Organic Photonics and Electronics

University of Strathclyde, UK

8–10 September 2014

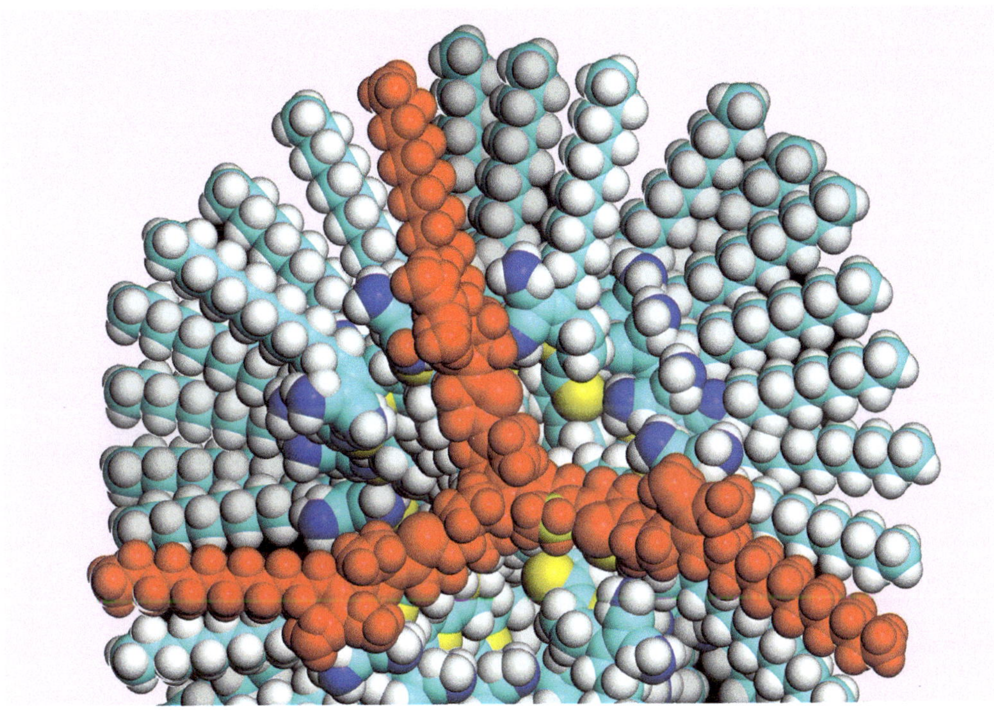

FARADAY DISCUSSIONS

Volume 174, 2014

ROYAL SOCIETY OF CHEMISTRY

The Faraday Division of the Royal Society of Chemistry, previously the Faraday Society, founded in 1903 to promote the study of sciences lying between Chemistry, Physics and Biology.

Faraday Discussions documents a long-established series of Faraday Discussion meetings which provide a unique international forum for the exchange of views and newly acquired results in developing areas of physical chemistry, biophysical chemistry and chemical physics.

Faraday Discussions (Print ISSN 1359-6640, Electronic ISSN 1364-5498) is published 8 times a year by the Royal Society of Chemistry, Thomas Graham House, Science Park, Milton Road, Cambridge, CB4 0WF, UK.

Volume 174 ISBN: 978-1-78262-048-8

2014 annual subscription price: print+electronic £827, US $1542; electronic only £785, US $1465. Customers in Canada will be subject to a surcharge to cover GST. Customers in the EU subscribing to the electronic version only will be charged VAT. All orders, with cheques made payable to the Royal Society of Chemistry, should be sent to RSC Order Department, Royal Society of Chemistry, Thomas Graham House, Science Park, Milton Road, Cambridge, CB4 0WF, UK.
Tel +44 (0) 1223 432398; E-mail www.orders@rsc.org

If you take an institutional subscription to any RSC journal you are entitled to free, site-wide web access to that journal. You can arrange access via Internet Protocol (IP) address at www.rsc.org/ip. Customers should make payments by cheque in sterling payable on a UK clearing bank or in US dollars payable on a US clearing bank.

Printed in the UK

Organic Photonics and Electronics

Faraday Discussions

www.rsc.org/faraday_d

A General Discussion on Organic Photonics and Electronics was held in Glasgow, UK on the 8th, 9th and 10th of September 2014.

CONTENTS

ISSN 1359-6640; ISBN: 978-1-78262-048-8

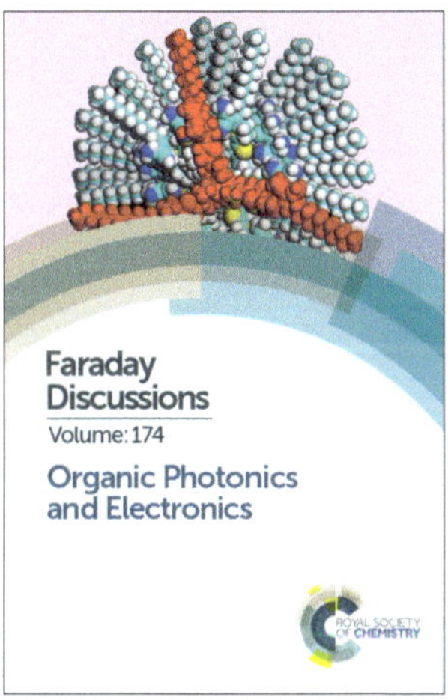

Cover
See S. A. Ponomarenko *et al.*, *Faraday Discuss.*, 2014, **174**, 313.

Molecular modelling of a column formed by star-shaped molecules N(Ph-2T-DCN-Dodec)$_3$ used as donor materials in efficient solution processable organic photovoltaics.

Image courtesy of S. A. Ponomarenko, Enikolopov Institute of Synthetic Polymeric Materials of Russian Academy of Sciences, Russia.

INTRODUCTORY LECTURE

PAPERS AND DISCUSSIONS

SPONSORS AND EXHIBITORS

BENTHAM

LOT
Quantum Design

SPECS™

CO-SPONSOR

IOP Institute of Physics
Quantum Electronics
and Photonics Group

CONCLUDING REMARKS

ADDITIONAL INFORMATION

Faraday Discussions

PAPER

Spiers Memorial Lecture

Organic electronics: an organic materials perspective

Fred Wudl*

Received 3rd October 2014, Accepted 13th October 2014
DOI: 10.1039/c4fd00191e

This Introductory Lecture is intended to provide a background to Faraday Discussion 174: "Organic Photonics and Electronics" and will consist of a chronological, subjective review of organic electronics. Starting with "ancient history" (1888) and history (1950–present), the article will take us to the present. The principal developments involved the processes of charge carrier generation and charge transport in molecular solids, starting with insulators (photoconductors) and moving to metals, to semiconductors and ending with the most popular semiconductor devices, such as organic light-emitting diodes (OLEDs), organic field effect transistors (OFETs) and organic photovoltaics (OPVs). The presentation will be from an organic chemistry/materials point of view.

1. Introduction

1.1 Why organic electronics?

The first question, of course, is whether one can fabricate organic devices that possess electronic and other materials properties analogous to, rivaling, or even surpassing those of traditional inorganic solids. Conventionally, most organic solids are lighter, more flexible and more elastic than their inorganic counterparts. For example, in terms of mechanical properties, one can compare a Kevlar® fiber to a thin steel wire, where the former is stronger than the latter, yet is lighter and more flexible. At the heart of organic solids is the element carbon, that distinguishes itself from the rest of the periodic table by producing long chains and rings involving double bonding with itself as well as nitrogen, oxygen, boron, sulfur and phosphorus. Carbon also forms stable triple bonds with itself and nitrogen. It stood (and stands) to reason, then, that one should be able to take advantage of the properties of carbon compounds but add to the list all the exciting possibilities of transporting charges, a property that until relatively recently was associated only with traditional semiconductors.

Department of Materials, University of California, Santa Barbara, CA 53706, USA. E-mail: wudl@chem.ucsb.edu

1.2 History: organic metals and superconductors

For this article, ancient history is defined as events in physics and chemistry of organic solids that predate the Rutherford model of the atom (1907). At first, scientists in the XIX century concerned themselves with achieving (observing) electrical conductivity in organic solids. According to Gutmann and Lyons (G&L),[1] who quoted Vartanian,[2] the first paper dedicated to the observation of a photovoltage by an organic dye was by the Russian N. Stoletov in 1888. In Western Europe, according to G&L, Pochettino was the first to observe photoconductivity in anthracene in 1906.[3] Between 1948 (ref. 4) and 1953, dark resistivities of organic solids were found to be on the order of 10^{11}–10^{18} Ω cm^{-1}.

A very important event took place in 1951, when Inokuchi and his group[5] reported a resistivity of 10^{-1} W cm^{-1} for an organic charge-transfer complex. This was followed by the DuPont work on charge-transfer salts of TCNQ[6] with resistivities on the order of 10^{-2} W cm^{-1}. In the 1970s, with the advent of TTF–TCNQ, the first organic solids exhibiting metallic conductivity were developed.[7] Research on these compounds quickly led to the discovery of organic superconductors.[8] At first the very low transition temperatures were achievable only under substantial pressure (25 kbar), but eventually the transition temperature was raised to *ca.* 13 K at atmospheric pressure[8] and ultimately 40 K for a fullerene salt.[9] This was a rather high value, especially considering that before the appearance of the high T_c cuprate superconductors, the highest ever observed T_c was only in the range of 22–23 K. In summary, before the end of the XX century, not only were charges generated in organic solids but they could also be transported with minimal resistance (metallic) to zero resistance (superconducting). The lessons learned from these developments were that the requirements for charge generation and transport were π-bonded carbon compounds containing sulfur and nitrogen. Further, these molecules either readily donated electrons (donors, D) to molecules that would easily accept them (acceptors, A). These D–A solids formed stacks of planar molecules, where the stacks consisted of like molecules; *i.e.*, segregated D and A stacks. In addition, the charge carriers in D–A ensembles delocalized preferentially along stacks, rather than between stacks. Finally, the stacks needed to be partially charged; *i.e.*, carrying approximately one charge (+ or −) for every two molecules within a stack.

1.3 History: organic semiconductors

Whereas metals exhibit non-activated conductivity, that is, their conductivity increases with decreasing temperature, semiconductors show activated conductivity. The charge carriers in the former are in a partially filled conduction band (CB), whereas in semiconductors, charge carriers are either thermo- or photo-excited from the filled valence band (VB) to the empty CB. In the case of organic semiconductors, it is relatively easy to remove electrons from the valence band of a D or add electrons to the empty conduction band of an A. These processes have a dramatic effect on the electronic properties of the semiconductor. Thus, electron removal from D, accompanied by the introduction of a counter anion to the resulting hole, is considered p-doping, and addition of an electron to the A with the addition of a counter cation to the electron is labeled n-doping. The VB of organic semiconductors is the π molecular orbital (MO) and the CB is the π* MO. The semiconductor gap of organic semiconductors can be relatively easily engineered to vary from larger than 3 eV to *ca.* 0.5 eV. This translates to interband light absorption wavelengths ranging from the UV to the NIR.

2. Classes of organic semiconductors

2.1 Small molecules

The most popular family of small molecule semiconductors are the polycyclic aromatic hydrocarbons and, of these, pentacene and some of its derivatives are the most studied in terms of device applications. Pentacene has a band gap of 2.2 eV. It is a p-type semiconductor with a hole mobility slightly higher than that of amorphous silicon, on the order of 5 $cm^2\ V^{-1}\ s^{-1}$.[10] Many dozens of organic π-bonded molecules have been examined in OFETs and a few of these have made it into J. Anthony's review.[10] These molecules contain sulfur as well as nitrogen. The overwhelming majority are p-type semiconductors with relatively low conductivities (10^{-12}–10^{-4} S cm^{-1}), particularly when compared to organic metals (10^2–10^3 S cm^{-1}). There are considerably fewer n-type semiconductor molecules. The salient examples are the acene imides such as naphthalene tetracarboxylic acid diimides (NTCDI)[11] and perylene tetracarboxylic acid diimide (PTCDI).[11] A method for the conversion of normally p-type molecules to n-type is multiple fluorination, thus, perfluoropentacene is an n-type material.[12] A more recent advent to small molecules is the introduction of specifically synthesized conjugated oligomers, to be described below. Small molecule organic semiconductors have had a major impact in OLEDs and OFETs. In the former application, metal complexes of organic ligands, particularly aluminum and iridium, are the most important. A small molecule that has had a very important place in organic photovoltaics (OPVs) is the fullerene C_{60}, particularly in the form of two derivatives to be described elsewhere in this article.

The main advantages of small molecules are the ease of synthesis and, especially, purification and characterization. Another advantage is the ease of processing. Small molecules can be processed into devices from solution or the vapor phase. The main disadvantage of small molecules in devices is that most small molecules tend to crystallize, a serious disadvantage for light emitting devices, where crystallites tend to scatter light as well as cause pin hole formation and loss of contact with addressing electrodes. Another potential problem is migration through the various device layers.

2.2 Conjugated polymers

Macromolecules have been examined for their electronic properties throughout the XX century. In the first half of the century, several researchers examined solids resulting from the pyrolysis of organic polymers. These were generally poorly characterized materials in terms of their chemical structure and composition. The first conducting polymer to be fully characterized was actually the inorganic poly(thiazyl) (SN_x) which, to this day, is the only superconducting polymer, albeit with a very low transition temperature (T_c 0.2–0.5 K).[13] Another polymer was poly(aniline) (PANI). These were mere curiosities until 1977, when Shirakawa, McDiarmid and Heeger, with their coworkers, reported the remarkable increases in conductivity of polyacetylene (PA) upon doping. Their report could be considered the beginning of organic conducting polymers.[14] Though PA turned out to be, essentially, the "fruit fly" of conducting polymers, it did not lead to any applications because it is an unstable solid, reacting exothermically with air and moisture. Nonetheless, p-doping eventually led to a solid with a conductivity on the order of 10^4 S cm^{-1}, a remarkable achievement for a solid consisting only of carbon and hydrogen. In quick succession the polymers based on thiophene poly(thiophene), PT, poly(3-alkylthiophene, PAT) and phenylene vinylene (PPV) were shown to be stable and dopable to conductivities on the order of 10^3 S cm^{-1}. Because the paradigm of the day was still a remnant of research on organic metals, the goal was to achieve higher and higher conductivities until in 1990 Friend and his group reported the electroluminescence of PPV[15] and in 1992 Sariciftci *et al.*[16] reported the first bulk heterojunction (BHJ) OPV, thus the emphasis switched to the semiconductor properties of undoped conjugated polymers. Further, all the physics and electronic engineering aspects of organic electronic devices were emphasized from that point on with the chemistry of the materials being guided by device physics and engineering. As is the case with small molecules, the number of p-type polymeric semiconductors is much larger than of the n-type conjugated polymers.

The advantages of polymers *vs.* small molecules are that polymers are natural film-formers with very low tendencies to migrate or crystallize. The major disadvantage is that polymers are very difficult to produce reproducibly with the same molecular weight and polydispersity. They are also more difficult to purify and characterize. Small oligomers, in principle, have the advantages of small molecules and polymers.

3. Basic concepts

3.1 The solid state

Organic semiconductors are molecular solids, in contrast to inorganic semiconductors, which are extended solids. The most attractive feature of molecular solids is that the properties of the solid are essentially those of its constituent molecules. Hence, if one were to increase (or decrease) the π–π^* gap of the molecule, the semiconductor energy gap of the solid would follow the same trend. This feature makes the design of organic semiconductors considerably easier than would otherwise be possible. Fortunately this idea can be extended to the design of polymers, where changes in the electronic character of the monomer are reflected directly in the properties of the polymer and the molecular solid derived from the ensemble of polymer chains. Thus, when it was desired to decrease the E_g of poly(thiophene) (E_g *ca.* 2.1 eV), decreasing the π–π^* gap of the monomer resulted in the decrease of the E_g of the resulting polymeric solid to 1.1 eV.[17] Unfortunately the energies involved in forming molecular solids are very small and, as a consequence, molecular solids have a propensity to produce many polymorphs for a given molecule. For example, a flat polycyclic aromatic molecule can form at least three different polymorphs: stacks, herringbone or discrete dimers. The situation becomes even more complicated in the case of polymers. Most polymers form amorphous solids but due to the backbone rigidity of conjugated polymers, they can have crystalline regions interspersed in an amorphous matrix. The crystalline segments can be manipulated by annealing (or tempering) either by heat-treatment or a very small amount of an additive. Polymers can also be coaxed into a relatively ordered state by external fields:

either mechanical (doctor blade, stretch-alignment, rubbed surface), or electric/magnetic fields for liquid crystalline systems. Ideally all active substrates in organic electronics devices would be crystalline with the molecular ensembles in preferred orientations relative to external influences such as electric field, magnetic field and light. However, in reality, a new molecule could be designed with, say, improvement in the power conversion efficiency (PCE) of an OPV but if its constituent molecules, be they small or macromolecules, refuse to form the proper morphology (polymorph) and cannot be coaxed to order in a desired form, the whole effort can, and will, be derailed. Over the last few decades, the rules of self-assembly of molecules into pre-determined structures have been evolving but, so far, incorporation of these to achieve particular goals in organic electronics and optoelectronics have had very minimal success.

3.2 Excitons, electrons and holes

Charges can be generated in organic semiconductors from the loosely held electrons of π molecular orbitals by heat, irradiation or an electric field. As shown in Fig. 1, an OLED is the microscopic reverse of an OPV; the principal species is an electron–hole pair or exciton. In the light emitter, the electrons and holes are generated at the electrodes, migrate to form excitons and combine, usually at an exciton trapping site, giving off a photon with energy less than the π–π* gap. Conversely, in an OPV, to separate the hole and the electron of an exciton, an acceptor is added to facilitate the charge separation and discourage electron–hole recombination. The excitons are generated by the absorption of photons, the excitons migrate to the D–A interface, where charges are separated, the electron delocalizes along the A component and the positive charge migrates along the D component to their corresponding electrodes.

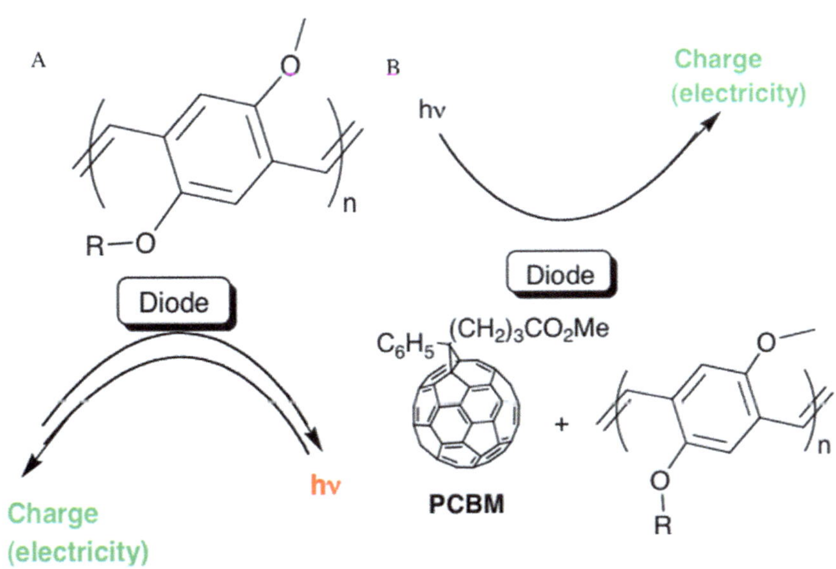

Fig. 1 (A) Schematic representation of the light emission and photovoltage generation being the reverse of each other; (B) improvement of the photovoltaic process by incorporation of the electron acceptor PCBM.

In p-type OFETs, the field effect results in the ionization of π-electrons (HOMO orbital) with concomitant hole-generation for charge transport. In n-type OFETs, the field effect produces a thin electron layer at the accessible sites in the π^* (LUMO orbital) for charge transport.

4. Specific devices

4.1 Molecular electronics

After the first organic metal, TTF–TCNQ, was discovered in 1973, Aviram and Ratner[18] proposed the first molecular rectifier, giving birth to the field. Over the years, with the advent of scanning probe microscopes, particularly break-junction (SMBJ) STM experiments, the examination of "single molecule" electronic properties was possible. Solomon discussed the thermoelectric properties as a function of π–π stacking in a molecular electronics experiment (DOI: 10.1039/c4fd00083h). Not surprisingly, the theory associated with single molecules is complicated and Bâldea showed the current thinking in terms of quantum chemistry may not be absolutely correct (DOI: 10.1039/c4fd00101j). Another aspect of theoretical concern is charge transfer and Venkatramani discussed this aspect of molecular electronics (DOI: 10.1039/c4fd00106k). In the meantime, Venkataraman discussed the implementation of quantum interference to interpret semiconductor properties in molecular electronics such as the molecular dependence of resistance ratios (DOI: 10.1039/c4fd00093e). Ottosson discussed the importance of the excited state electronic structure on SMBJs (DOI: 10.1039/c4fd00084f). He proposed, based on computational results, that photochemical rearrangements that could occur in a SMBJ could be desirable. Xu discussed his results of contact effects between the electrode and molecule on the properties of the "devices" (DOI: 10.1039/c4fd00080c). In this respect, Xu concluded: "Although an individual molecule could perform as a conductor, the produced current is far less than what is required for commercial use. Challenges exist in improving and amplifying the current. Doping special chemicals, metal ions and nanoparticles into individual molecules have been suggested to effectively increase the conductivity of a SMBJ. Attempts towards this direction have just started."

4.2 Light emitting devices

Though the first light emitting devices were reported in the early 1960s,[19] they were rather inefficient; it was not until 1987, when Tang and VanSlyke[20] reported the first organic thin film light emitting device based on an aluminum organometallic green light emitter, when the efficiency was improved considerably by energy level matching to the cathode and incorporating an organic hole transporting layer (HTL) at the anode to lower the hole transport barrier. With the exception of the addition of an electron-transporting layer (ETL), this architecture remains essentially the same in current devices. Three years later, the first polymer based OLED, a PLED, was reported by the Cambridge group led by Richard Friend. The architecture was considerably simpler, consisting of a thin (*ca.* 100 nm) film of the organic conjugated polymer sandwiched between the anode and cathode. Electrons were supplied through an Al contact and holes were collected at an ITO surface. Whereas it was relatively easy to obtain various hues of red and green as well as orange, deep blue was a challenge. Because blue was not so accessible, white organic LED light was also not achievable in the early days, but in the recent past several research groups and corporations have achieved white light emission with the well established stratagem of using the high energy blue emission as both the color and pump for the other lower energy colors. Fig. 2 is a depiction of a typical "modern" OLED architecture.

Finally, Forrest and Thompson made a most important contribution by looking at the relative contribution of triplet and singlet excited states to luminescence. They posited, and proved, that devices which exhibited phosphorescence should have dramatically higher efficiencies.[21] It was well known that organic complexes of heavy transition metals such as Pt, Pd or Ir were efficient phosphorescers. Incorporation of these luminescent molecules into OLEDs was a key step that was required to make these devices commercially important, leading to today's brilliant color Samsung smart phone and television displays. There have been a large number of important contributions, particularly in the recent past, to the successful development of organic light-emitting devices, and we see several of these during the Faraday Discussion. Lemmer discussed an intriguing fabrication improvement through a laser-assisted replication on an inexpensive poly(methyl methacrylate) substrate (DOI: 10.1039/c4fd00077c). Zysman-Colman discussed a novel approach to improve the quantum yield of blue emitters by proper design of the ligand for the transition metal complex (DOI: 10.1039/c4fd00107a). Leo, through his assistant Robert Brückner, discussed the effect of scattering, refractive or diffractive elements on the improvements of light-emitting devices (DOI: 10.1039/c4fd00094c). Sasaki discussed the possibility of dynamic hologram formation in ferroelectric liquid crystal displays (DOI: 10.1039/c4fd00068d); though these are not OLEDs, their display properties are based on a combination of photovoltaic and electro-optic effects. Dennany discussed the application of a variant of OLEDs: electrochemical luminescent (ECL) devices (DOI: 10.1039/c4fd00090k). With a new active star-shaped active organic layer they made very sensitive sensors. Laurand discussed an optically-pumped OLED as a very sensitive microsensor for biological systems that should be useful as label-free biosensing platforms (DOI: 10.1039/c4fd00091a).

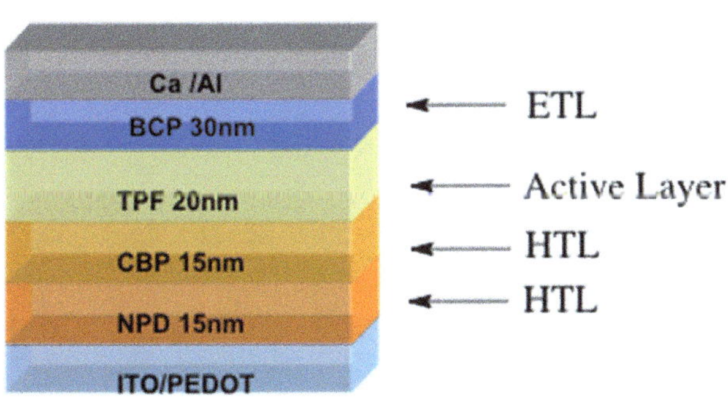

Fig. 2 Schematic of a typical OLED, where more than one HTL was required to lower the barrier to hole transport. In this case TPF is a blue light emitter and the device was fabricated by vacuum layer deposition.

4.3 Field effect transistors

In 1983, the first OFET based on polyacetylene[22] and in 1988 the same device based on polythiophene was reported. These devices, though behaving like classical transistors, showed very low hole mobilities. The first reasonable device was reported by Garnier in 1989 (ref. 23) and it was based on a small oligomer of polythiophene, sexythiophene, and in 1990 the same group reported an all organic device with the SiO_2 insulator replaced by an organic insulator and the glass substrate replaced by an organic plastic film. For vapor deposited thin organic films, the record is 35 $cm^2 V^{-1} s^{-1}$, held by Jurchescu[24] for pentacene, and for thin single crystals, the record for rubrene is 5–40 $cm^2 V^{-1} s^{-1}$.[25] For highly oriented thin polymeric films, oriented on a modified surface, the record is 36.3 $cm^2 V^{-1} s^{-1}$.[26] Fig. 3 shows a depiction of the most popular device fabrications. Most organic semiconductors lend themselves to bottom gate architectures, though in some cases the best results are obtained with top gate construction. The vast majority of organic semiconductors are p-type but n-type and ambipolar semiconductors have become more numerous in the very recent past.[27] The more interesting of these are the latter, where rather high inverter device gains of 35 (ref. 28) and 86 (ref. 29) have been reported. Functional OFETs are another more recent entry, also represented by ambipolar devices, particularly in the form of light-emitting devices (LEOFETs),[30] recently reviewed.[28] Though there have been a large number of important contributions, particularly in the recent past, to the successful development of organic transistors, and we will see several of these as part of this Faraday Discussion, there remain a number of challenges. Most have to do with chemistry. Though the community has been touting the advantages of OFETs such as processing simplicity, light weight and flexibility over the traditional electronics devices, a critical evaluation of the actual costs involved in mass production, including switching from current synthetic methods to more environmentally friendly methods and to more efficient syntheses of the more complicated molecules, oligomers and polymers, has yet to appear.[31] Torsi discussed a hybrid type of thin film transistor (TFT), the electrolyte-gated TFT (EG-TFT) (DOI: 10.1039/c4fd00081a). In these devices, the electrolyte serves as a high dielectric layer between the active layer and the gate. By using ZnO as a transparent electrode, Torsi's group has made excellent biocompatible devices. Fraboni discussed organic semiconductor single crystal (OSSC) devices and their application in radiation detection, particularly X-rays (DOI: 10.1039/c4fd00102h). McDouall discussed a theoretical approach to examine the hole mobility in OFETs as a function of structure, particularly the surface

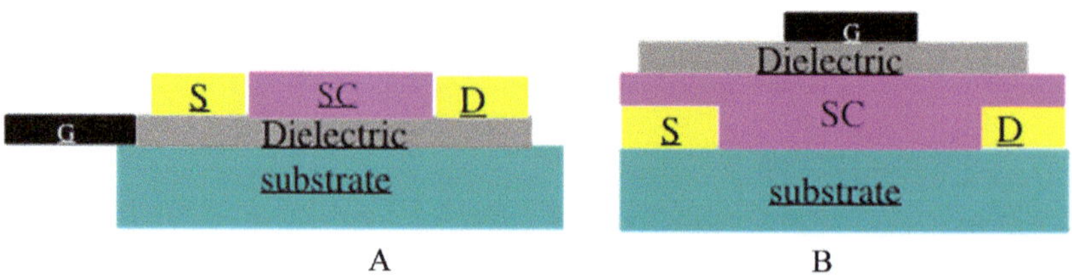

Fig. 3 (A) "Bottom gate" and (B) "Top gate" OFET architectures. D, drain; G, gate; S, source; SC, semiconductor.

structure effects (DOI: 10.1039/c4fd00082j). His results should help the engineering of FETs. Perepichka discussed a very successful crystal engineering result in the design of the active layer of OTFTs (DOI: 10.1039/c4fd00133h). They based their solid state engineering on complementary hydrogen bonding between the donor and acceptor components of the crystal, resulting in ambipolar devices. Melzer discussed a system already mentioned by Torsi (above) (DOI: 10.1039/c4fd00095a). In these devices, commercial poly(3-hexylthiophene) is used as the active layer and they find that the hole mobility is largely enhanced.

4.4 Organic photovoltaics (OPVs, plastic solar cells)

As discussed in relation to Fig. 1, the organic photovoltaic effect is essentially the reverse of the organic light-emitting electroluminescence effect. The main difference is that, in the photovoltaic device, the electron and hole of the exciton produced by photoexcitation need to be separated before they have a chance to recombine and emit. To achieve this, an electron acceptor is blended into the donor. In the early days of the field, the charge recombination was an apparently insurmountable problem, until the fullerenes were discovered. Nature's fullerene gift has two extremely beneficial properties: (i) fullerenes form good n-type semiconductor devices with relatively good electron mobilities, and (ii) the radical anion resulting from electron acceptance is very stable, providing a substantial barrier to electron hole recombination. The early attempts to form bilayer devices with a donor polymer–fullerene interface gave relatively low power conversion efficiencies (PCEs).[16] However, blends with a more soluble derivative of C_{60}, PCBM, soon improved the PCE to 2.9%.[32] These blends formed a phase segregated system that naturally formed a bulk heterojunction (BHJ). These BHJ devices are the mainstay of OPVs. In the meantime the PCE has been increased to 11.1%.[33] The main theoretical model construct for BHJ was given by Scharber, where they correlated the LUMO of the donor polymer with the band gap of the donor and extracted a theoretical PCE. The maximum predicted PCE of a BHJ cell based on PCBM as acceptor was 11%.[34] Nelson discussed a parameter free calculation result of the subgap density of states in poly(3-hexylthiophene) (DOI: 10.1039/c4fd00153b). One of the main conclusions is that intrachain torsional disorder is much more significant than the interchain coupling disorder in the non-crystalline regions of the solid. Kim discussed organic–inorganic hybrid based solar cells, particularly the stability (or lack thereof) of the hybrid structures toward annealing temperatures (DOI: 10.1039/c4fd00141a). Ponomarenko discussed the design, synthesis and applications of three-arm star-shaped molecules in OPV devices (DOI: 10.1039/c4fd00142g). These triphenylamine-based systems can produce an impressive PCE of up to 5.4%.

4.5 Other devices for the future

The future of organic electronics and photonics is bright, indeed. The new areas into which the science and engineering of organic condensed matter devices are expanding are spintronics and thermoelectrics. In the former, Vardeny[35] recently showed the fabrication of an effective spin valve using Alq_3 as the organic spacer between two ferromagnetic layers, and Ando with Sirringhaus recently showed a very exciting spin-charge converter by the use of highly conducting PEDT-PSS.[36]

The area of organic thermoelectrics is heating up, with the number of publications increasing from 1 in the 1980s to 70 in 2013.[37]

5. Conclusion and outlook

Research in organic electronics and photonics is a very active field indeed. The number of publications related to organic electronics involving physics, chemistry, materials science, engineering and biology in the last two decades is astounding and very difficult to measure due to the basic interdisciplinary nature of the endeavour. Fundamental research still dominates the field but several inroads have been made into technical applications and some will be presented as part of this Faraday Discussion. The salient features of the processes in the various devices, *viz.*, molecular electronics, OFETs, LEFETs, OLEDs and OPVs are now relatively well understood. However, many details are still to be worked out, particularly with regards to the end product, namely consumer electronics. For example, in the case of OPVs, Cao and Xue conclude:[38]

"Last, but not least, the stability of OPV devices is a critical hurdle that should be tackled before their commercialization. Although OPV devices with a lifetime over seven years have been demonstrated,[39] a more robust encapsulation technology, as well as more detailed study about the degradation mechanism, is required to improve the device stability. In particular, the phase separated bulk heterojunction structure generally is not at equilibrium; thus its morphological stability during operation may be a concern that needs to be better suited".

Some of these concerns can be directly translated from OPVs to the other devices described above and in this Faraday Discussion. For example, for OFETs, Bao states:[27]

"On the other hand, there are a few challenges the research community has to overcome before OFETs can take off for practical applications. Organic circuits are generally featured as low cost electronics, even though there is still the lack of a comprehensive and accessible analysis of cost-structures of OFET devices for various applications. From the aspect of synthetic chemistry, there are a few cost drivers, namely, reaction steps and yields, reagents and solvents, as well as purification approaches...Unfortunately, the majority of high performance OFET materials are currently processed from chlorinated solvents. It is thus important for chemists to develop OFET materials that can be processed from non-chlorinated solvents, while retaining high performance". Bao further concludes:[24]

"Some other issues for OFETs include (1) the long-term storage and operation stability of the devices, (2) the reliability and reproducibility of the devices, and (3) the patterning and integration of OFETs in integrated circuits. To create solutions for these challenges, not only are OSC (organic semiconductors) design and processing important, but the advancement of dielectric and encapsulation materials is also crucial. In addition, environmental assessment of OSCs is needed, since they are aromatic compounds and likely to be carcinogens. This is especially critical, if OFETs are ever used in disposable devices".

References

1 F. Gutmann and L. E. Lyons, *Organic Semiconductors*, Wiley, New York, 1967, pp. 632–650.

2 A. T. Vartanian, *Izv. Akad. Nauk SSSR, Ser. Fiz.*, 1952, **16**, 169.
3 A. Pochettino, *Atti Accad. Naz. Lincei, Cl. Sci. Fis., Mat. Nat., Rend.*, 1906, **15**, 355.
4 D. D. Eley, *Nature*, 1948, **162**, 819.
5 H. Akamatu, H. Inokuchi and T. Handa, *Nature*, 1951, **168**, 520.
6 R. G. Kepler, P. E. Biersted and R. E. Merrifield, *Phys. Rev. Lett.*, 1960, **5**, 503.
7 H. J. Keller, *Chemistry and Physics of One Dimensional Metals*, Plenum, New York, 1976, pp. 1–412.
8 J. M. Williams, J. R. Ferraro, R. J. Thorn. K. D. Karlson, U. Geiser, H. H. Wang, A. M. Kini and M. H. Whangbo, *Organic Superconductors (Including Fullerenes)*, Prentice, New Jersey, 1992, pp. 1–391.
9 T. T. M. Palstra, O. Zhou, Y. Iwasa, P. E. Sulewski, R. M. Fleming and B. R. Zegarski, *Solid State Commun.*, 1995, **93**, 327.
10 J. E. Anthony, *Chem. Rev.*, 2006, **106**, 5028.
11 Y. Zhao, Y. Guo and Y. Liu, *Adv. Mater.*, 2013, **25**, 5372.
12 Y. Sakamoto, T. Suzuki, M. Kobayashi, Y. Gao, Y. Fukai, Y. Inoue, F. Sato and S. Tokito, *J. Am. Chem. Soc.*, 2004, **126**, 8138.
13 H. J. Keller, *Chemistry and Physics of One Dimensional Metals*, Plenum, New York, 1976, pp. 167–195.
14 H. Shirakawa, E. J. Lewis, A. G. MacDiarmid, C. K. Chiang and A. J. Heeger, *J. Chem. Soc., Chem. Commun.*, 1977, 578.
15 J. H. Burroughes, D. D. C. Bradley, A. R. Brown, R. N. Marks, K. MacKay, R. H. Friend, P. L. Burns and A. B. Holmes, *Nature*, 1990, **347**, 539.
16 N. S. Sariciftci, L. Smilowitz, A. J. Heeger and F. Wudl, *Science*, 1992, **258**, 1474.
17 F. Wudl, M. Kobayashi and A. J. Heeger, *J. Org. Chem.*, 1984, **49**, 3382; M. Kobayashi, N. Colaneri, M. Boysel, F. Wudl and A. J. Heeger, *J. Chem. Phys.*, 1985, **82**, 5717.
18 A. Aviram and M. A. Ratner, *Chem. Phys. Lett.*, 1974, **29**, 277–283.
19 W. Helfrich and W. G. Schneidere, *Phys. Rev. Lett.*, 1965, **14**, 229; W. Helfrich and W. G. Schneidere, *J. Chem. Phys.*, 1965, **14**, 2902.
20 C. W. Tang and S. A. VanSlyke, *Appl. Phys. Lett.*, 1987, **51**, 913.
21 M. A. Baldo, D. F. O'Brien, Y. You, A. Shoustikov, S. Sibley, M. E. Thompson and S. R. Forrest, *Nature*, 1998, **395**, 151.
22 F. Ebisawa, T. Kurokawa and S. Nara, *J. Appl. Phys.*, 1983, **54**, 3255.
23 F. Garnier, G. Horowitz, X. Peng and D. Fichou, *Adv. Mater.*, 1990, **2**, 592.
24 O. D. Jurchescu, J. Baas and T. T. M. Palstra, *Appl. Phys.Lett.*, 2004, **84**, 3061.
25 D. Braga and G. Horowitz, *Adv. Mater.*, 2009, **21**, 1473–1486; J. Takeya, M. Yamagishi, Y. Tominari, R. Hirahara, Y. Na-kazawa, T. Nishikawa, T. Kawase, T. Shimoda and S. Ogawa, *Appl. Phys. Lett.*, 2007, **90**, 102120.
26 C. Luo, A. K. Ko Kyaw, L. A. Perez, S. Patel, M. Wang, B. Grimm, G. C. Bazan, E. J. Kramer and A. J. Heeger, *Nano Lett.*, 2014, **14**, 2764.
27 J. Mei, Y. Diao, A. L. Appleton, L. Fang and Z. Bao, *J. Am. Chem. Soc.*, 2013, **135**, 6724.
28 J. Fan, J. D. Yuen, M. Wang, J. Seifter, J.-H. Seo, A. R. Mohebbi, D. Zakhidov, A. Heeger and F. Wudl, *Adv. Mater.*, 2012, **24**, 2186.
29 Z. Chen, H. Lemke, S. Albert-Seifried, M. Caironi, M. M. Nielsen, M. Heeney, W. Zhang, I. McCulloch and H. Sirringhaus, *Adv. Mater.*, 2010, **22**, 2371.
30 A. Hepp, H. Heil, W. Weise, M. Ahles, R. Schmechel and H. von Seggern, *Phys. Rev. Lett.*, 2003, **91**, 157406.

31 S. Hotta, T. Yamao, S. Z. Bisri, T. Takenobu and Y. Iwasa, *J. Mater. Chem. C*, 2014, **2**, 965.
32 G. Yu, J. Gao, J. C. Hummelen, F. Wudl and A. J. Heeger, *Science*, 1995, **270**, 1789.
33 W. Cao and J. Xue, *Energy Environ. Sci.*, 2014, **7**, 2123.
34 M. C. Scharber, D. Muhlbacher, M. Koppe, P. Denk, C. Waldhauf, A. J. Heeger and C. J. Brabec, *Adv. Mater.*, 2006, **18**, 789.
35 Z. H. Xiong, D. Wu, Z. Valy Vardeny and J. Shi, *Nature*, 2004, **427**, 821.
36 K. Ando, S. Watanabe, S. Mooser, E. Saito and H. Sirringhaus, *Nat. Mater.*, 2013, **12**, 622.
37 P. J. Taroni, N. Stingelin, M. Heeney and E. Bilotti, *Isr. J. Chem.*, 2014, **54**, 534.
38 W. Cao and J. Xue, *Energy Environ. Sci.*, 2014, **7**, 2123.
39 Q. Gan, F. J. Bartoli and Z. H. Kafafi, *Adv. Mater.*, 2013, **25**, 2385.

Faraday Discussions

PAPER

Exploring coherent transport through π-stacked systems for molecular electronic devices

Qian Li and Gemma C. Solomon*

Received 28th April 2014, Accepted 2nd June 2014
DOI: 10.1039/c4fd00083h

Understanding electron transport across π-stacked systems can help to elucidate the role of intermolecular tunneling in molecular junctions and potentially with the design of high-efficiency molecular devices. Here we show how conjugation length and substituent groups influence the electron transport and thermoelectric response in the π-stacked structures by investigating five representative stacked molecular junctions. We found that a π-stacked system of two substituted anthracenes exhibits good thermopower and a high power factor, suggesting that increased conjugation can enhance the thermoelectric response. The fully eclipsed structure of quinhydrone exhibits a high power factor at the minimum energy structure and could thus be a better candidate in a thermoelectric device compared with the other π-stacked systems considered.

1 Introduction

The starting point for the chemists' vision of molecular electronics is a single molecule bound between electrodes, where chemical functionality within the molecule has been carefully designed to yield a particular electrical function. While such a design sounds simple, achieving this goal has been anything but simple. Leaving aside for a moment the challenge of developing a particular electrical function from chemical functionality, it is a nontrivial task to bind a single molecule between electrodes. Most of the molecules targeted have been large conjugated molecules and consequently aggregation is always a possibility. Within the junction, π-stacked structures can form by accident or by design. Within such structures, there is always the possibility of intermolecular tunnelling, introducing a degree of ambiguity in the interpretation of the results. The role of intermolecular π–π interactions in mediating transport has not been extensively explored and warrants further investigation.

π–π stacking is efficiently used by nature to achieve directed long-range electron transport, such as DNA base pairs[1,2] and amino acid residues. Careful

Nano-Science Center and Department of Chemistry, University of Copenhagen, Universitetsparken 5, 2100 Copenhagen, Denmark. E-mail: gsolomon@nano.ku.dk

experiments have found that the intermolecular π–π stacking interaction can form an electrical junction and lead to molecular junctions with significant conductance.[3,4] An STM break-junction experiment also measured single-molecule conductance through π–π stacked benzene rings and found hydrocarbons form direct electrical contacts with gold electrodes.[5] Understanding electron transport across π-stacked structures will help to elucidate the role of intermolecular tunneling in the molecular junction and allow us to use these interactions to design better molecular devices.

Returning to the other challenge in this problem, achieving particular electrical functionality, there has been considerable interest in recent years in quantum interference as a means to make conjugated molecules more than just wires. Destructive quantum interference has been shown to occur in π-stacked systems and can be modulated with changing spatial dislocation between the stacked elements.[6,7] Beyond traditional electrical components, there has also been an interest in molecular thermoelectrics. Thermoelectric effects may be dramatically enhanced near the transmission node that arises as a result of this destructive interference.[8] Several groups have reported the measurement of thermopower, S, of molecular junctions.[9–13] The thermopower enters the figure of merit, $ZT = S^2GT/\kappa$, which is used to describe the efficiency of a thermoelectric device. Here S is the thermopower, G is the electronic conductance, T is temperature and κ is the thermal conductance with contribution from electrons and phonons. As π-stacked systems are electronically coupled through-space, we anticipate that the thermal conductance will be low. Consequently, in this situation the magnitude of the figure of merit is mainly determined by the power factor S^2G. Although the conductance is low at energies where destructive interference dominates, the power factor (S^2G) can have a high value in the case where S^2 is much larger than G. In the case where an interference feature is in close proximity to a resonant peak (for example a fano-resonance), however, it is possible to obtain both high thermopower and relatively high conductance.

In this paper, we investigate electron transport and the thermoelectric response through five representative π-stacked systems (three systems with differing conjugation length and two systems with different substituents). For each system, the total transmission at the selected geometries is calculated to illustrate the change in transmission with the spatial dislocation. We examine how conjugation length and substituents influence the electronic transmission, thermopower and power factor in these π-stacked systems. Finally, the binding energy is calculated to show the differing stability of the structures.

The five π-stacked systems considered in this paper are shown in Fig. 1. As we expect an edge to face structure to have weak electronic coupling, resulting in low conductance, we only consider the face to face structure for π-stacked dimers in our study. The three systems A, B and C are π-stacked systems with different conjugation lengths. System D has a substituted cyano group and system E comprises a quinone and a hydroquinone. All of the junctions are formed by one molecule chemisorbed on the left electrode, and another one chemisorbed on the right electrode, as shown at the bottom of Fig. 1.

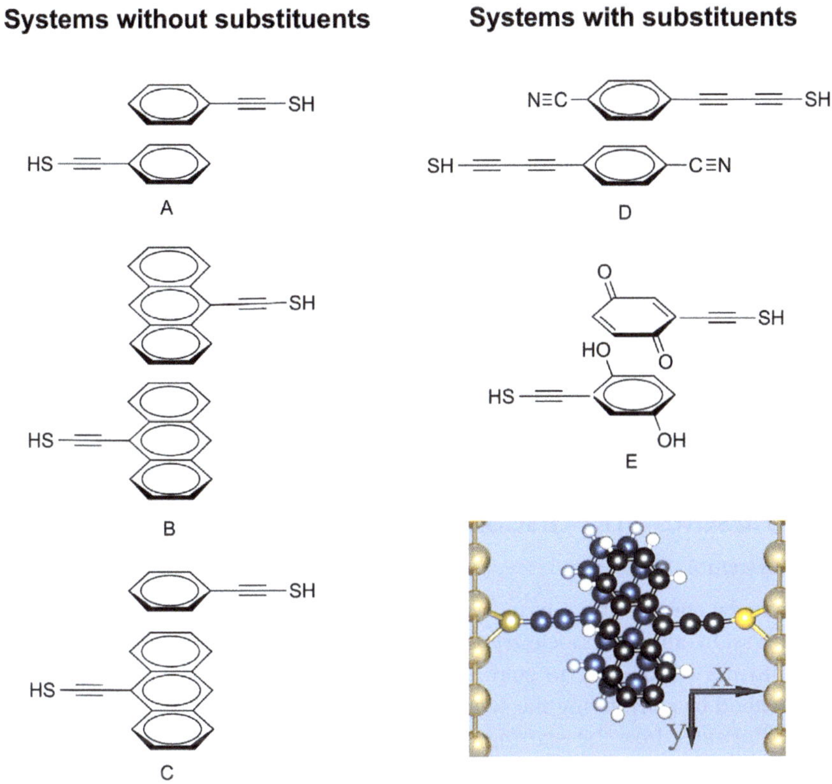

Fig. 1 Five different π-stacked molecules we consider in this work. A, B and C are π-stacked systems without substituents while system D is substituted with cyano groups and system E is the combination of a quinone and a hydroquinone. Bottom (right): an illustration of B when chemisorbed in between gold electrodes as used in our transport calculations.

2 Background

As inelastic transport is usually strongly suppressed at room temperature in single molecule junctions, we only consider the elastic transport. The current through a molecule is calculated with a Green's function implementation of the Landauer approach:[14]

$$I(V) = \frac{2e}{h}\int_{-\infty}^{\infty} \mathrm{d}E[f_{\mathrm{L}}(E,V) - f_{\mathrm{R}}(E,V)]T(E,V) \tag{1}$$

where $f_{\mathrm{L}}(E,V)$ and $f_{\mathrm{R}}(E,V)$ are the Fermi functions for the left and right electrodes, $T(E,V)$ is the energy and bias dependent transmission. Here, we report the transmission that underlies the zero-bias conductance. The Fermi energy is set to −5 eV. The thermopower is calculated by:

$$S = -\frac{\pi^2 k_{\mathrm{B}}^2 T}{3eT(E)} \left.\frac{\partial T(E)}{\partial E}\right|_{E=E_{\mathrm{F}}} \tag{2}$$

This equation indicates that a high thermopower is achieved when the slope of the transmission is steep.

3 Computational methods

The geometries of the isolated molecules were optimized using GPAW[15] (a grid-based real-space DFT code) using PBE exchange functional with van der Waals correction (TS09)[16] and the stacked structures were formed by combining two of these structures to calculate the transport and binding energy. The molecule–molecule separation in the z-direction (perpendicular to the molecular plane) was set to the value that gave the minimum energy structure. The molecules were chemisorbed on an FCC hollow site of the Au electrodes with terminal hydrogen atoms removed. All the transport calculations were performed using gDFTB[17,18] with the geometries obtained from GPAW.

4 Results and discussion

4.1 System A

We start by considering a basic system to illustrate the basic properties of the transmission and thermoelectric response: stacked benzene rings with thiol substituents for binding to gold electrodes. The transmission features can be understood by considering the system as an even alternate hydrocarbon.[19] Fig. 2 (top) illustrates how the starring assignment works in the π-stacked systems: mark every second atom with a star and continue across the nearest-neighbour interactions in the stack. When the two sulphur atoms are both marked with

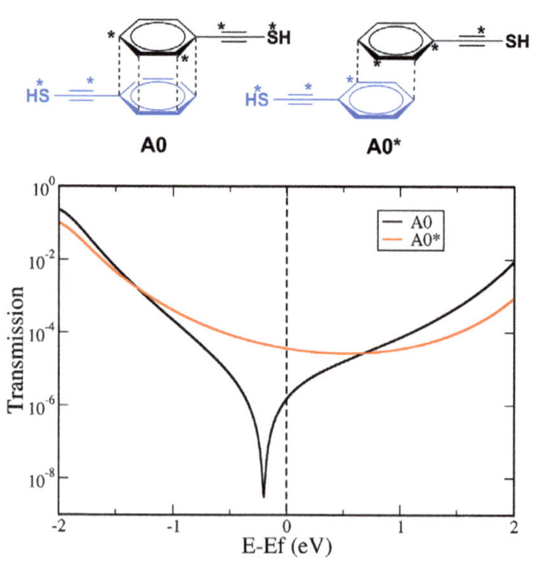

Fig. 2 Top: two selected structures of system A. The molecules are even alternate hydrocarbons and are marked with stars to show when alike (low) or disjoint (high) coupling is expected. Bottom: transmission through the two selected structures of system A.

stars, we say there is alike coupling and destructive interference and when only one is starred we have disjoint coupling and no interference.[7] Fig. 2 (bottom) shows that the transmission through the fully eclipsed structure has a destructive interference near the Fermi energy, which results in lower transmission, while the dislocated structure has higher transmission near Fermi energy without interference, in agreement with previous results.[6]

In addition, the full range of the transmission, thermopower and power factor are calculated as a function of translation along the long (x) and short (y) axes of the molecule. There is low transmission around (0.0, 0.0) in the transmission map, caused by the destructive interference as well as around (2.0, 0.0). The high values of thermopower always arise in the vicinity of the destructive interference. So, the whole thermopower map is almost the complementary map of transmission, shown in Fig. 3(b). The sign of thermopower changes with the translation of molecules (positive in red and negative in green). Due to the large

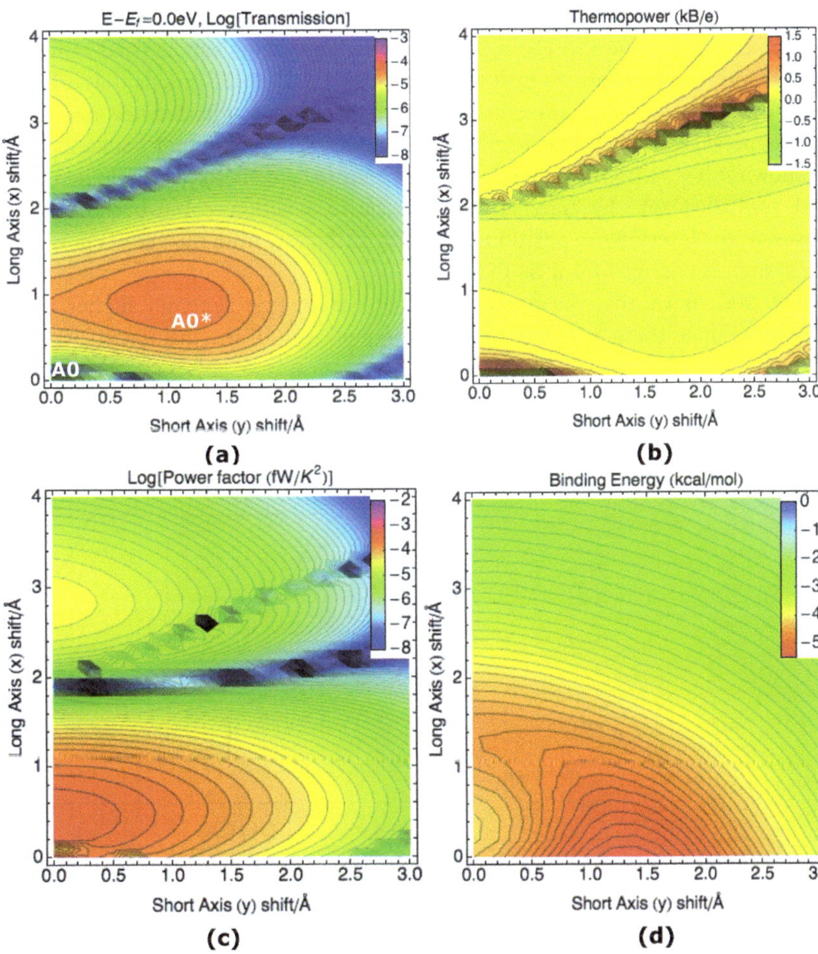

Fig. 3 Full spatial maps of the transmission (a), thermopower (b), power factor (c), binding energy (d) as a function of translation at the Fermi energy for A.

thermopower, the areas near (0.0, 0.0) which show low transmission, now have a high power factor S^2G.

The binding energies were calculated as the difference between the total energies of the π-stacked molecules and the isolated molecules, with van der Waals correction included. As we focus on the π–π stacking interaction, we only translate molecules along the positive direction of x and y-axis, avoiding transport from the triple bond on one molecule into the ring of the other. Consequently, the binding energy map is not the global potential surface. Fig. 3(d) shows that a local minimum is around (0.0, 1.4). Although the thermopower S and power factor S^2G have relatively high values at the (0.0, 0.0) position, the fully eclipsed structure is not the most stable geometry.

4.2 System B

In an effort to improve the transmission and thermoelectric properties, the conjugation length of molecules is increased by using anthracene rings instead of benzene rings in the π-stacked system. There are now three groups of fully eclipsed structures and dislocated structures as the rings are shifted with respect to each other, as shown in Fig. 4 (left). Starring arguments can be used again to predict interference features near the Fermi energy. All three groups that exhibit alike coupling have low transmission from destructive interference near the Fermi energy, while the structures with disjoint coupling have high transmission as shown in Fig. 4 (right). The interference features are effectively "switched on" and "switched off" by translating the relative positions of two molecules. However, the interference positions gradually shift away from the Fermi energy as the system is changed from B0 to B1, B2.

Fig. 5(a) shows that the transmission of system B is generally higher than system A, since there are larger π–π ring couplings. All the disjoint structures

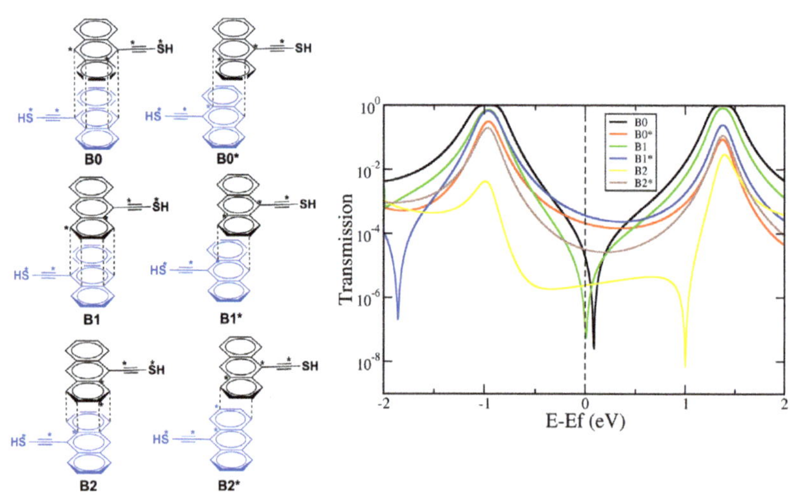

Fig. 4 Left: three groups of selected structures of system B. The molecules are even alternate hydrocarbons and are marked with stars to show when alike (low) or disjoint (high) coupling is expected. Right: transmission through these selected structures of system B.

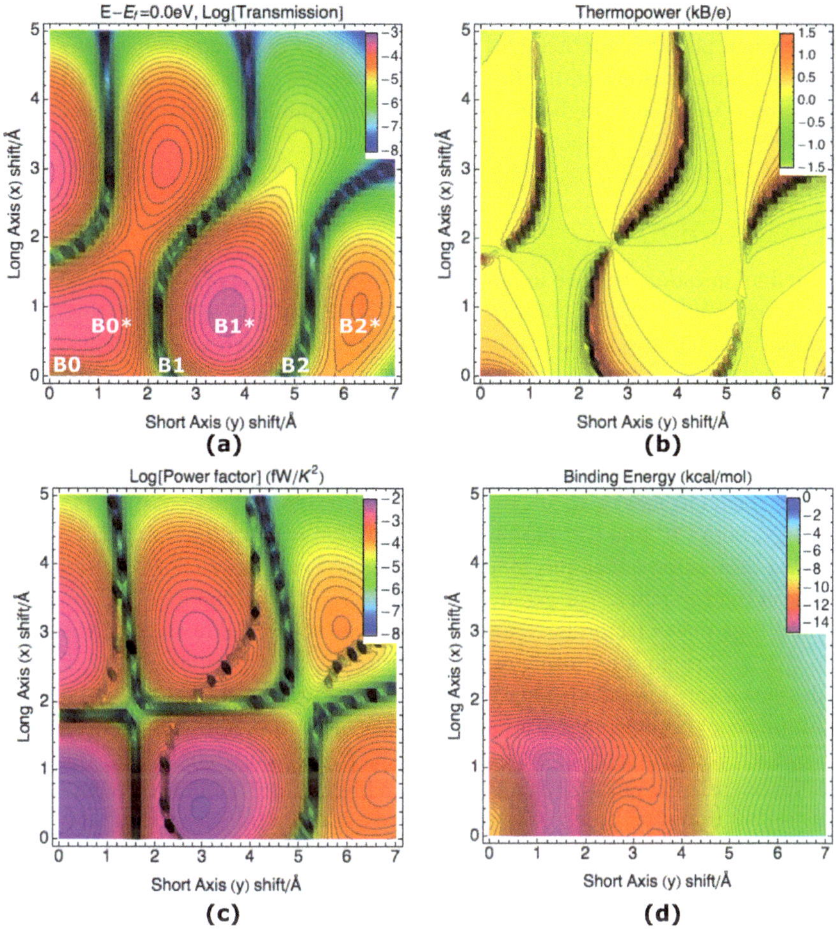

Fig. 5 Full spatial maps of the transmission (a), thermopower (b), power factor (c), and binding energy (d) as a function of translation at the Fermi energy for system B.

(B0*, B1*, B2*) can be found in the local maxima in the full transmission map, while the "alike" couplings (B0, B1, B2) are located in the low transmission areas. Fig. 5(b) shows more regions with a high thermopower as there are more geometries dominated by destructive interference effects. The sign of the thermopower at these regions switches from positive to negative, as the interference positions shift with the translation. However, the thermopower at the geometries around the fully eclipsed structure of system B all exhibit positive value, which is different from the results of system A. The whole power factor map is dramatically enhanced in the system of anthracene rings, especially at the geometries around the fully eclipsed structure, shown in Fig. 5(c). As the positions of the two molecules are translated along the y-direction, the power factor is modulated and periodically presents local maxima. Interestingly, the power factor exhibits very low values along $y = 1.6$, which is obviously problematic if we are looking for a large thermoelectric response. The binding energy of system B approaches a local minimum at (0.0, 1.4), consistent with what is observed for system A. Again, this minimum binding energy position is not the optimal position for the thermopower and power factor. Actually, the minimum energy structure has a very low power factor. In contrast, the fully eclipsed structure of system B has good thermopower value and high power factor and exhibits a good thermoelectric response.

4.3 System C

The final question we have in our investigation of the role of conjugation length is: does a mixed combination of an anthracene ring and a benzene ring work better? Fig. 6 (left) shows how the starring assignment works in system C. The transmission results change due to the symmetry breaking. There is destructive interference between the resonances from the highest occupied orbital and lowest unoccupied orbital for the structure C0, as predicted by the starring rules, but the position is shifted away from the Fermi energy. As the systems are translated along the short axis (y) to C1, the interference position shifts back to the Fermi energy. Both C0* and C1* have high transmission without interference.

Fig. 7(a) shows that the total transmission through system C is generally weaker than system B. The geometries around (0.0, 0.0) exhibit high transmission and a high power factor, but a low thermopower as the interference feature is no longer so close to the Fermi energy. Interestingly, the power factor for system C also has a very low value along $y = 1.6$, which is similar to system B. The local minimum of binding energy for system C is around (0.0, 0.8), shown in Fig. 7(d). However, the geometry of system C at (0.0, 0.0) is more stable than the fully eclipsed structure of system A and exhibits a high transmission and power factor.

4.4 System D

Another way to modify the conductance and change the binding energy is to add substituents. Here, we investigate the conductance and thermoelectric response of π-stacked systems substituted by cyano groups or the combination of a hydroquinone and a quinone. The components of the transmission between pairs

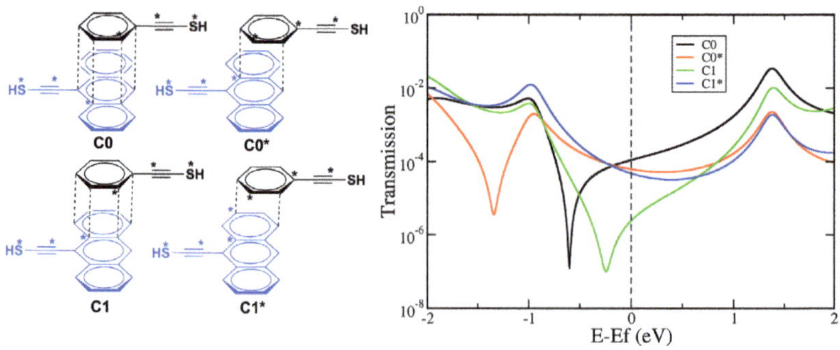

Fig. 6 Left: two groups of selected structures of system C. The molecules are even alternate hydrocarbons and are marked with stars to show when alike (low) or disjoint (high) coupling is expected. Right: transmission through these selected structures of system C.

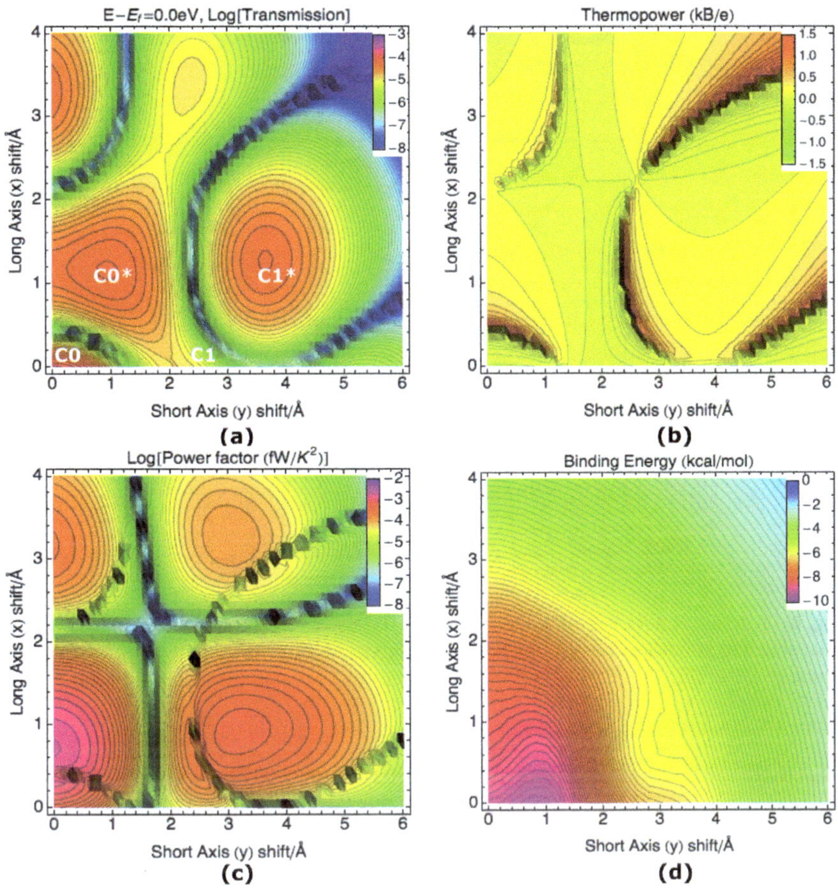

Fig. 7 Full spatial maps of the transmission (a), thermopower (b), power factor (c), and binding energy (d) as a function of translation at the Fermi energy for system C.

of atoms in the molecule, known as the local transmission[20] is calculated, in order to illustrate the roles of the substituents in the electron transport.

The starring rules cannot be used in the structures with substituents, however, we still consider the two typical geometries in the substituted systems for comparison with system A. Fig. 8 shows that the total transmission of the fully eclipsed structure exhibits destructive interference, similar to the result of system A0. However, at the Fermi energy, the total transmission through system D0 is higher than system A0. When the local transmission at the Fermi energy is calculated, it is found that several extra paths are added by the cyano group, and these contribute to the total transmission. The dislocated structure (D0*), with two sites eclipsed, exhibits high transmission near the Fermi energy and an absence of destructive interference. However, we see less influence of the cyano group, in terms of extra paths in the local transmission through D0* (Fig. 8).

The full spatial map of transmission (Fig. 9) at the Fermi energy shows high values around (0.0, 0.0), unlike that seen for system A. There are also more interference areas in the transmission map, compared with system A. This means

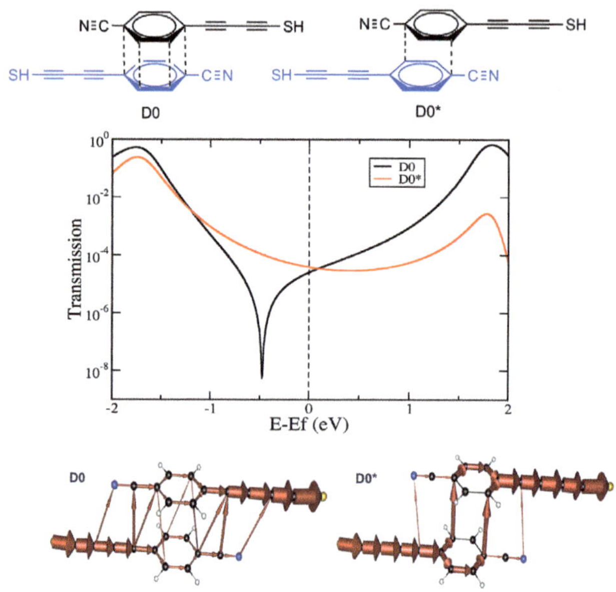

Fig. 8 Top: two selected structures of system D. Middle: total transmission through the two selected structures of system D. Bottom: the local transmission at the Fermi energy through D0 and D0*.

that there are also more regions with high thermopower. The most stable position of system D is located at (0.0, 1.4) in the binding energy map, similar to system A. There is, however, an additional local minimum at (1.6, 0.0). The power factor is low at both of these two minima areas, but has a high value at the geometries around the fully eclipsed structure, just like system A.

4.5 System E

The final system we consider comprises hydroquinone as a donor and quinone as an acceptor. Two selected structures E0 and E0* are shown in Fig. 10 (top), in order to compare with system A. The total transmission of the fully eclipsed structure E0 and the dislocated structure E0* are calculated and shown in Fig. 10 (middle). The transmission exhibits a high value in the structure of E0, and a destructive interference feature is present far from the Fermi energy. E0* shows high transmission without interference features across a wide energy range. The local transmission of system E doesn't indicate any extra paths from hydroquinone and quinone, shown in Fig. 10 (bottom). However, the local transmission of E0 and E0* both show several sites favoured by the overlap of the structure, resulting in their high transmission at the Fermi energy.

There is a region around the structure E0 which exhibits very high transmission in the full spatial map (Fig 11(a)). Fig. 11(b) shows that almost all the geometries exhibit a negative thermopower, except some limited regions with a high positive thermopower. The power factor at the geometries around E0 is significantly enhanced and form local maxima, shown in Fig. 11(c). The binding energies are shown in Fig. 11(d) as a function of (x, y) displacement. Interestingly,

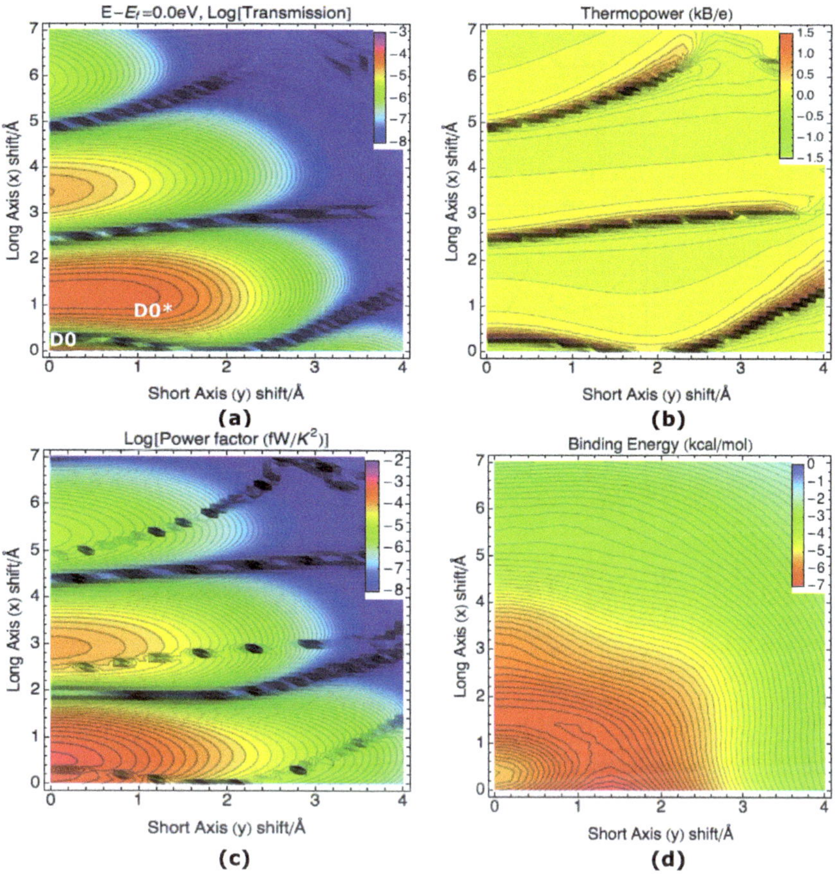

Fig. 9 Full spatial maps of the transmission (a), thermopower (b), power factor (c), and binding energy (d) as a function of translation at the Fermi energy for system D.

the minimum binding energy lies at (0.0, 0.0). The fully eclipsed structure of quinhydrone E0 is the most stable structure in all of these geometries, which is different from the former π-stacked systems. The transmission and power factor of quinhydrone both have high values at (0.0, 0.0). The fully eclipsed structure of quinhydrone that has the minimum energy and exhibits a high power factor, making it a possible candidate for a thermoelectric device. While the quinhydrone is a relatively simple donor–acceptor system, it is envisaged that other donor–acceptor pairs could also be explored by this method.

4.6 Discussion

While the systems considered here are somewhat artificial, they neatly illustrate a number of points for understanding transport through π-stacked systems. The conclusions that should be drawn are different for systems where π-stacked structures form by accident, from those where π-stacked structures are formed by design.

First we consider the case where aggregation in the junction results in the accidental formation of π-stacked structures. In a system without substituents,

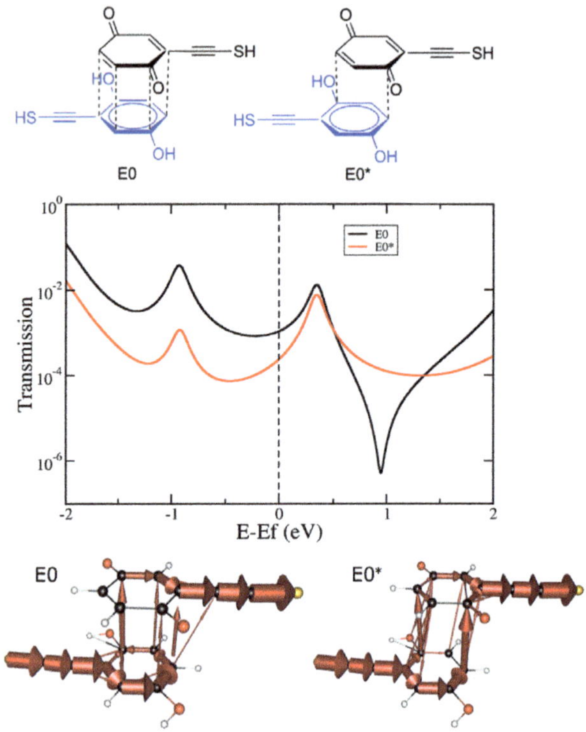

Fig. 10 Top: two selected structures of system E. Middle: total transmission through the two selected structures of system E. Bottom: the local transmission at the Fermi energy through E0 and E0*.

the binding energy indicates that the fully eclipsed structures are unlikely to form preferentially but rather some dislocated structure. This means that interference effects are unlikely to dominate and consequently the transport is measurable and the thermopower is not at a maximum. It may be that as these structures are pulled, for example in an STM-break junction experiment, the structure might move closer to or further from an interference feature and thus the transmission decreases or increases and the thermopower increases or decreases.

Substituents can have three effects, they can simply shift the transport resonances (the effect most commonly considered when they are described as electron donating or withdrawing groups), but they can also introduce additional pathways for transport and change the relative stability of different configurations. The case of the system D illustrated how cyano groups can provide an extra pathway and shift interference features. This case is important for considering the transport properties of accidentally formed π-stacked structures where the molecules involved have two binding groups. In this case, one binding group will be bound to each electrode and the other can be providing additional pathways and shifting interference features.

Using substituents to change the relative stability of different geometries becomes relevant when we seek to design optimal π-stacked structures. Modifying the system to a quinone and hydroquinone clearly (and somewhat

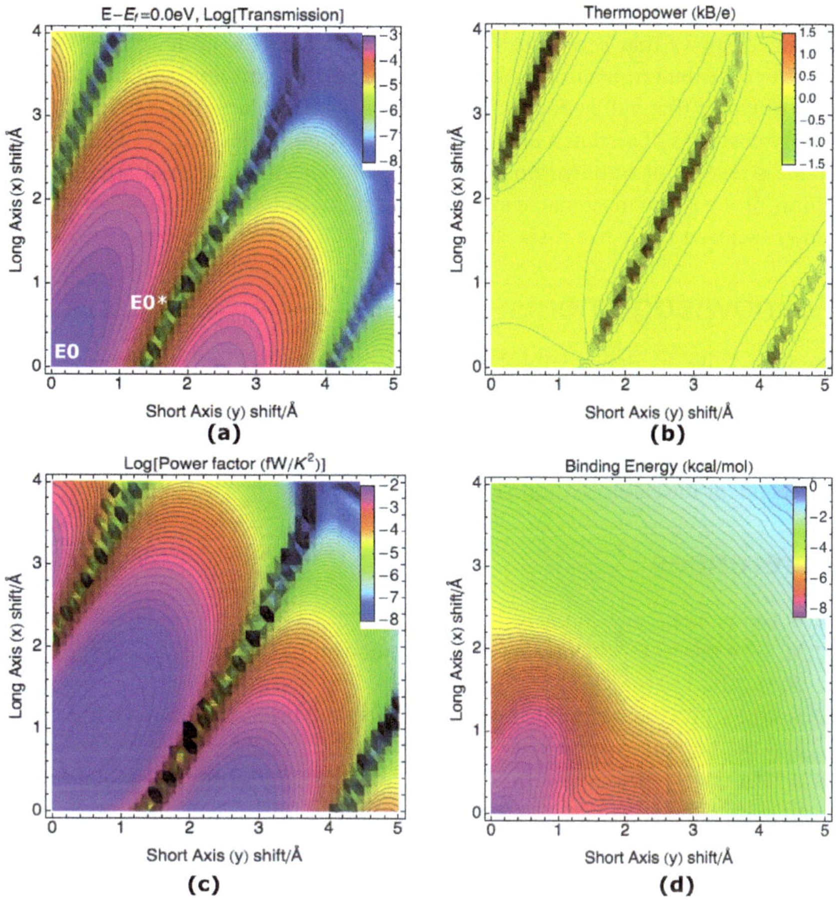

Fig. 11 Full spatial maps of the transmission (a), thermopower (b), power factor (c), and binding energy (d) as a function of translation at the Fermi energy for system E.

obviously) results in the fully eclipsed structure being favoured as two hydrogen bonds can form and we then can achieve the result that the most stable structure also maximises the thermoelectric response. The fully eclipsed structures could also be formed by binding the two sides together in a cyclophane; however, in this case the potential advantage of a non-bonded system for reducing heat transport is lost. While it was not explored in this paper, it has been shown that substitution can shift the minima across a considerable region of the map in conjugated systems,[21] suggesting that we really can pick out regions of interest and design molecules to form these structures.

5 Conclusions

We found that the transmission and power factor (S^2G) are both enhanced by increasing the conjugation length or adding substituent groups. In particular, the π-stacked system of two anthracene rings exhibits a good thermopower value and high power factor, predicting that an increased conjugation length can dramatically enhance the thermoelectric response. The local transmission of system D shows that several extra paths are added by the cyano group, which increases the total transmission at the Fermi energy. Substituents can also change the structures that will be formed in junctions, with the local minimum binding energy positions of system A and system B both lying at (0.0, 1.4), while the most stable structure of quinhydrone is at (0.0, 0.0). The fully eclipsed structure of quinhydrone is the most stable system that could exhibit high transmission and power factor in these five π-stacked systems.

Acknowledgements

We would like to thank Mikkel Strange, Jacob Lykkebo and Tim Hansen for helpful discussions. This work was supported through funding from the European Research Council under the European Union's Seventh Framework Program (FP7/2007- 2013)/ERC Grant Agreement no. 258806 and The Danish Council for Independent Research Natural Sciences.

References

1 B. Giese, *Annu. Rev. Biochem.*, 2002, **71**, 51–70.
2 M. D. Watson, F. Jckel, N. Severin, J. P. Rabe and K. Mllen, *J. Am. Chem. Soc.*, 2004, **126**, 1402–1407.
3 S. Wu, M. T. Gonzalez, R. Huber, S. Grunder, M. Mayor, C. Schonenberger and M. Calame, *Nat. Nanotechnol.*, 2008, **3**, 569–574.
4 S. Martin, I. Grace, M. R. Bryce, C. Wang, R. Jitchati, A. S. Batsanov, S. J. Higgins, C. J. Lambert and R. J. Nichols, *J. Am. Chem. Soc.*, 2010, **132**, 9157–9164.
5 S. T. Schneebeli, M. Kamenetska, Z. Cheng, R. Skouta, R. A. Friesner, L. Venkataraman and R. Breslow, *J. Am. Chem. Soc.*, 2011, **133**, 2136–2139.
6 G. C. Solomon, C. Herrmann, J. Vura-Weis, M. R. Wasielewski and M. A. Ratner, *J. Am. Chem. Soc.*, 2010, **132**, 7887–7889.
7 G. C. Solomon, J. Vura-Weis, C. Herrmann, M. R. Wasielewski and M. A. Ratner, *J. Phys. Chem. B*, 2010, **114**, 14735–14744.
8 J. P. Bergfield and C. A. Stafford, *Nano Lett.*, 2009, **9**, 3072–3076.
9 P. Reddy, S.-Y. Jang, R. A. Segalman and A. Majumdar, *Science*, 2007, **315**, 1568–1571.
10 K. Baheti, J. A. Malen, P. Doak, P. Reddy, S.-Y. Jang, T. D. Tilley, A. Majumdar and R. A. Segalman, *Nano Lett.*, 2008, **8**, 715–719.
11 J. A. Malen, P. Doak, K. Baheti, T. D. Tilley, R. A. Segalman and A. Majumdar, *Nano Lett.*, 2009, **9**, 1164–1169.
12 S. K. Yee, J. A. Malen, A. Majumdar and R. A. Segalman, *Nano Lett.*, 2011, **11**, 4089–4094.
13 A. Tan, J. Balachandran, S. Sadat, V. Gavini, B. D. Dunietz, S.-Y. Jang and P. Reddy, *J. Am. Chem. Soc.*, 2011, **133**, 8838–8841.
14 S. Datta, *Quantum Transport: Atom to Transistor*, Cambridge University Press, New York, 1st edn, 2005.
15 J. Enkovaara, C. Rostgaard, J. J. Mortensen, J. Chen, M. Duak, L. Ferrighi, J. Gavnholt, C. Glinsvad, V. Haikola, H. A. Hansen, H. H. Kristoffersen, M. Kuisma, A. H. Larsen, L. Lehtovaara, M. Ljungberg, O. Lopez-Acevedo,

P. G. Moses, J. Ojanen, T. Olsen, V. Petzold, N. A. Romero, J. Stausholm-Mller, M. Strange, G. A. Tritsaris, M. Vanin, M. Walter, B. Hammer, H. Hkkinen, G. K. H. Madsen, R. M. Nieminen, J. K. Nrskov, M. Puska, T. T. Rantala, J. Schitz, K. S. Thygesen and K. W. Jacobsen, *J. Phys.: Condens. Matter*, 2010, **22**, 253202.
16 A. Tkatchenko and M. Scheffler, *Phys. Rev. Lett.*, 2009, **102**, 073005.
17 T. Frauenheim, G. Seifert, M. Elstner, T. Niehaus, C. Khler, M. Amkreutz, M. Sternberg, Z. Hajnal, A. D. Carlo and S. Suhai, *J. Phys.: Condens. Matter*, 2002, **14**, 3015.
18 A. Pecchia and A. D. Carlo, *Rep. Prog. Phys.*, 2004, **67**, 1497.
19 L. Salem, *The Molecular Orbital Theory of Conjugated Systems*, ed. W. A. Benjamin, New York, 1st edn, 1966.
20 G. C. Solomon, C. Herrmann, H. Thorsten, V. Mujica and M. A. Ratner, *Nat. Chem.*, 2010, **2**, 223–228.
21 J. Vura-Weis, M. A. Ratner and M. R. Wasielewski, *J. Am. Chem. Soc.*, 2010, **132**, 1738–1739.

Faraday Discussions

PAPER

A quantum chemical study from a molecular transport perspective: ionization and electron attachment energies for species often used to fabricate single-molecule junctions

Ioan Bâldea†*

Received 8th May 2014, Accepted 21st May 2014
DOI: 10.1039/c4fd00101j

The accurate determination of the lowest electron attachment (EA) and ionization (IP) energies for molecules embedded in molecular junctions is important for correctly estimating, for example, the magnitude of the currents (I) or the biases (V) where an I–V curve exhibits significant non-Ohmic behavior. Benchmark calculations for the lowest electron attachment and ionization energies of several typical molecules utilized to fabricate single-molecule junctions characterized by n-type conduction (4,4′-bipyridine, 1,4-dicyanobenzene and 4,4′-dicyano-1,1′-biphenyl) and p-type conduction (benzenedithiol, biphenyldithiol, hexanemonothiol and hexanedithiol) based on the EOM-CCSD (equation-of-motion coupled-cluster singles and doubles) state-of-the-art method of quantum chemistry are presented. They indicate significant differences from the results obtained within current approaches to molecular transport. The present study emphasizes that, in addition to a reliable quantum chemical method, basis sets much better than the ubiquitous double-zeta set employed for transport calculations are needed. The latter is a particularly critical issue for correctly determining EAs, which is impossible without including sufficient diffuse basis functions. The spatial distribution of the dominant molecular orbitals (MOs) is another important issue, on which the present study draws attention, because it sensitively affects the MO energy shifts Φ due to image charges formed in electrodes. The present results cannot substantiate the common assumption of a point-like MO midway between electrodes, which substantially affects the actual Φ-values.

1 Introduction

Electron or hole injection into molecules embedded between two electrodes represents an important issue in the fabrication of molecular devices. The

Theoretische Chemie, Universität Heidelberg, Im Neuenheimer Feld 229, D-69120 Heidelberg, Germany. E-mail: ioan.baldea@pci.uni-heidelberg.de

† Also at National Institute for Lasers, Plasma and Radiation Physics, Institute of Space Sciences, Bucharest, Romania.

efficiency of the charge injection and transport in molecular junctions is controlled by the highest occupied or lowest unoccupied molecular orbital (HOMO or LUMO, respectively), whichever is closest to the electrodes' Fermi energy E_F, and the key quantity is the energy offset $\varepsilon_0 = \min(E_F - E_{HOMO}, E_{LUMO} - E_F)$. It can be compared to a tunneling barrier, which charge carriers have to overcome to generate a current. Ultraviolet photoelectron spectroscopy (UPS),[1] thermopower[2–4] and transition voltage spectroscopy (TVS)[5–10] represent current methods to estimate the relative alignment of the dominant molecular orbital from experimental data.

Recent analysis of a variety of transport data demonstrated that full current–voltage *I–V* curves beyond the ohmic regime can be quantitatively reproduced by assuming that molecular transport is dominated by a single level (to be identified with the HOMO or LUMO).[7,9,11–14] This is an enormous simplification. Still, the correct description of the relative alignment ε_0 ($=E_{LUMO} - E_F$ or $E_F - E_{HOMO}$) remains an important challenge for *ab initio* approaches to charge transport through single-molecule junctions. It is a challenge particularly because of the high accuracy needed. Values directly determined in ultraviolet photoelectron spectroscopy (UPS) experiments amount to $\varepsilon_0 \sim 1$ eV.[1] Results based on TVS by using a model able to excellently reproduce current–voltage ($I - V$) curves measured in single-molecule junctions demonstrate that ε_0 can be even smaller ($\varepsilon_0 \simeq 0.6$ eV (ref. 13)). It should be clear that, in view of such low ε_0 values, estimates for ionization energies (IPs) and electroaffinities (EAs) with errors $\simeq 0.5$ eV, typical for quantum chemical methods of moderate accuracy, are unacceptable. Noteworthily, the quantity ε_0 is important not only because it determines the magnitude of the currents, but also because it indicates the biases beyond which an *I–V* curve becomes significantly nonlinear.[12]

An important message, which the present study aims to convey, is that the accurate determination of the HOMO and LUMO energies (or, more precisely, the lowest IP and EA with reversed sign) is a non-trivial issue, even for isolated species of interest for molecular transport. Obviously, this is a minimum requirement for any molecular transport approach.

In the present paper we report the results of ongoing work focusing on several prototypical molecular species, which are mostly utilized in the fabrication of single-molecule junctions, and examine the reliability of the results for the IP ($\approx -E_{HOMO}$) and EA ($\approx -E_{LUMO}$) as obtained within the current methods employed in molecular transport with accurate estimates obtained within well-established quantum chemical methods. To avoid misunderstandings, it is worth explicitly mentioning that we will restrict ourselves throughout to the lowest IP and EA; in fact, if at all, for the molecules analyzed below it is only the lowest EA which corresponds to a stable anion (EA > 0).

The molecular species to be considered will include, besides 4,4′-bipyridine (44BPY), a molecule recently considered from a TVS perspective,[13,15] molecules often used to fabricate two classes of molecular junctions. One class is represented by molecules characterized by n-type (LUMO-mediated) conduction, and the other class comprises molecules characterized by p-type (HOMO-mediated) conduction. As specific examples belonging to the first class, we will consider 44BPY,[3] BDCN (1,4-dicyanobenzene)[16] and 2BDCN (4,4′-dicyano-1,1′-biphenyl).

From the second class, we will examine oligophenylene dithiols[4,8] and alkanemono- and dithiols.[17]

Because of the well-documented shortcomings of ubiquitous methods based on density functional theory (DFT), emphasis will be on post-DFT methods. Besides the methods already utilized in approaches to molecular transport (*e.g.*, Δ-SCF,[18] GW[19] and MP2 (ref. 20)), we will consider truly *ab initio* methods used in quantum chemistry: outer valence Green's functions (OVGF),[21,22] second-order algebraic-diagrammatic constructions [ADC(2)],[23,24] and equation-of-motion (EOM) coupled-cluster (CC)[25] approaches. The importance of using appropriate quantum chemical methods will be emphasized. As is well known, calculating EAs is a very delicate problem in quantum chemistry.[26]

2 Methods

Active molecules embedded in molecular junctions can be treated at various levels of theory, ranging from tight-binding (extended Hückel) to post-DFT. In order to facilitate understanding the message which the results reported below aim to convey, we will briefly present the methods utilized in this study.

(i) If the picture based on the *self-consistent field* (SCF) were valid (or, equivalently, electron correlations were absent), the energy of the highest occupied Hartree–Fock (HF) orbital (*i.e.*, HOMO) with reversed sign would represent the lowest ionization energy, IP $= -E_{\mathrm{SCF,HOMO}}$ (Koopmans theorem). The physical meaning of the virtual (unoccupied) orbital is controversial.[27,28] Virtual HF orbitals might have physical meaning if descriptions based on small basis sets succeeded (at least semi-) quantitatively, but this is often not the case. Table 8 illustrates this failure, where relevant results in connection with existing MP2-based transport approaches and minimal (STO-6G) basis sets[29] are presented. At the other extreme, it is also well known that the HF LUMO energy goes to zero in the complete basis set limit.[30,31] This is a reason why attempts to "improve" the quality of a transport approach by using larger and larger basis sets end up with unphysical results.[32] For large atomic orbital (AO) basis sets, the virtual HF orbitals have mathematical rather than physical meaning, namely in providing an expansion manifold for the physical states of interest. In some cases, individual virtual or unoccupied HF orbitals (the LUMO in particular) can reasonably describe, *e.g.*, anionic bound or resonance states semi-quantitatively, provided the size of the AO basis used is not too large.[13]

(ii) Within *DFT approaches*, the single particle solutions of the Kohn–Sham (KS) equations are handled as if the corresponding eigenvalues/eigenfunctions are real orbital energies/wave functions. The corresponding implementation in a Landauer-NEGF formalism is straightforward because the DFT description is mathematically a single-particle description. This is why the DFT approaches to molecular transport are by far the most popular to date. The drawbacks of such DFT approaches are well documented. The drastic underestimation of the HOMO–LUMO gap and the related lineup problem (HOMO–LUMO energies too close to the electrodes' Fermi energy) are the issues most frequently mentioned and are not at all surprising; as is well known,[18] KS "orbitals" are mathematical objects rather than physical orbitals.

The quantum chemical methods used in the present paper to compute the lowest EA and IP energies are as follows.

(iii) The *outer valence Green's function (OVGF) method*[22,33] represents the most elaborate quantum chemical approach based on a single-particle picture. To the best of our knowledge, the OVFG approach has not yet been utilized in molecular transport studies. The OVGF method is a way to approximately include the contribution of the electron–electron interaction beyond HF. Details can be found in ref. 22 and 33. As explained below, this method is superior to the MP2-like approximation used recently[20] (see Section 4.3). The OVGF method[22,33] exactly treats the full second- and third-order terms in the self-energy entering the Dyson equation for the one-electron Green function, and is augmented by a geometrical approximation to also include further higher-order corrections.[34] Let us briefly note at this point that the self-energy is quantity specific to the many-body theory; see, *e.g.*, ref. 35. The results for the electron affinities and ionization energies obtained by considering the second- and third-order terms are shown in the tables presented below (labeled as "2nd-order pole" and "3rd-order pole", respectively) along with those of the full OVGF. The corresponding pole strengths[22] are also indicated (percentages in parentheses).

(iv) *The algebraic-diagrammatic construction (ADC)* is based on a diagrammatic perturbation expansion. ADC(n) defines an approximate scheme of infinite partial summations exactly up to the n-th order of perturbation theory.[23,24] Its second-order version, ADC(2), is superior to the so-called *GW* approximation,[19] since only bubble contributions are included within the GW and they are not all second-order terms; expressing the self-energy Σ by the product of the single-electron Green's function G and the effective interaction W ($\Sigma \sim$ GW, thence the name GW), vertex corrections[35] are neglected within GW.

(iv) *Equation-of-motion (EOM) coupled-cluster (CC) approaches* restricted to singles and doubles (CCSD),[36] hybrid (CC2)[37] and perturbative [CCSD(2)][38,39] levels will be extensively applied in this paper. In the CC-terminology, "singles", "doubles" and "triples" mean that the cluster operator is truncated to single, double and triple excitations. Corrections due to triples [EOM-CCSD(T)] will be also considered; the fact that these corrections are altogether negligible is an indication of the accuracy of the state-of-the-art EOM-CCSD method.

(vi) *Energy difference* (Δ-) methods will be utilized for all of the aforementioned cases. Within these methods, the lowest IP and EA are estimated as the differences between the ground state energies E of the corresponding molecular charge species (M = SCF, DFT, MP2, CCSD, CC2) at the equilibrium geometry of the neutral molecule:

$$\mathrm{IP} \rightarrow \Delta^{i}_{\mathrm{M}} = \mathscr{E}_{\mathrm{M,cation}} - \mathscr{E}_{\mathrm{M,neutral}}, \tag{1}$$

$$\mathrm{EA} \rightarrow \Delta^{a}_{\mathrm{M}} = \mathscr{E}_{\mathrm{M,neutral}} - \mathscr{E}_{\mathrm{M,anion}}. \tag{2}$$

3 Computational details

The results of the SCF, DFT/B3LYP, MP2 and OVGF calculations reported below were computed with Gaussian 09.[40] Coupled-cluster calculations of the IP, EA, total energies of the various charge species and excitation energies were performed with CFOUR.[41] Calculations within the so-called regular (strict) ADC(2) reported here were made with the fully parallelized PRICD-$\Sigma(2)$ code,[42] which is interfaced to MOLCAS.[43] As amply documented by the extensive work of the Heidelberg theoretical chemistry group, the results based on the strict ADC(2) are comparable to the second-order approximate coupled cluster singles and doubles model (CC2).[37] Augmented with extra terms in an extended version [ADC(2)x],[44] the results are comparable to the equation of motion-coupled cluster singles and doubles method (EOM-CCSD).[36] Unfortunately, a code enabling computations for molecular sizes of interest for molecular transport is not yet available.

The inspection of the tables presented below reveals that the results obtained *via* the state-of-the-art IP- and EA-EOM-CCSD method[36,45] and the aug-cc-pVDZ (Dunning augmented correlation consistent double zeta) sets can be trusted. This is illustrated both by the good agreement between the EOM-CCSD and the Δ-CCSD values and by the fact that corrections due to triples [CCSD(*T*)] yield changes which are irrelevant within numerical errors. For understanding the impact of polarization and diffuse functions, we also present results obtained by using other basis sets: cc-pVDZ and cc-pVTZ, Pople basis sets (6-31G*, 6-311G* 6-311++G(d,p)), Dunning–Huzinaga double- and triple-zeta (DZ (DZ95 in ref. 40), DZP, TZ2P), Karlsruhe basis sets (svp, dzp, tzp, qz2p) and (merely to compare with earlier MP2-based transport calculations, *cf.* Table 8) STO-6G.

4 Results and discussion

Because the tables presented in this paper, which contain very detailed information on both the methods and the basis sets employed, are self-explanatory, below we will only briefly emphasize the main aspects related to the lowest electron affinity (EA) and ionization (IP) energies of the molecules of interest. Still, as a technical remark, let us mention that, in view of the fact that double zeta (DZ) sets are ubiquitous in transport studies, among other basis sets, we have always included DZ-based results in the relevant tables.

4.1 Lowest electron attachment energies

Tables 1, 2 and 3 report the results of the lowest electron attachment energies for 4,4′-bipyridine ($C_6H_4N_2$, 44BPY), 1,4-dicyanobenzene (NC–C_6H_4–CN, BDCN) and 4,4′-dicyano-1,1′-biphenyl (NC–$(C_6H_4)_2$–CN, 2BDCN).

Table 1 presents very detailed numerical results for 44BPY, a showcase molecule,[13,15,46–48] in order to illustrate the main issues which we have encountered in calculations of the electron affinities for molecules utilized in molecular electronics. The EA-EOM-CCSD method predicts a weakly bound anion $44BPY^{\bullet-}$ (EA $\gtrsim$ 0). The essential conditions for this is the inclusion of a sufficient number of diffuse basis functions. As emphasized recently,[47] it is not merely the basis set size that matters: as is visible in Table 1, the basis set CC-pVTZ, which is larger than aug-cc-pVDZ, cannot stabilize the anion, simply because diffuse functions are missing.

Notice also that a correct description requires a proper treatment of electron correlations. Even including sufficient diffuse basis functions (aug-cc-pVDZ), the SCF description (both Koopmans theorem and Δ-SCF) is even qualitatively inadequate; the anion is predicted to be unstable. This is already known from

Table 1 The lowest EA of 4,4′-bipyridine (44BPY) computed by various methods, and the exciton binding energy (EBE). The geometry of the neutral molecule was optimized at the DFT/B3LYP/aug-cc-pVDZ level

44BPY/method	Basis set	No. functions	EA (eV)	EBE (eV)
EOM-CCSD	DZ	136	−0.684	—
EOM-CCSD	6-31G**	208	−0.686	—
EOM-CCSD	svp	208	−0.421	—
EOM-CCSD	cc-pVDZ	208	−0.484	—
EOM-CCSD	dzp	220	−0.353	—
EOM-CCSD	6-311G**	264	−0.340	—
EOM-CCSD	tzp	276	−0.290	—
EOM-CCSD	TZ2P	360	−0.0247	—
EOM-CCSD	qz2p	416	−0.0123	—
EOM-CCSD	cc-pVTZ	472	−0.438	—
EOM-CCSD	aug-cc-pVDZ	348	0.0322	4.487
EOM-CCSD(*T*)	aug-cc-pVDZ	348	0.0293	—
EOM-CC2	aug-cc-pVDZ	348	0.360	—
ADC(2)	aug-cc-pVDZ	348	0.370	—
Δ-SCF	aug-cc-pVDZ	348	−0.564	—
Δ-MP2	aug-cc-pVDZ	348	−0.500	—
Δ-CCSD	aug-cc-pVDZ	348	0.0043	—
Δ-DFT/B3LYP	DZ	136	0.185	—
Δ-DFT	6-31G*	196	−0.126	—
Δ-DFT/B3LYP	svp	208	0.124	—
Δ-DFT	6-31G**	208	−0.115	—
Δ-DFT	cc-pVDZ	208	0.0290	—
Δ-DFT	6-31+G(d)	244	0.396	—
Δ-DFT	6-311G**	264	0.188	—
Δ-DFT	6-31+G(d,p)	268	0.407	—
Δ-DFT	6-31++G(d,p)	276	0.407	—
Δ-DFT/B3LYP	cc-pVTZ	472	0.281	—
Δ-DFT	aug-cc-pVDZ	348	0.444	—
Δ-DFT	aug-cc-pVTZ	736	0.467	—
-KS-LUMO	aug-cc-pVDZ	348	2.068	—
Koopmans theorem	aug-cc-pVDZ	348	−0.832	—
2nd-order pole (98.3%)	aug-cc-pVDZ	348	−0.466	—
3rd-order pole (98.6%)	aug-cc-pVDZ	348	−0.562	—
OVGF (98.5%)	aug-cc-pVDZ	348	−0.540	—

earlier work.[47,49,50] As is visible in Table 1, we found that electron correlations (which, by definition, measure the deviation from the SCF) cannot be adequately included *via* MP2 and the OVGF:[21,22] the $44BPY^{\bullet-}$ anion remains unstable.

On the other hand, while agreeing among themselves, the CC2, ADC(2) and Δ-DFT methods all overestimate the anion stability. With respect to CC2, one can remark that Δ-CC2 performs better than EOM-CC2.

The conclusions formulated above by analyzing the electron affinity of 44BPY also hold for the other two molecules (1,4-dicyanobenzene (BDCN) and 4,4′-dicyano-1,1′-biphenyl (2BDCN)) investigated (see Tables 2 and 3). Quantitatively, there is an important difference: the $BDCN^{\bullet-}$ and $2BDCN^{\bullet-}$ anions are substantially more stable: their electron affinities are by about 0.7 eV larger than that of the $44BPY^{\bullet-}$ anion.

Table 2 The lowest EA of 1,4-dicyanobenzene (BDCN) computed by various methods, and the exciton binding energy (EBE). The geometry of the neutral molecule was optimized at the DFT/B3LYP/aug-cc-pVDZ level

BDCN/method	Basis set	No. functions	EA (eV)	EBE (eV)
EOM-CCSD	DZ	108	−0.0833	—
EOM-CCSD	6-31G**	170	0.0893	—
EOM-CCSD	aug-cc-pVDZ	266	0.717	4.614
EOM-CCSD(*T*)	aug-cc-pVDZ	266	0.717	—
EOM-CC2	aug-cc-pVDZ	266	1.047	—
ADC(2)	aug-cc-pVDZ	266	1.107	—
Δ-MP2	aug-cc-pVDZ	266	0.394	—
Δ-CC2	aug-cc-pVDZ	266	0.602	—
Δ-CCSD	aug-cc-pVDZ	266	0.678	—
Δ-DFT/B3LYP	aug-cc-pVDZ	266	1.127	—
Δ-DFT/B3LYP	aug-cc-pVTZ	552	1.155	—
Δ-SCF	aug-cc-pVDZ	266	0.350	—
-KS-LUMO	aug-cc-pVDZ	266	2.918	—
Koopmans theorem	aug-cc-pVDZ	266	−0.536	—
2nd-order pole (88.6%)	aug-cc-pVDZ	266	1.034	—
3rd-order pole (91.1%)	aug-cc-pVDZ	266	0.253	—
OVGF (90.6%)	aug-cc-pVDZ	266	0.445	—

To summarize, the anions considered above can be accurately described at the CCSD level of theory provided that the basis set employed includes sufficient diffuse functions: triples corrections (*i.e.*, CCSD(*T*)) merely yield modifications of the EA values within numerical errors, and the EA-EOM-CCSD values agree well with the Δ-CCSD values.

Table 3 The lowest EA of 4,4′-dicyano-1,1′-biphenyl (2BDCN) computed by various methods, and the exciton binding energy (EBE). The geometry of the neutral molecule was optimized at the DFT/B3LYP/aug-cc-pVDZ level

2BDCN/method	Basis set	No. functions	EA (eV)	EBE (eV)
EOM-CCSD	DZ	176	−0.113	—
EOM-CCSD	6-31G**	264	0.0677	—
EOM-CCSD	aug-cc-pVDZ	440	0.697	3.567
EOM-CCSD(*T*)	aug-cc-pVDZ	440	0.697	—
EOM-CC2	aug-cc-pVDZ	440	1.033	—
Δ-MP2	aug-cc-pVDZ	440	0.0227	—
Δ-CCSD	aug-cc-pVDZ	440	0.601	—
Δ-CC2	aug-cc-pVDZ	440	0.611	—
Δ-DFT/B3LYP	6-311++G(d,p)	408	1.228	—
Δ-SCF	aug-cc-pVDZ	440	0.121	—
Δ-SCF	6-311++G(d,p)	408	0.117	—
-KS-LUMO	aug-cc-pVDZ	440	2.685	—
Koopmans theorem	aug-cc-pVDZ	440	−0.617	—
2nd-order pole (97.9%)	aug-cc-pVDZ	440	−0.171	—
3rd-order pole (98.3%)	aug-cc-pVDZ	440	−0.330	—
OVGF (98.2%)	aug-cc-pVDZ	440	−0.292	—

4.2 Lowest ionization energies

As representatives of molecules embedded in nanojunctions exhibiting a HOMO-mediated (p-type) conduction, we have studied and present detailed results for benzenedithiol (HS–C_6H_4–SH, BDT, Table 4) and related molecules (S–C_6H_4–S and S–C_6H_3–S, Table 8), dibenzenedithiol (HS–$(C_6H_4)_2$–SH, 2BDT, Table 5), 1,6-hexanemonothiol (H–$(CH_2)_6$–SH, C6MT, Table 6) and 1,6-hexanedithiol (HS–$(CH_2)_6$–SH, C6DT, Table 7).

As is visible in these tables, the lowest ionization energies can be estimated with a good relative accuracy ($\lesssim$4%), which is satisfactory for quantum chemical calculations for many purposes, within IP-EOM-CCSD calculations by using rather modest basis sets. Although not dramatically large, the corresponding absolute error ($\lesssim$0.4 eV) is still non-negligible from a molecular transport perspective in view of the rather small energy offset of the dominant molecular orbital relative to the electrodes' Fermi level. Hence, good basis sets are required

Table 4 The lowest IP of benzenedithiol (BDT) computed by various methods, and the exciton binding energy (EBE). The geometry of the neutral molecule was optimized at the DFT/B3LYP/aug-cc-pVDZ level

BDT/method	Basis set	No. functions	IP (eV)	EBE (eV)
EOM-CCSD	DZ	108	7.492	—
EOM-CCSD	6-31G**	150	7.669	—
EOM-CCSD	svp	150	7.671	—
EOM-CCSD	cc-pVDZ	150	7.642	—
EOM-CCSD	dzp	166	7.632	—
EOM-CCSD	cc-pVTZ	332	7.922	—
EOM-CCSD	aug-cc-pVDZ	246	7.816	3.868
EOM-CCSD(*T*)	aug-cc-pVDZ	246	7.816	—
EOM-CCSD(2)	aug-cc-pVDZ	246	7.883	—
EOM-CC2	aug-cc-pVDZ	246	7.497	—
Δ-MP2	aug-cc-pVDZ	246	8.114	—
Δ-CCSD	6-31G**	150	7.590	—
Δ-CCSD	aug-c-pVDZ	246	7.858	—
Δ-CC2	aug-cc-pVDZ	246	7.874	—
Δ-DFT/B3LYP	DZ	108	7.690	—
Δ-DFT/B3LYP	6-31G*	140	7.527	—
Δ-DFT/B3LYP	svp	150	7.559	—
Δ-DFT/B3LYP	cc-pVDZ	150	7.549	—
Δ-DFT/B3LYP	6-31G**	158	7.531	—
Δ-DFT/B3LYP	6-311G*	196	7.659	—
Δ-DFT/B3LYP	6-31++G(d,p)	196	7.750	—
Δ-DFT/B3LYP	6-311+G(d,p)	228	7.759	—
Δ-DFT/B3LYP	6-311++G(d,p)	234	7.770	—
Δ-DFT/B3LYP	cc-pVTZ	332	7.640	—
Δ-DFT/B3LYP	aug-cc-pVDZ	246	7.669	—
Δ-DFT/B3LYP	aug-cc-pVTZ	514	7.671	—
Δ-SCF	aug-cc-pVDZ	246	7.149	—
-KS-HOMO	aug-cc-pVDZ	246	5.865	—
Koopmans theorem	aug-cc-pVDZ	246	8.038	—
2nd-order pole (87.4%)	aug-cc-pVDZ	246	7.457	—
3rd-order pole (90.1%)	aug-cc-pVDZ	246	7.960	—
OVGF (89.3%)	aug-cc-pVDZ	246	7.750	—

Table 5 The lowest IP of dibenzenedithiol (2BDT) computed by various methods, and the exciton binding energy (EBE). The geometry of the neutral molecule was optimized at the DFT/B3LYP/6-31++G(d,p) level[64]

2BDT/method	Basis set	No. functions	IP (eV)	EBE (eV)
EOM-CCSD	DZ	176	7.294	—
EOM-CCSD	6-31G**	254	7.340	—
EOM-CCSD	svp	254	7.483	—
EOM-CCSD	dzp	276	7.437	—
EOM-CCSD	cc-pVDZ	254	7.433	—
EOM-CCSD	aug-cc-pVDZ	420	7.593	3.470
EOM-CCSD(*T*)	aug-cc-pVDZ	420	7.593	—
EOM-CCSD(2)	aug-cc-pVDZ	420	7.747	—
EOM-CC2	aug-cc-pVDZ	420	7.303	—
Δ-MP2	6-31G*	254	8.629	—
Δ-MP2	6-311++G(d,p)	394	8.831	—
Δ-CCSD	6-31G*	254	7.436	—
Δ-CCSD	aug-cc-pVDZ	420	7.714	—
Δ-CC2	aug-cc-pVDZ	420	7.888	—
Δ-DFT/B3LYP	DZ	176	7.255	—
Δ-DFT/B3LYP	6-31G*	238	7.092	—
Δ-DFT/B3LYP	svp	254	7.172	—
Δ-DFT/B3LYP	cc-pVDZ	254	7.149	—
Δ-DFT/B3LYP	6-31G**	268	7.097	—
Δ-DFT/B3LYP	6-311G*	298	7.248	—
Δ-DFT/B3LYP	6-311G**	328	7.256	—
Δ-DFT/B3LYP	cc-pVTZ	568	7.242	—
Δ-DFT/B3LYP	aug-cc-pVDZ	420	7.269	—
Δ-SCF	aug-cc-pVDZ	420	6.671	—
Δ-SCF	6311++G(d,p)	394	6.819	—
-KS-HOMO	6-311++G(d,p)	394	5.825	—
Koopmans theorem	6-311++G(d,p)	394	7.778	—
2nd-order pole (86.5%)	6-311++G(d,p)	394	7.260	—
3rd-order pole (89.8%)	6-311++G(d,p)	394	7.629	—
OVGF (88.8%)	6-311++G(d,p)	394	7.522	—

not only for EAs, but also for IPs. Even for cations, the various lower-level many-body approximations (MP2, CC2, ADC(2), CCSD(2) and their Δ-versions), deviating by up to ~0.4 eV from the EOM-CCSD approach, are still not too satisfactory.

In view of the present results, the Δ-DFT method cannot be recommended, as the deviations from the IP-EOM-CCSD estimate can be very large; the example presented in Table 7 indicates an error of ~0.8 eV.

A special mention is reserved for the OVGF approximation, which appears to provide IP estimates closest to EOM-CCSD; the differences are smaller than 0.1 eV.

4.3 Comparison with results of previously utilized many-body methods

The foregoing analysis emphasised that both the quantum chemical method and the basis set utilized are important for correct EA and IP estimations. Employing small basis sets is particularly tempting for truly *ab initio* approaches, which are

Table 6 The lowest IP of hexanemonothiol (C6MT) computed by various methods, and the exciton binding energy (EBE). The geometry of the neutral molecule was optimized at the DFT/B3LYP/aug-cc-pVDZ level

C6MT/method	Basis set	No. functions	IP (eV)	EBE (eV)
EOM-CCSD	DZ	106	8.543	—
EOM-CCSD	6-31G**	172	8.741	—
EOM-CCSD	svp	172	8.806	—
EOM-CCSD	dzp	183	8.785	—
EOM-CCSD	cc-pVDZ	172	8.803	—
EOM-CCSD	cc-pVTZ	291	9.107	—
EOM-CCSD	aug-cc-pVDZ	291	8.966	4.130
EOM-CCSD(*T*)	aug-cc-pVDZ	291	8.966	—
EOM-CCSD(2)	aug-cc-pVDZ	291	8.799	—
EOM-CC2	aug-cc-pVDZ	291	8.806	—
ADC(2)	aug-cc-pVDZ	291	8.623	—
Δ-MP2	aug-cc-pVDZ	291	9.0448	—
Δ-CCSD	aug-cc-pVDZ	291	8.922	—
Δ-CC2	aug-cc-pVDZ	291	9.012	—
Δ-DFT/B3LYP	DZ	106	8.989	—
Δ-DFT/B3LYP	svp	172	8.925	—
Δ-DFT/B3LYP	6-31G*	137	8.930	—
Δ-DFT/B3LYP	cc-pVDZ	172	8.918	—
Δ-DFT/B3LYP	6-31G**	179	8.926	—
Δ-DFT/B3LYP	6-311G*	176	9.019	—
Δ-DFT/B3LYP	cc-pVTZ	410	8.975	—
Δ-DFT/B3LYP	aug-cc-pVDZ	291	8.994	—
Δ-DFT/B3LYP	aug-cc-pVTZ	648	8.990	—
Δ-SCF	aug-cc-pVDZ	291	7.988	—
-KS-HOMO	aug-cc-pVDZ	291	6.490	—
Koopmans theorem	aug-cc-pVDZ	291	9.575	—
2nd-order pole (90.6%)	aug-cc-pVDZ	291	8.632	—
3rd-order pole (91.1%)	aug-cc-pVDZ	291	9.109	—
OVGF (90.7%)	aug-cc-pVDZ	291	9.058	—

otherwise impractical, as they require much more RAM memory, disc space and computational time than ubiquitous DFT methods.

In this subsection, we will scrutinize the reliability of the results obtained by the post-DFT approaches reported in two earlier studies.[29,51] To assess the validity of those methods, we will present a comparison with the results of the present methods also using the same small basis sets of ref. 29 and 51.

Ref. 29 reports results obtained within an MP2-like approach for two molecules, namely S–C_6H_4–S and S–C_6H_3F–S, which are similar to BDT(=HS–C_6H_4–SH). In Table 8, we present results for these molecules obtained within the methods described above, along with those extracted from ref. 29. Following ref. 29, we refer to the latter results as "MP2-based". Still, for clarity, we should note that in the present terminology, the method of ref. 29 coincides with that labeled "2nd-order pole" here, as it corresponds to the second-order correction in the electron–electron interaction to the electronic self-energy.[21,22] So, the results labeled "MP2-based" and "2nd-order pole" in Table 8 should coincide. However, they do not coincide; or, more precisely, they are substantially different. We are

Table 7 The lowest IP of hexanedithiol (C6DT) computed by various methods, and the exciton binding energy (EBE). The geometry of the neutral molecule was optimized at the DFT/B3LYP/aug-cc-pVDZ level

C6DT/Method	Basis set	No. functions	IP (eV)	EBE (eV)
EOM-CCSD	DZ	124	8.591	—
EOM-CCSD	6-31G*	148	8.748	—
EOM-CCSD	svp	190	8.847	—
EOM-CCSD	dzp	206	8.829	—
EOM-CCSD	cc-pVDZ	190	8.846	—
EOM-CCSD	cc-pVTZ	444	9.162	—
EOM-CCSD	aug-cc-pVDZ	318	9.026	4.129
EOM-CCSD(*T*)	aug-cc-pVDZ	318	9.026	—
EOM-CCSD(2)	aug-cc-pVDZ	318	8.859	—
EOM-CC2	aug-cc-pVDZ	318	8.482	—
ADC(2)	aug-cc-pVDZ	318	8.696	—
Δ-MP2	aug-cc-pVDZ	318	9.092	—
Δ-CCSD	aug-cc-pVDZ	318	8.974	—
Δ-CC2	aug-cc-pVDZ	318	9.063	—
Δ-DFT/B3LYP	DZ	124	8.209	—
Δ-DFT/B3LYP	svp	190	8.111	—
Δ-DFT/B3LYP	cc-pVDZ	190	8.136	—
Δ-DFT/B3LYP	6-31G**	198	8.167	—
Δ-DFT/B3LYP	6-311G**	244	8.287	—
Δ-DFT/B3LYP	cc-pVTZ	444	8.246	—
Δ-DFT/B3LYP	aug-cc-pVDZ	318	8.273	—
Δ-DFT/B3LYP	aug-cc-pVTZ	698	8.279	—
Δ-SCF	aug-cc-pVDZ	318	8.042	—
-KS-HOMO	aug-cc-pVDZ	318	6.535	—
Koopmans theorem	aug-cc-pVDZ	318	9.628	—
2nd-order pole (90.7%)	aug-cc-pVDZ	318	8.712	—
3rd-order pole (91.1%)	aug-cc-pVDZ	318	9.158	—
OVGF (90.7%)	aug-cc-pVDZ	318	9.116	—

unable to understand these large differences ($\gtrsim$1 eV) visible in Table 8. The only thing not specified in ref. 29 is the molecular geometry utilized in the calculations. However, as expected, the results presented in Table 8 reveal that differences in the (optimized) geometries have a considerably smaller impact (maybe ~0.2 eV). Most importantly, as seen in Table 8, it is the combined effect of an inaccurate method and a too-small basis set that results in very large errors (~3 eV) for IPs.

By using the same (DZ) basis sets as in ref. 51, Table 9 demonstrates that the results deduced within the GW method are not adequate to estimate the energies of the frontier orbitals with the accuracy required for molecular transport studies. The differences of ~0.5 eV between the GW and EOM-CCSD in Table 9 are too large, given that for this molecule (BDT) the HOMO energy offset directly measured by ultraviolet photoelectron spectroscopy (UPS) amounts to $\simeq$1 eV.[1] Drawbacks of GW-based transport approaches were previously pointed out.[52]

4.4 Spatial distribution of the frontier molecular orbitals

As highlighted above, both the method and the basis sets employed are essential to properly estimate the energy of the frontier orbitals of an isolated molecule. Still, however important, the energy of the frontier molecular orbitals of an

Table 8 MP2-based results for the lowest IP of the benzenedithiol (BDT) and BDT-like molecules considered in ref. 29, along with results obtained within methods utilized in this paper

Molecule/optimization	Method	Basis set	No. functions	IP (eV)
BDT				
Optimized B3LYP/aug-cc-pVDZ	2nd-order pole (87.4%)	aug-cc-pVDZ	246	7.457
	3rd-order pole (90.1%)	aug-cc-pVDZ	246	7.960
	OVGF (89.3%)	aug-cc-pVDZ	246	7.750
	2nd-order pole (91.9%)	STO-6G	54	4.702
	3rd-order pole (91.5%)	STO-6G	54	4.775
	OVGF (91.4%)	STO-6G	54	4.754
S–C_6H_4–S				
Ref. 29	MP2-based	STO-6G	52	6.5267
Optimized B3LYP/aug-cc-pVDZ	EOM-CCSD	aug-cc-pVDZ	228	8.920
	2nd-order pole (87.3%)	STO-6G	52	5.139
	3rd-order pole (83.6%)	STO-6G	52	5.319
	OVGF (82.9%)	STO-6G	52	5.353
Optimized MP2/STO-6G	2nd-order pole (88.3%)	STO-6G	52	5.382
	3rd-order pole (85.0%)	STO-6G	52	5.499
	OVGF (84.4%)	STO-6G	52	5.521
S–C_6H_3F–S				
Ref. 29	MP2-based	STO-6G	56	6.8019
Optimized B3LYP/aug-cc-pVDZ	EOM-CCSD	aug-cc-pVDZ	242	8.229
	2nd-order pole (87.4%)	STO-6G	56	5.235
	3rd-order pole (83.7%)	STO-6G	56	5.370
	OVGF (83.0%)	STO-6G	56	5.395
Optimized MP2/STO-6G	2nd-order pole (88.2%)	STO-6G	56	5.444
	3rd-order pole (85.1%)	STO-6G	56	5.522
	OVGF (84.5%)	STO-6G	56	5.534

Table 9 Results for the lowest IP, EA, and HOMO–LUMO gap obtained by using several methods and the same double zeta (DZ) basis sets. Notice that the GW estimates taken from ref. 51 significantly differ from the EOM-CCSD values, which are very accurately estimated within the OVGF. The fact that all methods predict a stable anion (EA > 0) is an artefact of the small (DZ) basis set utilized in this table

Method	IP (eV)	EA (eV)	HOMO–LUMO gap (eV)
EOM-CCSD	7.492	−2.016	9.500
OVGF	7.560	−2.073	9.633
GW	6.9	−2.2	9.1

isolated molecule is not the whole issue. In a molecular junction, the active molecule is linked to electrodes, which yield shifts in energy *via* image charge effects. This effect, which is well established in surface science,[53] was embodied in recent studies on molecular electronics in a simplified form, namely by assuming point-like molecular orbitals midway between electrodes.[3,54] This assumption is indeed a comfortable approximation, as it can readily be implemented in one-shot DFT + Σ transport calculations.[3,55]

The obvious critical point here, on which ref. 13 has recently drawn attention, is to what extent is it legitimate to approximate a real molecular orbital as a point charge. Within classical electrostatics, the interaction energy of an electron located at z, with the image charges created in two infinite planar electrodes, can be exactly expressed by:[56,57]

$$\phi_i(z) = \frac{e^2}{4d}\left[-2\psi(1) + \psi\left(\frac{z - z_s}{d}\right) + \psi\left(\frac{z_t - z}{d}\right)\right]. \tag{3}$$

Here, z_s and z_t are the positions of the image planes (which are slightly shifted from the real electrodes, see, *e.g.*, ref. 13 and citations therein), $d \equiv z_t - z_s$, and $\psi(z) \equiv d \log \Gamma(z)/dz$ is the digamma function. For a real molecular orbital, the image-driven energy shift should be computed by weighting eqn (3) with the MO-spatial density $\rho_{MO} = |\Psi_{MO}|^2$, which is determined by its wave function Ψ_{MO}:

$$\Phi = \int_{z_s}^{z_t} dz \rho_{1D}^{MO}(z)\phi_i(z), \tag{4}$$

$$\rho_{1D}^{MO}(z) = \iint dx dy |\Psi_{MO}(x,y,z)|^2. \tag{5}$$

Close to the electrodes, *e.g.*, $z \gtrsim z_s$, eqn (3) recovers the classical expression

$$\phi_i(z)^{z \gtrsim z_s} \approx \phi_{i,cl}(z) = -\frac{e^2}{4(z - z_s)} \tag{6}$$

for a single image plane. This demonstrates that, in cases of molecular orbitals with significant spatial extension, the main contributions to the image-driven energy shift Φ come from regions close to the electrodes.

Since, except for ref. 13, the spatial extension of the dominant molecular orbitals did not receive consideration in previous studies, we have decided to systematically investigate this aspect for typical molecules of interest for nano-transport. Towards this aim, inspecting spatial densities of (completely unphysical LUMO) Kohn–Sham orbitals makes little sense, and (especially LUMO) HF orbitals may represent too crude an approximation. Therefore, as in ref. 26, we have calculated the natural orbital expansion of the corresponding reduced density matrices at the EOM-CCSD level.

This is the most reliable approach to characterize the spatial distribution of the extra electron or hole in molecules with n-type (LUMO-mediated) or p-type (HOMO-mediated) conduction, and we are not aware of any similar studies conducted in conjunction with molecular transport at this level of theory. For all the molecules considered, by inspecting the natural orbital expansion, we found that the extra electron or hole is almost entirely (>97%) concentrated in a single natural orbital. It is the density of this natural orbital ρ_{MO}, which we have used in the calculations based on eqn (4).

Most importantly from the present standpoint, we did not even find a molecule whose dominant MO resides in a very narrow spatial region around the center.

Rather than being strongly localized close to the center, in cases of n-conduction, we found that the natural orbital of the extra electron is more or less uniformly spread over the whole molecule. This is illustrated by the examples depicted in Fig. 1 and 2 (we employed Gabedit[58] to generate the figures presented

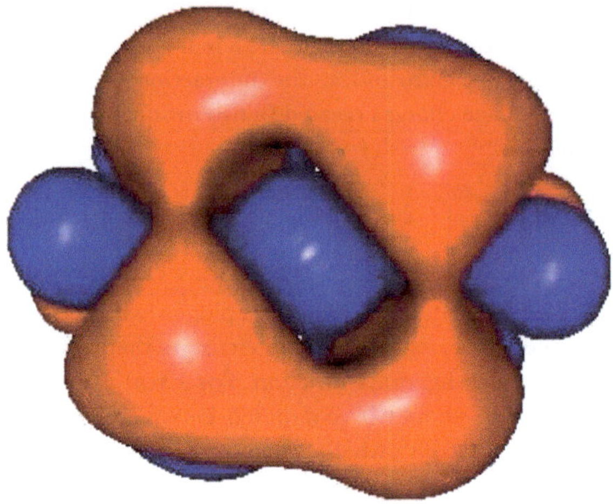

Fig. 1 The almost singly occupied natural orbital corresponding to the extra electron of the 44BPY$^{\bullet-}$ anion ("LUMO") is delocalized over the whole molecule, as obtained *via* EA-EOM-CCSD/aug-cc-pVDZ calculations.

Fig. 2 The almost singly occupied natural orbital of the extra electron in the 2BDCN anion ("LUMO") is delocalized over the whole molecule, as obtained *via* EA-EOM-CCSD/aug-cc-pVDZ calculations.

in this paper.) Therefore, the difference between the value $\Phi_{\mathrm{loc}} = -e^2 \ln 2/d$ obtained by setting $z = (z_s + z_t)/2$ in eqn (3), which corresponds to a point-like MO located in the middle of two infinite metallic plates,[3] is substantially different from the value Φ_r deduced *via* eqn (4) using the realistic natural orbital density. For the 44BPY molecule, the values thus obtained for the image-driven LUMO shifts are $\Phi_r = -2.13$ eV and $\Phi_{\mathrm{loc}} = -1.16$ eV. A difference of about 1 eV is indeed a large effect.

We have also computed spatial distributions of the natural orbital of the extra hole ("HOMO") in cation species relevant for molecules exhibiting p-type conduction. The examples presented in Fig. 3 and 4 illustrate two different behaviors, which we found to be characteristic for HOMO distributions. On one side, we found HOMOs like that presented in Fig. 3, which are similar to the LUMOs discussed above; they are delocalized over the whole molecule. On the other side, we found HOMOs strongly localized on the anchoring groups. This behavior is typical for alkanemono- and dithiols, as illustrated in Fig. 4.

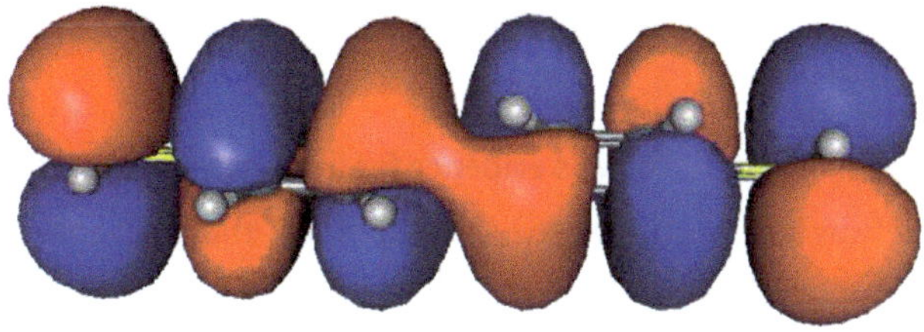

Fig. 3 The almost singly occupied natural orbital of the extra hole ("HOMO") in the 2BDT cation is delocalized over the whole molecule, as obtained *via* IP-EOM-CCSD/aug-cc-pVDZ calculations.

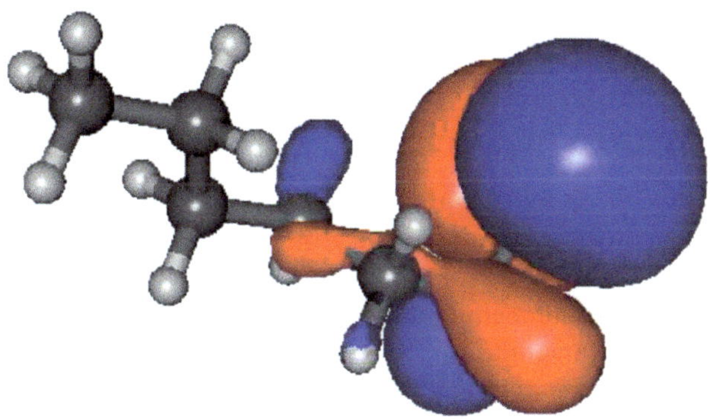

Fig. 4 The almost singly occupied natural orbital of the extra hole ("HOMO") in the hexanemonothiol cation is localized in the vicinity of the sulfur atom, as obtained *via* IP-EOM-CCSD/aug-cc-pVDZ calculations.

To obtain the above value Φ_r ($= -2.16$ eV) for 44BPY, we applied the cutoff procedure near electrodes, as described in detail in ref. 13. The results of preliminary calculations with metal atoms linked to the molecules with LUMO-mediated conduction considered here indicate that, as in ref. 13, the spatial density LUMO (single occupied natural orbital corresponding to the extra electron) does not substantially penetrate into the electrodes. This is important: corrections due to image charges are not dramatically affected by the cutoff procedure close to the electrodes. This behavior contrasts to that of the HOMOs. Whether delocalized (like that of Fig. 3) or localized (like that from Fig. 4) on the terminal groups of the *isolated* molecules, we found that the HOMO distributions substantially penetrate into the electrodes. Although we only checked that this happens for molecules of the types discussed above, we believe that this is a general HOMO property which ensures stable molecule–electrode bonds. A cutoff procedure is needed to eliminate spurious divergences of the classical expression of the interaction energy with image charges at $z = z_{s,t}$ [*cf.* eqn (3) and (6)], which ignores quantum mechanical effects and the electrodes' atomistic structure. A possible cutoff procedure consists of multiplying the classical expression with factors $1 - \exp(-\mu|z - z_{s,t})$ ($\propto |z - z_{s,t}|$ as $z \to z_{s,t}$); see, *e.g.*, ref. 13 and citations therein. Cutoff procedures make sense only if the results do not sensitively depend on the value of the cutoff parameter μ, and this cannot be the case if the

(HO)MO densities have non-negligible values in spatial regions close to the electrodes ($z \sim z_{s,t}$).

4.5 The exciton binding energy as evidence for important electron correlations

As a possible way to quantify electron correlations, the solid-state community employs the difference between the so-called charge gap Δ_c and the optical gap (= lowest excitation energy) Δ_o; see *e.g.* ref. 46 and 59 and citations therein. The charge gap, which is what molecular physicists normally call the HOMO–LUMO gap, can be expressed as $\Delta_c = \text{IP} - \text{EA}$. Loosely speaking (because in reality the single-particle (MO) picture breaks down in the exciton problem), the difference between the charge gap and the optical gap is that, in determining the EA, both the HOMO and LUMO are occupied. By contrast, in the lowest optical excitation, the HOMO becomes empty as the LUMO becomes occupied. Δ_c should be larger than Δ_o because of the (negative) attraction energy between the oppositely charged electron (LUMO) and hole (HOMO). This difference is referred to as the exciton binding energy EBE (see, *e.g.*, ref. 35)

$$\text{EBE} = \Delta_c - \Delta_o. \tag{7}$$

The various tables of the present paper only include results for EBE obtained within the most accurate method utilized (EOM-CCSD). The EBE values shown there are substantial, amounting to up to ~50% of the charge gap. As expected for molecules with aromatic units and delocalized electrons, the EBE decreases with increasing molecular size; compare the EBE values for BDCN (Table 2) and 2BDCN (Table 3), and for BDT (Table 4) and 2BDT (Table 5). The EBE values (EBE ~ 3.5–4.6 eV) estimated for all of the molecules analyzed in the present paper are substantially larger than, *e.g.*, for π-conjugated organic thin films (EBE ~ 0.6–1.4 eV).[60]

So, one should conclude that, for species of interest for molecular electronics, electron correlations are very strong. This aspect may be quite relevant for developing correlated transport approaches;[61] *e.g.*, even if the charge transport is dominated by the LUMO, an electron traveling through the molecule can interact with the HOMO.

5 Conclusion

In this paper, we have presented benchmark quantum chemical calculations for the lowest electron attachment and ionization energies of several isolated molecular species of interest for molecular electronics. In assessing the importance of the accuracy to be achieved by estimating EAs and IPs, it is worth mentioning that a proper understanding of the charge transport at the nanoscale does not only mean to reproduce the (order of) magnitude of the currents (which can be adjusted by "manipulating" both ε_0 and the broadening functions Γ not considered here), but also the biases V characterizing a non-Ohmic regime, which is determined by ε_0 ($eV \sim \varepsilon_0$, see ref. 12).

The main results presented above can be summarized as follows:

(i) For all molecules, the differences between the HF-MO energy (Koopmans theorem) and the Δ-SCF values are large. This demonstrates that orbital relaxation is substantial. Electron correlations are also important, as revealed by important departures from the SCF results as well as by substantial differences between the various post-SCF methods considered.

(ii) The present results demonstrate the need both for accurate methods and good basis sets beyond those currently utilized in transport approaches. In particular, employing basis sets with sufficient diffuse functions is essential to correctly describe electron affinities.

(iii) Kohn–Sham orbital energies can by no means be used to estimate IPs and EAs. Even with Δ-corrections, DFT-based methods do not appear to achieve the desired accuracy of estimating the relevant MO energy offsets ε_0. As visible in the various tables, because Δ-DFT estimates are much more weakly dependent on the basis sets that those of EOM-CCSD, Δ-DFT may convey a false impression on the importance of the basis sets to be utilized in calculations.

(iv) MP2-based methods appear to be completely inadequate for describing anions. For cations, they may still yield substantially different results; *e.g.*, compare the results *via* Δ-MP2 and the second-order correction to self-energy ("2nd-order pole) (>1.6 eV in Table 5). Examples showing that different methods, to include second-order terms in the electron–electron interaction, yield results that are significantly different, have been discussed previously.[62]

(v) For the presently investigated molecules exhibiting p-type conduction, the OVGF method represents an excellent compromise in terms of computational effort and the accuracy of IP estimates. Unlike the other diagrammatic methods considered here (GW and ADC), the OVGF method does not require to self-consistently solve a Dyson equation (which is an integral equation); the electron self-energy $\Sigma(\varepsilon)$ can be expressed in closed analytical form, and what needs to be solved is a nonlinear algebraic equation for ε.[22] Let us also mention that, for the presently considered molecules that form junctions characterized by n-type conduction, the OVGF method turned out to be totally inadequate.

(vi) The spatial distribution of the frontier orbitals plays an important role to reliably estimate image-driven shifts of the relevant MO energies. To the best of our knowledge, this is the first systematic study on the spatial distribution of the extra electron or hole in molecular species of interest for molecular transport at this (EOM-CCSD natural orbital expansion) level of theory. None of the molecules considered in this paper were found to possess point-like frontier molecular orbitals, a fact that contradicts the common assumption made in the field.

Acknowledgements

The author thanks Shachar Klaiman and Evgeniy Gromov for invaluable help in performing the quantum chemical calculations, and Jochen Schirmer for useful discussions. Thanks are also due to Vitja Vysotskiy, whose fully parallelized PRICD-Σ(2) code has been employed to obtain the ADC(2) results for electron affinity and ionization energies reported here. The calculations for this work have been partially done on the high performance bwGrid cluster.[63] Financial support from the Deutsche Forschungsgemeinschaft (grant BA 1799/2-1) is gratefully acknowledged.

Notes and references

1 B. Kim, S. H. Choi, X.-Y. Zhu and C. D. Frisbie, *J. Am. Chem. Soc.*, 2011, **133**, 19864–19877.

2 K. Baheti, J. A. Malen, P. Doak, P. Reddy, S.-Y. Jang, T. D. Tilley, A. Majumdar and R. A. Segalman, *Nano Lett.*, 2008, **8**, 715–719.
3 J. R. Widawsky, P. Darancet, J. B. Neaton and L. Venkataraman, *Nano Lett.*, 2012, **12**, 354–358.
4 S. Guo, G. Zhou and N. Tao, *Nano Lett.*, 2013, **13**, 4326–4332.
5 J. M. Beebe, B. Kim, J. W. Gadzuk, C. D. Frisbie and J. G. Kushmerick, *Phys. Rev. Lett.*, 2006, **97**, 026801.
6 M. Araidai and M. Tsukada, *Phys. Rev. B: Condens. Matter Mater. Phys.*, 2010, **81**, 235114.
7 I. Bâldea, *Chem. Phys.*, 2010, **377**, 15–20.
8 S. Guo, J. Hihath, I. Diez-Pérez and N. Tao, *J. Am. Chem. Soc.*, 2011, **133**, 19189–19197.
9 I. Bâldea, *J. Am. Chem. Soc.*, 2012, **134**, 7958–7962.
10 T. K. Tran, K. Smaali, M. Hardouin, Q. Bricaud, M. Oafrain, P. Blanchard, S. Lenfant, S. Godey, J. Roncali and D. Vuillaume, *Adv. Mater.*, 2013, **25**, 427–431.
11 I. Bâldea, *Phys. Rev. B: Condens. Matter Mater. Phys.*, 2012, **85**, 035442.
12 I. Bâldea, *Chem. Phys.*, 2012, **400**, 65–71.
13 I. Bâldea, *Nanoscale*, 2013, **5**, 9222–9230.
14 I. Bâldea, *J. Phys. Chem. C*, 2013, **117**, 25798–25804.
15 I. Bâldea, *Electrochem. Commun.*, 2013, **36**, 19–21.
16 H. Song, Y. Kim, Y. H. Jang, H. Jeong, M. A. Reed and T. Lee, *Nature*, 2009, **462**, 1039–1043.
17 H. Song, M. A. Reed and T. Lee, *Adv. Mater.*, 2011, **23**, 1583–1608.
18 R. O. Jones and O. Gunnarsson, *Rev. Mod. Phys.*, 1989, **61**, 689–746.
19 F. Aryasetiawan and O. Gunnarsson, *Rep. Prog. Phys.*, 1998, **61**, 237–312.
20 T. Shimazaki and K. Yamashita, *Int. J. Quantum Chem.*, 2006, **106**, 803–813.
21 L. S. Cederbaum, *J. Phys. B: At. Mol. Phys.*, 1975, **8**, 290.
22 W. von Niessen, J. Schirmer and L. S. Cederbaum, *Comput. Phys. Rep.*, 1984, **1**, 57–125.
23 J. Schirmer, *Phys. Rev. A*, 1982, **26**, 2395–2416.
24 J. Schirmer, *Phys. Rev. A*, 1991, **43**, 4647–4659.
25 J. F. Stanton and R. J. Bartlett, *J. Chem. Phys.*, 1993, **98**, 7029–7039.
26 S. Klaiman, E. V. Gromov and L. S. Cederbaum, *J. Phys. Chem. Lett.*, 2013, **4**, 3319–3324.
27 J. M. Schulman, J. W. Moskowitz and C. Hollister, *J. Chem. Phys.*, 1967, **46**, 2759–2764.
28 W. J. Hunt and W. A. G. III, *Chem. Phys. Lett.*, 1969, **3**, 414–418.
29 T. Shimazaki and K. Yamashita, *Int. J. Quantum Chem.*, 2009, **109**, 1834–1840.
30 J. Garza, J. A. Nichols and D. A. Dixon, *J. Chem. Phys.*, 2000, **113**, 6029–6034.
31 G. Zhang and C. B. Musgrave, *J. Phys. Chem. A*, 2007, **111**, 1554–1561.
32 C. Herrmann, G. C. Solomon, J. E. Subotnik, V. Mujica and M. A. Ratner, *J. Chem. Phys.*, 2010, **132**, 024103.
33 L. S. Cederbaum and W. Domcke, *Adv. Chem. Phys.*, Wiley, New York, 1977, vol. 36, pp. 205–344.
34 Notice that, *via* the Dyson equation, each term included in the self-energy amounts to include an infinite sub-series of terms in the perturbation expansion.

35 G. D. Mahan, *Many-Particle Physics*, Plenum Press, New York and London, 2nd edn, 1990.

36 M. Nooijen and R. J. Bartlett, *J. Chem. Phys.*, 1995, **102**, 3629–3647.

37 O. Christiansen, H. Koch and P. Jørgensen, *Chem. Phys. Lett.*, 1995, **243**, 409–418.

38 J. F. Stanton and J. Gauss, *J. Chem. Phys.*, 1995, **103**, 1064–1076.

39 M. Nooijen and J. G. Snijders, *J. Chem. Phys.*, 1995, **102**, 1681–1688.

40 M. J. Frisch, G. W. Trucks, H. B. Schlegel, G. E. Scuseria, M. A. Robb, J. R. Cheeseman, G. Scalmani, V. Barone, B. Mennucci, G. A. Petersson, H. Nakatsuji, M. Caricato, X. Li, H. P. Hratchian, A. F. Izmaylov, J. Bloino, G. Zheng, J. L. Sonnenberg, M. Hada, M. Ehara, K. Toyota, R. Fukuda, J. Hasegawa, M. Ishida, T. Nakajima, Y. Honda, O. Kitao, H. Nakai, T. Vreven, J. A. Montgomery, Jr, J. E. Peralta, F. Ogliaro, M. Bearpark, J. J. Heyd, E. Brothers, K. N. Kudin, V. N. Staroverov, T. Keith, R. Kobayashi, J. Normand, K. Raghavachari, A. Rendell, J. C. Burant, S. S. Iyengar, J. Tomasi, M. Cossi, N. Rega, J. M. Millam, M. Klene, J. E. Knox, J. B. Cross, V. Bakken, C. Adamo, J. Jaramillo, R. Gomperts, R. E. Stratmann, O. Yazyev, A. J. Austin, R. Cammi, C. Pomelli, J. W. Ochterski, R. L. Martin, K. Morokuma, V. G. Zakrzewski, G. A. Voth, P. Salvador, J. J. Dannenberg, S. Dapprich, A. D. Daniels, O. Farkas, J. B. Foresman, J. V. Ortiz, J. Cioslowski and D. J. Fox, *Gaussian 09, Revision B.01*, Gaussian, Inc., Wallingford CT, 2010.

41 CFOUR, Coupled-Cluster techniques for Computational Chemistry, a quantum-chemical program package by J. F. Stanton, J. Gauss, M. E. Harding and P. G. Szalay with contributions from A. A. Auer, R. J. Bartlett, U. Benedikt, C. Berger, D. E. Bernholdt, Y. J. Bomble, L. Cheng, O. Christiansen, M. Heckert, O. Heun, C. Huber, T.-C. Jagau, D. Jonsson, J. Jusélius, K. Klein, W. J. Lauderdale, D. A. Matthews, T. Metzroth, L. A. Mück, D. P. O'Neill, D. R. Price, E. Prochnow, C. Puzzarini, K. Ruud, F. Schiffmann, W. Schwalbach, C. Simmons, S. Stopkowicz, A. Tajti, J. Vázquez, F. Wang, J. D. Watts and the integral packages MOLECULE (J. Almlöf and P. R. Taylor), PROPS (P. R. Taylor), ABACUS (T. Helgaker, H. J. A. Jensen, P. Jørgensen and J. Olsen), and ECP routines by A. V. Mitin and C. van Wüllen. For the current version, see http://www.cfour.de.

42 V. P. Vysotskiy and L. S. Cederbaum, *J. Chem. Phys.*, 2010, **132**, 044110.

43 F. Aquilante, L. D. Vico, N. Ferre, G. Ghigo, P. A. Malmqvist, P. Neogrady, T. B. Pedersen, M. Pitonak, M. Reiher, B. O. Roos, L. Serrano-Andres, M. Urban, V. Veryazov and R. Lindh, *J. Comput. Chem.*, 2010, **31**, 224.

44 I. Bâldea, B. Schimmelpfennig, M. Plaschke, J. Rothe, J. Schirmer, A. Trofimov and T. Fanghänel, *J. Electron Spectrosc. Relat. Phenom.*, 2007, **154**, 109–118.

45 J. F. Stanton and J. Gauss, *J. Chem. Phys.*, 1994, **101**, 8938–8944.

46 I. Bâldea, *Europhys. Lett.*, 2012, **99**, 47002.

47 I. Bâldea, H. Köppel and W. Wenzel, *Phys. Chem. Chem. Phys.*, 2013, **15**, 1918–1928.

48 I. Bâldea, *J. Phys. Chem. C*, 2014, **118**, 8676–8684.

49 L. Ould-Moussa, O. Poizat, M. Castellà-Ventura, G. Buntinx and E. Kassab, *J. Phys. Chem.*, 1996, **100**, 2072–2082.

50 M. Castellà-Ventura and E. Kassab, *J. Raman Spectrosc.*, 1998, **29**, 511–536.

51 M. Strange, C. Rostgaard, H. Häkkinen and K. S. Thygesen, *Phys. Rev. B: Condens. Matter Mater. Phys.*, 2011, **83**, 115108.
52 C. D. Spataru, M. S. Hybertsen, S. G. Louie and A. J. Millis, *Phys. Rev. B: Condens. Matter Mater. Phys.*, 2009, **79**, 155110.
53 M.-C. Desjonqueres and D. Spanjaard, *Concepts in Surface Physics*, Springer Verlag, Berlin, Heidelberg, New York, 1996.
54 J. B. Neaton, M. S. Hybertsen and S. G. Louie, *Phys. Rev. Lett.*, 2006, **97**, 216405.
55 H. J. Choi, M. L. Cohen and S. G. Louie, *Phys. Rev. B: Condens. Matter Mater. Phys.*, 2007, **76**, 155420.
56 A. Sommerfeld and H. Bethe, *Handbuch der Physik*, Julius-Springer-Verlag, Berlin, 1933, vol. 24(2), p. 446.
57 I. Bâldea and H. Köppel, *Phys. Status Solidi B*, 2012, **249**, 1791–1804.
58 A.-R. Allouche, *J. Comput. Chem.*, 2011, **32**, 174–182.
59 I. Bâldea and L. S. Cederbaum, *Handbook of Nanophysics*, CRC Press, Taylor & Francis, Boca Raton, 2010, vol. 4 (Nanotubes and Nanowires), ch. 42, pp. 1–15.
60 C. D. Zangmeister, S. W. Robey, R. D. van Zee, Y. Yao and J. M. Tour, *J. Phys. Chem. B*, 2004, **108**, 16187–16193.
61 I. Bâldea, H. Köppel, R. Maul and W. Wenzel, *J. Chem. Phys.*, 2010, **133**, 014108.
62 L. J. Holleboom and J. G. Snijders, *J. Chem. Phys.*, 1990, **93**, 5826–5837.
63 bwgrid, http://www.bw-grid.de, member of the German d-grid initiative, funded by the Ministry for Education and Research (Bundesministerium für Bildung und Forschung) and the Ministry for Science, Research and Arts Baden-Württemberg (Ministerium für Wissenschaft, Forschung und Kunst Baden-Württemberg).
64 We used the larger 6-311++G(d,p) basis set for this molecule instead of the aug-cc-pVDZ basis set employed in all other cases, merely because it created convergence problems.

Faraday Discussions

PAPER

Breaking the simple proportionality between molecular conductances and charge transfer rates

Ravindra Venkatramani,*[ab] Emil Wierzbinski,[c] David H. Waldeck*[c] and David N. Beratan*[bd]

Received 11th May 2014, Accepted 27th May 2014

DOI: 10.1039/c4fd00106k

A theoretical framework is presented to describe and to understand the observed relationship between molecular conductances and charge transfer rates across molecular bridges as a function of length, structure, and charge transfer mechanism. The approach uses a reduced density matrix formulation with a phenomenological treatment of system–bath couplings to describe charge transfer kinetics and a Green's function based Landauer–Buttiker method to describe steady-state currents. Application of the framework is independent of the transport regime and includes bath-induced decoherence effects. This model shows that the relationship between molecular conductances and charge transfer rates follows a power-law. The nonlinear rate–conductance relationship is shown to arise from differences in the charge transport barrier heights and from differences in environmental decoherence rates for the two experiments. This model explains otherwise puzzling correlations between molecular conductances and electrochemical kinetics.

The last decade has witnessed a growing interest in understanding the relationship between charge transfer (CT) kinetics and molecular conductance.[1–4] Substantial literature exists for unimolecular CT rates in donor–bridge–acceptor (D–B–A) structures comprised of charge donor (D) and acceptor (A) moieties that are connected by a bridging molecular unit (B).[5–9] In addition, the chemical donor and acceptor groups may be replaced with electrodes (M) to measure unimolecular electrochemical CT rates[10–13] and molecular conductances.[14–19] One expects unimolecular CT rates, molecular conductances, and electrochemical rates to be linked because the conduit (bridge) for charge transmission can be

[a]Department of Chemical Sciences, Tata Institute of Fundamental Research, Homi Bhabha Road, Colaba, Mumbai 400 005, India. E-mail: ravi.venkatramani@tifr.res.in

[b]Department of Chemistry, Duke University, Durham, NC, 27708, USA

[c]Department of Chemistry, University of Pittsburgh, Pittsburgh, PA, 15260, USA. E-mail: dave@pitt.edu

[d]Departments of Biochemistry and Physics, Duke University, Durham, NC, 27708, USA. E-mail: david.beratan@duke.edu

made chemically identical in the different experiments. Yet, recent experimental studies find fundamental differences among the transport characteristics in these experiments, and these differences may lead to nonlinear relationships among the measured quantities.[20]

Nitzan highlighted the fundamental connection between steady-state CT rates in D–B–A systems and their molecular conductances over a decade ago.[21,22] He proposed a linear relationship between the nonadiabatic, unimolecular CT rate constant in D–B–A structures and the zero bias molecular conductance in M–B–M junctions, assuming identical molecular bridges and a superexchange CT mechanism.[1] Subsequently, he described hopping CT through structures with identical repeating bridging units, and found a similar linear correlation between CT rates and molecular conductances for long bridges with high tunneling barriers.[2] Berlin and Ratner extended that treatment to the hopping CT regime with more general molecular bridge models, while retaining the large-barrier assumption.[3] In an alternative approach, Lewis and coworkers used Fermi's golden rule to relate unimolecular CT rates (in D–B–A molecules), electrochemical rate constants (in M–B–A assemblies), and molecular conductances (in M–B–M junctions),[4] treating the donor–acceptor electronic coupling as a parameter. For D–B–A systems, the CT was assumed to occur between a single electronic state of the donor and a single electronic state on the acceptor; for M–B–A and M–B–M systems, the CT was allowed to occur between multiple electronic states on the electrode and a molecular acceptor (M–B–A), or with multiple electronic states on a second electrode (M–B–M). If each electronic state on an electrode is considered analogous to the single state on the donor–acceptor molecule, then CT across M–B–A/M–B–M junctions can be described by the effective electronic coupling per electronic state times a density-of-states factor for the electrodes. Assuming that the D–B–A electronic coupling is equal to the M–B–A/M–B–M electronic coupling per charge carrying state of the electrode, Traub *et al.* predicted the resistances of M–B–M structures based on available CT rate data for alkanethiolate, oligophenylene, and DNA bridges in D–B–A or M–B–A experiments.[4] Each of these earlier studies[1–4] predicted a linear correlation between the tunneling or hopping CT rate and the conductance for a high barrier molecular bridge. Yet experiments indicate that molecular conductances and CT rates do not satisfy simple linear correlations (*vide infra*, ref. 20).

Fig. 1 plots experimental conductances (obtained from scanning tunneling microscope-controlled break junction (STM-BJ) experiments) *versus* electrochemical CT rates for the same molecular bridges.[20] In all cases, molecules were attached (at both ends in the conductance experiments and at one end in the electrochemical experiments) to the gold electrode through thiol linkers. In the electrochemical experiments, ferrocene tethered to the molecule's free terminus was used as a redox reporter for rate measurements. Data are presented for three families of bridges: (i) alkanes, (ii) single stranded peptide nucleic acids (ss PNA), and (iii) double stranded peptide nucleic acids (ds PNA). For each of the three families of bridges, the conductance and rate data are plotted as a function of bridge lengths. These data reveal two dramatic features that are at odds with previous theoretical predictions.[1–4] First, the conductance–rate data for each type of bridge are best fit by a power law relationship of the form: $\sigma \propto (k^0)^m$. Assuming that each measurement decays exponentially with distance ($k^0 \propto \exp(-\beta_{k^0} \times L)$ and $\sigma \propto \exp(-\beta_\sigma \times L)$), the exponent is $m = \beta_\sigma/\beta_{k^0}$.[20] Thus, a linear correlation

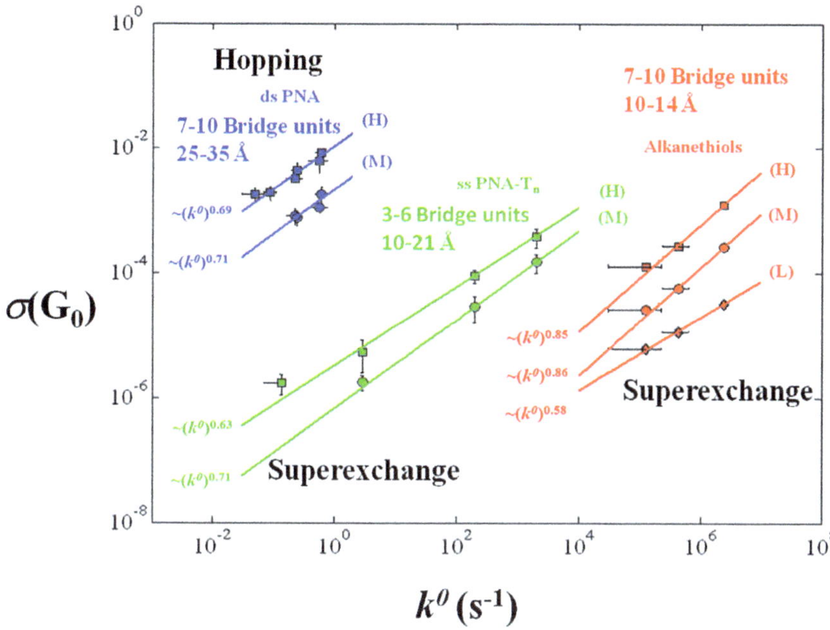

Fig. 1 Dependence of the single molecule conductance, σ, on the electrochemical charge transfer rate constant, k^0, for alkanes (red symbols), ss PNA constructed of thymine nucleotides (ss PNA-T_n; $n = 3$–6) (green symbols), and ds PNAs (blue symbols). The lines represent the least squares fits in the form $\sigma = A \times (k^0)^m$. Electrochemical rate constant values are adapted from ref. 10 for alkanethiols, and ref. 11 and 12 for ss PNAs. Conductance values for alkanedithiols are adapted from ref. 19. Note that the same type of linker was used to attach the ss- and ds PNA molecules to the gold surface (Cys), whereas the alkyl chains are directly coupled to gold with thiol groups, rather than cysteines. In the conductance measurements, several values of single-molecule conductance were determined for each molecule, depending on the details of the Au–S covalent bond geometry. H, M, and L represent high, medium, and low families of reported conductances, respectively. Figure adapted from ref. 20.

between rates and conductances is recovered only when $\beta_\sigma = \beta_{k^0}$. However, m values that are extracted from the data in Fig. 1 range between ~0.6–0.9 for alkanethiols and between ~0.6–0.7 for the nucleic acid bridges.[20] Secondly, the value of m is not transferrable for rate–conductance relationships among different molecular bridges. For example, 7–10 base pair ds PNA (~25–35 Å length) shows both lower CT rates and higher molecular conductances than 10–18 Å ss PNA sequences and 10–14 Å alkane chains. In this manuscript we develop a theoretical framework to identify the physical parameters governing the complex rate–conductance relationship shown in Fig. 1.

As previous theoretical models[1–4] predict a linear correlation between CT rate and molecular conductance, assumptions inherent to these models must be violated for the data in Fig. 1. First, the expressions in ref. 1–4 were derived for specific mechanisms of charge transport, *i.e.*, superexchange[1,4] or incoherent hopping with large donor/electrode-to-bridge energy gaps.[2,3] In general, CT may include admixtures from different coherent mechanisms as well as from incoherent variable-range hopping and nearest neighbor hopping.[23–25] Indeed, recent studies with nucleic acids (PNA, DNA) found that a mixture of charge-transport mechanisms operate as a consequence of thermal fluctuations of the structures.[25] Another simplifying assumption used in deriving the expressions in ref. 1–4 is that the CT energy barriers are equal (*i.e.*, the energy gaps between charge carrying states on the bridge and those on the donor–acceptor–electrode are equal) for D–B–A/M–B–A structures and for M–B–M junctions. As the CT energy barrier is a crucial determinant of CT mechanism(s), different CT barriers in conductance and rate measurements can lead to different CT mechanisms, or their admixtures. Third, the measurements are performed under different environmental conditions. That is, D–B–A rates are measured in solution, M–B–A rates are measured at electrochemical interfaces, and M–B–M conductances are measured in molecular junctions. These different environments can lead to differences in the relaxation rates for populations and coherences of the bridge states. Thus, even if the molecular bridge retains the same conformation and electronic structure in the different settings, differences in the molecule–bath interactions and differences in the energies of donor–electrode charge states could cause deviations from a linear conductance–rate relationship. These effects are examined in the models developed here.

Describing mixed mechanism charge-transport is challenging, and general theoretical expressions that link unimolecular charge-transfer rates with steady state molecular conductances arising from mixed mechanisms do not yet exist. Reduced density matrix approaches to calculate charge-transfer rates, including bath-induced decoherence, were formulated by several groups.[21,26,27] The theoretical rates may be linked to unimolecular CT rates measured in experiments[27] when excess charge population on the bridge is negligible (*i.e.*, for large tunneling barriers). A recent study by Yeganeh *et al.* describes an approach to calculate steady-state non-adiabatic CT rates in the framework of non-equilibrium Green's functions, which are typically used to calculate molecular conductances.[28] This approach is appealing as it enables the description of steady-state non-adiabatic CT rates and conductances in one formalism; however, the bridge is not explicitly treated in the prescription and the approach cannot be used to describe transient CT kinetics in D–B–A or M–B–A structures. In the present study, the molecular conductance through a bridge is modeled in a Green's function based Landauer–Buttiker framework, and the transient unimolecular CT rate (and the electrochemical rate constant) is treated with an initial value formulation of the density matrix. The framework presented here allows the computation of rate–conductance relationships for molecular bridges of different lengths, for different chemical species, and for mixed CT mechanisms. Our model calculations find that different CT energy barriers and different bath-induced decoherence effects (for molecular conductance *versus* CT rates) can produce nonlinearities in the rate–conductance relationships. These effects are not included in earlier theoretical studies of the link between rate and conductance,[1–4] which therefore predict linear correlations. The theoretical framework developed here helps to rationalize the complex relationship between the experimental rate and conductance data shown in Fig. 1.

Theoretical models

We aim to establish a framework for calculating CT rates and molecular conductance with a minimum number of transferable parameters. The computational approach is structured to be independent of the charge-transfer mechanism and can incorporate bath-induced decoherence. For both rate and conductance calculations, the bridge is represented with a tight-binding Hamiltonian ($\hat{H}_{\rm BR}$). There are N identical repeat units in the bridge with equal site energies ($E_{\rm B}$) and equal nearest-neighbor couplings ($t_{\rm B}$):

$$\hat{H}_{\rm BR} = \sum_{m=1}^{N} |m> E_{\rm B} <m| \; + \sum_{m=1}^{N-1} |m> t_{\rm B} <m+1| + |m+1> t_{\rm B} <m| \qquad (1)$$

As shown in Fig. 2, the tight-binding bridge (green lines) is then either coupled to molecular donor–acceptor sites with nearest neighbor couplings ($t_{\rm D1}/t_{N\rm A}$ terms in Fig. 2A) for CT rate calculations, or to donor–acceptor electrodes represented through renormalized complex broadening matrices ($\Gamma_{\rm L1}/\Gamma_{N\rm R}$). We next describe how the schemes in Fig. 2 can be used to describe unimolecular charge transfer rates, electrochemical rate constants, and molecular conductances. The theoretical approach presented here can be extended to more elaborate model structures and Hamiltonians.

Expression for the charge transfer rate

The CT rate is computed from the reduced density matrix of a donor–bridge–acceptor system coupled to a bath, which is obtained by solving the quantum Liouville equation. In our simple model, the molecular bridge is connected to the

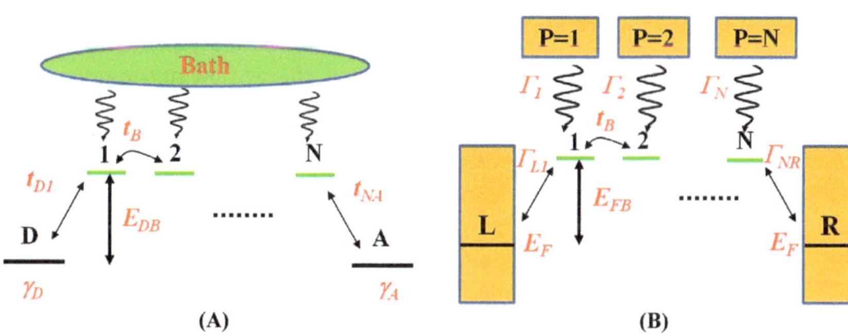

Fig. 2 Schemes depicting models used for calculating CT rates (left) and conductance (right). In both models, the bridge is defined by tight-binding couplings, t_{ij}, and equal site energies, $E_{\rm B}$. The bridge is coupled (couplings $t_{\rm D1}/t_{N\rm A}$) to donor–acceptor sites (energies $E_{\rm D} = E_{\rm A}$) in the CT rate model and to electrodes (couplings $\Gamma_{\rm L1}/\Gamma_{N\rm R}$) of Fermi energy, $E_{\rm F}$, in the conductance model. The CT energy barriers are given by $E_{\rm DB} = E_{\rm B} - E_{\rm D}$ and $E_{\rm FB} = E_{\rm B} - E_{\rm F}$ in the CT rate and conductance models, respectively. Bath effects are included in the CT rate model through donor–acceptor population relaxation terms, $\gamma_{\rm D}/\gamma_{\rm A}$, and through pure dephasing terms, γ_{mn} (not shown), of the system–bath Liouvillian (eqn (4)), which damp the coherences involving bridge states. Bath effects are included in the conductance model by coupling each bridge site to Buttiker probe electrodes (P = 1, …, N) with couplings given by $\Gamma_{\rm P}$.

donor (D) and acceptor (A). The donor with energy E_D is coupled to the first bridge site with interaction t_{D1}. An acceptor site (energy E_A) is connected to the Nth bridge site with coupling t_{NA}. The system Hamiltonian, $\hat{H}_S$, is:

$$\hat{H}_S = \hat{H}_{BR} + \sum_{m=D,A} |m> \varepsilon_m <m| + |D> t_{D1} <1| + |1> t_{D1} <D| + |A> t_{NA} <N| + |N> t_{NA} <A| \quad (2)$$

and the quantum Liouville equation is

$$\frac{d\rho}{dt} = -\frac{i}{\hbar}\hat{L}\rho = -\frac{i}{\hbar}\left(\hat{L}_S + \hat{L}_{SB}\right)\rho \quad (3)$$

here, the density matrix dynamics is determined by the Liouvillian, $\hat{L} = \hat{L}_S + \hat{L}_{SB}$, which is a sum of the real system Liouvillian, $\hat{L}_S = [\hat{H}_S, \rho]$, and the complex system–bath interaction Liouvillian, $\hat{L}_{SB}$, that accounts for the decay of the density matrix elements. Coupling of the D–B–A system electronic states to a bath introduces dissipative energy exchange between the system and the bath leading to population relaxation. Furthermore, interaction with the bath creates a distribution of electronic transition energies leading to a decay of electronic coherences (dephasing). The bath is not described explicitly (through a microscopic model), and its influence on the system density matrix (population relaxation and dephasing) is included phenomenologically in the system–bath Liouvillian, $\hat{L}_{SB}$:

$$\left[\hat{L}_{SB}\right]_{mn,mn} = -i\left(\frac{\gamma_m}{2} + \frac{\gamma_n}{2} + \gamma_{mn}[1 - \delta_{mn}]\right) \quad (4)$$

here, $\gamma_m/\hbar$ is the population relaxation rate from site m (inducing relaxation of the diagonal density matrix element, ρ_{mm}), and $\gamma_{mn}/\hbar$ is the pure dephasing rate that accounts for coherence loss between sites m and n (ρ_{mn}). With this formulation, population relaxation on sites m and n (γ_m/γ_n) produces dephasing that causes coherences $\rho_{mn}(m \neq n)$ to decay, even if $\gamma_{mn}(m \neq n)$ is set to zero in eqn (4). The time-dependent density matrix of eqn (3) describes charge flow in the system.

At time $t = 0$, the charge is localized on the donor ($\rho_{DD} = 1$ and all other diagonal/off-diagonal density matrix elements are zero). Eqn (3) is then solved to determine the charge population on the acceptor as a function of time, $\rho_{AA}(t)$. This computation is accessible in the eigenstate basis that diagonalizes the Liouvillian. Because the Liouvillian is complex, it has two sets of eigenvectors that connect the site basis to its eigenstates.[29,30] The right (ket) eigenvectors, X^R, are defined by:

$$\hat{L}X^R = \varepsilon^R X^R \quad (5)$$

and the left (bra) eigenvectors, X^L, are defined by:

$$X^L\hat{L} = \varepsilon^L X^L \quad (6)$$

In practice, we compute the left eigenvector from $X^L = (X')^\dagger$ where X' is the eigenvector of $\hat{L}^\dagger$. Both left and right eigenvectors correspond to the same set of complex eigenvalues, $\varepsilon = \varepsilon^R = \varepsilon^L$, and are normalized as:

$$\tilde{X}^{\mathrm{L}} = \frac{X^{\mathrm{L}}}{\sqrt{X^{\mathrm{L}}X^{\mathrm{R}}}}; \ \tilde{X}^{\mathrm{R}} = \frac{X^{\mathrm{R}}}{\sqrt{X^{\mathrm{L}}X^{\mathrm{R}}}} \tag{7}$$

These eigenvectors link the density matrix elements in the site basis to the density matrix elements in the eigenstate basis.

$$\rho_{mn}(t) = \sum_{a,b=1}^{N+2} \rho_{ab}(t)\tilde{X}^{\mathrm{R}}_{mnab} \tag{8}$$

The pairs of indices m, n and a, b enumerate density matrix elements in the site basis (m, n = D, A, 1, …, N) and in the eigenstate basis (a, b = 1, …, N + 2), respectively. The time evolution of the density matrix elements in the site basis may be written in terms of the complex eigenvalues of the Liouvillian $\varepsilon_{ab} = [\hat{L}]_{abab}$:

$$\rho_{mn}(t) = \sum_{a,b,k,l=1}^{N+2} \rho_{kl}(0)\tilde{X}^{\mathrm{L}}_{abkl}\tilde{X}^{\mathrm{R}}_{mnab}\mathrm{e}^{-\frac{i}{\hbar}\varepsilon_{ab}t} \tag{9}$$

The donor to acceptor CT rate constant, k_{CT}, is computed from the probability of charge (all on the donor site at time $t = 0$) being transferred to the acceptor site at time t. Eqn. (10) defines the CT rate constant, k_{CT}, as the inverse of the average time, $\langle\tau_{\mathrm{AA}}\rangle$, to populate the acceptor; namely[31]

$$k_{\mathrm{CT}} = \frac{1}{\langle\tau_{\mathrm{AA}}\rangle} = \frac{\int_0^\infty \rho_{\mathrm{AA}}(t)\mathrm{d}t}{\int_0^\infty t\rho_{\mathrm{AA}}(t)\mathrm{d}t} \tag{10}$$

The rate in eqn (10) is an activated rate, *i.e.* charge transfer occurs when the donor and acceptor are resonant. Therefore, eqn (10) does not include a description for nuclear relaxation following electron transfer. However, we now show that the rate constant, k_{CT}, in eqn (10) is proportional to the non-adiabatic charge transfer rate constant, k_{DA}, between the donor and acceptor when the bridge is energetically well separated from the donor and acceptor.[32] For large energy separations between the donor and bridge states, the Hamiltonian in eqn (2) may be transformed into an effective two-state Hamiltonian by a Löwdin partitioning scheme:[33–35]

$$\hat{H}_{\mathrm{S}} = \sum_{i=\mathrm{D,A}} |i> \varepsilon_i < i| + \ |\mathrm{D}> t_{\mathrm{DA}} < \mathrm{A}| + |\mathrm{A}> t_{\mathrm{DA}} < \mathrm{D}| \tag{11}$$

where t_{DA} is the effective electronic coupling between donor and acceptor. In the case of population relaxation only on the acceptor (*i.e.*, $\gamma_{\mathrm{D}} = 0$) and $t_{\mathrm{DA}} \ll \gamma_{\mathrm{A}}$, eqn (10) gives the analytical result $k_{\mathrm{CT}} \sim 2|t_{\mathrm{DA}}|^2/\hbar\gamma_{\mathrm{A}}$. Comparison of this result with the high-temperature non-adiabatic CT rate, given by

$$k_{\mathrm{DA}} = \frac{2\pi}{\hbar}|t_{\mathrm{DA}}|^2\frac{1}{\sqrt{4\pi\lambda k_{\mathrm{B}}T}}\exp\left(-\frac{(\Delta G^0 + \lambda)^2}{4\lambda k_{\mathrm{B}}T}\right) \tag{12}$$

where λ is the reorganization energy and ΔG^0 is the Gibbs free energy of reaction,[32] indicates that

$$k_{\mathrm{DA}} = \frac{\pi \gamma_{\mathrm{A}}}{\sqrt{4\pi\lambda k_{\mathrm{B}} T}} k_{\mathrm{CT}} \exp\left(-\frac{(\Delta G^0 + \lambda)^2}{4\lambda k_{\mathrm{B}} T}\right) \tag{13}$$

Thus, when the electronic states of the bridge are energetically distant from the donor–acceptor energies, the non-adiabatic charge transfer rate is given by k_{CT} in eqn (10), evaluated for resonant donor–acceptor states and weighted by the Marcus activation factor. Both the prefactor and the Marcus activation factor in eqn (13) depend only on properties of the donor–acceptor and are independent of the bridge (although λ and ΔG^0 are weakly distance dependent). Thus, the non-adiabatic charge transfer rate, k_{DA}, for donor–bridge–acceptor systems is linearly related to the rate constant, k_{CT}, that is defined by eqn (10), when the donor and acceptor are the same.

Expression for the electrochemical rate constant

In electrochemical experiments, the charge transfer occurs from an electrode surface to a molecular species (redox couple) such as ferrocene. The Gibbs free energy for the charge transfer reaction from an electrode state of energy ε is:

$$\Delta G^0 = \varepsilon_{\mathrm{F}} - \varepsilon + e\eta \tag{14}$$

where η is the electrochemical overpotential and ε_{F} is the Fermi function of the electrode. In a density of states picture,[36] the electrochemical charge transfer rate from the donor electrode to the redox couple may be written by generalizing eqn (13) as:

$$k_{\mathrm{EL}} = \frac{\pi \gamma_{\mathrm{A}}}{\sqrt{4\pi\lambda k_{\mathrm{B}} T}} \int_0^\infty k_{\mathrm{CT}}(\varepsilon) \exp\left(-\frac{(\varepsilon_{\mathrm{F}} - \epsilon + e\eta + \lambda)^2}{4\lambda k_{\mathrm{B}} T}\right) \rho_{\mathrm{EL}} f_{\mathrm{EL}}(\varepsilon) \mathrm{d}\varepsilon \tag{15}$$

where ρ_{EL} is the density of states and $f_{\mathrm{EL}}(\varepsilon) = [1 + \exp((\varepsilon - \varepsilon_{\mathrm{F}})/k_{\mathrm{B}}T)]^{-1}$ is the Fermi distribution function. The standard heterogeneous rate constant is then given by the electrochemical rate at $\eta = 0$

$$k^0 = \frac{\pi \gamma_{\mathrm{A}} \rho_{\mathrm{EL}}}{\sqrt{4\pi\lambda k_{\mathrm{B}} T}} \int_0^\infty k_{\mathrm{CT}}(\varepsilon) \exp\left(-\frac{(\varepsilon_{\mathrm{F}} - \epsilon + \lambda)^2}{4\lambda k_{\mathrm{B}} T}\right) f_{\mathrm{EL}}(\varepsilon) \mathrm{d}\varepsilon \tag{16}$$

In our rate calculations, we assumed $T = 300$ K, and $\lambda = 0.8$ eV, representative of the reorganization energy for the ferrocene redox couple that was used in the electrochemical measurements to determine the rates shown in Fig. 1.[37]

Expression for the molecular conductance

The steady-state current in molecular conductance measurements is often modeled using a Landauer–Buttiker formulation in terms of non-equilibrium Green's functions.[38,39] The conductance is modeled using the molecular bridge Hamiltonian of eqn (1) with site 1 connected to a source electrode (L) and site N connected to a drain electrode (R). In addition, a Buttiker probe (P) electrode is located on each of the N bridge sites (see Fig. 2B). The Buttiker probe electrodes absorb current from the bridge sites and re-inject it into the bridge with a random phase. This 'scattered' current can travel to the acceptor electrode or to other Buttiker probe electrodes, and it can also be reflected back to the source through independent coherent channels. Thus, the Buttiker probes simulate phase breaking and backscattering events on the bridge. The bridge Green's function, including the influence of electrode contacts, is:

$$\boldsymbol{G}(E) = \frac{1}{(E\boldsymbol{I} - \boldsymbol{H}) - \boldsymbol{\Sigma}_{\mathrm{L}} - \boldsymbol{\Sigma}_{\mathrm{R}} - \sum_{P=1}^{\mathrm{N}} \boldsymbol{\Sigma}_{\mathrm{P}}} \tag{17}$$

here, $\boldsymbol{H}$ is the molecular bridge Hamiltonian of eqn (1), and $\boldsymbol{I}$ is the identity matrix. The influence of electrodes L and R and of the Buttiker probes on the bridge are included in the bridge Green's function through the complex self-energy matrices $\boldsymbol{\Sigma}_{\mathrm{L}}$, $\boldsymbol{\Sigma}_{\mathrm{R}}$, and $\boldsymbol{\Sigma}_{\mathrm{P}}$, respectively. The real (imaginary) parts of the self-energy matrix elements add to the corresponding elements of the bridge Hamiltonian and manifest themselves through shifts (broadenings) of the molecular orbital energies in the local density of states (LDOS):[38,39]

$$\mathrm{LDOS}(E) = \frac{i}{2\pi}\mathrm{Tr}\left[\boldsymbol{G}(E) - \boldsymbol{G}^{\dagger}(E)\right] \tag{18}$$

where $\boldsymbol{G}^{\dagger}$ is the Hermitian adjoint of $\boldsymbol{G}$. Applying a potential bias across electrodes L and R causes a current to flow between the electrodes. This current $I(\mathrm{L} \to \mathrm{R})$ is

$$I(\mathrm{L}\to\mathrm{R}) = \frac{2e}{h}\int\left[T_{\mathrm{LR}}(E)[f_{\mathrm{L}}(E) - f_{\mathrm{R}}(E)] + \sum_{P=1}^{N} T_{\mathrm{PR}}(E)[f_{\mathrm{P}}(E) - f_{\mathrm{R}}(E)]\right]\mathrm{d}E \tag{19}$$

where $T(E)$ is the transmission function and $f(E)$ is the Fermi function. $T(E)$ between a pair of electrodes m and n (m, n = L, R, P) is

$$T_{mn}(E) = \mathrm{Tr}[\boldsymbol{\Gamma}_m \boldsymbol{G} \boldsymbol{\Gamma}_n \boldsymbol{G}^{\dagger}] \tag{20}$$

The $\boldsymbol{\Gamma}$ (broadening) matrices for each electrode m are defined as: $\boldsymbol{\Gamma}_m = [\boldsymbol{\Sigma}_m - (\boldsymbol{\Sigma}_m)^{\dagger}]$. The Fermi functions $f_m(E)$ describe the electron occupancy of the states in the m = L, R, P electrodes at energy E:

$$f_m(E) = [1 + \exp((E - \mu_m)/k_{\mathrm{B}}T)]^{-1} \tag{21}$$

The Fermi functions for the source (L) and drain (R) electrodes are defined by their chemical potentials: $\mu_{\mathrm{L}} = E_{\mathrm{F}} - eV/2$ and $\mu_{\mathrm{R}} = E_{\mathrm{F}} + eV/2$. Here, E_{F} (the Fermi energy of the electrodes) and V (the potential bias) are parameters. The chemical potential, μ_{P}, of a Buttiker probe at site P is unknown, but it can be determined from the constraint on each Buttiker probe P to absorb no net current:[40]

$$I_{\mathrm{P}} = 0 = \frac{2e}{h}\int\Big[\big[T_{\mathrm{LP}}(E)[f_{\mathrm{L}}(E) - f_{\mathrm{P}}(E)] + T_{\mathrm{RP}}(E)[f_{\mathrm{R}}(E) - f_{\mathrm{P}}(E)]\big] + \sum_{P'=1, P'\neq \mathrm{P}}^{N} T_{P'\mathrm{P}}(E)\left[f_{P'}(E) - f_{P'}(E)\right]\Big]\mathrm{d}E \tag{22}$$

By making the further simplifying assumption that the net current absorbed by each Buttiker probe at energy E is zero, the integrand in eqn (22) may be set to zero for each probe. This condition means that the scattering by each probe is elastic. The unknown Fermi functions, f_P, are obtained by solving the resulting N coupled differential equations on a discrete energy grid. In the Landauer–Buttiker approach described here, charges are exchanged with the system only at electrodes L and R. The Buttiker probe electrodes are elastic scattering centers for resonant current. Scattering of the current at the Buttiker probes leads to phase relaxation of the carriers without any momentum or energy relaxation.

In our numerical simulations, the steady-state current between electrodes was calculated using eqns (19) through to (22) for a potential bias $V = 0.1$ V and $T = 300$ K. In this model, both thermally activated, and tunneling currents can be computed, including bath-induced decoherence. However, inelastic current contributions arising from interactions of the charge with molecular vibrational modes[41–43] are not considered within the approximations of the model. The conductance (σ) is calculated as:

$$\sigma = I(\mathrm{L} \rightarrow \mathrm{R})/V \tag{23}$$

We end this section with a discussion of how population relaxation and decoherence effects are incorporated in the rate and conductance models, and of the assumptions that we make to enable a comparison of charge transport among the models. Both rate and conductance models include bath-induced relaxation of diagonal (populations) and off-diagonal (coherences) elements of the system density matrix. In the rate model, the bridge is coupled to the donor and acceptor (Fig. 2A) that are, in turn, connected to the bath continuum through the lifetime parameters, γ_D and γ_A. As given by eqn (4), these lifetime contributions (imaginary terms in the diagonal elements of the Hamiltonian) also contribute to the relaxation of coherences between bridge and donor–acceptor states. In the conductance model, the bridge is directly coupled (at sites 1 and N in Fig. 2B) to a continuum of electrode states through the broadening matrices, Γ_L and Γ_R. Analogous to the rate model, the broadening matrices add imaginary terms to the diagonal elements of the Hamiltonian (see eqn (17)) and hence contribute towards both population as well as coherence relaxation. Furthermore, in the conductance model, the Buttiker probe couplings, Γ_P, also add imaginary terms to the Hamiltonian which might appear to be equivalent to adding lifetime broadening terms to the bridge (γ_m, where m = bridge sites) in the CT rate model. However, there is a critical difference. In the conductance model, the conductance is computed with zero current absorption at each Buttiker probe. This is equivalent to eliminating the population relaxation contributions provided by Γ_P, while retaining its dephasing contributions. On the other hand, adding lifetime broadening couplings (γ_m) to the bridge in the CT rate model leads to irreversible population relaxation from the bridge. To avoid population loss from the bridge to the bath, we set all lifetime broadening terms as zero, $\gamma_m = 0$ (m = bridge sites), in the rate model. We then control the decoherence in the CT rate model through the pure dephasing term, γ_{mn}, of the system bath Liouvillian (see eqn (4)) which does not have an analog in the conductance model. Thus, while the decoherence in the conductance model arises from lifetime broadening (Γ_P) of bridge-state energies (see eqn (17)) because of its coupling to Buttiker probes (a variable parameter in our calculations), the decoherence in the CT rate model arises from lifetime broadening of the donor and acceptor state energies (fixed for a given donor–acceptor pair) and through γ_{mn} (variable parameter in our calculations). While this approach allows comparison of the decoherence effects in the CT rate and conductance models on a similar footing, it is based on phenomenological descriptions of decoherence in the rate and the conductance models which are not entirely equivalent. The extent of decoherence in the rate and conductance models can thus only be compared qualitatively, *e.g.*, how the decoherence influences the distance dependence of CT in the two models. In our numerical simulations, we demonstrate that using different models to describe rates and conductance does not limit our conclusions regarding the influence of dephasing rates and CT energy barriers on the rate–conductance correlation.

Numerical results and discussion

The theoretical framework developed above is now used to calculate the CT rate and the molecular conductance as a function of the bridge length, N. The change in CT rate and molecular conductance as a function of bridge length provides information on the underlying charge-transport mechanism. When the donor is energetically well separated from the bridge, charge transport occurs predominantly *via* superexchange for short bridge lengths, and the molecular conductance or CT rate drops approximately exponentially with the bridge length. Beyond a certain bridge length, determined by the donor–bridge energy gap and the temperature, thermal activation of the charge to populate bridge states becomes competitive with tunneling, and the distance dependence softens – approaching an ohmic ($1/N$) dependence for long bridges. Previous theoretical analysis[29,44] has found that the transition to a $1/N$ distance dependence is gradual, and that the fall off in CT rate may even appear distance independent, for intermediate bridge lengths.

Distance dependence of charge transport with decoherence

Fig. 3 compares the calculated distance dependence for the CT rate and the molecular conductance in a model system with identical CT energy barriers, $E_{DB} = E_{FB}$, but with variable bath-induced decoherence. The CT rates and conductances show an exponential distance dependence at short bridge lengths with a crossover to a softer distance dependence at longer bridge lengths. Interestingly, the distance dependences of the CT rate and the molecular conductances are not identical (note that in Fig. 3A both rates and conductance are normalized with respect to their values at 1 BU which allows us to compare their distance dependence). The distance dependence of CT rate and conductance coincide only at short bridge lengths with deviations appearing near the crossover region from exponential to soft distance dependence. Because thermally-activated transport competes with tunneling transport near the cross-over region, the differences in distance dependence of the CT rates and conductances are associated with model parameters that control resonant CT processes. Specifically, the difference occurs because of differences in the treatment of bath-induced decoherence in the rate and conductance models.

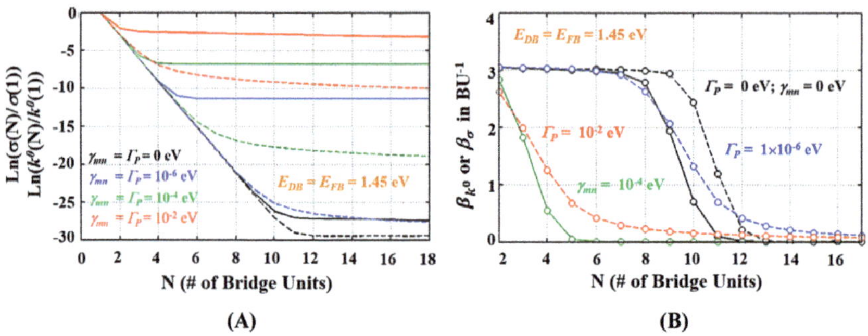

Fig. 3 (A) This plot shows the distance dependence of the electrochemical rate constant (solid line) and molecular conductance (dashed lines) normalized with respect to their values at 1 BU. Rate constants were calculated using eqn. 16 ($E_{DB} = 1.45$ eV; $V_{BB} = 0.3$ eV; $V_{D1} = 0.1$ eV; $V_{NA} = 0.1$ eV; $\gamma_D = 1 \times 10^{-16}$ eV; $\gamma_A = 0.1$ eV and $\gamma_m = 0$ eV m = bridge sites). Molecular conductances were calculated using eqn (23) ($E_{FB} = 1.45$ eV; $V_{BB} = 0.3$ eV; $\Gamma_{L1} = 0.1$ eV and $\Gamma_{NR} = 0.1$ eV). The model system parameters for the bridge most closely resemble alkanethiol chains, however the CT energy gap and dephasing parameters were chosen to facilitate the discussion and may not reflect values present in actual experiments. The colored lines show the effect of pure dephasing (included by varying γ_{mn} and Γ_P) on the distance dependence of the CT rate/molecular conductance. To minimize the number of parameters only γ_{mn} terms controlling the decay of coherences between donor–acceptor and bridge states were set to the values shown in the figure (*i.e.* $\gamma_{DA} = 0$, $\gamma_{mn} = 0$ when both m and n are bridge sites; trial calculations showed that non-zero values of these terms do not alter the general trends in distance dependence of rates considered here or our final conclusions). (B) The distance decay constants, β_{k^0} for CT rates (solid lines) and β_σ for molecular conductance (dashed lines), are plotted *versus* the number of bridge units for the different dephasing rates. The β values for bridge unit N were calculated by fitting the N dependence of $\ln(k^0)$ or $\ln(\sigma)$ for three values from $N-1$ to $N+1$.

In the conductance analysis, dephasing contributions arise from the coupling (Γ_P) of the bridge to the Buttiker probes (imaginary part of the self-energy, Σ_P, in eqn (17)). For CT, dephasing is introduced *via* two parameters in the system–bath Liouvillian (eqn (4)): (1) the population relaxation terms (γ_m) and (2) the pure dephasing, γ_{mn} ($m \neq n$), terms. This feature results in CT rate models that contain non-zero dephasing contributions from the donor–acceptor population relaxation ($\gamma_{D/A}$), even after zeroing the γ_{mn} and γ_m ($m \neq$ D, A) contributions for the bridge states. In contrast, the conductance model with $\Gamma_P = 0$ includes no dephasing contributions and leads to a conductance distance dependence which differs from that of the CT rates. The conductance computed for a Buttiker probe coupling value of 10^{-6} eV (dashed blue line) produces a distance dependence that closely matches the computed CT rate with $\gamma_{mn} = 0$ (solid black line). As the bridge parameters and the CT energy gaps are constrained to be the same for rate/conductance calculations, we infer qualitatively that dephasing (k^0) ~ dephasing (σ) for the rate model with $\gamma_{mn} = 0$ (solid black line) and the conductance model with $\Gamma_P = 10^{-6}$ eV (dashed blue line). This qualitative comparison between different phenomenological models for incorporating bath-induced decoherence in the CT rate and conductance is justified because both models of decoherence produce the same effect on the distance dependence of CT transport (see Fig. 3A and discussion below).

Earlier studies[26,6] found that bridge decoherence can significantly influence the distance dependence of CT rates/molecular conductances. Fig. 3 shows that, for large CT energy barriers, dephasing changes the distance dependence of both the CT rate and the molecular conductance in similar ways. For both rates and conductances, the crossover from an exponential distance dependence to a nearly distance-independent regime shifts to shorter bridge lengths for larger bridge decoherence (larger values of γ_{mn} and Γ_P). Bath-induced decoherence broadens the bridge state energies, effectively reducing the tunneling barrier and boosting both the superexchange and thermally activated transport rates.[45] Decoherence also introduces backscattering, which reduces ballistic (activated but coherent) CT.[45] Fig. 3A shows that increasing the decoherence makes the thermally activated transport mechanisms more competitive with the superexchange for shorter bridges, leading to the observed shift in the crossover region.

To draw a link between the data in Fig. 3A and the experimental observations in Fig. 1, we fit a decay length, β, to the distance dependence of rates/conductances. In experiments, β_σ and β_{k^0} are often extracted from linear fits to experimentally measured $\ln(k^0/\mathrm{s}^{-1})$ or $\ln(\sigma/\Omega^{-1})$ (typically 3–10 BU).[10,11,19,20] Accordingly, β_σ and β_{k^0} values were calculated as a function of the number of bridge units (N) by linearly fitting three data points for $\ln(k^0/\mathrm{s}^{-1})$ or $\ln(\sigma/\Omega^{-1})$ computed for $N-1$, N, and $N+1$ bridge units. The variation in β_σ and β_{k^0} as a function of N and with different levels of bath-induced decoherence is shown in Fig. 3B. With increasing bridge length, both β_σ and β_{k^0} switch over from a value of 3 BU^{-1} to $\sim$0 BU^{-1} signifying a change from superexchange to hopping CT. The increase in dephasing contributions tends to lower both β_σ and β_{k^0} at all bridge lengths (compare red dashed line with black dashed line or green solid line with black solid line), although at intermediate dephasing values the effects are more complex and exhibit a bridge length dependence (compare blue dashed line with black dashed line). Notably, the ratio β_σ/β_{k^0}, which dictates the power law in Fig. 1, approaches unity only when the dephasing (k^0) $\sim$ dephasing (σ).

To summarize, Fig. 3 indicates that bath-induced decoherence can significantly change the distance dependence of both the CT rate and the molecular conductance, even if the electronic structure of the bridge is only weakly perturbed. Thus, different distance dependences could be produced by experiments that measure molecular conductances (*e.g.*, STM-BJ) and CT rates (*e.g.*, measured with electrochemistry) for identical molecular bridges, because of different medium interactions. While the crossover from an exponential to a soft distance dependence has been observed in experiments for a number of CT rate studies, the molecular conductance measurements for the same set of molecules are often exponential within the same bridge-length range, exhibiting no crossover.

Distance dependence of charge transport with varying CT energy barriers

We now explore how the magnitude of the CT energy barriers (parameters E_{DB} and E_{FB}, from Fig. 2) influence β_σ and β_k. For superexchange, β should decrease as the barrier decreases;[9,34,46–48] and the crossover point from exponential to $1/N$ behavior should shift to shorter chain lengths for lower barriers.[49] The distance dependence of the CT rate and the molecular conductance were calculated for: (1) $E_{DB} = E_{FB} = 1.45$ eV, (2) $E_{DB} = E_{FB} = 1.15$ eV, and (3) $E_{DB} = E_{FB} = 0.75$ eV. The values of the parameters Γ_P and γ_{mn} were chosen so that dephasing (k^0) $\sim$ dephasing (σ)

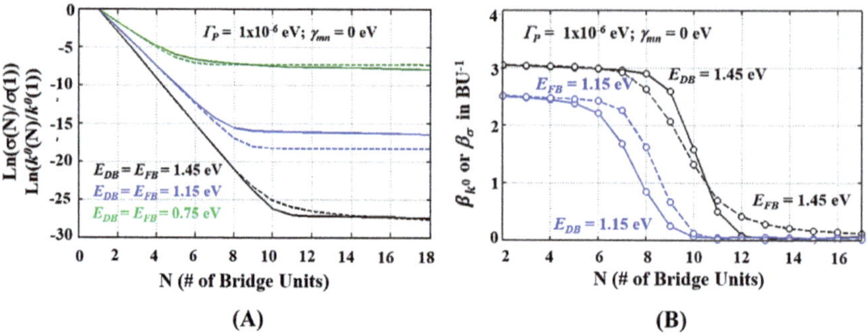

Fig. 4 (A) Dependence of CT rates (solid lines) and molecular conductance (dashed lines) on bridge length for different energy barrier heights (colored lines) and dephasing (k^0) ~ dephasing (σ) in the rate–conductance models. Rate–conductance values are normalized with respect to their values at 1 BU. (B) Variation of the distance decay constants, $\beta_{k^0}$for CT rates (solid lines) and β_σ for molecular conductance (dashed lines) with the number of bridge units. The calculations were performed for two different values of the CT barriers; all other parameters are the same as the model system described in Fig. 3.

based on the comparison of distance dependence of rates and conductances at $E_{DB} = E_{FB} = 1.45$ eV. Fig. 4A shows that for this minimal dephasing model ($\gamma_{mn} = 0$ eV; $\Gamma_P = 10^{-6}$ eV), decreasing the CT energy barrier shifts the crossover region (from an exponential to a soft distance dependence) to shorter bridge lengths for both CT and molecular conductance. Decreasing the barrier also decreases β (Fig. 4B), this is attributed to: (1) an increase in tunneling probability and (2) an increase in the probability of thermally activated CT. The slight dip in the conductance curves (dashed green) around 6 BU in Fig. 4A is not a crossover point (from exponential to soft distance dependence); rather the CT energy barrier ($E_{FB} = 0.75$ eV) is small enough that the conductance has thermally-activated contributions for all bridge lengths (showing soft distance dependence). In summary, the relative values of β_σ and β_{k^0} are significantly modulated by the CT energy barrier in conductance and rate measurements. The data in Fig. 4B show that $\beta_\sigma \neq \beta_{k^0}$ when $E_{DB} \neq E_{FB}$ (compare β values given by the solid and dashed lines of different colors). Thus, $E_{DB} = E_{FB}$ (compare β values given by the solid and dashed lines of the same color) is a necessary condition (along with dephasing (k^0) ~ dephasing (σ)) for the power law exponent, β_σ/β_{k^0}, in Fig. 1 to be unity.

Relationship between molecular conductance and the electrochemical rate constant

The results in the previous sections (Fig. 3B and 4B) indicate that the CT energy barrier and bath-induced decoherence strongly influence the distance dependence of rates and conductances. Previous theoretical studies for high barriers[1,2] found that the CT rates in both the superexchange and the hopping regimes are proportional to the conductances. In contrast, experiments (Fig. 1) find that the rate–conductance relationship follows a power law:[20]

$$\sigma = \sigma_{(L=0)} \times \left(\frac{k^0}{k^0_{(L=0)}}\right)^{(\beta_\sigma/\beta_{k^0})} \tag{24}$$

As the observables are measured under different experimental conditions, the bridge decoherence and the CT energy barriers are likely to be different, leading to $\beta_\sigma \neq \beta_{k^0}$.

Fig. 5 shows how decoherence rate differences between the measurements influence the correlation between σ and k_{CT} (for the data from Fig. 4A). We consider three cases of relative decoherence (as determined from the distance dependence data in Fig. 4A) in rate and conductance models: (1) dephasing (k^0) < dephasing (σ) (red symbols; $\gamma_{mn} = 0$ eV; $\Gamma_P = 10^{-2}$ eV), (2) dephasing (k^0) ~ dephasing (σ) (blue symbols; $\gamma_{mn} = 0$ eV; $\Gamma_P = 10^{-6}$ eV), and (3) dephasing (k^0) > dephasing (σ) (green symbols; $\gamma_{mn} = 10^{-4}$ eV; $\Gamma_P = 10^{-6}$ eV) where the CT energy barriers in the CT rate and the molecular conductance models are the same ($E_{DB} = E_{FB} = 1.45$ eV). The correlation between rates and conductance is plotted on a log–log scale, so that the slope of the correlation is the ratio β_σ/β_{k^0}. The black line represents a linear correlation (unit slope) between rate and conductance. The rate constant and the conductance are only linearly correlated for bridge lengths where their distance dependence is identical, *i.e.* $\beta_\sigma = \beta_{k^0}$. Fig. 5 shows that for fixed barrier heights, the correlation is linear ($m = \beta_\sigma/\beta_{k^0} = 1$) over a large range (over 10 orders of magnitude) of conductances and rates when dephasing (k^0) ~ dephasing (σ) (blue symbols in Fig. 5). Deviation from linearity occurs near the crossover point (above 8 BU in Fig. 5B) from exponential to soft distance dependence. Because the dephasing rates in the two models are similar but not identical, the distance dependence for rates and conductances near the cross-over point differs (compare solid black and dashed blue lines in Fig. 4) leading to non-linearity in the rate–conductance relation at long bridge lengths. In contrast, when the dephasing rates in the two models differ significantly, the non-linearity in the rate–conductance relation is evident over the entire range of data at all

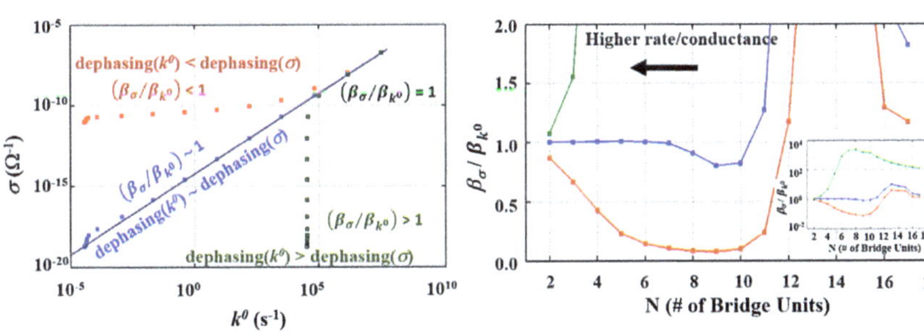

(A) (B)

Fig. 5 (A) Correlation between the CT rate and the molecular conductance for the three cases of relative dephasing rates on the bridge. The black line indicates a linear correlation between conductance and rate: $\sigma = A \times (k^0)^m$ (where $m = \beta_\sigma/\beta_{k^0} = 1$). (B) The exponent $m = \beta_\sigma/\beta_{k^0}$ calculated for the data in panel (A) as a function of bridge length. Red, blue and green lines/symbols correspond to the exponents obtained from fits to the red, blue and green symbols in panel (A), respectively. The inset shows the full range of exponents on a logscale (β_σ/β_{k^0} for the green line reaches very large values as $\beta_{k^0} \sim 0$) obtained at all bridge lengths. The data in panel (B) are restricted to exponent values between 0–2 and show that $m = 1$ is obtained over a large range of rates–conductances only when the dephasing rates in the rate and conductance models are similar. The rate–conductance is non-linear ($m \neq 1$) over the whole data range, when the dephasing rates in the rate and conductance models are different.

bridge lengths. When dephasing (k^0) < dephasing (σ) (red symbols in Fig. 5), the crossover from exponential to soft distance decay occurs much earlier for the conductance (compare solid black and dashed red lines in Fig. 4). As a result, the rate conductance correlation is close to linear only at short bridge lengths where superexchange CT dominates. Fig. 5B shows $m \neq 1$ even for rates–conductances across 1 BU. At longer bridge lengths, β_σ drops much more rapidly than β_{k^0} because of stronger dephasing in the conductance model to give a ratio $\beta_\sigma/\beta_{k^0} < 1$. A complementary decoupling of the rate and conductance is evident in the other limit: dephasing (k^0) > dephasing (σ) (dark green symbols in Fig. 5) where the crossover from an exponential to a softer distance decay occurs earlier for the rates (compare solid green and dashed blue lines in Fig. 4). As before, a near-linear correlation appears for short bridges where superexchange CT dominates. For longer bridges, β_{k^0} drops much faster than β_σ because of stronger dephasing in the rate model to give a ratio $\beta_\sigma/\beta_{k^0} > 1$. This analysis suggests that the power law exponent ($m = \beta_\sigma/\beta_{k^0} \sim 0.6$–$0.9 < 1$) obtained from fits to the experimental rate *versus* conductance data in Fig. 1 may arise from larger dephasing rates in conductance measurements than in the rate measurements (see red symbols in Fig. 5B). Note that the numerical values of β_σ/β_{k^0} from Fig. 5B were obtained for a model system (caption of Fig. 3). Thus, while the analysis in Fig. 5 provides general insights to understand the trends in the rate–conductance scaling exponents (β_σ/β_{k^0}) as a function of dephasing, the numerical values in Fig. 5B do not correspond to exponents of the experimental data in Fig. 1. The extent to which dephasing influences the non-linearity shown in Fig. 1 can be determined only through quantitative fits of the framework developed here to the experimental data, a task which we will undertake in a separate study. An important point is that the decay constants (β_σ and β_{k^0}) for pure superexchange transport are not influenced by changes in the dephasing rates (slopes of the exponential decay of rates–conductances with bridge length in Fig. 3 do not change with dephasing rates). This means that it is possible to find linear correlations ($m = \beta_\sigma/\beta_{k^0} \sim 1$) between rates and conductances at short bridge lengths, despite the significant differences in dephasing rates in the two models. The near linear correlation at short bridge lengths in Fig. 5A is a manifestation of this effect.

The magnitude of the CT barrier also affects the correlation between the CT rate and the molecular conductance. Fig. 6 shows a log–log plot of the rate–conductance correlation (as in Fig. 5), except for the three cases of relative CT energy barriers (see Fig. 6 caption) in the rate and the conductance models. Both models have similar dephasing rates ($\gamma_{mn} = 0$ eV; $\Gamma_P = 10^{-6}$ eV) as in case 2 of Fig. 5. The rate–conductance correlation for $E_{DB} = E_{FB}$ (blue symbols in Fig. 6) is the same as case 2 of Fig. 5, indicating a near linear correlation for a wide range of bridge lengths. As discussed above, the non-linearity for long bridge lengths arises because the dephasing in the rate and conductance models are not identical. In contrast, when the CT energy barriers in the two models are different, the rate–conductance correlation is non-linear at all bridge lengths. When $E_{DB} > E_{FB}$ (red symbols), the superexchange CT rate drops more rapidly and the crossover of the rate to a softer distance dependence occurs at longer bridge lengths than for the conductance. Consequently, β_{k^0} is larger than β_σ, and $\beta_\sigma/\beta_{k^0} < 1$ for all bridge lengths. The situation is reversed when $E_{DB} < E_{FB}$ (green symbols), as the conductance drops more rapidly than the rate in the superexchange regime with crossover to a weaker distance dependence occurring at longer bridge lengths

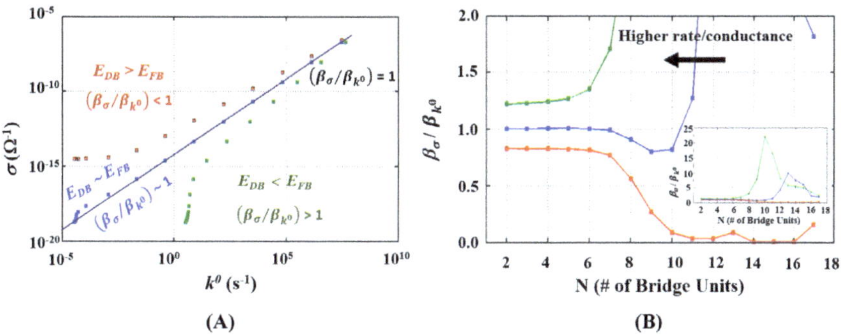

Fig. 6 (A) Correlation between the CT rate and the molecular conductance for three different CT energy barrier conditions: (i) E_{DB} (1.45 eV) > E_{FB} (1.15 eV), (ii) $E_{DB} = E_{FB}$ (1.45 eV), and (iii) E_{DB} (1.15 eV) < E_{FB} (1.45 eV). The black line indicates a linear correlation ($m = \beta_\sigma/\beta_{k^0} = 1$). (B) The exponent $m = \beta_\sigma/\beta_{k^0}$ calculated for the data in panel (A) as a function of bridge length. Red, blue and green lines/symbols correspond to the exponents obtained from fits to the red, blue and green symbols in panel (A), respectively. The inset shows the full range of exponents obtained at all bridge lengths. The data in panel (B) are restricted to exponent values between 0–2 and show that $m = 1$ is obtained over a large range of rates–conductances only when the energy barriers in the rate and conductance models are same. The rate–conductance is non-linear ($m \neq 1$) over the whole data range, when the energy barriers in the rate and conductance models are different.

than that for the rate. Consequently, β_σ is larger than β_{k^0} and the ratio $\beta_\sigma/\beta_{k^0} > 1$ for all bridge lengths. This analysis on our model system (parameters are given in the caption of Fig. 3B) indicates that the exponent in the power law fits ($\beta_\sigma/\beta_{k^0} \sim 0.6$–$0.9 < 1$) of Fig. 1 may arise from smaller CT barriers in the conductance measurements than in the rate measurements (red symbols in Fig. 6B). Note that the decay constants (β_σ and β_{k^0}) for pure superexchange transport are strongly dependent on the energy barriers for charge transport (slopes of the exponential decays of rates–conductances with distance in Fig. 4A change with the CT energy barrier). Thus, when energy barriers for charge transport are significantly different, the rate–conductance correlation can never be linear ($m = \beta_\sigma/\beta_{k^0} \neq 1$) as seen in Fig. 6A and B.

General role of dephasing and CT energy barriers in shaping the non-linear relationship between charge transport measurements

In this section, we show that our findings regarding the origin of the rate–conductance non-linearity are in fact generally applicable to the study of correlations between any two charge transport measures (rates or conductances). In other words, the same property (for example, two CT rates or two conductances) when measured subject to conditions where the environmental decoherence contributions and/or CT energy barriers are different, will exhibit a non-linear correlation. Fig. 7 shows the correlation between two CT rates (k_1^0 and k_2^0 in Fig. 7A) and two conductances (σ_1 and σ_2 in Fig. 7B) computed for identical charge transfer barriers ($E_{DB}(k_1^0) = E_{DB}(k_2^0)$ and $E_{FB}(\sigma_1) = E_{FB}(\sigma_2)$), but with three choices of the relative dephasing rates ($\gamma_{mn}(k_1^0) < \gamma_{mn}(k_2^0)$, $\gamma_{mn}(k_1^0) = \gamma_{mn}(k_2^0)$, $\gamma_{mn}(k_1^0) > \gamma_{mn}(k_2^0)$ and $\Gamma_P(\sigma_1) < \Gamma_P(\sigma_2)$, $\Gamma_P(\sigma_1) = \Gamma_P(\sigma_2)$, $\Gamma_P(\sigma_1) > \Gamma_P(\sigma_2)$). The data in Fig. 7 can be directly compared with that in Fig. 5 and show a non-linear rate–rate and

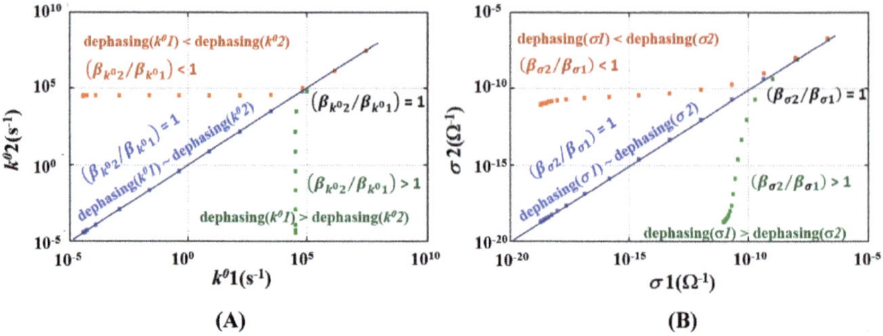

Fig. 7 (A) Correlation between two CT rates, k_1^0 and k_2^0, computed as a function of bridge length with identical CT energy barriers ($E_{DB}(k_1^0) = E_{DB}(k_2^0) = 1.45$ eV) but for three different conditions of relative dephasing: (i) $\gamma_{mn}(k_1^0) = 0$ eV < $\gamma_{mn}(k_2^0) = 10^{-4}$ eV, (ii) $\gamma_{mn}(k_1^0) = 0$ eV = $\gamma_{mn}(k_2^0)$ and (iii) $\gamma_{mn}(k_1^0) = 10^{-4}$ eV > $\gamma_{mn}(k_2^0) = 0$ eV. (B) Correlation between two conductances, σ_1 and σ_2, computed as a function of bridge length with identical CT energy barriers ($E_{FB}(\sigma_1) = E_{FB}(\sigma_2) = 1.45$ eV) but for three different conditions of relative dephasing: (i) $\Gamma_P(\sigma_1) = 10^{-6}$ eV < $\Gamma_P(\sigma_2) = 10^{-2}$ eV, (ii) $\Gamma_P(\sigma_1) = 10^{-6}$ eV = $\Gamma_P(\sigma_2)$ and (iii) $\Gamma_P(\sigma_1) = 10^{-2}$ eV > $\Gamma_P(\sigma_2) = 10^{-6}$ eV. The black lines in both panels indicate a linear correlation ($m = \beta_{k_1^0}/\beta_{k_2^0} = 1$) in (A) and ($m = \beta_{\sigma_1}/\beta_2 = 1$) in (B).

conductance–conductance correlation dependent on the relative dephasing rates, similar to the behavior exhibited by the rate–conductance correlation. The non-linear rate–conductance trends seen in Fig. 1 may also be seen for two series of CT rate measurements (or two conductance measurements) when environment-induced decoherence contributions in the two experimental setups are different. The calculations shown in Fig. 7 also help validate the robustness of our conclusions on how dephasing influences the rate–conductance correlation. The trends in the rate–conductance correlation shown in Fig. 5 were obtained by comparing two different model calculations (the CT rate model based on the reduced density matrix framework and the conductance model based on the Landauer–Buttiker framework). A concern arises that the non-linear rate–conductance correlation obtained in Fig. 5 may arise from differences in model assumptions and parameters. For instance, there is no theoretical expression to directly link the numerical values of dephasing in the CT rate models to those of the conductance models (see discussion following eqn (23) in the "Theoretical models" section). The strong agreement among the trends in the data of Fig. 7, where comparisons are within the same model framework (between CT rate models or between conductance models), and the data of Fig. 5, justifies our assumptions for comparing dephasing rates between rate and conductance models to extract the factors behind the rate–conductance non-linearity.

Our conclusions regarding the influence of the CT energy barrier on the rate–conductance correlation can also be extended to rate–rate and conductance–conductance correlations. Fig. 8 shows the correlation between two CT rates (k_1^0 and k_2^0 in Fig. 8A) and two conductances (σ_1 and σ_2 in Fig. 8B) computed for identical dephasing rates ($\gamma_{mn}(k_1^0) = \gamma_{mn}(k_2^0)$ and $\Gamma_P(\sigma_1) = \Gamma_P(\sigma_2)$) but with three cases of relative CT energy barriers ($E_{DB}(k_1^0) < E_{DB}(k_2^0)$, $E_{DB}(k_1^0) = E_{DB}(k_2^0)$, $E_{DB}(k_1^0) > E_{DB}(k_2^0)$ and $E_{FB}(\sigma_1) < E_{FB}(\sigma_2)$, $E_{FB}(\sigma_1) = E_{FB}(\sigma_2)$, $E_{FB}(\sigma_1) > E_{FB}(\sigma_2)$). The data in Fig. 8 show that the rate–rate and conductance–conductance correlations are non-

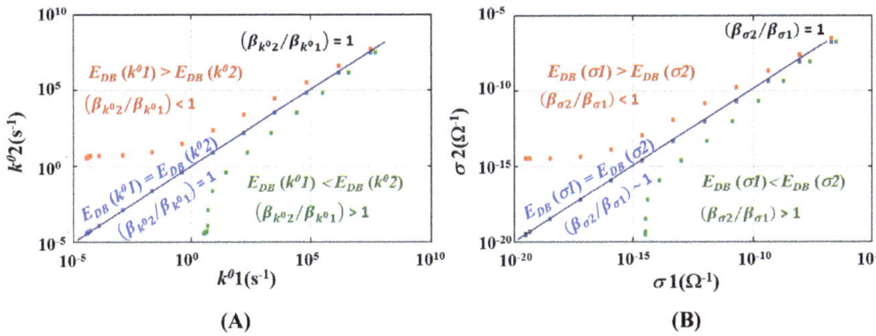

Fig. 8 (A) Correlation between two CT rates, k_1^0 and k_2^0, computed as a function of bridge length with identical dephasing rates ($\gamma_{mn}(k_1^0) = 0$ eV $= \gamma_{mn}(k_2^0)$) but for three different conditions of relative CT energy barriers: (i) $E_{DB}(k_1^0) = 1.15$ eV < $E_{DB}(k_2^0) = 1.45$ eV, (ii) $E_{DB}(k_1^0) = E_{DB}(k_2^0) = 1.45$ eV and (iii) $E_{DB}(k_1^0) = 1.45$ eV > $E_{DB}(k_2^0) = 1.15$ eV. (B) Correlation between two conductances, σ_1 and σ_2, computed as a function of bridge length with identical dephasing rates ($\Gamma_P(\sigma_1) = 10^{-6}$ eV $= \Gamma_P(\sigma_2)$) but for three different conditions of relative CT energy barriers: (i) $E_{FB}(\sigma_1) = 1.15$ eV < $E_{FB}(\sigma_2) = 1.45$ eV, (ii) $E_{FB}(\sigma_1) = 1.45$ eV $= E_{FB}(\sigma_2)$ and (iii) $E_{FB}(\sigma_1) = 1.45$ eV > $E_{FB}(\sigma_2) = 1.15$ eV. The black lines in both panels indicate a linear correlation ($m = \beta_{k_1^0}/\beta_{k_2^0} = 1$) in (A) and ($m = \beta_{\sigma_1}/\beta_2 = 1$) in (B).

linear, influenced by relative CT energy barriers similar to that seen for the rate–conductance correlation in Fig. 6. Thus, the non-linear rate–conductance trends observed in Fig. 1 are also observed for two series of CT rate measurements (or two conductance measurements) when the CT energy barriers in the two experimental setups are different (*e.g.*, obtained using different electrode binding linkers/redox reporters molecular designs or different electrode materials).

Conclusions

The analysis in this paper indicates that the CT energy barrier and decoherence on the bridge can independently, or in combination, influence the predicted rate–conductance relationship, producing non-linearity in the power law seen in Fig. 1. Specifically, our numerical simulations show that the scaling exponent ($m = \beta_\sigma/\beta_{k^0} \sim 0.6$–$0.9 < 1$) in the rate–conductance correlation of Fig. 1, may arise under conditions where dephasing (conductance) > dephasing (rate) and/or from $E_{DB} > E_{FB}$. Replacing a molecular donor–acceptor with an electrode introduces a continuum of electrode states, which provides an additional source of broadening for the bridge states. As an electrode provides irreversible decay channels for charge population on the bridge, population relaxation and dephasing rates on the bridge are expected to be greater when coupled to electrodes rather than to molecular donors and acceptors. Overall, decoherence effects are expected to increase with the number of electrodes connected to the bridge. Thus, we expect decoherence (D–B–A) < decoherence (M–B–A/D–B–M) < decoherence (M–B–M). However, other medium effects may vary among experimental schemes as well. Molecular bridges coupled to electrodes will experience stabilization of electronic states from electrode polarization in response to injection of charge on the bridge.[50,51] Image charge effects of this kind can lead to modified charge transfer barriers in the different experiments, even when the molecular bridges are identical. Finally, donor electronic state energies may differ from the work functions of metal electrodes. As such, different CT energy barriers appear likely in D–B–A, M–B–A/D–B–M, and M–B–M structures. A quantitative fit of the framework developed here to the experimental data in Fig. 1 is expected to constrain the CT barriers and decoherence values, and analysis of such fits will be presented elsewhere.

In summary, we have developed a theoretical framework to compute CT rates and molecular conductances without assuming a transport mechanism *a priori*. This approach reveals that a non-linear correlation exists between rates and conductances. The non-linearity arises from CT energy barrier and decoherence differences in the experiments. A decrease in the CT energy barrier and/or an increase in the decoherence rate shifts the position of the crossover from exponential to a weaker distance dependence of rates and conductances. In fact, by manipulating the CT energy barriers and bridge decoherence, a wide range of power law correlations can be accessed among CT rates and molecular conductances. This theoretical framework explains the observed non-linear rate–conductance relationships evident in experimental data.

Acknowledgements

R.V. acknowledges financial support from Tata Institute of Fundamental Research (Department of Atomic Energy, India). The authors gratefully acknowledge support from the National Science Foundation (CHE 1057953 to D.N.B. and CHE 1057981 to D.H.W.).

References

1 A. Nitzan, *J. Phys. Chem. A*, 2001, **105**(12), 2677–2679.
2 A. Nitzan, *Isr. J. Chem.*, 2002, **42**(2–3), 163–166.
3 Y. A. Berlin and M. A. Ratner, *Radiat. Phys. Chem.*, 2005, **74**(3–4), 124–131.
4 M. C. Traub, B. S. Brunschwig and N. S. Lewis, *J. Phys. Chem. B*, 2007, **111**(24), 6676–6683.
5 G. B. Schuster, *Long-Range Charge Transfer in DNA I and II*, Springer: Berlin, 2004, vol. 236–237.
6 W. B. Davis, M. R. Wasielewski, M. A. Ratner, V. Mujica and A. Nitzan, *J. Phys. Chem. A*, 1997, **101**(35), 6158–6164.
7 F. D. Lewis, T. F. Wu, Y. F. Zhang, R. L. Letsinger, S. R. Greenfield and M. R. Wasielewski, *Science*, 1997, **277**(5326), 673–676.
8 Y. A. Berlin, I. V. Kurnikov, D. Beratan, M. A. Ratner and A. L. Burin, DNA electron transfer processes: Some theoretical notions, in *Long-Range Charge Transfer in DNA II*, Springer-Verlag Berlin, Berlin, 2004, vol. 237, pp. 1–36.
9 F. D. Lewis, J. Q. Liu, W. Weigel, W. Rettig, I. V. Kurnikov and D. N. Beratan, *Proc. Natl. Acad. Sci. U. S. A.*, 2002, **99**(20), 12536–12541.
10 J. F. Smalley, H. O. Finklea, C. E. Chidsey, M. R. Linford, S. E. Creager, J. P. Ferraris, K. Chalfant, T. Zawodzinsk, S. W. Feldberg and M. D. Newton, *J. Am. Chem. Soc.*, 2003, **125**(7), 2004–2013.
11 A. Paul, R. M. Watson, E. Wierzbinski, K. L. Davis, A. Sha, C. Achim and D. H. Waldeck, *J. Phys. Chem. B*, 2010, **114**(45), 14140–14148.

12 A. Paul, R. M. Watson, P. Lund, Y. J. Xing, K. Burke, Y. F. He, E. Borguet, C. Achim and D. H. Waldeck, *J. Phys. Chem. C*, 2008, **112**(18), 7233–7240.
13 J. F. Smalley, S. W. Feldberg, C. E. D. Chidsey, M. R. Linford, M. D. Newton and Y. P. Liu, *J. Phys. Chem.*, 1995, **99**(35), 13141–13149.
14 J. He, O. Sankey, M. Lee, N. J. Tao, X. L. Li and S. Lindsay, *Faraday Discuss.*, 2006, **131**, 145–154.
15 D. Porath, A. Bezryadin, S. de Vries and C. Dekker, *Nature*, 2000, **403**(6770), 635–638.
16 H. Cohen, C. Nogues, D. Ullien, S. Daube, R. Naaman and D. Porath, *Faraday Discuss.*, 2006, **131**, 367–376.
17 X. F. Guo, A. A. Gorodetsky, J. Hone, J. K. Barton and C. Nuckolls, *Nat. Nanotechnol.*, 2008, **3**(3), 163–167.
18 A. V. Tivanski, Y. F. He, E. Borguet, H. Y. Liu, G. C. Walker and D. H. Waldeck, *J. Phys. Chem. B*, 2005, **109**(12), 5398–5402.
19 C. Li, I. Pobelov, T. Wandlowski, A. Bagrets, A. Arnold and F. Evers, *J. Am. Chem. Soc.*, 2008, **130**, 318–326.
20 E. Wierzbinski, R. Venkatramani, K. L. Davis, S. Bezer, J. Kong, Y. Xing, E. Borguet, C. Achim, D. N. Beratan and D. H. Waldeck, *ACS Nano*, 2013, **7**(6), 5391–5401.
21 D. Segal, A. Nitzan, W. B. Davis, M. R. Wasielewski and M. A. Ratner, *J. Phys. Chem. B*, 2000, **104**(16), 3817–3829.
22 D. Segal, A. Nitzan, M. Ratner and W. B. Davis, *J. Phys. Chem. B*, 2000, **104**(13), 2790–2793.
23 D. N. Beratan, S. S. Skourtis, I. A. Balabin, A. Balaeff, S. Keinan, R. Venkatramani and D. Q. Xiao, *Acc. Chem. Res.*, 2009, **42**(10), 1669–1678.
24 S. S. Skourtis, D. H. Waldeck and D. N. Beratan, *Annu. Rev. Phys. Chem.*, 2010, **61**, 461–485.
25 R. Venkatramani, S. Keinan, A. Balaeff and D. N. Beratan, *Coord. Chem. Rev.*, 2011, **255**, 635–658.
26 A. K. Felts, W. T. Pollard and R. A. Friesner, *J. Phys. Chem.*, 1995, **99**(9), 2929–2940.
27 A. Okada, V. Chernyak and S. Mukamel, *J. Phys. Chem. A*, 1998, **102**(8), 1241–1251.
28 S. Yeganeh, M. A. Ratner and V. Mujica, *J. Chem. Phys.*, 2007, **126**, 161103.
29 S. Skourtis and A. Nitzan, *J. Chem. Phys.*, 2003, **119**(12), 6271–6276.
30 S. S. Skourtis, D. N. Beratan, R. Naaman, A. Nitzan and D. H. Waldeck, *Phys. Rev. Lett.*, 2008, **101**, 238103.
31 S. S. Skourtis, A. J. R. Dasilva, W. Bialek and J. N. Onuchic, *J. Phys. Chem.*, 1992, **96**(20), 8034–8041.
32 R. A. Marcus and N. Sutin, *Biochim. Biophys. Acta*, 1985, **811**(3), 265–322.
33 P. O. Lowdin, *J. Math. Phys.*, 1962, **3**, 969–982.
34 S. Larsson, *J. Am. Chem. Soc.*, 1981, **103**(14), 4034–4040.
35 S. Larsson, *J. Chem. Soc., Faraday Trans. 2*, 1983, **79**, 1375–1388.
36 A. K. Mishra and D. H. Waldeck, *J. Phys. Chem. C*, 2011, **115**(42), 20662–20673.
37 A. M. Napper, H. Y. Liu and D. H. Waldeck, *J. Phys. Chem. B*, 2001, **105**(32), 7699–7707.
38 S. Datta, *Quantum Transport: Atom to Transistor*, Cambridge University Press, 2005.

39 S. Datta, *Electronic transport in mesoscopic systems*, Cambridge University Press, Cambridge, 1997.
40 J. Maassen, F. Zahid and H. Guo, *Phys. Rev. B: Condens. Matter Mater. Phys.*, 2009, **80**, 125423.
41 M. Galperin, A. Nitzan and M. A. Ratner, *Phys. Rev. B: Condens. Matter Mater. Phys.*, 2006, **74**, 075326.
42 R. Hartle, R. Volkovich, M. Thoss and U. Peskin, *J. Chem. Phys.*, 2010, **133**, 081102.
43 J. Lykkebo, A. Gagliardi, A. Pecchia and G. C. Solomon, *ACS Nano*, 2013, **7**(10), 9183–9194.
44 M. Bixon and J. Jortner, *Chem. Phys.*, 2002, **281**(2–3), 393–408.
45 R. Venkatramani, K. L. Davis, E. Wierzbinski, S. Bezer, A. Balaeff, S. Keinan, A. Paul, L. Kocsis, D. N. Beratan, C. Achim and D. H. Waldeck, *J. Am. Chem. Soc.*, 2011, **133**, 62–72.
46 H. McConnell, *J. Chem. Phys.*, 1961, **35**(2), 508–515.
47 J. N. Onuchic and D. N. Beratan, *J. Am. Chem. Soc.*, 1987, **109**(22), 6771–6778.
48 Y. Xing, T. H. Park, R. Venkatramani, S. Keinan, D. N. Beratan, M. J. Therien and E. Borguet, *J. Am. Chem. Soc.*, 2010, **132**(23), 7946–7956.
49 Y. A. Berlin, A. L. Burin and M. A. Ratner, *Chem. Phys.*, 2002, **275**(1–3), 61–74.
50 S. Kubatkin, A. Danilov, M. Hjort, J. Cornil, J. L. Bredas, N. Stuhr-Hansen, P. Hedegard and T. Bjornholm, *Nature*, 2003, **425**(6959), 698–701.
51 J. B. Neaton, M. S. Hybertsen and S. G. Louie, *Phys. Rev. Lett.*, 2006, **97**, 216405.

Faraday Discussions

PAPER

Molecular diodes enabled by quantum interference†

Arunabh Batra,[a] Jeffrey S. Meisner,‡[b] Pierre Darancet,[ac] Qishui Chen,[b] Michael L. Steigerwald,[b] Colin Nuckolls*[b] and Latha Venkataraman*[a]

Received 5th May 2014, Accepted 17th June 2014
DOI: 10.1039/c4fd00093e

We use scanning tunneling microscope break-junction (STM-BJ) measurements to study the low-bias conductance and high-bias current–voltage (*IV*) characteristics of a series of asymmetric *para*–*meta* connected diphenyl-oligoenes. From tight-binding calculations, we determine that the quantum interference features inherent in our molecular design result in a 'through-bond' coupling on the *para*-side, and through-space coupling on the *meta*-side. We show that these molecular junctions form single molecule diodes, and show that the rectification results from a difference in the voltage dependence of the coupling strength on the through-bond and the through-space side. The interplay between the applied voltage and the molecule–metal coupling results from the asymmetric polarizability of the conducting orbital under an external field.

Understanding charge transport across single molecule junctions as well as transfer at metal-organic interfaces is of critical importance in the field of molecular/organic electronics, such as light emitting devices (OLEDs) and photovoltaics (OPVs).[1–6] Organic materials as alternates to standard silicon based semiconductors are fast becoming commercially viable especially in applied areas including flexible electronics, sensors and photovoltaics.[7–10] However, the physical mechanisms by which charge transport and transfer occur in relation to the chemical structure in these systems are poorly understood. One route to understand and fully characterize such physical processes in organic electronics is to focus on nanoscale single-molecule devices where one can probe structure–function relations directly. The proposal to create molecular analogs of circuit components dates back to the work of Aviram and Ratner. In their seminal work from 1974,[11] they proposed using a single molecule as a diode circuit element in giving birth to the field of molecular electronics,[12] and research in this field has

[a]Department of Applied Physics and Applied Mathematics, Columbia University, New York, NY, USA. E-mail: lv2117@columbia.edu
[b]Department of Chemistry, Columbia University, New York, NY, USA. E-mail: cn37@columbia.edu
[c]Molecular Foundry, Lawrence Berkeley National Laboratory, Berkeley, California, USA

† Electronic supplementary information (ESI) available. See DOI: 10.1039/c4fd00093e

‡ Current address: Department of Chemistry, Duke University.

advanced tremendously since then.[13] For example, highly conductive molecular junctions have been realized,[14,15] and recent scientific developments have demonstrated molecular analogues of several conventional circuit components, such as transistors,[16–18] potentiometers,[19] diodes[20–23] and switches.[24–27] Furthermore, devices that utilize purely quantum mechanical aspects of charge transport – interference effects arising due to the wave nature of electrons – have been demonstrated;[28–30] such behavior has no counterpart in conventional electronic circuits.

Molecular diodes are of great interest in the nanoscale electronics community because they are the simplest example of non-trivial electronic behavior in a molecular system, and because they form the basis for more complicated device architectures.[31,32] Despite the long-standing interest in the production of these devices, their experimental realization has been difficult, with only a handful of studies showing rectification at the single molecule level.[20,21,23] This is because most theoretical designs, such as the Aviram–Ratner model of rectification,[11] rely on the complex interplay between many variables, such as the level alignments of the molecular components and the Fermi level of the metal electrodes.[33] For example, in the pioneering work of Elbing *et al.*,[20] the molecule studied consisted of two weakly coupled conjugated segments that were energetically shifted relative to the electrode Fermi level by using electron withdrawing and electron donating substituents. They showed experimentally that the current voltage characteristics across a molecular device was asymmetric with respect to the applied voltage, however, it was hard to quantitatively assess the effects of the substituents on the rectification ratio determined experimentally making the behavior of such diodes difficult to predict.[34,35] Simpler 'unimolecular' rectifier designs had been proposed,[36–38] but have been demonstrated primarily on thin-films, or many molecule junctions.[39–42] Theoretical work has also focused on the role of the electronic contact[43] in defining rectification behavior, and recent experimental evidence[23] has shown the efficacy of such an approach in developing tunable rectifiers. Here, we show that quantum interference effects[29,44,45] can be used to create asymmetrically coupled molecular junctions[46] and attribute the rectification to a novel mechanism that can in principle be applied to a variety of other molecular structures.

In this work, we measure conductance and current–voltage characteristics in a series of conjugated molecules of increasing length with linkers at the *para*-position on one side and at the *meta*-position on the other, and find that these molecular junctions rectify the current. From our experimental and theoretical results, we conclude that the quantum interference inherent in our molecular design effectively ensures that the molecular orbital dominating transport across the junction is strongly coupled to one electrode while allowing only weak electronic coupling to the other electrode. We show further that rectification results from a difference in the voltage dependence of the coupling strength between two sides of the molecule, an effect that depends on the molecular backbone length. The interplay between the applied voltage and the molecule–metal coupling results from the asymmetric polarizability of the conducting orbital under an external field.

The molecules used in this work are based on diphenyl-oligoene backbones[46] (Fig. 1A, molecules 1–5) of lengths ranging from 1.2 nm to 2.3 nm. The backbones are functionalized with two methylsulfide (SMe) linkers,[47,48] in the *para*-position

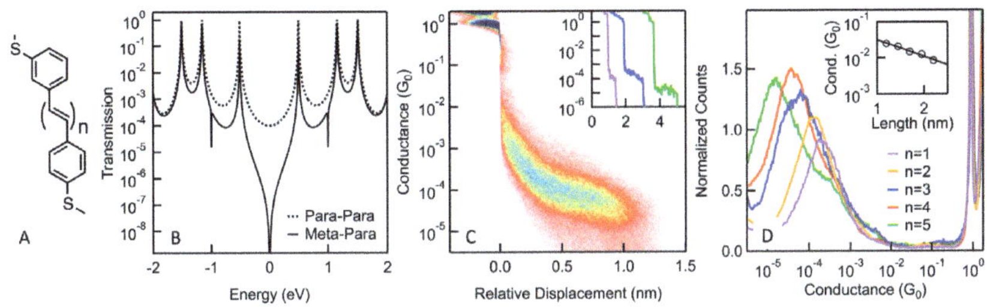

Fig. 1 (A) Molecular structures for molecules with $n = 1$–5. (B) Tight-binding transmission functions for molecular junctions with molecule 1 (black); this *para–meta* system shows negligible transmission probability near E_{Fermi} (0 eV). A *para–para* (dashed black) connected version of molecule 1 shows higher transmission. (C) Inset: sample conductance traces for molecules 1, 3 and 5 showing molecular plateaus. Two dimensional (2D) conductance histogram for molecule 3, showing characteristic downward slope in conductance with increasing relative displacement. (D) Conductance histograms for *para–meta* connected molecules 1–5 showing measurable conductance. Inset: logarithm of conductance *versus* molecule length obtained using *S–S* distances from DFT optimized structures of molecules 1–5 fit with a line. The decay constant determined from this least-squares fit is $\beta = 0.25 \pm 0.03$ Å^{-1}.

on one phenyl ring and in the *meta*-position on the other (see the ESI† for synthesis details).[45,46] To first understand what conductance we can expect to measure in this system, we carry out a simple tight-binding based theoretical model calculation for molecular junctions formed using such *para–meta* linked molecules. We focus on the stilbene molecule (1) and consider only the p_z orbitals on each carbon atom in our model. We assume an on-site energy of $\varepsilon = 0$ eV and a coupling strength to the nearest neighbor orbital, $\tau = 1$ eV. The molecule is coupled to electrodes through Γ_L and Γ_R at the carbon atom at the 4 position on one benzene ring (*para*) and at the 3′ (*meta*) or 4′ (*para*) position on the other benzene ring. For the purpose of this calculation we set $\Gamma_L = \Gamma_R = 0.04$. Transmission probability $T(E)$ is calculated by first determining the Hamiltonian matrix H (which in this case is a 14 × 14 matrix). The retarded Green's function for the model junction is then $G(E) = [EI - H]^{-1}$ and the transmission is given by $T(E) = \Gamma^2|G_{LR}|^2$ where G_{LR} denotes the component of the Green's function matrix that describes the propagation between the left and right electrodes.[49] Fig. 1B compares the tight-binding transmission functions for the *para–para* connected stilbene (dashed curve) and the *para–meta* connected version (solid curve). We see a clear non-zero transmission probability at zero energy for the *para–para* system while the *para–meta* curve shows close to zero transmission probability. The negligible transmission seen in the *para–meta* system is due to an anti-resonance which results from destructive interference at the Fermi energy.[29,45,50]

To determine the conductance from these transmission probabilities, we use the Landauer formalism,[51] which is perhaps the most well accepted theoretical framework for describing ballistic transport in one dimension. This method converts the electron transport problem into a scattering problem. In this formalism, the conductance G of a metal–molecule–metal junction is related to the probability of an electron from one electrode, emerging at the Fermi energy (modulo any applied bias), being transmitted across the molecular junction. The conductance G at zero applied bias is given by

$$G = G_0 T(E_F) = \frac{2e^2}{h} \sum_i T_i(E_F) \tag{1}$$

where e is the charge of an electron, h is Planck's constant, i is the index associated with a conducting channel, and $G_0 = 2e^2/h$ is the quantum of conductance. Applying the Landauer equation to the transmission function shown in Fig. 1B, we would expect a close to zero conductance for these *para–meta* coupled molecules at zero bias.

We measure the conductance and current–voltage characteristics of this series of molecules in ambient, room-temperature conditions with a scanning tunneling microscope (STM) in break-junction mode.[52–55] Conductance measurements under low-bias ($V_{bias} \sim 250$ mV) conditions are first carried out first using 1 mM solutions of the compounds in 1,2,4-trichlorobenzene (Sigma-Aldrich, >99% purity). A gold STM tip is brought into contact with a solution covered gold-on-mica substrate until a junction conductance greater than 5 G_0 ($G_0 = 2e^2/h = 77.6$ μS) is measured. The tip is then withdrawn at a rate of 15 nm s^{-1} while the current is measured at a constant bias. The inset of Fig. 1C shows sample conductance traces showing a prominent molecular conductance feature for molecules 1, 3 and 5, in contrast with the negligible conductance predicted by tight-binding.

Fig. 1C shows a two dimensional conductance histogram for molecule 3 which is created by overlaying thousands of conductance traces after first aligning them to have zero displacement at a conductance of 0.5 G_0. The conductance feature seen in this 2D histogram starts at 10^{-3} G_0 at zero relative displacements and goes down to $\sim 3 \times 10^{-5}$ G_0 by 1 nm relative displacement. The sloped nature of this feature indicates that the molecular conductances decrease as electrode separation is increased. Previous measurements with similar molecules have shown that such a sloped feature results from a non-specific π–Au molecule–electrode interaction.[19,56] This is in contrast to symmetric *para*-connected versions of the same molecules studied previously,[46] which show a flatter conductance feature typically seen when molecules are coupled through donor–acceptor bonds on both ends. We conclude that the asymmetric functionalization allows for one side (*para*) to be electronically coupled 'through-bond', and the other side (*meta*) to be coupled 'through-space'.[45,46] Fig. 1D shows logarithmically binned one-dimensional histograms of thousands of such traces for each molecule without data selection, which allows a characteristic conductance value to be statistically determined. On going from the shortest molecule, 1, through the series to the longest molecule, 5, the conductance decays exponentially (inset Fig. 1D), with a decay parameter $\beta = 0.25 \pm 0.03$ Å^{-1}. This is a similar decay constant to previous studies on the oligoene molecular system,[45,46] and suggests that the π–Au interaction occurs at end carbons on the phenyl ring near the *meta*-linker, rather than on the backbone.

We next measure the current–voltage (*IV*) characteristics of these molecules using procedures published previously.[54] Briefly, a gold STM tip is brought into contact with a gold-on-mica substrate. The tip is first withdrawn at a rate of 15 nm s^{-1} for 125 ms, and held at this displacement for 150 ms before being withdrawn for an additional 75 ms. During the 'hold' section when the tip/substrate distance is not changed, the voltage is ramped between ±1 V, while the current is measured (Fig. 2A). The *IV* procedure is repeated tens of thousands of times. Each measured trace is first analyzed to determine if a molecule was trapped during the hold section by comparing the conductance at the start and end of this section with the range determined from the conductance histogram shown in Fig. 1D. If the conductance corresponds to that of a molecular junction, we select the trace to analyze the *IV* data. We typically find that ~1–10% of all measured junctions show a stable, unbroken, molecular conductance signature during the *IV* measurement. Due to the instability of molecular junctions at higher biases at room temperature, measurements beyond ±1 V do not yield sufficient numbers of traces to analyze statistically. We then sort all selected *IV* traces into two sets based on the magnitude of current in a range of positive biases (+0.75 to +0.80 V) and corresponding negative voltages (−0.75 V to −0.80 V).[23] Traces that show a larger current at positive voltages compared with negative voltages correspond to molecular junctions in the forward orientation; traces that show smaller current at positive voltages compared with negative voltages correspond to molecular junctions in the reverse orientation. For each molecule, the *IV* portion of the selected traces from each set are binned logarithmically on the current axis, and added together to create two two-dimensional histograms of absolute current against voltage.

As an illustration, sorted 2D histograms from 4 are shown in Fig. 2B (reverse bias, top panel and forward bias, bottom panel). To make a quantitative comparison of the histograms, we fit each vertical slice of the histograms to a Gaussian distribution and extract the most frequently measured current at each voltage.[57] These values are plotted as the red and black *IV* curves overlaid in the top and bottom panels of Fig. 2B, respectively, and are also shown on a linear scale in Fig. 2C. The resulting curves represent the averaged *IV* curves for this molecule. The forward bias average *IV* (black curve, Fig. 2C) overlays the reverse bias *IV* (red curve), which has been mirrored in the voltage axis. The near-identical shape of these curves indicates that sorting recovers the inherent structure in the *IV* curves without biasing the data. A similar sorting algorithm applied to fully symmetric molecules further confirms this conclusion: the separated 2D

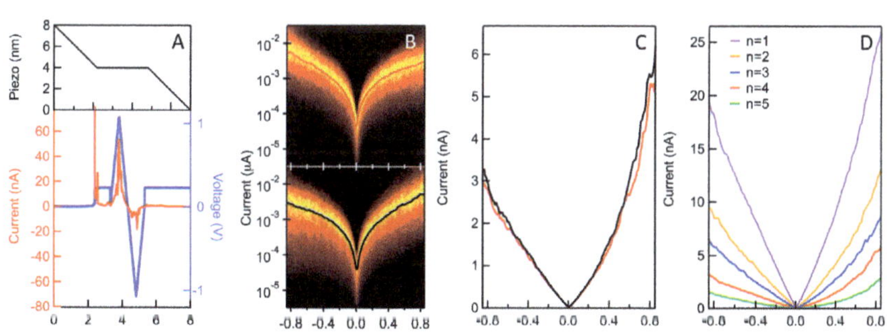

Fig. 2 (A) Representative current, voltage and displacement (piezo *z*-position) traces for a single break junction measurement for molecule 4. The *IV* ramp is applied when *z* is held constant. (B) Two-dimensional histograms created from thousands of *IV* curves for molecule 4 that are either in reverse bias (top panel) or forward bias (bottom panel). Average *IV* curves are overlaid in red and black, respectively. The counts in this 2D histogram range from 0 (black) to 1000 (yellow). (C) Average *IV* curves from B shown on a linear scale with reverse-bias (red) reflected about the *x*-axis. (D) Averaged *IV* curves for molecules 1–5 analyzed following the same procedure detailed in panel B and C.

histograms for these control molecules are identical and do not show rectification.[23] Fig. 2D compares statistically averaged *IV* curves for molecules 1–5. We find that the absolute current decreases with length. The rectification ratio for each molecule is determined from the ratio of the current at +0.85 V and −0.85 V. We see that the rectifying behavior increases slightly with increasing molecular length as will be discussed further below.

To understand the origin of rectification and its length dependence, we conduct density functional theory (DFT) calculations of molecules 1–5 in which we include an electric field applied parallel to the molecular axis. Molecules are structurally relaxed by performing spin-unrestricted calculations using the B3LYP exchange-correlation functional and 6-31G* basis set using the Q-Chem software suite.[58] Default grids and convergence thresholds are used for relaxation. Subsequently, single-point calculations are carried out using B3LYP/6-311G** either without field, or with a monopole field applied along the long axis of the molecule. We apply the field along the long axis of the molecule to simulate the field due to the applied bias on a molecule trapped between the tip and substrate.

Generally, the factors affecting rectification are voltage dependence in energy level alignment[59] (Stark shifts), or a voltage dependent coupling term Γ. For an applied voltage of ±1 V, the HOMOs of molecules 1–5 show negligible Stark shifts as shown in Fig. 3A, where we see that the energies of the HOMO orbitals do not differ by more than 0.1 eV under positive or negative voltage. This small Stark shift is not enough to explain the rectification effects observed experimentally. We therefore concentrate on the voltage dependence of the coupling terms on the *para* and *meta*-side, Γ_{para} and Γ_{meta}. Fig. 3B shows representative isosurface plots of the HOMO of molecule 5 under an applied voltage of +1 V (upper panel) and −1 V (lower panel). The weight of the HOMO on the carbon atoms adjacent to the *para* and *meta* linkers (arrows, Fig. 3B) changes significantly with field polarity. We can quantify this change in HOMO distribution using the projection of HOMO onto the p_z orbital of the *para* carbon of the 'through-bond' coupled side (Fig. 3C), and the projection of the p_z orbitals of all carbons on the phenyl ring of the 'through-space' *meta* connected side of the molecule (Fig. 3D). We show in Fig. 3C and D these projections for the entire series at 0 V and ±1 V.

The coupling Γ of the HOMO to each electrode is proportional to the square of the projections shown in Fig. 3C and D.[35,36] To determine the voltage dependence of the coupling terms Γ_{meta} and Γ_{para}, we square the projections shown in Fig. 3C and D and fit a line to the result to determine the change in the Γ values as a function of voltage. The voltage dependence on Γ is thus incorporated as:

$$\Gamma_{meta/para}(V) = \Gamma_{0,meta/0,para}(1 + \nu_{meta/para} \times V) \tag{2}$$

where $\Gamma_{0,meta/0,para}$ is the zero-field $\Gamma_{meta/para}$, and $\nu_{meta/para}$ is calculated from the slope of the voltage dependence of $\Gamma_{meta/para}$. In Fig. 4A we plot the difference between the −1 V and +1 V $\Gamma_{meta/para}$ normalized by the 0 V $\Gamma_{meta/para}$ ($\Delta\Gamma_{meta/para}/\Gamma_{0,meta/0,para}$) for all five molecules. We see first that longer molecules have a larger fractional change in $\Gamma_{meta/para}$ due to a higher molecular polarizability. More importantly, we see that this trend is more pronounced for Γ_{meta} than for Γ_{para}. This indicates that the coupling on the *meta*-side of the molecule changes much more with applied bias than that on the *para*-side. In other words, a positive or negative bias applied to the *para*-side has a much smaller effect on the charge

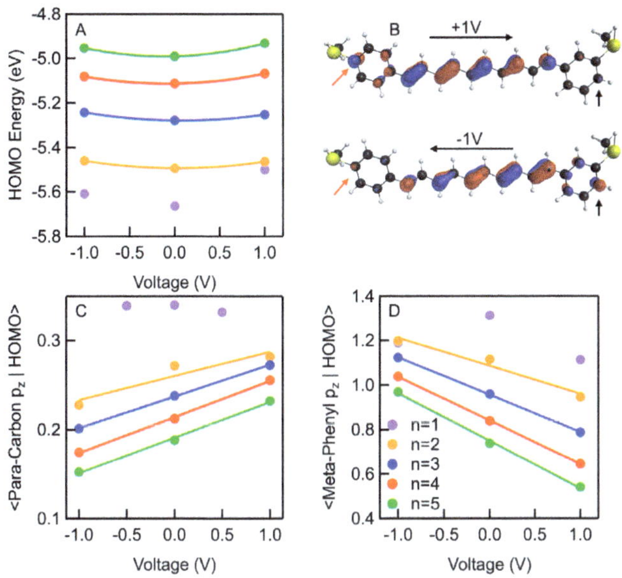

Fig. 3 (A) Stark shifts for HOMO as a function of constant field (voltage scaled with molecular length, with ±1 V applied to molecule 5). Molecules 2–5 (purple, yellow, blue, red, and green, respectively) show quadratic dependence on field, as expected for molecules without significant dipoles. Molecule 1 shows a slight stark shift due to a dipole moment along the length of the molecule. (B) DFT calculated HOMO for molecule 5 with an applied potential of +1 V (upper panel) and −1 V (lower panel). The red arrows point to the *para* carbon on the strongly coupled side. The black arrows point to the *para* carbon on the weakly coupled side. Isosurface value = $0.05/A^3$. (C) Projection of HOMO on the p_z orbitals of the carbon attached to the sulfur on the *para*-side of the molecule for constant voltage of ±1 V (for 1–5). (D) Projection of HOMO on the p_z orbitals of six ring carbons on the *meta*-side of the molecule for constant voltage of ±1 V (for 1–5). Molecules 2–5 all show a linear dependence of projection on voltage. Molecule 1 shows nonlinearities associated with a dipole moment along the length of the backbone.

density than the same voltage applied to the *meta*-side. We therefore attribute the rectification observed in these molecular junctions to the asymmetric polarizability of the conducting orbital in this system.

We now extend the tight binding model developed earlier to obtain current–voltage curves based on model transmission functions to understand if this change in coupling does indeed result in rectification for this system. To simplify the modeling, we use a single molecular orbital (the HOMO) coupled to the left and right electrode through coupling terms Γ_{para} and Γ_{meta} (Fig. 4B, inset). This reduces the tight binding Hamiltonian to a 1 × 1 matrix rather than the atomistic 14 × 14 Hamiltonian used to illustrate the interference effect above,[60,61] yielding a transmission function that has a single-Lorentzian form.[62] Including a voltage dependent Γ_{para} and Γ_{meta} given in eqn (2), the single-level model described results in a transmission function of the form:

$$T(E,V) = \frac{\Gamma_{para}\Gamma_{meta}\left(1+\nu_{para}V\right)\left(1+\nu_{meta}V\right)}{(E-E_0)^2 + \frac{1}{4}\left(\Gamma_{para}\left(1+\nu_{para}V\right)+\Gamma_{meta}\left(1+\nu_{meta}V\right)\right)^2} \tag{3}$$

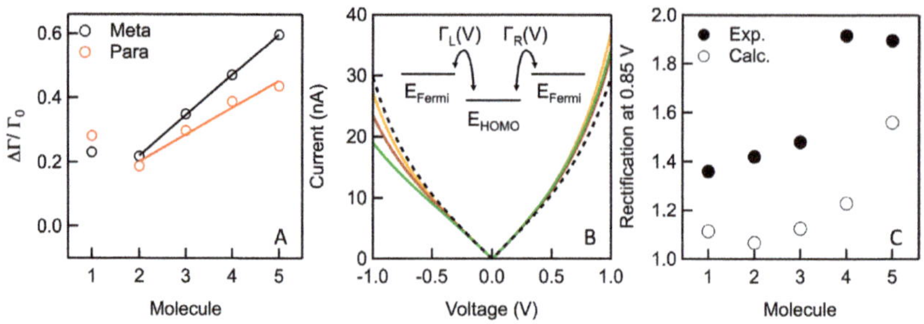

Fig. 4 (A) Fractional change in coupling Γ_{meta} (black) and Γ_{para} (red), estimated as the square of projection of the HOMO on the p_z atomic basis of the *para* atom (Γ_{para}) or the *meta*-connected phenyl ring (Γ_{meta}). Fractional change shown is an average of change for $V = +1$ V and $V = -1$ V. (B) Calculated *IV* curves for a single-level model (inset) with $E_0 = -0.8$ eV, $\Gamma_{para} = 0.04$, $\Gamma_{meta} = 0.004$, with Γ_{meta} and Γ_{para} set to vary linearly with voltage (see main text). The yellow, red, and green curves are calculated using voltage dependent Γ_{para} and Γ_{meta} with $\nu_{para} = 0.19, 0.39, 0.43$ and $\nu_{meta} = -0.22, -0.47, -0.60$ for molecules 2, 4, and 5, respectively (panel A). These curves show the effect of increasing differences in $\Delta\Gamma_{para}$ and $\Delta\Gamma_{meta}$ on the *IV* line shape. The dashed line shows a non-rectifying case where Γ_{meta} and Γ_{para} have the same voltage dependence. (C) Rectification ratio calculated (solid markers) from the ratios shown in (A), compared with experimental data (open circles). Both ratios are calculated by taking the ratio of the current at 0.85 V and −0.85 V. Rectification increases with the increase in molecular length.

IV curves are generated from the transmission function using the voltage dependent Landauer formula

$$I(V) = \frac{2e}{h}\int \mathrm{d}E\, T(E, V)\left[f\left(E + \frac{eV}{2}\right) - f\left(E - \frac{eV}{2}\right)\right] \quad (4)$$

where f is the Fermi distribution function. For simplicity, we take the temperature to be 0 K, although results from using $T = 300$ K are not very different.

The calculated model *IV* curves determined from eqn (4) above while varying the voltage dependences of Γ_{para} and Γ_{meta} are shown in Fig. 4B. We see that rectification increases with the difference in $\Delta\Gamma_{para}$ and $\Delta\Gamma_{meta}$, and when $\Delta\Gamma_{para}$ and $\Delta\Gamma_{meta}$ are equal, no rectification occurs (dashed *IV* curve). For each *IV* curve, we determine a rectification from the ratio of the current at 0.85 V to −0.85 V. These are plotted in Fig. 4C along with the experimentally determined rectifications. Our model explains the length dependence of our data, and suggests that rectification occurs not due to a difference between Γ_{para} and Γ_{meta}, but due to a difference in $\Delta\Gamma_{meta/para}$ which we attribute to the different nature of the electronic coupling on the *para* (through-bond) and *meta* (through-space) sides. Further, our model shows that increasing the length of the molecule can enhance rectification by increasing the asymmetric polarizability of the conducting orbital (in this case, the HOMO).

In summary, we have used the unique properties of quantum interference to create a series of asymmetrically coupled molecular diodes. We find that rectification ratios increase with molecular length, and model this trend using a tight-binding model taking into account the asymmetric coupling in these molecular junctions. We show that the increase in rectification is due to differences in the voltage-dependence of 'through-space' and 'through-bond' coupling terms with molecular length.

Acknowledgements

This work has been supported in part by the NSF Award (DMR-1206202) and the Packard Foundation. A.B. acknowledges support from the National Science Foundation Graduate Research Fellowship under grant no. DGE-07-07425. Part of this work was carried out at the Center for Functional Nanomaterials, Brookhaven National Laboratory, supported by the DOE Office of Basic Energy Sciences (contract no. DE-AC02- 98CH10886). The synthetic portions of this work was supported as part of the Center for Re-Defining Photovoltaic Efficiency Through Molecular-Scale Control, an Energy Frontier Research Center funded by the U.S. Department of Energy (DOE), Office of Science, Office of Basic Energy Sciences under award number DE-SC0001085.

References

1 Y. Loo and I. McCulloch, *MRS Bull.*, 2008, **33**, 653–662.
2 G. Yu, J. Gao, J. C. Hummelen, F. Wudl and A. J. Heeger, *Science*, 1995, **270**, 1789.
3 J. J. M. Halls, C. A. Walsh, N. C. Greenham, E. A. Marseglia, R. H. Friend, S. C. Moratti and A. B. Holmes, *Nature*, 1995, **376**, 498–500.
4 S. Beaupré, P.-L. T. Boudreault and M. Leclerc, *Adv. Mater.*, 2010, **22**, E6–E27.
5 Q. Pei, G. Yu, C. Zhang, Y. Yang and A. J. Heeger, *Science*, 1995, **269**, 1086–1088.
6 T. Mitsumori, L. M. Campos, M. A. Garcia-Garibay, F. Wudl, H. Sato and Y. Sato, *J. Mater. Chem.*, 2009, **19**, 5826–5836.
7 M. Bendikov, F. Wudl and D. F. Perepichka, *Chem. Rev.*, 2004, **104**, 4891–4946.
8 C. N. Hoth, P. Schilinsky, S. A. Choulis and C. J. Brabec, *Nano Lett.*, 2008, **8**, 2806–2813.
9 J. E. Anthony, M. Heeney and B. S. Ong, *MRS Bull.*, 2008, **33**, 698–705.
10 H. E. Katz, Z. Bao and S. L. Gilat, *Acc. Chem. Res.*, 2001, **34**, 359–369.
11 A. Aviram and M. A. Ratner, *Chem. Phys. Lett.*, 1974, **29**, 277–283.
12 N. J. Tao, *Nat. Nanotechnol.*, 2006, **1**, 173–181.
13 S. V. Aradhya and L. Venkataraman, *Nat. Nanotechnol.*, 2013, **8**, 399–410.
14 Z. L. Cheng, R. Skouta, H. Vazquez, J. R. Widawsky, S. Schneebeli, W. Chen, M. S. Hybertsen, R. Breslow and L. Venkataraman, *Nat. Nanotechnol.*, 2011, **6**, 353–357.
15 W. Chen, J. R. Widawsky, H. Vázquez, S. T. Schneebeli, M. S. Hybertsen, R. Breslow and L. Venkataraman, *J. Am. Chem. Soc.*, 2011, **133**, 17160–17163.
16 J. Park, A. N. Pasupathy, J. I. Goldsmith, C. Chang, Y. Yaish, J. R. Petta, M. Rinkoski, J. P. Sethna, H. D. Abruna, P. L. McEuen and D. C. Ralph, *Nature*, 2002, **417**, 722–725.
17 W. J. Liang, M. P. Shores, M. Bockrath, J. R. Long and H. Park, *Nature*, 2002, **417**, 725–729.
18 H. Song, M. A. Reed and T. Lee, *Adv. Mater.*, 2011, **23**, 1583–1608.
19 J. S. Meisner, M. Kamenetska, M. Krikorian, M. L. Steigerwald, L. Venkataraman and C. Nuckolls, *Nano Lett.*, 2011, **11**, 1575–1579.
20 M. Elbing, R. Ochs, M. Koentopp, M. Fischer, C. von Hanisch, F. Weigend, F. Evers, H. B. Weber and M. Mayor, *Proc. Natl. Acad. Sci. U. S. A.*, 2005, **102**, 8815–8820.

21 I. Diez-Perez, J. Hihath, Y. Lee, L. Yu, L. Adamska, M. A. Kozhushner, I. I. Oleynik and N. Tao, *Nat. Chem.*, 2009, **1**, 635–641.
22 E. Lortscher, B. Gotsmann, Y. Lee, L. P. Yu, C. Rettner and H. Riel, *ACS Nano*, 2012, **6**, 4931–4939.
23 A. Batra, P. T. Darancet, Q. Chen, J. Meisner, J. R. Widawsky, J. B. Neaton, C. Nuckolls and L. Venkataraman, *Nano Lett.*, 2013, **13**, 6233–6237.
24 J. E. Green, J. Wook Choi, A. Boukai, Y. Bunimovich, E. Johnston-Halperin, E. DeIonno, Y. Luo, B. A. Sheriff, K. Xu, Y. Shik Shin, H.-R. Tseng, J. F. Stoddart and J. R. Heath, *Nature*, 2007, **445**, 414–417.
25 F. Moresco, G. Meyer, K. H. Rieder, H. Tang, A. Gourdon and C. Joachim, *Phys. Rev. Lett.*, 2001, **86**, 672–675.
26 E. Lortscher, J. W. Ciszek, J. Tour and H. Riel, *Small*, 2006, **2**, 973–977.
27 S. Y. Quek, M. Kamenetska, M. L. Steigerwald, H. J. Choi, S. G. Louie, M. S. Hybertsen, J. B. Neaton and L. Venkataraman, *Nat. Nanotechnol.*, 2009, **4**, 230–234.
28 M. Mayor, H. B. Weber, J. Reichert, M. Elbing, C. von Hanisch, D. Beckmann and M. Fischer, *Angew. Chem., Int. Ed.*, 2003, **42**, 5834–5838.
29 C. M. Guédon, H. Valkenier, T. Markussen, K. S. Thygesen, J. C. Hummelen and S. J. van der Molen, *Nat. Nanotechnol.*, 2012, **7**, 305–309.
30 H. Vazquez, R. Skouta, S. Schneebeli, M. Kamenetska, R. Breslow, L. Venkataraman and M. S. Hybertsen, *Nat. Nanotechnol.*, 2012, **7**, 663–667.
31 J. C. Ellenbogen and J. Christopher Love, *Proc. IEEE*, 2000, **88**, 386–426.
32 M. A. Reed and J. M. Tour, *Sci. Am.*, 2000, **282**, 86–93.
33 K. Stokbro, J. Taylor and M. Brandbyge, *J. Am. Chem. Soc.*, 2003, **125**, 3674–3675.
34 V. Mujica, M. A. Ratner and A. Nitzan, *Chem. Phys.*, 2002, **281**, 147–150.
35 S. K. Yee, J. B. Sun, P. Darancet, T. D. Tilley, A. Majumdar, J. B. Neaton and R. A. Segalman, *ACS Nano*, 2011, **5**, 9256–9263.
36 C. Joachim, J. K. Gimzewski and A. Aviram, *Nature*, 2000, **408**, 541–548.
37 R. M. Metzger, *Acc. Chem. Res.*, 1999, **32**, 950–957.
38 R. M. Metzger, *Chem. Rev.*, 2003, **103**, 3803–3834.
39 M. A. Reed, C. Zhou, C. J. Muller, T. P. Burgin and J. M. Tour, *Science*, 1997, **278**, 252–254.
40 M. L. Chabinyc, X. X. Chen, R. E. Holmlin, H. Jacobs, H. Skulason, C. D. Frisbie, V. Mujica, M. A. Ratner, M. A. Rampi and G. M. Whitesides, *J. Am. Chem. Soc.*, 2002, **124**, 11730–11736.
41 J. G. Kushmerick, C. M. Whitaker, S. K. Pollack, T. L. Schull and R. Shashidhar, *Nanotechnology*, 2004, **15**, S489–S493.
42 B. Larade and A. M. Bratkovsky, *Phys. Rev. B: Condens. Matter*, 2003, **68**, 235305.
43 J. Taylor, M. Brandbyge and K. Stokbro, *Phys. Rev. Lett.*, 2002, **89**, 1383011–1383014.
44 S. H. Ke, W. T. Yang and H. U. Baranger, *Nano Lett.*, 2008, **8**, 3257–3261.
45 S. V. Aradhya, J. S. Meisner, M. Krikorian, S. Ahn, R. Parameswaran, M. L. Steigerwald, C. Nuckolls and L. Venkataraman, *Nano Lett.*, 2012, **12**, 1643–1647.
46 J. S. Meisner, S. Ahn, S. V. Aradhya, M. Krikorian, R. Parameswaran, M. Steigerwald, L. Venkataraman and C. Nuckolls, *J. Am. Chem. Soc.*, 2012, **134**, 20440–20445.

47 Y. S. Park, A. C. Whalley, M. Kamenetska, M. L. Steigerwald, M. S. Hybertsen, C. Nuckolls and L. Venkataraman, *J. Am. Chem. Soc.*, 2007, **129**, 15768–15769.
48 Y. S. Park, J. R. Widawsky, M. Kamenetska, M. L. Steigerwald, M. S. Hybertsen, C. Nuckolls and L. Venkataraman, *J. Am. Chem. Soc.*, 2009, **131**, 10820–10821.
49 S. Datta, *Electronic Transport in Mesoscopic Systems*, Cambridge University Press, 1995.
50 T. Markussen, R. Stadler and K. S. Thygesen, *Phys. Chem. Chem. Phys.*, 2011, **13**, 14311–14317.
51 M. Büttiker, Y. Imry, R. Landauer and S. Pinhas, *Phys. Rev. B: Condens. Matter Mater. Phys.*, 1985, **31**, 6207–6215.
52 B. Q. Xu and N. J. Tao, *Science*, 2003, **301**, 1221–1223.
53 L. Venkataraman, J. E. Klare, I. W. Tam, C. Nuckolls, M. S. Hybertsen and M. L. Steigerwald, *Nano Lett.*, 2006, **6**, 458–462.
54 J. R. Widawsky, M. Kamenetska, J. Klare, C. Nuckolls, M. L. Steigerwald, M. S. Hybertsen and L. Venkataraman, *Nanotechnology*, 2009, **20**, 434009.
55 M. Kamenetska, M. Koentopp, A. Whalley, Y. S. Park, M. Steigerwald, C. Nuckolls, M. Hybertsen and L. Venkataraman, *Phys. Rev. Lett.*, 2009, **102**, 126803.
56 S. M. Wu, M. T. Gonzalez, R. Huber, S. Grunder, M. Mayor, C. Schonenberger and M. Calame, *Nat. Nanotechnol.*, 2008, **3**, 569–574.
57 J. R. Widawsky, M. Kamenetska, J. Klare, C. Nuckolls, M. L. Steigerwald, M. S. Hybertsen and L. Venkataraman, *Nanotechnology*, 2009, **20**, 434009.
58 Y. Shao, L. F. Molnar, Y. Jung, J. Kussmann, C. Ochsenfeld, S. T. Brown, A. T. B. Gilbert, L. V. Slipchenko, S. V. Levchenko, D. P. O'Neill, R. A. DiStasio Jr, R. C. Lochan, T. Wang, G. J. O. Beran, N. A. Besley, J. M. Herbert, C. Yeh Lin, T. Van Voorhis, S. Hung Chien, A. Sodt, R. P. Steele, V. A. Rassolov, P. E. Maslen, P. P. Korambath, R. D. Adamson, B. Austin, J. Baker, E. F. C. Byrd, H. Dachsel, R. J. Doerksen, A. Dreuw, B. D. Dunietz, A. D. Dutoi, T. R. Furlani, S. R. Gwaltney, A. Heyden, S. Hirata, C.-P. Hsu, G. Kedziora, R. Z. Khalliulin, P. Klunzinger, A. M. Lee, M. S. Lee, W. Liang, I. Lotan, N. Nair, B. Peters, E. I. Proynov, P. A. Pieniazek, Y. Min Rhee, J. Ritchie, E. Rosta, C. David Sherrill, A. C. Simmonett, J. E. Subotnik, H. Lee Woodcock Iii, W. Zhang, A. T. Bell, A. K. Chakraborty, D. M. Chipman, F. J. Keil, A. Warshel, W. J. Hehre, H. F. Schaefer Iii, J. Kong, A. I. Krylov, P. M. W. Gill and M. Head-Gordon, *Phys. Chem. Chem. Phys.*, 2006, **8**, 3172–3191.
59 P. Darancet, J. R. Widawsky, H. J. Choi, L. Venkataraman and J. B. Neaton, *Nano Lett.*, 2012, **12**, 6250–6254.
60 A. Batra, G. Kladnik, H. Vazquez, J. S. Meisner, L. Floreano, C. Nuckolls, D. Cvetko, A. Morgante and L. Venkataraman, *Nat. Commun.*, 2012, **3**, 1086.
61 L. Viani, Y. Olivier, S. Athanasopoulos, D. A. da Silva Filho, J. Hulliger, J.-L. Brédas, J. Gierschner and J. Cornil, *ChemPhysChem*, 2010, **11**, 1062–1068.
62 J. R. Widawsky, W. Chen, H. Vazquez, T. Kim, R. Breslow, M. S. Hybertsen and L. Venkataraman, *Nano Lett.*, 2013, **13**, 2889–2894.

Faraday Discussions

PAPER

Measurement and control of detailed electronic properties in a single molecule break junction

Kun Wang, Joseph Hamill, Jianfeng Zhou, Cunlan Guo and Bingqian Xu*

Received 24th April 2014, Accepted 16th May 2014
DOI: 10.1039/c4fd00080c

The lack of detailed experimental controls has been one of the major obstacles hindering progress in molecular electronics. While large fluctuations have been occurring in the experimental data, specific details, related mechanisms, and data analysis techniques are in high demand to promote our physical understanding at the single-molecule level. A series of modulations we recently developed, based on traditional scanning probe microscopy break junctions (SPMBJs), have helped to discover significant properties in detail which are hidden in the contact interfaces of a single-molecule break junction (SMBJ). For example, in the past we have shown that the correlated force and conductance changes under the saw tooth modulation and stretch–hold mode of PZT movement revealed inherent differences in the contact geometries of a molecular junction. In this paper, using a bias-modulated SPMBJ and utilizing emerging data analysis techniques, we report on the measurement of the altered alignment of the HOMO of benzene molecules with changing the anchoring group which coupled the molecule to metal electrodes. Further calculations based on Landauer fitting and transition voltage spectroscopy (TVS) demonstrated the effects of modulated bias on the location of the frontier molecular orbitals. Understanding the alignment of the molecular orbitals with the Fermi level of the electrodes is essential for understanding the behaviour of SMBJs and for the future design of more complex devices. With these modulations and analysis techniques, fruitful information has been found about the nature of the metal–molecule junction, providing us insightful clues towards the next step for in-depth study.

Introduction

The concept of molecular electronics started with the idea of wiring an individual molecule to two metal electrodes as an analogy of single electronic components in commercial microelectronic devices.[1,2] Stimulated by the high degree of diversity

Single Molecule Study Laboratory, College of Engineering and Nanoscale Science and Engineering Center, University of Georgia, Athens, GA 30602, USA. E-mail: bxu@engr.uga.edu

and functionality of molecules and the nanometre scale dimensions of such systems, both experimental and theoretical investigations in this direction have been in steady progress for more than two decades, although with many obstacles.[3–11] Among the techniques enabling the sandwiching of an individual molecule between two electrodes, single-molecule break junctions (SMBJs) provide a robust and tuneable platform to probe the inherent properties and gain a fundamental understanding of molecular-scale electronic elements.[10,12] The key feature lies with the electronic transport properties of a SMBJ.

A SMBJ is a simple yet very complex system. To depict the simplicity, a SMBJ only consists of two parts: a nanoscopic molecule and two macroscopic electrodes. Multiple metal materials have been studied as SMBJ electrodes, such as Au, Pt, Ag and Pd.[7,13–15] Similarly, molecular species spanning from simple saturated molecules[16–18] to intricate conjugated molecules[19–22] have been tested for various purposes. Within a SMBJ, the metal electrodes consist of a very condensed set of continuous energy states, the Fermi level of which varies from material to material, but the isolated molecule in the center of the junction contains a discrete set of energy levels, with a highest-occupied molecular orbital (HOMO) and a lowest unoccupied molecular orbital (LUMO). This last quality is what distinguishes SMBJs from classical p–n junctions which can be modelled with continuous bands throughout. Bridging an individual molecule between two metal electrodes leads to the broadening of discrete energy levels, causing a quasicontinuum density of states due to the hybridization with the metal wave functions. In other words, the charge transport properties across a SMBJ are essentially determined by the alignment of continuous electrode energy levels and discrete molecular energy levels. The interface connecting the molecule and electrodes is the source of the complexity of a SMBJ system. Unlike in a semiconductor system where the contact parts are ignored because of their Ohmic behaviour, the contact interfaces in a SMBJ have a strong influence, and sometimes even dominate the electron transport. Apart from the electrodes, chemical ligands, the other factor involved in the interfaces which are used to wire a molecule to the electrodes, have been widely studied. These include amines ($-NH_2$), thiols (–SH), carboxyls (–COOH), dimethyl phosphines ($-PMe_2$) and, recently, selenols (–SeH).[23–29] The environment where electrical measurements are conducted, often in a liquid buffer, also has a non-negligible impact on the properties of a SMBJ.[30,31] Especially for DNA molecules, this influence is critical.[32,33]

Challenges emerge in both experiments and simulations. Since experimental data always reflect a mixed result of various effects that are involved in a SMBJ system, it is necessary to elaborately define each factor that may influence the final result. Without enough understanding of them, no conclusion can be drawn convincingly. Experimental difficulties rest with how to control those factors separately and explore them one by one at a scale far beyond human eyesight. Similarly, development in simulations is limited by the size of the system and oversimplification.[34] Even so, to date great efforts have been made experimentally and theoretically.[10,35–41] Investigations into different combinations of molecules and electrodes, and multiple controls of the contact effect and environmental conditions, have been reported over the past decade.[42–44] To clear the uncertainty in this field, researchers are driven to keep innovating experimental designs and simulation methods. We have been focusing on experimental modulations and designs using the SPMBJ technique in order to gain deep insight towards a comprehensive understanding of the whole SMBJ system. In the following, we will briefly discuss our previous work in tuning the measurement process, and report on the controls of contact interfaces and relevant theoretical analysis.

Results and discussion

1. Experimental modulations

This section will briefly discuss multiple experimental modulations that we previously developed and the detailed electrical and mechanical properties discovered by using these modified techniques.

The power of the SPM (STM/CAFM) experimental setup comes from the flexibility it offers the experimentalist: (1) the piezoelectric transducer (PZT) provides precise displacement of the SPM tip, allowing for the gap between the tip and substrate to be adjusted continuously (A, B and C modulation modes in Fig. 1a); and (2) the SPM allows for a bias to be applied and modulated between the tip and substrate (D modulation mode in Fig. 1a). Both of these variables can be controlled by a computer program and a feedback mechanism.

1.1 Continuous-stretch mode. The first experiments using the SPM design created an ensemble of transient junctions by rapidly and repeatedly approaching the tip to the substrate, and retracting it in a continuous motion, controlled by the PZT.[10] The conductance of the junction during the retraction phase firstly revealed quantized Landauer steps[45] corresponding to the Au–Au junction, and then once the final Au–Au junction broke, quantized steps caused by individual molecules in the junction (Fig. 2a). This second set of quantized steps in the conductance was associated with one molecule, two molecules, *etc.* present in the junction. The introduction of this technique resolved most discrepancies caused by the defective formation of molecular junctions.

1.2 Dual-mode feedback. At the single-molecule level, one of the difficulties in experimental measurements is to maintain complete molecular junctions over thousands of repeated measurement processes. In order to achieve multiple electromechanical modulations but still keep the junction system intact enough, we developed a comprehensive controlling system.[46] With a dual-mode feedback

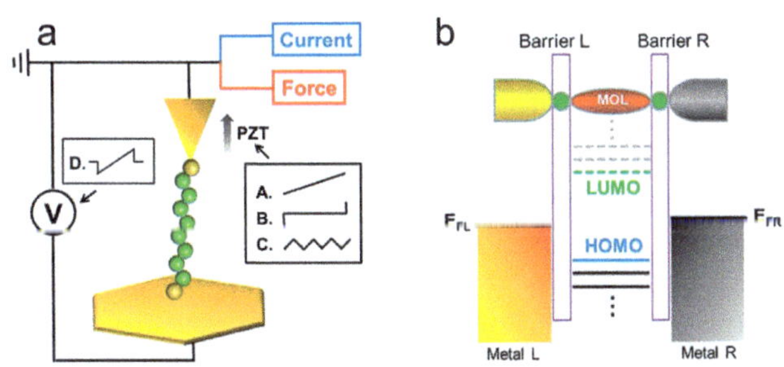

Fig. 1 (a) Schematic of the SPMBJ setup. Continuous-stretch A, stretch–hold B, and saw tooth C are possible PZT displacement modulation modes. D is a possible bias modulation mode. (b) Energy profiles of a SMBJ, including the continuous energy states of the metal electrodes and discrete energy levels of the molecule (HOMO in solid blue and LUMO in dashed green).

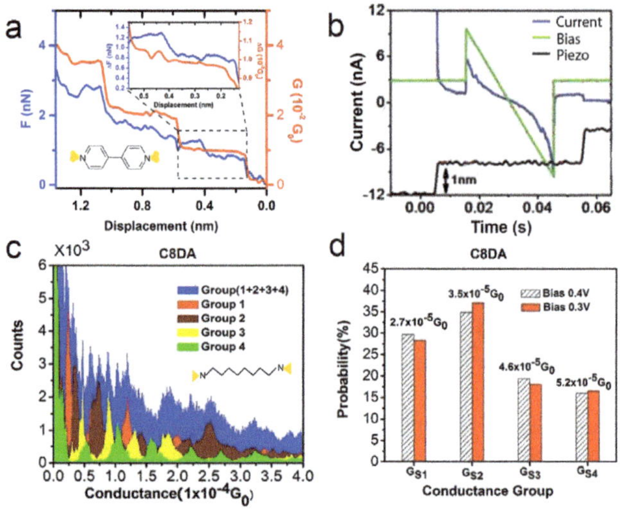

Fig. 2 (a) Continuous-stretch mode force (blue) and conductance (red) traces from the 4,4′-bipyridine SPM break junctions showing quantized steps for 1, 2 and 3 conducting molecules in the junction. (b) The black curve is the displacement of the PZT during stretch–hold mode; the green curve is the applied bias modulation during the step of the stretch–hold modulation; the blue curve is the induced current when the bias across the molecule is swept from 1 V to −1 V. (c) Conductance histograms from the C8DA SPM break junctions showing four different sets of conductance. (d) Conductance values measured under 0.3 V and 0.4 V for the four sets in (c). (c) and (d) are reprinted from ref. 47.

control on traditional conducting atomic force microscopy (CAFM) techniques, this highly integrated SPM system realized the tuneable contact strength and many other modulations. Based on this system, the tip was first controlled by the "current-feedback" mode due to its high sensitivity to tip–substrate separation. However, the possible peeling of the conductive coating on the tip, which results in decreased conductance, may blind the engaging process by pushing the tip until the limit of the PZT, which in most cases will crash the CAFM tip. The core of our design is that the system normally runs under the "current-feedback" mode, but if the current exceeds a higher current set point, the system will automatically switch to the "force-feedback" mode to avoid the crashing of the CAFM tip. If the current resumes, the system can automatically turn back to the "current-feedback" mode. The experimental tests on octanedithiol (C8DT) molecules proved the robustness of this system.

1.3 Stretch–hold mode. The steady linear retractions of continuous-stretch mode SPM measurements created single molecular junctions across which single molecular conductance was measured, but the junctions were short-lived and the histograms missed significant details which are important for understanding the SMBJ systems they were measuring. The solution was to stair-step the retraction process as the B modulation mode in Fig. 1a shows, so that the system paused momentarily and allowed the junction to settle into a quasi-relaxed state.[47] This allows one to eliminate, or at least minimize, variations in the experimental conditions. The retraction–pause–retraction was controlled by a computer program prescribing the displacement of the PZT controlling the SPM tip.

1.4 Bias modulation mode. When the stretch–hold mode SPM pauses during retraction, it provides a unique opportunity for the molecule to be perturbed while in a quasi-steady state junction. For example, the bias can be swept through a range of values to create an *I–V* curve similar to those used to study other electronic devices.[48] Fig. 2b shows a typical *I–V* curve responding to the bias sweep applied on a stretch–hold mode single-molecule conductance plateau. The characteristics of the *I–V* curve can reveal information about the symmetry of the junction and the molecular orbitals within the junction. Indeed, one of the first goals after creating SMBJs was to create rectifying asymmetric single molecular junctions.[1] Since then it has been shown that even geometrically symmetric junctions can show rectifying behaviour in SPM break junctions.

1.5 Saw tooth modulations. The displacement of the SPM tip as it is retracted from the surface can be controlled in numerous ways, limited only by the programmability of the PZT in control of the tip. Besides the stretch–hold modulation, the tip can also be induced to cycle through saw tooth displacements while it is being retracted. In one cycle, the tip performs a compression movement, followed by an elongation movement. Modulations in this manner influence the specific geometric orientation of the molecule with the electrodes, and therefore alter the conductance histograms. The SPM mechanism can be modified to measure force simultaneously with conductance using CAFM, and when this saw tooth modulation was applied to CAFM SPM measurements, the effects of the contact geometry and bond distance were isolated in the conductance signal.

1.6 Detailed properties. Using the modulations described above, we have discovered in detail the significant properties of a SMBJ. The stretch–hold mode conductance histograms yielded much finer peaks and also successfully identified the small variation due to the molecule–electrode contact configuration differences of the Au–C8DT/C8DA–Au molecular junctions (Fig. 2c and d). The saw tooth modulation applied on the continuous-stretch mode conductance measurements revealed regular fluctuations (Fig. 3a) on the conductance traces with a 180° phase shift (Fig. 3b) in relation to the PZT modulation. The saw tooth modulation applied on the free-holding process of the stretch–hold mode measurements of Au–C8DT/C8DA–Au illustrated more detailed phenomena: (1) conductance switching between different conductance sets were observed on some conductance traces (Fig. 3c), which is suggested to be caused by the switching among the different adsorption sites of Au atoms induced by the mechanical modulation; (2) for C8DA, the conductance switching accompanied with a force change close to the Au–NH_2 binding force strongly suggested that the responsive conductance changes are a combination of the switching between different adsorption sites and the dissociation of the molecule from the electrode with the bond broken under modulations (Fig. 3d). In addition, using the bias modulation, single molecule junctions composed of a thiol-terminated Ru(II) bis-terpyridine (Ru(tpy–SH)$_2$) molecule sandwiched between two gold electrodes was observed to have a contact-specific negative differential resistance (NDR) at a positive low-bias range.[48] The NDR was observed to occur only for the hollow-top Au–S binding configuration. The differential conductance profiles showed extra peaks, which we suggested to be caused by a small HOMO–LUMO gap and hybridization among the transition metal center, organic molecule backbone and

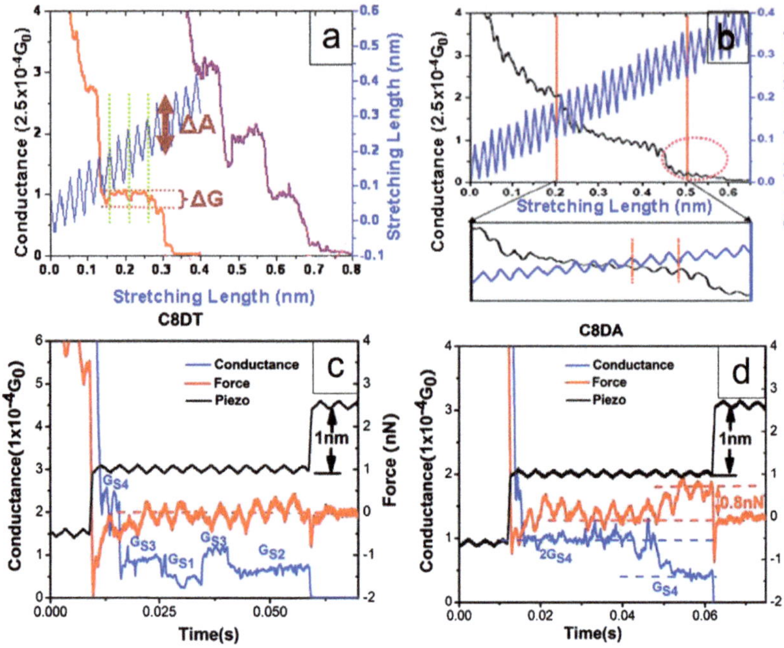

Fig. 3 Representative traces indicating various specific characteristic features. The conductance traces of C8DT under the saw-tooth modulations of the continuous retraction process show a correspondent conductance fluctuation of ΔG (a) and out-of-phase (180° phase shift) correspondence to the modulation (b). (c) and (d) show the conductance and force traces under the saw tooth modulation applied under the stretch–hold mode. (c) The trace with conductance switching between different single molecular conductance sets but without an obvious change in force. (d) The trace with conductance changes with the dissociation of molecules from the electrodes caused by the breakage of the contact bond. GS1–GS4 (C8DT, 4.8×10^{-5}, 7.0×10^{-5}, 9.0×10^{-5} and 24.9×10^{-5} G_0) in (c) and GS4 (C8DA, 5.2×10^{-5} G_0) in (d) are the single molecular conductance sets reported in ref. 47. (a) is reprinted from ref. 49. (c) and (d) are reprinted from ref. 50.

the Au electrodes. Simultaneous force measurements suggested the bias-induced molecule–electrode coupling change as the major cause. In such a strongly coupled molecular junction, we proposed that the specific contact configuration dominated the NDR effect.

1.7 **Barrier model.** To understand the experimental phenomena observed under applied mechanical modulations, we established a modified multiple tunneling barrier model incorporating a contact resistance term (Fig. 4).[49] In this model, the molecule core was defined as a rigid body which remained intact under junction extension Δd. The extension parameter Δd is to simulate the elongation movement of the tip in the saw tooth modulation. The molecule–electrode contacts were considered as a rectangular potential barrier where the contact decay constant β_C (the height of the potential barrier) and the width of the potential barrier d were used as two crucial indices to describe the effects of the contact conformation change. Thus, the final conductance $G = A\exp(-\beta_M) \times \exp(-\beta_C)$. Using this model, we identified the static contact resistances for stable C8DT and C8DA molecular junctions, which were divided into four conductance

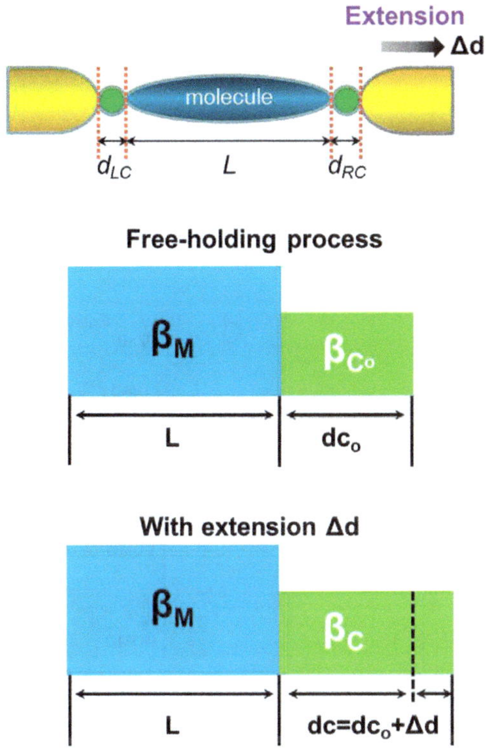

Fig. 4 Modified multiple tunneling barrier models.

sets. By monitoring the decay constant changes and force changes under mechanical modulation/extensions, we found that they changed differently not only with molecule–electrode contact bonds (Au–S and Au–NH_2) but also with their conductance sets.

These experimental modulations and theoretical models have either yielded new detailed phenomena or further improved existed theories, providing improved physical understanding about the missing essence of a molecular junction.

2. Control of contact variables

In this section we will present the new experimental measurements in controlling the contact effect and the relevant data analysis.

As the bias-sweep modulation on the stretch–hold mode SPMBJ enables us to study the *I*–*V* characteristics of a single-molecule junction, the *I*–*V* characteristics of potential functional molecular devices, such as molecular rectifiers, are our main focus. One of the less-understood issues in the rectification behaviour of a SMBJ is the role of the contact interfaces. To gain insights into the rectification behaviour induced by asymmetric contacts in a SMBJ, we performed a comprehensive investigation of the charge transport properties of molecular junctions with systematic controls on both anchoring groups and electrodes (Fig. 5a). Combining the stretch–hold and bias sweep modification of the SPMBJ, *I*–*V*

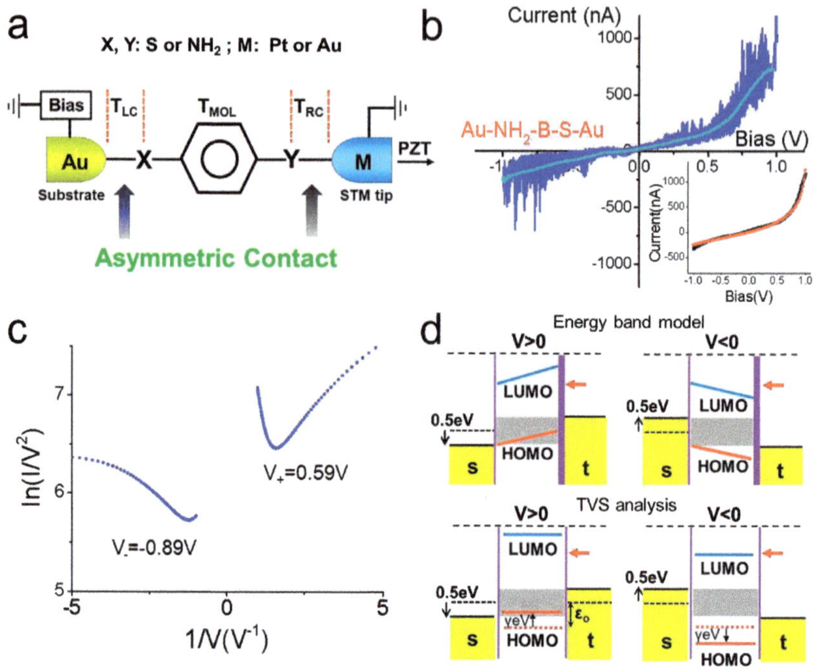

Fig. 5 (a) Schematic of benzene-based molecular junctions with changeable electrodes and anchoring groups; (b) I–V characteristics and Landauer fitting (inset red curve) of the Au–NH_2–B–S–Au junction; (c) $\ln(I/V^2)$ *vs.* $1/V$ plot for the I–V curves in (b), transition voltages (V_+ and V_-) are labeled; (d) energy band model and TVS analysis schematics.

characteristics were measured for molecules consisting of a central benzene ring (B) and alternating thiol (–SH) and amine (–NH_2) anchoring groups. The electrodes consist of an Au substrate and a STM tip which alternates between Au and Pt. In total, five different molecular junctions were measured. Due to the symmetric and rigid inherent structure of the benzene ring, any asymmetric electronic behaviour should come from the two contacts. As the experimental result revealed, pronounced rectification behaviour occurred when an asymmetric component (an asymmetric anchoring group, an asymmetric electrode, or both in combination) was introduced into the molecular junction. Example I–V curves for the Au–NH_2–B–SH–Au junction are shown in Fig. 5b. Using an energy band model,[50] where molecular orbitals at the strong coupled contact tend to shift together with the chemical potential of the corresponding electrode, rectification was accessed due to the asymmetric shift of the molecular frontier orbitals with respect to the Fermi levels of the electrodes, which resulted from the coupling strength asymmetry at two contact interfaces (Fig. 5d). Comparing the binding energies of different anchoring group–electrode combinations,[14,51] the differences in coupling strength were determined. It turned out that our experimental results agreed well with the prediction of the energy band model. The degree of asymmetry of the coupling strength at the two contacts determined the rectification ratio of the molecular junction. For example, the Au–NH_2–B–SH–Pt junction, which possesses the most asymmetric contact coupling strength, should have the

Table 1 Landauer fitting and TVS analysis results for the different molecular junctions

	Landauer fitting			TVS	
Molecular junction[a]	Γ_L (eV)	Γ_R (eV)	ε_{fit} (eV)	ε_{TVS} (eV)	γ
Au–S–B–S–Au	7.63×10^{-4}	7.69×10^{-4}	0.71	0.61	−0.01
Au–NH_2–B–NH_2–Au	3.11×10^{-4}	3.18×10^{-4}	0.71	0.69	−0.01
Au–NH_2–B–S–Au	3.19×10^{-4}	6.60×10^{-4}	0.87	0.62	−0.06
Au–S–B–S–Pt	5.89×10^{-4}	7.10×10^{-4}	0.75	0.49	−0.03
Au–S–B–NH_2–Pt	1.15×10^{-3}	5.09×10^{-4}	0.92	0.57	0.08

[a] Molecular junctions in the first column are expressed in the format of substrate–molecule–tip.

largest rectification ratio. This was in accordance with our experimental results. Otherwise, we believe that the external bias played a significant role in triggering the rectifying behaviour.

2.1 *I–V* curve calculations. Possibly the most valuable information that experimentalists hope to discover using SMBJs is the energy gap between the Fermi level of the electrodes and the nearest conducting orbital of the molecule in the break junction. This parameter determines the turn-on voltage necessary to yield a high current in a molecular electronics device. This orbital is either the HOMO or LUMO, and usually the Fermi level of the electrodes is closer to one or the other, which means when a bias is applied, the nearest orbital is reached first, resulting in a drastic increase in conductance. However, experimental junctions often break down before this voltage can be reached, so the energy gap must be derived from more lengthy methods. Two such methods are easily calculated from the *I–V* curves measured from bias-modulated mode SMBJs. Each method has strengths and weaknesses.

2.1.1 I–V curve fitting. An *I–V* curve can be fitted to a simplified version of the Landauer formula, involving only three fitting parameters, using a Levenberg–Marquardt least squares fitting algorithm.[52] The three parameters correspond to the gap ε_{fit} described above, and the degree of coupling, Γ_L and Γ_R, to each separate electrode. When this fitting method was applied to the five different junctions with variable anchoring groups and electrodes, the results agreed with observations made from the qualitative analysis of the *I–V* curves. The symmetric junctions yielded symmetric values for Γ_L and Γ_R, and yielded unequal values when the junctions were asymmetric (see Table 1). This provides quantitative confirmation of the observation that the asymmetric junctions yielded asymmetric *I–V* curves due to unequal coupling to the molecule. Furthermore, the energy levels when Pt was used as one of the electrodes were considerably altered compared to those of Au–molecule–Au junctions.

2.1.2 Transition voltage spectroscopy (TVS). Another interpretation of the Landauer formula yields information about the energy gap ε_{TVS} of the SMBJ from a Fowler–Nordheim (FN) plot of the *I–V* curve. Specifically, a FN plot of an *I–V* curve measured from a SMBJ can yield minima at specific transition voltages (Fig. 5c). The corresponding bias voltage at a specific minimum indicates the transition voltage where a trapezoid-shaped tunneling barrier turns to a triangle-shaped barrier. It is necessary to notice that this transition voltage is different from the bias voltage, enabling the alignment of the frontier molecular orbital of the molecule with the chemical potential of the metal electrodes. Namely, the transition voltage is smaller than the voltage potential necessary for resonant tunneling. Using these minima, it is possible to calculate the energy gap between the Fermi level of the electrodes and the nearest conducting molecular orbital, and a coefficient γ which determines the degree of relative shift of the nearest molecular orbital under a certain bias.[6,53] By comparing these two core parameters, ε_{TVS} and γ, we found: (1) comparing the Au–SH–B–SH–Au and Au–NH_2–B–NH_2–Au junctions, the latter has a bigger ε_{TVS} but smaller γ, which resulted in a lower current within the applied bias range. As the binding energy calculation suggested, this difference should be a result of the difference between the Au–SH and Au–NH_2 contact bonds. (2) Comparing Au–SH–B–SH–Au and Au–NH_2–B–SH–Au, when an asymmetric anchoring group (weaker coupling Au–NH_2 bond) was introduced, ε_{TVS} became larger, but a greater γ effect was also introduced. Notably, ε_{TVS} for the Au–NH_2–B–SH–Au junction is between those of the Au–SH–B–SH–Au and Au–NH_2–B–NH_2–Au junctions, suggesting it is a hybridization of those two junctions. (3) Comparing Au–SH–B–SH–Pt with Au–SH–B–SH–Au, when an asymmetric electrode component was introduced, the whole system was altered since the Fermi level of one electrode is offset from that of the other. ε_{TVS} becomes much smaller and γ almost doubles. The smaller ε_{TVS} of Au–SH–B–SH–Pt could be caused by the asymmetric Fermi level at the two electrodes. (4) For junction Au–SH–B–NH_2–Pt, there is a similar ε_{TVS} compared to the Au–SH–B–SH–Au junction, although these two junctions are based on two different molecules, which suggests a similar bonding strength between Au–SH and Pt–NH_2, as confirmed by previous binding energy calculations. This junction has the largest γ, which is believed to be the main reason for the rectification behavior observed at a negative bias (see Table 1).

Overall, based on TVS, whenever an asymmetric component was introduced it increased the value of γ, intensifying the degree of the relative HOMO shift. Thus, it is possible that the frontier molecular orbital shifted into the transmission window, or the field emission transport was induced under high bias. Besides, when the Pt electrode was involved, the ε_{TVS} energy offset becomes smaller (the results from the Landauer fit, namely ε_{fit}, do not display this behavior—this will be discussed in the following paragraph), which could be induced by the asymmetric Fermi level positions of Au and Pt. Compared with the concept of strong affinity between the frontier molecular orbitals and the electrode Fermi level described in the energy band model, the γ value introduced in TVS analysis could be understood in a similar way. Take junction Au–NH_2–B–SH–Au as an example (Fig. 4d), based on TVS, the relative shift of the frontier molecular orbitals always tries to catch up with the movement of the chemical potential of the electrode at the strong coupling interface, which results in the sharp current increase under a high positive bias.

To yield precise calculation results from the experimental data, theoretical calculations such as Landauer fitting and TVS analysis often require smooth and rather pronounced rectifying *I–V* curves, which rarely occur in the bias range where reasonable electrical signals are obtained in real experiments. This could be the major reason why different calculation algorithms gave different results. Another reason lies with the difficulty in obtaining well-defined minima in the

TVS plot. For instance, the ill-defined minimum at the negative bias regime in Fig. 5c may bring a greater calculation error than the well-defined minima under the positive bias. The common feature the two calculation methods suggested is that resonant tunneling was not reached within our bias sweep range (−1 to 1 V), and that the asymmetric coupling, asymmetric Γs in the Landauer fitting and an asymmetric large γ in TVS are responsible for the resulting rectification *I–V* characteristics. Both methods rely on simplifications to theoretical models for SMBJs, and therefore do not provide highly accurate results. Their value rests in the relative ease at which important parameters can be calculated from the experimental data. Furthermore, the length dependence and effects of screening have been documented for TVS, and represent some of the uncertainty in the results of the TVS analysis.[54] However, it is possible that this length dependence provides more, not less, information about the details of the conducting molecular orbitals.[55] On the other hand, the results from a Landauer fit are uncertain because any theoretically derived Landauer formula relies on many more variables than the coupling strength γ and energy gap ε_{fit}. The simplifications needed to provide a functional form with a limited number of parameters necessitates that the results be interpreted with this in mind: the value of the Landauer fit is the relative ease at which it can be applied, and the importance of the parameters it yields, not the precision of the results. Precision is still best provided by more complicated simulation methods.

Further experimental and theoretical efforts are needed to test and complete our proposed mechanisms concluded from the responsive experimental signals towards detailed controls, as reported above.

Conclusions

Expected to be the revolutionary technology for next generation electronic devices, molecular electronics are currently at a fledgling stage. To date, only a small section of this field has been touched using the existing techniques. We have explored the responses of the electronic properties of a molecular junction to multiple mechanical and electrical modulations and used recently developed data analysis techniques to calculate important physical parameters from the data. The experimental results reflect the inherent impact of the contact interface of the electrode and molecule, which is probably the most critical factor in a SMBJ. We have shown that by tuning the contact distance for alkane molecules, responsive changes in force and conductance indicated the essential difference in contact configurations, which was also explained using a modified multiple tunneling barrier model. By controlling the anchoring groups coupling the molecule to the electrode and the electrodes themselves, we significantly altered how the HOMO of benzene related to the conductance band of the Au electrodes. Specifically, we have shown that the most dramatic effect to the *I–V* characteristics of a SMBJ occurs with a combination of asymmetric electrodes and asymmetric end groups. In this case, the change in the HOMO and electrode Fermi level gap is significant, but the most dramatic effect is in the relative shift of the HOMO due to the modulation of the bias.

However, a thorough physical and chemical understanding of the electronic properties of a SMBJ requires finer controls which could minimize the influences from most of the irrelevant factors, and more advanced data analysis techniques.

For example, careful control of the orientation of the molecule sandwiched between the electrodes could provide essential information.[56] In this direction, experimental attempts have been made using molecules consisting of benzene rings,[57,58] but obviously more are needed. Similarly, the fabrication and characterization of molecular junctions with the third gate electrode are of critical significance. From the simulation point of view, methods enabling the simulation of long and complex molecules and calculations requiring less simplification and assumptions will provide more precise prediction and constructive guidance for experimentalists.

Although an individual molecule could perform as a conductor, the produced current is far less than what is required for commercial use. Challenges exist in improving and amplifying the current. Doping special chemicals, metal ions and nanoparticles into individual molecules have been suggested to effectively increase the conductivity of a SMBJ. Attempts towards this direction have just started.

Other physical properties involved in a molecular junction, such as optoelectronics,[59,60] thermoelectrics,[61,62] spintronics[63–65] and quantum interference,[40] deserve more attention. The responsive signal from them may inspire novel experimental designs and activate new research branches.

Acknowledgements

The authors thank the U. S. National Science Foundation for funding this work (ECCS 0823849, ECCS 1231967).

Notes and references

1 A. Aviram and M. A. Ratner, *Chem. Phys. Lett.*, 1974, **29**, 277–283.
2 R. L. Carroll and C. B. Gorman, *Angew. Chem., Int. Ed.*, 2002, **41**, 4378–4400.
3 A. Nitzan and M. A. Ratner, *Science*, 2003, **300**, 1384–1389.
4 J. C. Love, L. A. Estroff, J. K. Kriebel, R. G. Nuzzo and G. M. Whitesides, *Chem. Rev.*, 2005, **105**, 1103–1169.
5 N. J. Tao, *Nat. Nanotechnol.*, 2006, **1**, 173–181.
6 J. M. Beebe, B. Kim, J. W. Gadzuk, C. D. Frisbie and J. G. Kushmerick, *Phys. Rev. Lett.*, 2006, **97**, 026801.
7 C. Li, I. Pobelov, T. Wandlowski, A. Bagrets, A. Arnold and F. Evers, *J. Am. Chem. Soc.*, 2008, **130**, 318–326.
8 C. Toher, I. Rungger and S. Sanvito, *Phys. Rev. B: Condens. Matter Mater. Phys.*, 2009, **79**, 205427.
9 H. Dalgleish and G. Kirczenow, *Nano Lett.*, 2006, **6**, 1274–1278.
10 B. Q. Xu and N. J. Tao, *Science*, 2003, **301**, 1221–1223.
11 C.-C. Kaun and T. Seideman, *Phys. Rev. B: Condens. Matter Mater. Phys.*, 2008, **77**, 033414.
12 R. H. M. Smit, Y. Noat, C. Untiedt, N. D. Lang, M. C. van Hemert and J. M. van Ruitenbeek, *Nature*, 2002, **419**, 906–909.
13 C. M. Kim and J. Bechhoefer, *J. Chem. Phys.*, 2013, **138**, 014707.
14 M. Kiguchi, S. Miura, T. Takahashi, K. Hara, M. Sawamura and K. Murakoshi, *J. Phys. Chem. C*, 2008, **112**, 13349–13352.

15 J. M. Beebe, B. Kim, C. D. Frisbie and J. G. Kushmerick, *ACS Nano*, 2008, **2**, 827–832.
16 X. D. Cui, A. Primak, X. Zarate, J. Tomfohr, O. F. Sankey, A. L. Moore, T. A. Moore, D. Gust, G. Harris and S. M. Lindsay, *Science*, 2001, **294**, 571–574.
17 V. B. Engelkes, J. M. Beebe and C. D. Frisbie, *J. Am. Chem. Soc.*, 2004, **126**, 14287–14296.
18 M. A. Reed, C. Zhou, C. J. Muller, T. P. Burgin and J. M. Tour, *Science*, 1997, **278**, 252–254.
19 R. E. Holmlin, R. Haag, M. L. Chabinyc, R. F. Ismagilov, A. E. Cohen, A. Terfort, M. A. Rampi and G. M. Whitesides, *J. Am. Chem. Soc.*, 2001, **123**, 5075–5085.
20 E. G. Emberly and G. Kirczenow, *Phys. Rev. B: Condens. Matter Mater. Phys.*, 2001, **64**, 235412.
21 S. H. Choi, B. Kim and C. D. Frisbie, *Science*, 2008, **320**, 1482–1486.
22 S. O. Kelley, N. M. Jackson, M. G. Hill and J. K. Barton, *Angew. Chem., Int. Ed.*, 1999, **38**, 941–945.
23 L. Venkataraman, J. E. Klare, I. W. Tam, C. Nuckolls, M. S. Hybertsen and M. L. Steigerwald, *Nano Lett.*, 2006, **6**, 458–462.
24 W. Haiss, S. Martin, E. Leary, H. van Zalinge, S. J. Higgins, L. Bouffier and R. J. Nichols, *J. Phys. Chem. C*, 2009, **113**, 5823–5833.
25 R. J. Nichols, W. Haiss, S. J. Higgins, E. Leary, S. Martin and D. Bethell, *Phys. Chem. Chem. Phys.*, 2010, **12**, 2801–2815.
26 M. Kamenetska, M. koentopp, A. C. Whalley, Y. S. Park, M. L. Steigerwald, C. Nuckolls, M. S. Hybertsen and L. Venkataraman, *Phys. Rev. Lett.*, 2009, **102**, 126803.
27 E. Adaligil, Y.-S. Shon and K. Slowinski, *Langmuir*, 2010, **26**, 1570–1573.
28 Y. S. Park, A. C. Whalley, M. Kamenetska, M. L. Steigerwald, M. S. Hybertsen, C. Nuckolls and L. Venkataraman, *J. Am. Chem. Soc.*, 2007, **129**, 15768–15769.
29 F. Chen, X. Li, J. Hihath, Z. Huang and N. Tao, *J. Am. Chem. Soc.*, 2006, **128**, 15874–15881.
30 D. P. Long, J. L. Lazorcik, B. A. Mantooth, M. H. Moore, M. A. Ratner, A. Troisi, Y. Yao, J. W. Ciszek, J. M. Tour and R. Shashidhar, *Nat. Mater.*, 2006, **5**, 901–908.
31 V. Fatemi, M. Kamenetska, J. B. Neaton and L. Venkataraman, *Nano Lett.*, 2011, **11**, 1988–1992.
32 R. N. Barnett, C. L. Cleveland, A. Joy, U. Landman and G. B. Schuster, *Science*, 2001, **294**, 567–571.
33 Y. A. Mantz, F. L. Gervasio, T. Laino and M. Parrinello, *Phys. Rev. Lett.*, 2007, **99**, 058104.
34 S. M. Lindsay and M. A. Ratner, *Adv. Mater.*, 2007, **19**, 23–31.
35 M. Brandbyge, J. L. Mozos, P. Ordejon, J. Taylor and K. Stokbro, *Phys. Rev. B: Condens. Matter Mater. Phys.*, 2002, **65**, 165401.
36 Y. Selzer, L. T. Cai, M. A. Cabassi, Y. X. Yao, J. M. Tour, T. S. Mayer and D. L. Allara, *Nano Lett.*, 2005, **5**, 61–65.
37 F. Chen, J. He, C. Nuckolls, T. Roberts, J. E. Klare and S. Lindsay, *Nano Lett.*, 2005, **5**, 503–506.
38 I. Diez-Perez, J. Hihath, Y. Lee, L. Yu, L. Adamska, M. A. Kozhushner, I. I. Oleynik and N. Tao, *Nat. Chem.*, 2009, **1**, 635–641.
39 H. Vazquez, R. Skouta, S. Schneebeli, M. Kamenetska, R. Breslow, L. Venkataraman and M. S. Hybertsen, *Nat. Nanotechnol.*, 2012, **7**, 663–667.

40 C. M. Guedon, H. Valkenier, T. Markussen, K. S. Thygesen, J. C. Hummelen and S. J. van der Molen, *Nat. Nanotechnol.*, 2012, **7**, 305–308.

41 M. Taniguchi, M. Tsutsui, R. Mogi, T. Sugawara, Y. Tsuji, K. Yoshizawa and T. Kawai, *J. Am. Chem. Soc.*, 2011, **133**, 11426–11429.

42 M. Kiguchi and S. Kaneko, *Phys. Chem. Chem. Phys.*, 2013, **15**, 2253–2267.

43 C. Jia and X. Guo, *Chem. Soc. Rev.*, 2013, **42**, 5642–5660.

44 R. L. McCreery, H. Yan and A. J. Bergren, *Phys. Chem. Chem. Phys.*, 2013, **15**, 1065–1081.

45 V. Rodrigues, T. Fuhrer and D. Ugarte, *Phys. Rev. Lett.*, 2000, **85**, 4124–4127.

46 F. Chen, J. Zhou, G. Chen and B. Xu, *IEEE Sens. J.*, 2010, **10**, 485–491.

47 J. Zhou, F. Chen and B. Xu, *J. Am. Chem. Soc.*, 2009, **131**, 10439–10446.

48 J. Zhou, S. Samanta, C. Guo, J. Locklin and B. Xu, *Nanoscale*, 2013, **5**, 5715–5719.

49 J. Zhou, C. Guo and B. Xu, *J. Phys.: Condens. Matter*, 2012, **24**, 164209.

50 J. Zhao, C. Yu, N. Wang and H. Liu, *J. Phys. Chem. C*, 2010, **114**, 4135–4141.

51 M. Tsutsui, M. Taniguchi and T. Kawai, *J. Am. Chem. Soc.*, 2009, **131**, 10552–10556.

52 B. M. Briechle, Y. Kim, P. Ehrenreich, A. Erbe, D. Sysoiev, T. Huhn, U. Groth and E. Scheer, *Beilstein J. Nanotechnol.*, 2012, **3**, 798–808.

53 I. Baldea, *Phys. Rev. B: Condens. Matter Mater. Phys.*, 2012, **85**, 035442.

54 F. Mirjani, J. M. Thijssen and S. J. van der Molen, *Phys. Rev. B: Condens. Matter Mater. Phys.*, 2011, **84**, 115402.

55 M. C. Lennartz, N. Atodiresei, V. Caciuc and S. Karthaeuser, *J. Phys. Chem. C*, 2011, **115**, 15025–15030.

56 A. V. Malyshev, *Phys. Rev. Lett.*, 2007, **98**, 096801.

57 L. Venkataraman, J. E. Klare, C. Nuckolls, M. S. Hybertsen and M. L. Steigerwald, *Nature*, 2006, **442**, 904–907.

58 T. Kim, P. Darancet, J. R. Widawsky, M. Kotiuga, S. Y. Quek, J. B. Neaton and L. Venkataraman, *Nano Lett.*, 2014, **14**, 794–798.

59 S. Lara-Avila, A. V. Danilov, S. E. Kubatkin, S. L. Broman, C. R. Parker and M. B. Nielsen, *J. Phys. Chem. C*, 2011, **115**, 18372–18377.

60 S. Battacharyya, A. Kibel, G. Kodis, P. A. Liddell, M. Gervaldo, D. Gust and S. Lindsay, *Nano Lett.*, 2011, **11**, 2709–2714.

61 J. R. Widawsky, P. Darancet, J. B. Neaton and L. Venkataraman, *Nano Lett.*, 2012, **12**, 354–358.

62 K. Baheti, J. A. Malen, P. Doak, P. Reddy, S.-Y. Jang, T. D. Tilley, A. Majumdar and R. A. Segalman, *Nano Lett.*, 2008, **8**, 715–719.

63 L. Bogani and W. Wernsdorfer, *Nat. Mater.*, 2008, **7**, 179–186.

64 C. Iacovita, M. V. Rastei, B. W. Heinrich, T. Brumme, J. Kortus, L. Limot and J. P. Bucher, *Phys. Rev. Lett.*, 2008, **101**, 116602.

65 R. Vincent, S. Klyatskaya, M. Ruben, W. Wernsdorfer and F. Balestro, *Nature*, 2012, **488**, 357–360.

Faraday Discussions

PAPER

A computational study of potential molecular switches that exploit Baird's rule on excited-state aromaticity and antiaromaticity†

H. Löfås,[a] B. O. Jahn,[b] J. Wärnå,[a] R. Emanuelsson,[b] R. Ahuja,[ac] A. Grigoriev[a] and H. Ottosson*[b]

Received 28th April 2014, Accepted 23rd June 2014
DOI: 10.1039/c4fd00084f

A series of tentative single-molecule conductance switches which could be triggered by light were examined by computational means using density functional theory (DFT) with non-equilibrium Green's functions (NEGF). The switches exploit the reversal in electron counting rules for aromaticity and antiaromaticity upon excitation from the electronic ground state (S_0) to the lowest $\pi\pi^*$ excited singlet and triplet states (S_1 or T_1), as described by Hückel's and Baird's rules, respectively. Four different switches and one antifuse were designed which rely on various photoreactions that either lead from the OFF to the ON states (switches **1**, **2** and **4**, and antifuse **5**) or from the ON to the OFF state (switch **3**). The highest and lowest ideal calculated switching ratios are 1175 and 5, respectively, observed for switches **1** and **4**. Increased thermal stability of the **1-ON** isomer is achieved by benzannulation (switch **1B-OFF/ON**). The effects of constrained electrode–electrode distances on activation energies for thermal hydrogen back-transfer from **1-ON** to **1-OFF** and the relative energies of **1-ON** and **1-OFF** at constrained geometries were also studied. The switching ratio is strongly distance-dependent as revealed for **1B-ON/OFF** where it equals 711 and 148 when the ON and OFF isomers are calculated in electrode gaps with distances confined to either that of the OFF isomer or to that of the ON isomer, respectively.

1 Introduction

Molecules composed of a few dozens of atoms constitute the smallest objects that can be designed in an essentially unlimited number of versions and which also

[a]*Department of Physics and Astronomy, Uppsala University, Box 516, SE-75120, Uppsala, Sweden*
[b]*Department of Chemistry – BMC, Uppsala University, Box 576, SE-75123, Uppsala, Sweden. E-mail: henrik.ottosson@kemi.uu.se*
[c]*Applied Materials Physics, Department of Materials and Engineering, Royal Institute of Technology (KTH), SE-10044, Stockholm, Sweden*

† Electronic supplementary information (ESI) available: Interference of **1-OFF**, full ref. 41, energy tables and Cartesian coordinates. See DOI: 10.1039/c4fd00084f

are stable at ambient conditions. For example, it has been estimated that 10^{40} different molecules with sizes of a regular drug molecule can be designed based on the most common elements.[1] Thus, the combination of molecular chemistry and electronics provides for an exceptional wealth of design possibilities for the smallest possible memory and logic devices that are of such high stabilities that they potentially can be used in everyday applications.[2,3] The molecular switch is therefore a component that has received significant attention within the past decades, and a variety of different molecular switches have been reported,[3–5] activated by different kinds of external stimuli, such as light, force, heat or, as recently shown, a current running through a junction.[6] Light is a particularly attractive stimulus due to its fast response time and compatibility with already existing experimental setups.[4] The existing molecular switches can roughly be divided into two categories: isomerization switches,[7–19] where the three-dimensional structure (connectivity, configuration, or conformation) of the molecular system is changed; or redox/polarization-switches,[20–22] where the molecule (or part of it) takes up or loses an electron. Our tentative molecular switches, studied through quantum chemical calculations reported herein, fall into the first category and they could presumably be stimulated by light in one direction and by heat in the other.

A major drawback for some of the previously studied molecular switches, *e.g.*, azobenzenes, is the fact that they change length during switching, making them unsuitable in solid-state devices.[7,8,11] Furthermore, isomerization switches should ideally have the ON state represented by a planar compound with a linearly conjugated path with maximal p_{π}-orbital overlap. The compound representing the OFF state, on the other hand, should have a disruption in its π-conjugation, and such a disruption could be accomplished by a saturated molecular segment (*structural isomerization*) or by a large twist in the conjugated path (*stereo-isomerization*), in the optimal case not changing the length. Furthermore, it should be possible to switch the molecule between the two states with external stimuli.

We argue that quantum chemical calculations should be efficient tools in the development of completely new classes of molecular-switch compounds. In the switch design we exploit Baird's rule as a qualitative guiding tool.[23–25] This rule states that the numbers of π-electrons needed for aromaticity and antiaromaticity, respectively, in the lowest ππ* excited singlet and triplet states (S_1 and T_1) of cyclic fully conjugated hydrocarbons (annulenes) are opposite to that of the electronic ground state (S_0), as described by Hückel's rule,[26–29] *i.e.*, annulenes with $4n$ π-electrons are aromatic and those with $4n + 2$ π-electrons are antiaromatic in the T_1 and S_1 states. This rule was derived by Colin Baird in 1972 through use of perturbation molecular orbital (PMO) theory.[23] Since then it has been verified through a series of quantum chemical calculations using different aromaticity indices, and it has been shown to be valid also in the S_1 state.[30,31] Yet, it is very seldom applied in the design of optically- and photochemically-active compounds.[24,25]

Following Baird's rule, compounds which are aromatic in S_0 (*e.g.*, benzene) become antiaromatic upon excitation. Thus, they change character from being exceptionally stable and become more reactive, and for this reason we recently described benzene as a molecular "Dr Jekyll and Mr Hyde".[25] Indeed, the alleviation of the antiaromatic character of a benzene ring in its S_1 and T_1 states should be a driver for various photorearrangements in benzene derivatives, and this could be exploited for the identification of new compounds that may be applicable in single-molecule photoswitches. At this point it should be noted that our switch candidates either have essentially the same lengths in the ON and the OFF isomers, or they are conformationally highly flexible whereby only small energy penalties are needed for adjustment to electrode gaps that are not ideal for a specific isomer.

Two situations which are useful for molecular switches can be identified and imply either a decreased degree of antiaromaticity or an increased degree of aromaticity along the reaction coordinates in both S_1 and S_0 (Fig. 1). First are compounds which are S_0 aromatic and S_1 antiaromatic (Baird-antiaromatic), and that as a result of the latter are prone to photorearrange to a non-aromatic isomer. The second class of compounds are nonaromatic in both their S_0 and S_1 states but can photorearrange so as to attain aromaticity in S_1 (Baird-aromaticity). After decay to S_0 these species may thermally rearrange back to the start upon regain in S_0 aromaticity and/or reduction in S_0 antiaromaticity.

Photorearrangements of benzene derivatives could be useful in the design of new optical conductance switches provided that the switches can be tailored so that rearrangements lead from low-conducting to more highly conducting isomers. Photoinduced hydrogen shifts could be attractive in this context because with properly designed switching compounds the reaction may lead to only small length differences between the ON- and the OFF-isomers (states), as earlier concluded by Benesch *et al.*[9] Additionally, if the hydrogen back-transfer is hampered by the immobilization between the electrodes to the extent that back-transfer to the starting isomer (state) does not occur spontaneously at ambient temperatures then the isomer pair could represent a new molecular switch type.

We have now investigated the transport properties of new types of tentative light-triggered molecular switches, where systems **1** and **2** are based on photochemical hydrogen-shifts (Scheme 1)[32–35] and systems **3** and **4** exploit two different pericyclic photoreactions (Scheme 2).[36–39] We also propose an "antifuse", **5**, based on a double Norrish Type I reaction leading to a photodecarbonylation (Scheme 3).[40] This optically activated "antifuse" is the opposite of a fuse as its electrically conductive

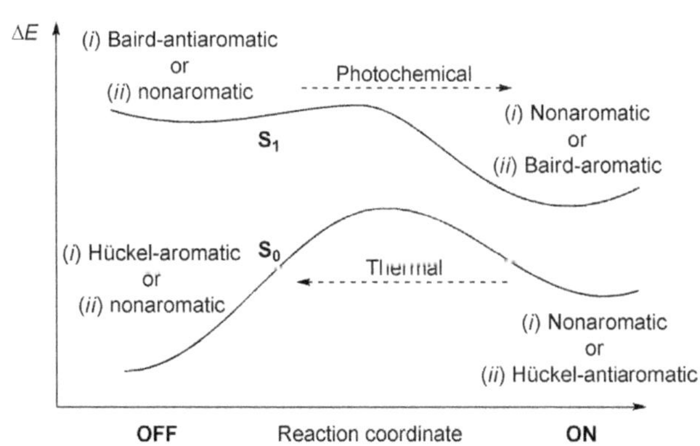

Fig. 1 The general shapes of the S_0 and S_1 potential energy surfaces and the expected influence of aromaticity, antiaromaticity and non-aromaticity of the central unit at the various minima in the S_0 and S_1 states.

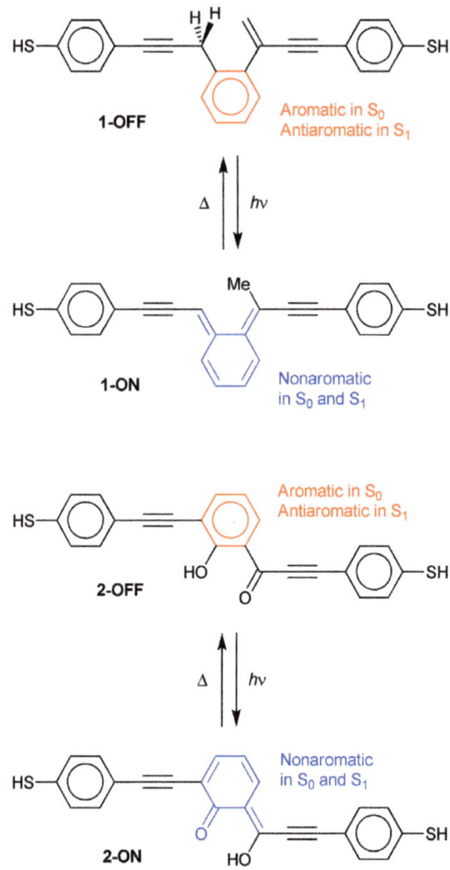

Scheme 1 Potential molecular conductance switches based on photochemical hydrogen shifts in which the S_1 state antiaromaticity is diminished through the photoreaction. The benzene ring which is the core of the photorearrangement of the OFF-state is shown in red, and the *ortho*-xylylene fragments found in the ON-states are shown in blue.

path is opened when triggered by light of the appropriate wavelength. The general feature of switches **1**, **2** and the antifuse **5** is that the photoreactions induced by an S_1 antiaromatic benzene ring lead to molecular fragments of *ortho*-xylylene type, providing for linearly π-conjugated paths through the molecules, and eliminated aromaticity in the S_0 state. This should lead to a particularly high conductivity, in line with recent findings on the conductivities of compounds in which the degree of aromaticity of the central unit could be varied.[41] In the tentative switches **3** and **4** it is the gain in excited-state $4n$ π-electron aromaticity that drives the reactions forward.

At this point it should be noted that the presently reported switch candidates primarily represent templates for further design. Still, it should be remarked that all of these tentative switches are based on molecules and reactions that have been studied experimentally in other contexts.[32–40] The photoreactions upon which the switch function in the specific candidates rely represent only a few of the reactions that likely can be used. Thus, the present investigation represents a

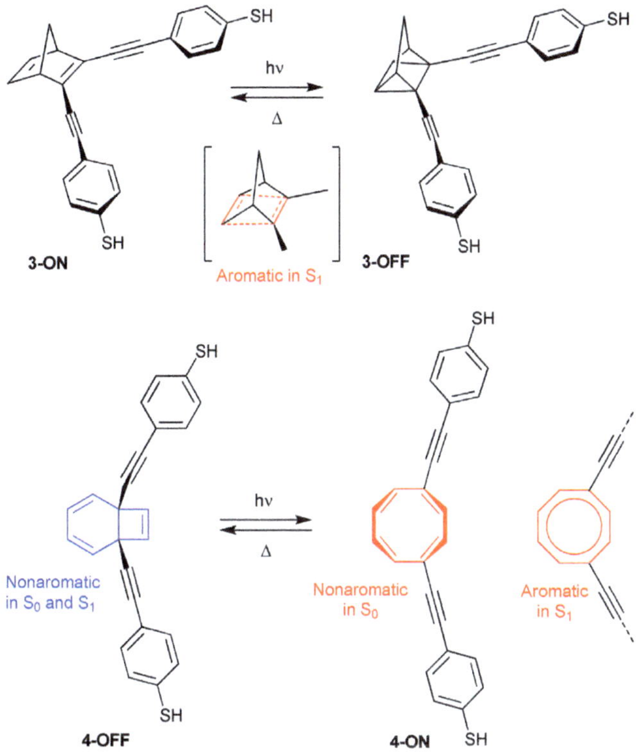

Scheme 2 Potential molecular conductance switches based on photochemical cycloadditon and electrocyclic ring-opening reactions in which the S_1 state aromaticity is gained, in the case of **3** leading to a photocycloaddition.

start on the design of optical switches which exploit the opposite rules for aromaticity and antiaromaticity in the ground state *vs.* the lowest $\pi\pi^*$ excited states.

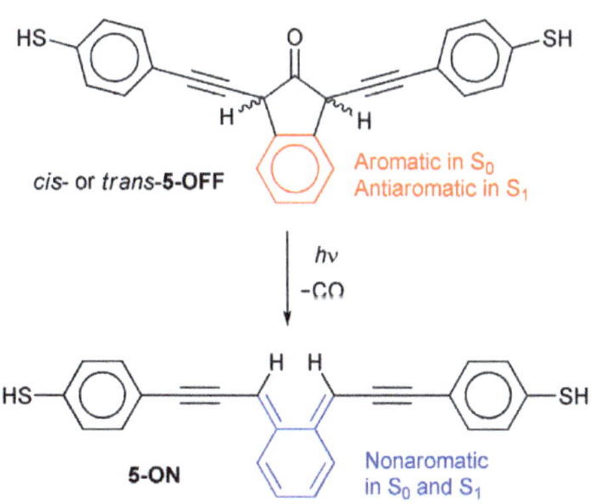

Scheme 3 A potential molecular antifuse based on a Norrish Type I reaction in which the S_1 state antiaromaticity of the benzene ring marked in red is diminished through photodecarbonylation leading to the formation of an *ortho*-xylylene fragment marked in blue.

2 Computational methods

The isolated molecules were first optimized by use of the B3LYP hybrid density functional theory with the 6-31G(d) valence double-zeta basis set.[42–44] These calculations were performed using the Gaussian09 program package.[45] The activation barriers for the thermal back-reactions were optimized without geometry constraints as well as at constrained S···S distances corresponding to those of either the ON or the OFF isomer.

The modelled device consists of three parts: left electrode, molecule, and right electrode. First, the relaxed molecular structures shown in Schemes 1–3 were inserted between two Au(111)-surfaces and relaxed with three gold layers on each side. To remove interactions between periodic images in the lateral dimensions, a 6×6 supercell (17.5×17.5 Å) was used. For transport calculations three more layers of gold are added on each side at experimental bulk positions. All relaxations are performed at the DFT-level with the SIESTA package[46,47] and core electrons are modeled using Troullier-Martins[48] soft norm-conserving pseudopotentials. The valence electrons are expanded in a basis set of local orbitals using a double-zeta plus polarization orbital (DZP) basis set for electrons in the molecule and a single-zeta plus polarization orbital (SZP) basis set for electrons in the gold electrode. The generalized gradient approximation (GGA) of Perdew–Burke–Ernzerhof was used for the exchange-correlation functional.[49] Transport calculations were carried out from first principles with a method based on non-equilibrium Green's functions (NEGF) combined with DFT as implemented in the TranSIESTA package.[50] The zero-bias transmission, $T = T(E)$, was calculated at the Fermi energy E_F and sampled over a 12×12 $k_{||}$-mesh.

3 Results and discussion

To analyze the switching behaviour in the reactions outlined in Schemes 1–3, we first calculated the electronic transport of the ON and OFF isomer pairs of the various switch candidates, attached to gold electrodes, in the framework of DFT–NEGF. In these first calculations the most extended conformations of the molecules were chosen. The zero-bias transmissions for the different compounds are

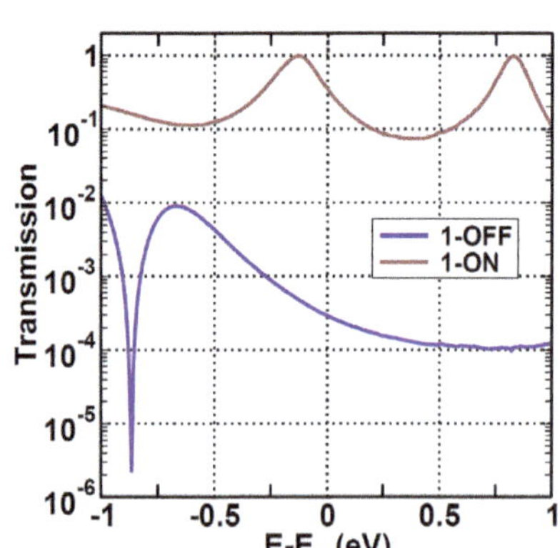

Fig. 2 Transmission spectra ($V = 0$) for switch **1** shown in Scheme 1.

shown in Fig. 2, 4, 6, 8 and 9 where each panel shows the transmission for the ON and the OFF isomers (states) for the different switch candidates. The results for the zero-bias conductance together with switching ratios (SRs) at $V = 0$ and the length in the junction of the most extended conformation of the compounds (S···S distances) are collected in Table 1.

Subsequently, we analyze how realistic the switches are. In particular we examined (i) if the activation energies for thermal back-reactions are sufficiently high to allow the ON (or OFF) isomer to be persistent at ambient temperatures, and (ii) what effect the constrained electrode–electrode distances have on the transport properties and SRs of the systems found to have the highest idealized SRs. We also explored further derivatizations of the switches which could lead to higher thermal stability of the ON as well as the OFF isomers.

3.1 Switches based on photoinduced hydrogen-transfer

Switch **1** is based on a photochemically-induced antarafacial [1,5]-sigmatropic H-shift, and this reaction converts **1-OFF**, having one saturated sp^3 C atom and a cross-conjugated segment, into **1-ON** having a linearly conjugated path spanning the complete molecule. For this switch there are large differences between the two isomers. For **1-OFF** the HOMO-level is positioned far from the E_F and there is a large antiresonance in the transmission function about 0.9 eV below E_F. On the other hand, with the extended conjugation in **1-ON** the HOMO shifts closer to E_F and, at the same time, the HOMO–LUMO gap is decreased to give high transmission over the whole energy range considered in Fig. 2.

The differences for the HOMO and LUMO levels in the ON and OFF states are clearly seen in the plot of the spatial distribution of the density of states in the junction (see Fig. 3). For **1-OFF** the HOMO-level is about 0.5 eV below E_F and mostly localized to the two conjugated phenylethynyl segments with a very weak connection through the central part of the molecule. In contrast, the HOMO

Table 1 Conductance characteristics of switches

Compound	Conductance[a] (G_0)	SR[b]	S···S distance[c]
1-ON	3.4×10^{-1}	1175	20.6
1-OFF	2.9×10^{-4}		20.7
2-ON	6.6×10^{-2}	548	20.9
2-OFF	1.2×10^{-4}		20.8
3-ON	2.8×10^{-1}	47	12.0
3-OFF	6.1×10^{-3}		11.7
4-ON	5.9×10^{-3}	5	11.4
4-OFF	1.1×10^{-3}		12.2
5-ON	2.8×10^{-1}	623	20.4
cis-**5-OFF**	4.5×10^{-4}		19.4
trans-**5-OFF**	6.3×10^{-4}	450	19.4

[a] Conductance at zero bias (0 V). [b] Conductance ON/OFF switching ratio at zero bias (0 V). [c] Distance between thiol end-groups [Å].

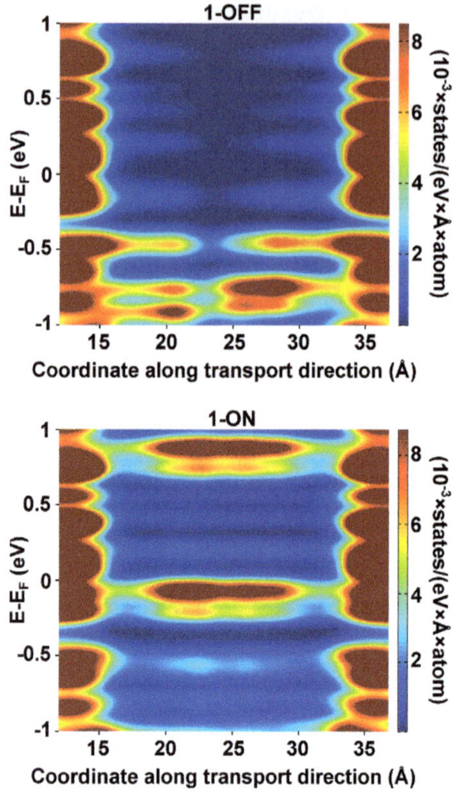

Fig. 3 PDOS for **1-OFF** and **1-ON**.

of **1-ON** is shifted close to E_F and it is delocalized over the complete molecule. Also, the LUMO level is now visible ~1 eV above E_F. Thus, for **1** there is both a shift of the most conducting level closer to E_F as well as a delocalization of the level over the complete molecule when going from the OFF- to the ON-state. This has been shown earlier to give a high switching ratio,[15] and this is also the case here as

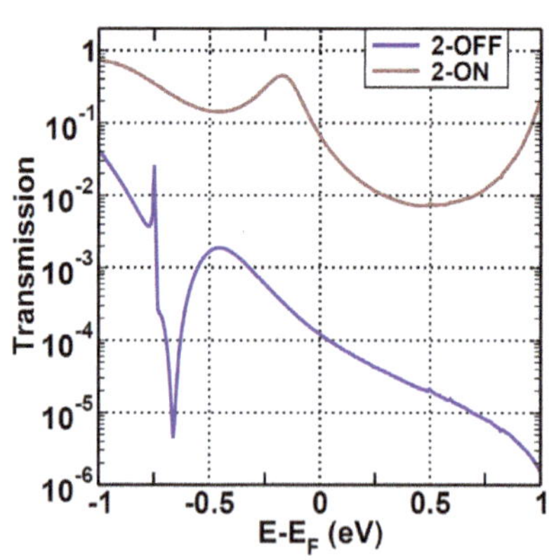

Fig. 4 Transmission spectra ($V = 0$) for switch **2** shown in Scheme 1.

an idealized switching ratio of nearly 1200 was calculated (Table 1). It is noteworthy that the large anti-resonance in **1-OFF** can be traced back to destructive interference introduced by cooperative effects of the cross-conjugated unit and the central S_0-state aromatic benzene ring, seen as a reversal of the local currents in this segment at the energy of the antiresonance (see the ESI†).

The second switch based on an intramolecular hydrogen-shift is **2**. The **2-OFF** isomer has a low conductance (Fig. 4) which should be due to (i) the cross-conjugated connectivity around the carbonyl group, and (ii) the *meta*-substitution pattern around the S_0 aromatic benzene ring. Similarly as with switch **1**, the benzene ring promotes a H-shift lowering the S_1 antiaromaticity. This leads to a central (non-aromatic) *ortho*-xylylene derivative **2-ON** which has a linearly conjugated path between the two electrodes. The compounds have previously been studied experimentally; however, the corresponding ON-state is a short-lived transient species.[35] Thus, usage of these or similar compounds in molecular-switch applications need further tailoring.

The charge transport characteristic of **2** is very similar to that of **1**. The overall conductances of both **2-OFF** and **2-ON** are lower compared to the corresponding isomers of **1**, yet, it is known that *meta*-substituted benzene rings give lower transmissions.[51,52] Similarly as with **1**, the aromatic ring in **2** which is *meta*-substituted together with the cross-conjugated segment around the carbonyl

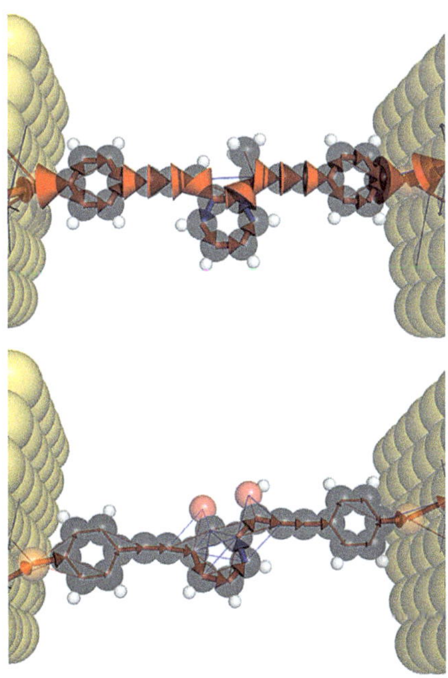

Fig. 5 Local currents in **1-ON** (top) and **2-ON** (bottom), where the cross-sectional area of the cylinder is proportional to the current density. Red currents represent a positive transport direction, and blue currents represent a negative direction. The currents are calculated at the Fermi energy (($\mu_L + \mu_R$)/2), but a small positive bias voltage is assumed for computational purposes. For a full derivation see Okabayashi *et al.*[55] and Paulsson *et al.*[56] In order to become visible, the local currents in **2** are magnified by a factor of two compared with **1**.

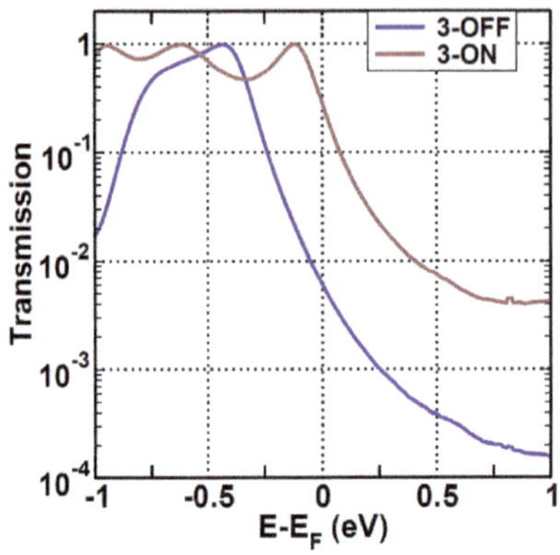

Fig. 6 Transmission spectra ($V = 0$) for the switch **3** shown in Scheme 2.

group lowers the transmission in **2-OFF** significantly compared to the fully conjugated **2-ON**.[53,54] The **2-OFF** isomer also has a deep antiresonance about 1 eV above E_F introduced by interference in the *meta*-substituted benzene which leads to an even further decrease in the transmission at E_F. Moreover, the lower conductance of **2-ON** when compared to **1-ON** can be explained by the cross-conjugated carbonyl moiety which reduces the number of linearly conjugated paths through the *ortho*-xylylene unit. This can be seen in Fig. 5 where the local currents in **1-ON** and **2-ON** are shown. For **1-ON** there are two well-conducting linearly conjugated paths through the central *ortho*-xylylene unit, one short butadiene path and one long octatetraene path. In contrast, the shorter path in **2-ON** is disrupted by the cross-conjugated carbonyl segment and the current follows the longer conjugated path. Despite this, **2** has a reasonably high switching ratio of approximately 500 (Table 1).

3.2 Switches based on ring-closure/opening reactions

Two different photochemical reactions that lead to/from cyclic (partially) saturated molecular segments were also examined as bases for tentative switches; (i) the [2 + 2] photocycloaddition leading from a norbornadiene to a quadricyclane, and (ii) a photochemical electrocyclic ring-opening leading from a bicyclo[4.2.0]octa-2,4,7-triene to a cyclooctatetraene. It is of note that, while the photoreactions of **1**, **2**, **4** and **5** lead from an OFF state to an ON state, the photoreaction of **3** leads from the ON to the OFF state.

We first consider **3** which is based on the norbornadiene to quadricyclane photoisomerization. In **3-ON**, one of the C=C double bonds of the norbornadiene unit provides a conjugated path while the quadricyclane unit of the **3-OFF** is a fully saturated hydrocarbon which ideally should disrupt the π-conjugation. It is noteworthy that the electron transfers through norbornadiene and quadricyclane moieties have been examined experimentally earlier through intervalence electron transfer studies using a bis(pentaamine–ruthenium(II)) complex of dicyanonorbornadiene.[36,37] The effective coupling through the dicyanonorbornadiene ligand was determined to 0.023 eV; however, no intervalence transition could be found for the corresponding quadricyclane complex.

Here, the transmissions of **3-ON** and **3-OFF** show similar behaviours with broad peaks in the DOS from HOMOs close to E_F, which give high transmissions for both isomers (Fig. 6). The thermal cycloreversion which leads from **3-OFF** to the **3-ON** shifts the HOMO level even closer to E_F, making it almost resonant. Nonetheless, due to the high transmission of the OFF-isomer this results in a moderate switching ratio of about 50 for the **3-ON/OFF** isomer pair. The high transmission of **3-OFF** is possibly due to the two strained

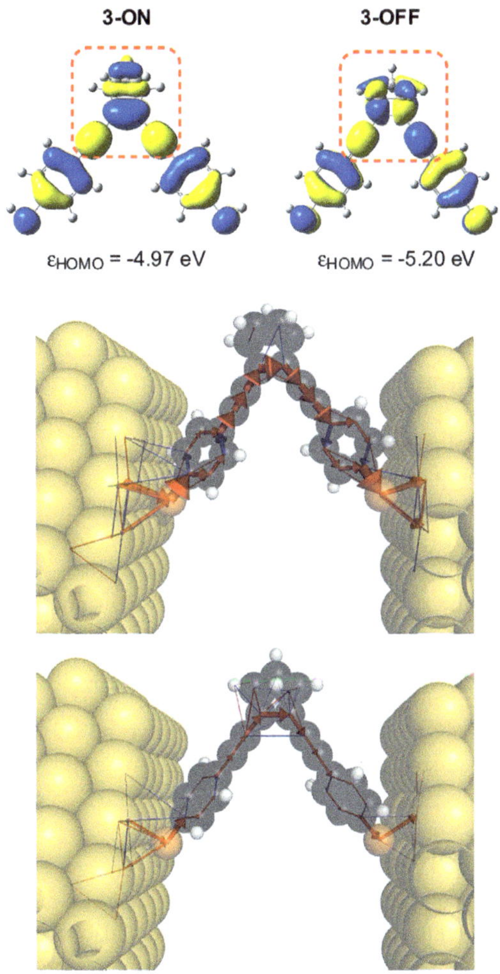

Fig. 7 (top) Orbital plots of the HOMOs of **3-OFF** and **3-ON** at the B3LYP/6-31G(d) level with the central norbornadiene (**3-ON**) or quadricyclane (**3-OFF**), and the two adjacent ethynyl segments, encircled in red. The symmetry notations of the HOMOs are *b* (**3-ON**) and *a* (**3-OFF**), respectively, in the C_2 point group (top frame). Local currents in **3-ON** (middle) and **3-OFF** (bottom), where the cross-sectional area of the cylinder is proportional to the current density. Red currents represent a positive transport direction, and blue currents represent a negative direction. The currents are calculated at the Fermi energy $((\mu_L + \mu_R)/2)$, but a small positive bias voltage is assumed for computational purposes. For a full derivation see Okabayashi *et al.*[55] and Paulsson *et al.*[56] In order to become visible, the local currents in **3-OFF** are magnified by a factor of 20 compared to **3-ON**.

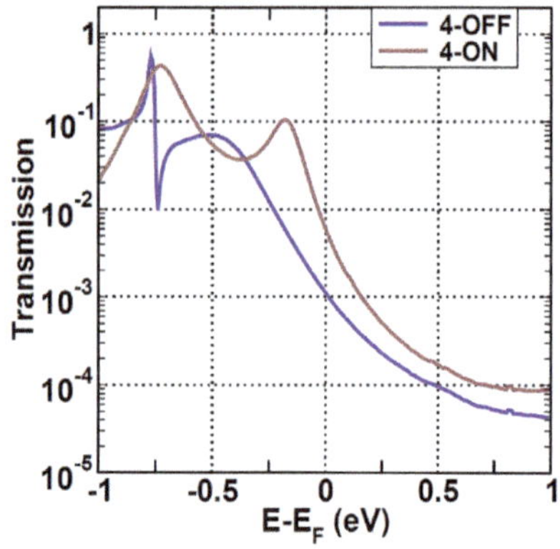

Fig. 8 Transmission spectra ($V = 0$) for the switch **4** shown in Scheme 2.

cyclopropane rings in the quadricyclane fragment providing bent bond orbitals which can overlap with the π-orbitals of the phenylethynyl arms. Indeed, when viewing the HOMOs of the isolated **3-ON** and **3-OFF** (Fig. 7, top), respectively, one can note that the central norbornadiene/quadricyclane core contributes significantly in both cases and that the orbital energies differ by only 0.23 eV, in approximate agreement with the difference in the positions of the resonance transmission in Fig. 6. The local currents at the Fermi level are also significant through the central quadricyclane moiety of **3-OFF** (Fig. 7, bottom) and increases as the energy is off-set.

The next switch, **4**, has a similar behavior to **3** with a broad DOS close to E_F from the HOMO level (Fig. 8), with a slight shift between **4-ON** and **4-OFF**. The introduction of the two cyclic units in the bicyclo[4.2.0]octatri-2,4,7-ene moiety creates destructive interference in **4-OFF** showing up as an antiresonance about 0.7 eV below E_F. Here the interference can be traced back to a reversal of the currents in the four-membered ring-segment including the central C–C single bond. The moderate conductance of **4-ON** gives a very low ON/OFF ratio of merely five, and this is likely due to the poor conjugative p_π-overlap because of the puckered, tub-shaped structure of the cyclooctatetraene ring.

3.3 An antifuse based on photodecarbonylation

The last compound pair considered, **5** (Scheme 3), is not a molecular switch but an antifuse. The **5-OFF** and **5-ON** compounds are also not isomers, and the photoreaction is irreversible as carbon monoxide leaves. Here, the OFF-state is a molecule with the current path disrupted by two sp^3-hybridized C atoms. The photoinitiated decarbonylation leading to the conducting **5-ON** state can be viewed as a result of (i) the double α-cleavage (Norrish Type I) of the C–C bonds adjacent to the carbonyl group, as well as (ii) a way for the S_1 state benzene ring to alleviate its destabilizing antiaromaticity. The S_1-antiaromatic character of the benzene ring presumably facilitates the decarbonylation. The linear conjugation introduced in **5-ON** increases the transmission (Fig. 9), and the result is very close to **1-ON**, except that the LUMO is shifted minutely closer to E_F as a result of the cross-conjugated $C{=}CH_2$ instead of the C=CHMe moiety.

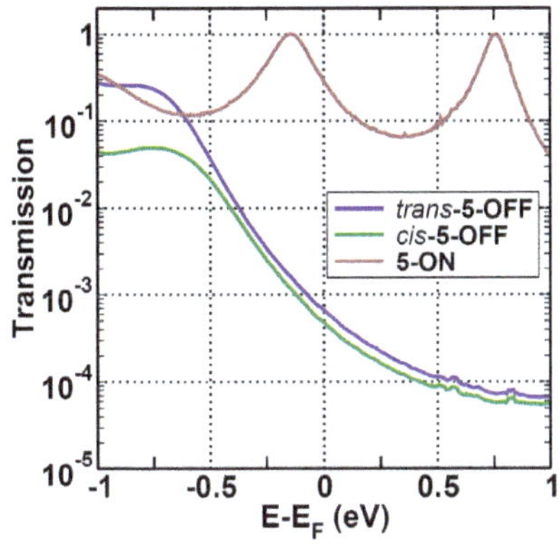

Fig. 9 Transmission spectra ($V = 0$) for the antifuse **5** shown in Scheme 3.

Note that, depending on whether one considers the *cis*- or the *trans*-isomer of **5-OFF**, the ON/OFF ratio is either about 600 or 450. The *cis*-**5-OFF** isomer is the least conducting according to our calculations. However, the *cis*-**5-OFF** isomer is not suitable because when the formation of **5-ON** in a junction with the electrode–electrode distance constrained to the *cis*-**5-OFF** is studied computationally it becomes clear that a **5-ON** isomer formed from *cis*-**5-OFF** would close instantaneously to a low-conducting benzocyclobutene moiety with two saturated (insulating) C atoms.

3.4 Conclusions from transmissions at ideal situations

Among the five switch/antifuse candidates investigated only **1**, **2** and **5** have a sufficiently high SR, and these compounds were analyzed further. None of these are, however, perfect as a molecular switch because the *ortho*-xylylene isomers (**1-ON**, **2-ON**, and **5-ON**) will most likely only be transient species. Immobilization between electrodes may decrease their potential to undergo dimerization reactions; however, this may also facilitate the back-reactions to the corresponding OFF isomers. In addition, the experimental situation as such is complicated by a series of factors. The electrode–electrode distance is unclear, and the effects of stabilizing features in **1**, **2** and **5** on the conductance are also rather unclear.

3.5 Thermal interconversion processes

The realization of the isomer pairs to operate as molecular switches relies on the existence of two stable isomers. As we want the switching process to progress by light activation in one direction and heat activation in the other there should also be an energetic drive towards the starting isomer in the S_0 state (OFF for **1**, **2** and **4**, and ON for **3**). Thus, we first examined the activation barriers for the thermal back-rearrangements from **1-ON** and **2-ON** to **1-OFF** and **2-OFF**, respectively. We also probed how to raise the activation barriers for these rearrangements above 120 kJ mol^{-1}, which should lead to a stable isomer at ambient temperatures. In addition to the parent switch-pairs **1-ON/OFF** and **2-ON/OFF** we therefore examined modified switch-pairs with either benzannulation to the central rearranging moiety (**1-ON/OFF**, Scheme 4) or with oxygen atoms exchanged with sulfur (**2-ON/**

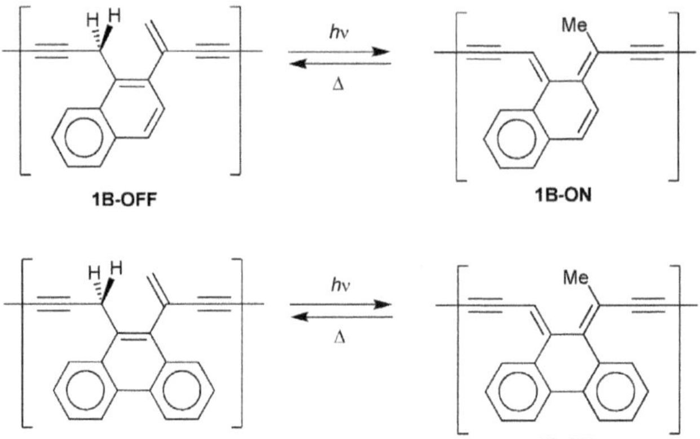

Scheme 4 The modified switches of type **1-ON/OFF** with benzannulation to increase the thermal stabilities of the OFF-isomers (states).

OFF). For a full list of the investigated switches see the ESI.† One conclusion from these calculations is that the free energies of activation for the thermal hydrogen shift from **1-ON** to **1-OFF** can be modified extensively: from 88.4 kJ mol^{-1} in **1-ON/OFF** to 104.5 and 126.4 kJ mol^{-1} in the switch pairs **1B-ON/OFF** and **1C-ON/OFF**, respectively, shown in Scheme 4. The corresponding transition-state structure for the hydrogen shift from **1-OFF** to **1-ON** is displayed in Fig. 10, top panel.

In contrast, the barriers for the hydrogen back-transfers in all derivatives of **2-ON/OFF** switches are too low (within the range 17.8–57.5 kJ mol^{-1}). Also, the transition-state structures correspond to rotation about the C–O bond to the hydroxyl group rather than a shift of the H atom (Fig. 10, bottom). The only substitution which has a substantial effect is the exchange of the hydroxyl group of **2-ON/OFF** with a methyl group and exchange of the carbonyl group with a vinyl group because this leads to an ON state with a significantly higher activation barrier for hydrogen back-transfer (~80 kJ mol^{-1}). However, this is still lower than our desired barrier height. Interestingly, and in contrast to the observation for **1B-ON/OFF** and **1C-ON/OFF**, further benzannulation lowered the barrier. For that reason, focus in the following text is placed exclusively on derivatives of **1-ON/OFF**.

Furthermore, the restriction with fixed electrode–electrode distances will also influence the activation barriers for the hydrogen back-transfer reactions. Yet, rather than regarding the complete set of the various derivatives of switches we considered the change in the activation energy of switch **1-ON/OFF** at electrode–electrode distances which are optimal for either the ON isomer or the OFF-isomer (Fig. 10, middle structures). When the S···S distance is kept at the optimal OFF distance (17.191 Å) then $\Delta G^{\ddagger}(298\ \text{K}) = 116.2$ kJ mol^{-1}, whereas at the optimal ON distance (20.303 Å) $\Delta G^{\ddagger}(298\ \text{K}) = 145.1$ kJ mol^{-1}. The lower activation energy for the shorter constrained S···S distance is in line with Hammond's postulate as this transition-state structure is more OFF-like, and thus, lower in relative energy than the more ON-like transition-state structure at more extended S···S distances. The energy needed to adjust **1-OFF** to an S···S distance corresponding to that of the ON-isomer is merely 3.5 kJ mol^{-1}. Hence, the procedure in a tentative experiment should be to insert **1-OFF** into an electrode gap which is larger by ~3 Å than the length of **1-OFF** at its most stable conformer.

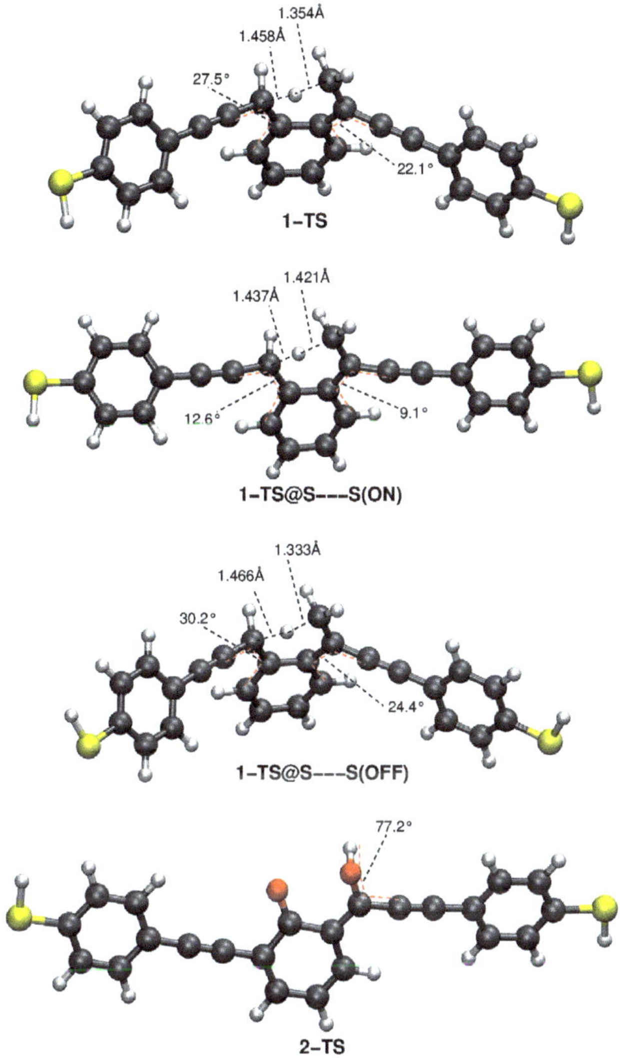

Fig. 10 Transition-state structures for thermal hydrogen back-transfer from **1-ON** to **1-OFF**, and from **2-ON** to **2-OFF**. Calculations at the B3LYP/6-31G(d) level.

The energy of **1-ON** relative to that of **1-OFF** is 42.3 kJ mol^{-1} at the optimal S···S ON distance but just 5.5 kJ mol^{-1} at the optimal OFF distance. Again, it is apparent that it is ideal to keep the **1-ON/OFF** at the S···S distance of the ON isomer as this creates the largest energetic drive for return to the OFF isomer in the S_0 state.

3.6 Effects on transport upon derivatization and electrode–electrode distance variation

Since both the **1-ON** and **1-OFF** isomers, as well as the corresponding isomers of **1B** and **1C**, are conformationally flexible they should potentially be able to fit into electrode–electrode gaps which have slightly different gap distances

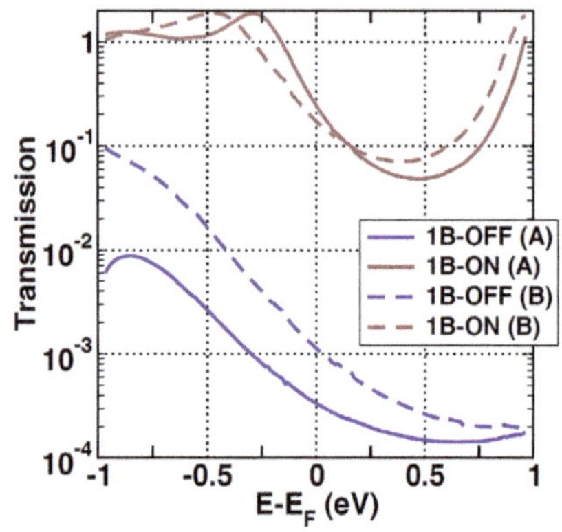

Fig. 11 Transmissions through **1B-OFF** and **1B-ON** at electrode–electrode distances which are optimal for (A) **1B-OFF** and (B) **1B-ON**.

than ideal for a certain isomer. Here we particularly examined switch **1B-ON/OFF** as this should have a higher thermal stability for the ON isomer than the parent **1-ON/OFF** switch (activation energy for hydrogen back-transfer is 104.5 kJ mol^{-1}). For this switch the ΔH(298 K) needed to compress **1B-ON** to the S···S distance of **1B-OFF** is 6.8 kJ mol^{-1}, and the ΔH(298 K) to elongate **1B-OFF** to that of **1B-ON** is 11.4 kJ mol^{-1}. We also examined the transport properties of **1B-ON** and **1B-OFF** when confined to electrode gaps that are ideal for either the ON or the OFF isomer (Fig. 11). From these computations it becomes clear that the transmission at zero-bias voltage of the OFF isomer is more geometry-dependent than the transmission of the ON isomer. The zero-bias transmissions of both the ON and OFF isomers are highest when the isomers are kept at lengths which are ideal for the opposite isomer, and these transmissions are higher by factors of 1.4 and 3.4, respectively, than when the isomers are kept at their optimal lengths. A high switching ratio is clearly possible for **1B-ON/OFF** when kept at the OFF-distance (SR = 711) while that at the ON-distance is moderate (SR = 148). Here, a comparison to other optically activated molecular conductance switches should be made. A class of switches which has been examined both experimentally and computationally is various diarylethene switches.[10–15,18,19] The experimentally determined single-molecule switching ratios for various derivatives are found in the range 6–300,[13,15] and the calculated ones vary from being negligible to a few hundreds.[18,19] Thus, **1B-ON/OFF** displays a high ratio which should motivate future experimental investigations. Yet, the SR is still far below that of the recent dimethyldihydropyrene/cyclophanediene (DHP/CPD) derivative studied experimentally by Royal, Wandlowski and co-workers and which have SR values of $>10^4$.[17]

Indeed, it should be noted that the DHP/CPD switch can be turned photochemically in both directions, and both processes can be rationalized by an alleviation of excited-state antiaromaticity.[25] In the OFF→ON state by alleviation of 14π-electron S_1 antiaromaticity and in the ON→OFF state by alleviation of the 6π-electron S_1 antiaromaticity of the two benzene rings.

Conclusions and outlook

Aromaticity and antiaromaticity effects in the lowest $\pi\pi^*$ excited singlet- or triplet-states (S_1 and T_1) of fully conjugated carbocycles, as given by Baird's rule [aromaticity (antiaromaticity): $4n$ ($4n + 2$) π-electrons], can be applied for the design of new classes of photochemically-triggered molecular conductance switches. The tendency of an excited-state antiaromatic benzene ring to rearrange its substituent pattern so as to break the cyclic π-conjugation into an acyclic linearly π-conjugated *ortho*-xylylene, or a heteroderivative, is exploited in the first two switches, **1** and **2**, as well as in the antifuse **5**. These systems display calculated ideal **ON/OFF** switching ratios in the range 550–1175. This is, however, the maximal switching ratio, and lower ratios are to be expected as the molecules when in a junction presumably will adopt conformations with reduced p_π-overlap and thus weaker π-conjugation in the ON isomer (state). For this reason we analyzed the transmission of a benzannulated derivative of switch **1** (**1B-ON/OFF**), and found in particular that the transmission of the OFF isomer was strongly dependent on the geometry.

We also investigated two tentative molecular switches where we exploit the favourable involvement of excited-state aromatic $4n$ π-electron cycles. Such cycles are formed during the photochemical [2 + 2] cycloaddition of norbornadiene to quadricyclane, leading from **3-ON** to **3-OFF**, and in photochemical retrocyclizations of bicyclo[4.2.0]octa-2,4,7-triene to cyclooctatetraene, leading from **4-OFF** to **4-ON**. In both of these switches the OFF state has two saturated sp^3 carbon atoms which are converted to sp^2-hybridized C atoms; however, it should be noted that the photorearrangements lead from the ON to the OFF state in the first switch and from the OFF to the ON state in the second one. Still, neither **3** nor **4** possess high ON/OFF ratios.

To conclude, molecular switches of type **1-ON/OFF** should be particularly interesting for experiments. Such experimental studies should resolve both the photophysical and photochemical properties of the molecular switches, the thermal stabilities, and the charge transport characteristics.

Although this present investigation is exclusively computational, some general guidelines can be drawn based on our findings. Besides the fact that rearrangements leading from isomers that contain cross-conjugated and/or saturated molecular segments to isomers with linearly conjugated paths are desirable, a few additional generalizations can also be made. Saturated cyclopropyl units bridging two π-conjugated segments may only be saturated in a formal sense as the local Walsh orbitals of the cyclopropyl groups can interact with the π-orbitals of the unsaturated segments. Moreover, large annulene moieties which pucker, *e.g.*, a cyclooctatetraene unit, should be avoided in the ON state for the apparent reason that the π-conjugation through such units is weakened. Finally, it could be desirable if the OFF isomer is conformationally flexible, as in **1-OFF**, so that it can adjust to an electrode–electrode gap with some variation in the gap distance with only a small energy penalty.

Acknowledgements

We gratefully acknowledge Uppsala University through funding of the U3MEC KoF07-initiative, the Swedish Research Council (Vetenskapsrådet), Carl Tryggers

Stiftelse för Vetenskaplig Forskning, the Wenner-Gren Foundations, the Swedish Energy Agency, and Stiftelsen för Strategisk Forskning (SSF (SUSBATT)) for financial support. The calculations were performed on resources provided by the Swedish National Infrastructure for Computing (SNIC) at C3SE and NSC.

Notes and references

1 P. Ball, *Nature*, 2006, **442**, 500–502.
2 H. Song, M. A. Reed and T. Lee, *Adv. Mater.*, 2011, **23**, 1583–1608.
3 S. J. van der Molen and P. Liljeroth, *J. Phys.: Condens. Matter*, 2010, **22**, 133001.
4 N. Fuentes, A. Martín-Lasanta, L. Álvarez de Cienfuegos, M. Ribagorda, A. Parra and J. M. Cuerva, *Nanoscale*, 2011, **3**, 4003.
5 M. Galperin and A. Nitzan, *Phys. Chem. Chem. Phys.*, 2012, **14**, 9421–9438.
6 C. Schirm, M. Matt, F. Pauly, J. C. Cuevas, P. Nielaba and E. Scheer, *Nat. Nanotechnol.*, 2013, **8**, 645–648.
7 M. Alemani, M. V. Peters, S. Hecht, K.-H. Rieder, F. Moresco and L. Grill, *J. Am. Chem. Soc.*, 2006, **128**, 14446–14447.
8 B.-Y. Choi, S.-J. Kahng, S. Kim, H. Kim, H. Kim, Y. Song, J. Ihm and Y. Kuk, *Phys. Rev. Lett.*, 2006, **96**, 156106.
9 C. Benesch, M. F. Rode, M. Cizek, R. Härtle, O. Rubio-Pons, M. Thoss and A. L. Sobolewski, *J. Phys. Chem. C*, 2009, **113**, 10315–10318.
10 S. J. van der Molen, J. Liao, T. Kudernac, J. S. Agustsson, L. Bernard, M. Calame, B. J. van Wees, B. L. Feringa and C. Schönenberger, *Nano Lett.*, 2009, **9**, 76–80.
11 K. Smaali, S. Lenfant, S. Karpe, M. Oçafrain, P. Blanchard, D. Deresmes, S. Godey, A. Rochefort, J. Roncali and D. Vuillaume, *ACS Nano*, 2010, **4**, 2411–2421.
12 S. Lara-Avila, A. V. Danilov, S. E. Kubatkin, S. L. Broman, C. R. Parker and M. B. Nielsen, *J. Phys. Chem. C*, 2011, **115**, 18372–18377.
13 Y. Kim, T. J. Hellmuth, D. Sysoiev, F. Pauly, T. Pietsch, J. Wolf, A. Erbe, T. Huhn, U. Groth, U. E. Steiner and E. Scheer, *Nano Lett.*, 2012, **12**, 3736–3742.
14 Y. Kim, A. Garcia-Lekue, D. Sysoiev, T. Frederiksen, U. Groth and E. Scheer, *Phys. Rev. Lett.*, 2012, **109**, 226801.
15 C. Jia, J. Wang, C. Yao, Y. Cao, Y. Zhong, Z. Liu, Z. Liu and X. Guo, *Angew. Chem., Int. Ed.*, 2013, **52**, 8666–8670.
16 H. Löfås, A. Orthaber, B. O. Jahn, A. M. Rouf, A. Grigoriev, S. Ott, R. Ahuja and H. Ottosson, *J. Phys. Chem. C*, 2013, **117**, 10909–10918.
17 D. Roldan, V. Kaliginedi, S. Cobo, V. Kolisoska, C. Bucher, W. Hong, G. Royal and T. Wandlowski, *J. Am. Chem. Soc.*, 2013, **135**, 5974–5977.
18 A. Odell, A. Delin, B. Johansson, I. Rungger and S. Sanvito, *ACS Nano*, 2010, **4**, 2635–2642.
19 Y. Tsuji and R. Hoffmann, *Angew. Chem.*, 2014, **126**, 4177–4181.
20 F. Chen, J. He, C. Nuckolls, T. Roberts, J. E. Klare and S. Lindsay, *Nano Lett.*, 2005, **5**, 503–506.
21 L. Bogani and W. Wernsdorfer, *Nat. Mater.*, 2008, **7**, 179–186.
22 J. Liao, J. S. Agustsson, S. Wu, C. Schönenberger, M. Calame, Y. Leroux, M. Mayor, O. Jeannin, Y.-F. Ran, S.-X. Liu and S. Decurtins, *Nano Lett.*, 2010, **10**, 759–764.
23 N. C. Baird, *J. Am. Chem. Soc.*, 1972, **94**, 4941–4948.

24 H. Ottosson, *Nat. Chem.*, 2012, **4**, 969–971.
25 M. Rosenberg, C. Dahlstrand, K. Kilså and H. Ottosson, *Chem. Rev.*, 2014, **114**, 5379–5425.
26 E. Hückel, *Grundzüge der Theorie Ungesättiger und Aromatischer Verbindungen*, Verlag Chemie GmbH, Berlin, 1938.
27 V. I. Minkin, M. N. Glukhovtsev and B. Y. Simkin, *Aromaticity and antiaromaticity: Electronic and Structural Aspects*, John Wiley and Sons Ltd, New York, 1994.
28 P. v. R. Schleyer, *Chem. Rev.*, 2001, **101**, 1115–1118.
29 N. Martín, M. M. Haley and R. R. Tykwinski, *Chem. Commun.*, 2012, **48**, 10471.
30 P. B. Karadakov, *J. Phys. Chem. A*, 2008, **112**, 12707–12713.
31 P. B. Karadakov, *J. Phys. Chem. A*, 2008, **112**, 7303–7309.
32 M. Moreno, R. Gelabert, J. M. Lluch and J. M. Ortiz Sánchez, *Phys. Chem. Chem. Phys.*, 2013, **15**, 20236–20246.
33 A. C. Pratt, *J. Chem. Soc., Chem. Commun.*, 1974, 183–184.
34 S. V. Kessar, T. Singh and A. K. S. Mankotia, *J. Chem. Soc., Chem. Commun.*, 1989, 1692.
35 S. J. Formosinho and L. G. Arnaut, *J. Photochem. Photobiol., A*, 1993, **75**, 21–48.
36 P. Lainé, V. Marvaud, A. Gourdon, J.-P. Launay, R. Argazzi and C.-A. Bignozzi, *Inorg. Chem.*, 1996, **35**, 711–714.
37 S. Fraysse, C. Coudret and J.-P. Launay, *Eur. J. Inorg. Chem.*, 2000, 1581–1590.
38 D. Bryce-Smith and J. E. Lodge, *Proc. Chem. Soc.*, 1961, 333.
39 J. E. Garbe and V. Boekelheide, *J. Am. Chem. Soc.*, 1983, **105**, 7384–7388.
40 K. K. De Fonseka and J. J. McCullough, *J. Am. Chem. Soc.*, 1979, **101**, 3277–3282.
41 W. Chen, H. Li, J. R. Widawsky, C. Appayee, L. Venkataraman and R. Breslow, *J. Am. Chem. Soc.*, 2014, **136**, 918.
42 A. D. Becke, *J. Chem. Phys.*, 1993, **98**, 1372–1377.
43 P. J. Stephens, F. J. Devlin, C. F. Chabalowski and M. J. Frisch, *J. Phys. Chem.*, 1994, **98**, 11623–11627.
44 P. C. Hariharan and J. A. Pople, *Theor. Chim. Acta*, 1973, **28**, 213–222.
45 M. J. Frisch *et al.*, *Gaussian 09, Revision C.01 and D.01*, Gaussian Inc. Wallington CT, 2009; for the complete reference, see ESI†
46 J. M. Soler, E. Artacho, J. D. Gale, A. Garcia, J. Junquera, P. Ordejon and D. Sánchez-Portal, *J. Phys.: Condens. Matter*, 2002, **14**, 2745–2779.
47 E. Artacho, E. Anglada, O. Diéguez, J. D. Gale, A. García, J. Junquera, R. M. Martin, P. Ordejon, J. M. Pruneda, D. Sanchez-Portal and J. M. Soler, *J. Phys.: Condens. Matter*, 2008, **20**, 064208.
48 N. Troullier and J. L. Martins, *Phys. Rev. B: Condens. Matter Mater. Phys.*, 1991, **43**, 1993–2006.
49 J. P. Perdew, K. Burke and M. Ernzerhof, *Phys. Rev. Lett.*, 1996, **77**, 3865–3868.
50 M. Brandbyge, J.-L. Mozos, P. Ordejon, J. Taylor and K. Stokbro, *Phys. Rev. B: Condens. Matter Mater. Phys.*, 2002, **65**, 165401.
51 M. Mayor, H. B. Weber, J. Reichert, M. Elbing, C. von Hänisch, D. Beckmann and M. Fischer, *Angew. Chem., Int. Ed.*, 2003, **42**, 5834–5838.
52 T. Hansen, G. C. Solomon, D. Q. Andrews and M. A. Ratner, *J. Chem. Phys.*, 2009, **131**, 194704.
53 M. Ernzerhof, H. Bahmann, F. Goyer, M. Zhuang and P. Rocheleau, *J. Chem. Theory Comput.*, 2006, **2**, 1291–1297.

54 G. C. Solomon, C. Herrmann, T. Hansen, V. Mujica and M. A. Ratner, *Nat. Chem.*, 2010, **2**, 223–228.
55 N. Okabayashi, M. Paulsson, H. Ueba, Y. Konda and T. Komeda, *Phys. Rev. Lett.*, 2010, **104**, 077801.
56 M. Paulsson and M. Brandbyge, *Phys. Rev. B: Condens. Matter Mater. Phys.*, 2007, **76**, 115117.

Faraday Discussions

DISCUSSIONS

Molecular electronics: general discussion

Justin Hodgkiss, Eli Zysman-Colman, Simon Higgins, Gemma Solomon, Ioan Bâldea, Ifor Samuel, Latha Venkataraman, Fred Wudl, Bingqian Xu, Ravindra Venkatramani, Henrik Ottosson, Dmitrii Perepichka, Uli Lemmer, Peter Skabara, Andrew Mount and Donal Bradley

DOI: 10.1039/c4fd90049a

Dmitrii Perepichka opened the discussion of the introductory lecture by Fred Wudl: The photovoltaic devices based on your open-shell materials show no sensitivity in the near-IR region, where the band-gap is. Does this imply that the observed photoeffect in the UV-Vis region involves charge separation of the higher excited state ("hot" exciton)?

Fred Wudl answered: Thanks for your question. Yes, that could be the case. The long wavelength absorption may be entirely due to transitions of the open shell species that do not result in carrier generation.

Peter Skabara said: Do you have crystal structures of any of the BBT molecules and, if so, what observations did you make concerning conformation and packing?

Fred Wudl answered: Thanks for your question. Unfortunately we do not have crystal structures for the small oligomers mentioned in the talk. In our hands, they either gave fine powders or films that we did not investigate further by powder pattern techniques.

Uli Lemmer remarked: You very nicely pointed out the historical development in the field and highlighted the role of charge transfer salts. What was the main reason that these materials never really made it into an application?

Fred Wudl answered: Thank you for the question, Prof. Lemmer. Those days were about 20 or more years before the advent of "nano", and the materials available were very tiny, brittle crystals. Prof. Qichun Zhang from NTU in Singapore has been relatively successful in obtaining nanoscale objects of TCNQ charge transfer salts. The heyday of charge transfer salts was also before the introduction of the important concept of supramolecular control, the way that Prof. Perepichka's paper[1] shows it.

1 H. T. Black, H. Lin, F. Bélanger-Gariépy and D. F. Perepichka, *Faraday Discuss.*, 2014, DOI: c4fd00133h.

Dmitrii Perepichka commented: You mentioned lack of processability as one of the chief reasons for the faded interest in organic metals, as compared to organic semiconductors. I wonder, if the processability issue was resolved, what key applications would you envisage for such materials?

Fred Wudl responded: Thanks for your question. It is similar to Uli Lemmer's earlier question. Please see the reply to Uli Lemmer's question.

Donal Bradley stated, in response to Dmitrii Perepichka's question: There are conductors that are widely used but are not ideal and that we would like to be able to replace. In particular for indium tin oxide (ITO), achieving a high work function is problematic and it varies significantly with different environmental conditions. ITO is also brittle and therefore suffers from micro-cracking on flexible substrates which increases the sheet resistance. It is also relatively rough, leading to the need for a planarizing layer.

Ifor Samuel remarked: You mentioned that you think organic photovoltaics are now a topic for industry. This contrasts with a discussion at the recent NREL-organised meeting in Telluride, where in a discussion session, delegates started to list the basic things we do not understand in organic photovoltaics (including, for example, aspects of charge separation) so I think there is still very considerable need for work by academics and national research labs as well as in industry for the field to advance.

Fred Wudl responded: Of course you are right. Scientists should refrain from making predictions or making sweeping generalizations. By my comment I was expressing a general sentiment of the academic community working in the field.

Latha Venkataraman opened the discussion of the paper by Gemma Solomon: Your paper discusses an interesting idea relating to the difference in conductance and thermopower for molecular junctions formed through the overlap of two pi-systems. My question relates mainly to whether the theory could be tested experimentally? Given that very small changes in the overlap between the two pi-systems result in such a large difference in the thermopower and conductance, I am not clear how one could observe such a change in an experiment.

Gemma Solomon replied: I am not sure that it is realistic to deliberately probe the subtle geometric changes that we calculate in an experimental setup. Our purpose here is to examine what the likely conformations are in junctions that spontaneously form of this kind (for example Wu *et al.*[1]) and what sort of variation we can expect if we can perturb the nature of the minimum energy structures.

1 S. Wu, M. T. González, R. Huber, S. Grunder, M. Mayor, C. Schönenberger and M. Calame, *Nat. Nanotechnol.*, 2008, **3**, 569–574.

Ioan Bâldea asked: Some of your results refer to molecular geometries, which are not the most stable geometries. So, those molecules are under stress. Did you estimate the corresponding forces and check whether they are comparable to those in STM junctions under mechanical stretching (F~nN, *cf. e.g.* Tao *et al.*[1])?

1 B. Xu, X. Xiao and N. J. Tao, *J. Am. Chem. Soc.*, 2003, **125**, 16164–16165.

Gemma Solomon answered: We did not calculate the forces in these cases. We were not imagining that the STM experiment would be able to map the structures that we explored, but rather we wanted to see what the variation was with conformation. In other pi-stacked systems, substituents can be used to tune the minimum energy structures[1] and we imagined that this approach could possibly be used to control the minimum energy structures.

1 J. Vura-Weis, M. A. Ratner and M. R. Wasielewski, *J. Am. Chem. Soc.*, 2010, **132**, 1738–1739.

Ioan Bâldea asked: You have studied various molecular stacks, wherein the two thiol groups are at opposite ends. For an STM setup this would correspond to a (say) left thiol group of the lower half (covalently) bound to the STM substrate and to a right thiol group of the upper half (covalently) bound to the STM tip. I wonder whether such conformations could occur in a real situation, given the fact that in real situations a self-assembled monolayer (SAM) is first prepared on the STM substrate, and then the STM tip is approached/retracted. I think that a more realistic conformation would be with both thiol groups of the stack chemisorbed on the substrate while the opposite end of the stack is physisorbed on the tip. What would be the impact of this (perhaps more realistic) conformation?

Gemma Solomon answered: The conformation you suggest is certainly likely to occur in some instances; however, in that case I would expect to simply see the conductance of two molecules in parallel. There will be an environmental effect of multiple molecules in a monolayer but I would not expect this to be as large as the effects we see here. The conformations we suggest with one molecule bound to the tip and the other to the surface have been observed experimentally,[1] so they are not unrealistic.

1 S. Wu, M. T. González, R. Huber, S. Grunder, M. Mayor, C. Schönenberger and M. Calame, *Nat. Nanotechnol.*, 2008, **3**, 569–574.

Henrik Ottosson said: In your calculations of the geometries you used a dispersion-corrected DFT method, and then these geometries were applied in the transport calculations. However, if dispersion could be taken into consideration also in the transport calculations, what do you think the result would be? Also, has anyone already studied such effects?

Gemma Solomon replied: Our method most certainly underestimates the transport across the pi-stack, not because of the missing dispersion necessarily, but more importantly because of the finite spatial extent of the parameter set we use for the DFTB. We would expect to see higher conductance but the same qualitative trends in a method that had longer-range coupling elements.

Henrik Ottosson stated: It might be interesting to look at couplings between molecules with heteroaromatic rings instead of phenyl or anthryl groups. First, pyridyl groups linked to the ethynyl group at their 4-positions would possibly display a repulsion between the two N-atoms, which could lead to a stronger preference for the type of conformation displayed at A0 in Fig. 2 in your paper.[1]

Secondly, it may also be interesting to investigate the coupling between two compounds with N-linked pyrroles as the aromatic groups. They would be non-alternate, yet could provide (in one and the same structural arrangement) both a path with both sulfur atoms starred and a path with a star on only one S (depending if one takes one or two routes around the pyrrole ring before "jumping" to the other one). This same feature would be displayed by non-alternate groups such as the azulenyl group. However, it may not be interesting as it just becomes something in between the transmission displayed by conformations A0 and A0*.

1 Q. Li and G. C. Solomon, *Faraday Discuss.*, 2014, DOI: c4fd00083h.

Gemma Solomon answered: We feel that there are certainly many possibilities with self-assembled non-bonded systems of this kind and hope that the future will bring some interesting investigations in this direction, both theoretically and experimentally.

Ravindra Venkatramani said: The variation of thermoelectric conductance features with molecular stacking geometries described in your paper is very interesting. In the present formulation, it appears that the effect of temperature on conductance is included only through the Fermi functions, is this correct? The electrode–molecule couplings (broadening factors) should also be sensitive to temperature. There could also be interesting effects on the molecular geometry due to the temperature gradient across the molecule. Can you comment on how these effects might influence your results (particularly the interference near the Fermi energy)? What temperature gradients are typically applied in the thermoelectric conductance measurements?

Gemma Solomon replied: In the present formulation, the variation of the junction geometry (and thereby electronic coupling elements) with temperature is not included. Typically, the temperature gradients that are applied in experiments are up to 30 K. We would expect the energetic position of the interference features to shift; however, we would not expect to see dephasing effects remove the destructive interference. For some calculations exploring these effects, see Andrews *et al.*[1]

1 D. Q. Andrews, G. C. Solomon, R. H. Goldsmith, T. Hansen, M. R. Wasielewski, R. P. Van Duyne and M. A. Ratner, *J. Phys. Chem. C*, 2008, **112**, 16991–16998.

Fred Wudl asked: What is the effect of interplanar distance within a pi-stack on conductivity, does the conductivity decrease with increasing distance?

Gemma Solomon replied: The conductance will decrease significantly with increasing the inter planar distance across the pi-stack.

Fred Wudl responded: In real life crystals the highest conductivity is along slip-stacks, that is, the molecules within a stack do not sit exactly on top of each other.

Gemma Solomon replied: I was not aware of this, but it is interesting to hear that this is also the case in other systems. We found that the explanation for our systems was that the symmetry breaking effect of the binding groups meant that the fully eclipsed structures were not the most favourable for transport.

Simon Higgins asked: Could you elaborate on why the regions of high thermopower in your 2-dimensional plots coincide with regions of low transmittance? Would it be possible to design a system where both occurred at the same point, as needed for a particularly high figure of merit, as defined by the equation on page 2 in your paper?[1]

1 Q. Li and G. C. Solomon, *Faraday Discuss.*, 2014, DOI: c4fd00083h.

Gemma Solomon responded: The thermopower can be related to the derivative (slope) of the electronic transmission with respect to energy.[1] This means that when the transmission varies a lot with changing the injection energy of the tunnelling electron, the Seebeck coefficient (S) will be large. In this case, we achieve a very steep slope near features of destructive interference (nodes) in the transmission. With this strategy, we will always see low transmission associated with a high S. We would have to introduce a nearby transmission resonance (*i.e.* not just destructive interference) in order to achieve higher levels of transport whilst maintaining the higher values for S.

1 M. Paulsson and S. Datta, *Phys. Rev. B*, 2003, **67**, 241403(R).

Uli Lemmer queried: What is the role of vibrational heat transfer as compared to heat transfer by charges?

Gemma Solomon answered: Generally, we would expect vibrational heat transfer to be suppressed in these junctions due to the phonon/vibration mismatch between the metal and the molecule. Consequently heat transfer by charges is likely to dominate.

Bingqian Xu asked: The work is insightful in understanding the coherent transport in molecular junction devices, yet the design of the measuring system seems extremely challenging. How can we make such a molecular junction and maintain the molecule in the junction, especially the "through-space" part? In ultra high vacuum (UHV) and low temperature with the thermal fluctuation mitigated, it may be possible. However, how could we then apply the temperature gradient to measure the thermal power?

Gemma Solomon replied: Here we model the pi-stack as two separate molecules in order to probe a range of configurations and see the effect of geometry on the conductance properties. In order to construct devices where the electronic properties are controlled by a pi-stacked structure, it may be necessary to use cyclophanes where the two elements are bound together through a saturated linker.

Simon Higgins communicated: In your quinone–dihydroquinone system, as well as pi-pi overlap there will be a degree of charge transfer interaction between the electron-deficient quinone and the electron-rich dihydroquinone. Do your calculations take this into account?

Gemma Solomon communicated in response: The hydrogen bonding between these two species is what shifts the minimum energy structure to the fully eclipsed position (0,0 in our plots). The extent to which this is modelled correctly is limited by the theoretical methods we use (DFT for the geometries, DFTB for the transport). We would expect that the most significant qualitative effects of these substituents are accounted for with this approach, but the details are certainly not quantitatively correct.

Gemma Solomon opened the discussion of the paper by Ioan Bâldea: Herrmann *et al.* have investigated increasing the size of the basis set used in density functional theory calculations of coherent transport and found that sometimes large basis sets can make the results worse.[1] Do you have any suggestions for what are sensible larger basis sets in the particular context of transport?

1 C. Herrmann, G. C. Solomon, J. E. Subotnik, V. Mujica and M. A. Ratner, *J. Chem. Phys.*, 2010, **132**, 024103.

Ioan Bâldea replied: The case considered in the paper by Herrmann *et al.*[1] is a typical illustration for one of the main ideas expressed in my paper: KS "orbitals" are not true molecular orbitals. Any implementation based on NEGF+KS-DFT (including that used by Herrmann *et al.*[1]) uses them as if they were real orbitals. Albeit merely from a technical/computational point of view, virtual orbitals (LUMO *et al.*) have been nicely identified by Herrmann *et al.* as a source of difficulty when increasing the basis set size. Attempting to achieve (nearly) convergent results by enlarging the basis set makes sense only in approaches that are conceptually sound. The basis set merely provides an expansion manifold used to compute, for example, the poles (which provide ionizations and affinities) of the electronic Green's function within the OVGF or ADC(n) approaches; there, their physical meaning is guaranteed (Lehmann representation). Parenthetically, convergence is not the only issue. Based on earlier literature, I mentioned in my paper[2] that, in the limit of a complete basis set, the energy of the HF-LUMO does converge; still, it converges to zero, which is unphysical. If we knew the exact exchange correlation functional, the KS-HOMO would converge to the lowest ionization energy with a reversed sign (−IP). However, all of the examples presented in my paper obtained by using the widely employed functional B3LYP (this is not the only one that I studied) demonstrate large differences between KS-HOMO and −IP. So, we are still far from knowing the exact functional. This may also explain why convergence upon increasing the basis set size is problematic not only when "nonvariational" properties are computed within DFT (see ref. 62 cited by Herrmann *et al.*).

1 C. Herrmann, G. C. Solomon, J. E. Subotnik, V. Mujica and M. A. Ratner, *J. Chem. Phys.*, 2010, **132**, 024103.
2 I. Bâldea, *Faraday Discuss.*, 2014, DOI: c4fd00101j.
3 J. Nobel, S. Trickey, J. R. Sabin and J. Oddershede, *Chem. Phys.*, 2005, **309**, 89–94.
4 M. N. Paddon-Row and K. D. Jordan, Through-Bond and Through-Space Interactions in Unsaturated Hydrocarbons: Their Implications for Chemical Reactivity and Long-Range Electron Transfer, in *Modern Models of Bonding and Delocalization*, ed. J. F. Liebman and A. Greenberg, Wiley VCH, New York, 1989.

Dmitrii Perepichka asked: In your presentation, you often characterized the results of calculations with different methods as being reasonable or not reasonable. I wonder what are the main criteria to tell "reasonable" results from "unreasonable" results, without comparing them to the experimental data? Why are such comparisons not discussed in the paper?

Ioan Bâldea responded: By chance, questionable theoretical methods can sometimes "reproduce" experimental data. What can we learn from such "agreements" between theory and experiment? I emphasized throughout the need for a sound, reliable theoretical description. In my paper,[1] I always referred to the (theoretical) results at the EOM-CCSD level using aug-cc-pVDZ basis sets; this is the reference for assessing "reasonable" or "unreasonable" results. The fact that these results insignificantly differ from those obtained (i) *via* Δ-CCSD, (ii) by including corrections due to triples [CCSD(T)], or (iii) by augmenting the basis sets (by using aug-cc-pVTZ) represents an important internal self-consistency test within the realm of theory. The fact that at the molecular size considered, experimental results on the lowest ionization and electron attachment energies can be accurately reproduced at this level of theory is amply documented. Even for larger molecules like C_{60}, the differences between theory and experiment are ~0.1 eV (*cf.* Ortiz *et al.*[2]).

1 I. Bâldea, *Faraday Discuss.*, 2014, DOI: c4fd00101j.
2 V. G. Zakrzewski, O. Dolgounitcheva and J. V. Ortiz, *J. Phys. Chem. A*, 2014, 118, 7424–7429.

Ravindra Venkatramani asked: You have examined the band gap of molecules with several high level electronic structure calculations. Such systematic studies are very useful to this field of molecular electronics. Can you comment on the effects of solvation and solvent polarization on the calculated band gap? How were solvation effects included in the calculations?

Ioan Bâldea answered: Calculations which are not reported here (*e.g.* at the CC2, SAC-CI, or TD-DFT level) indicate that solvent effects on the band gap are very weak. On the contrary, the charge (transport) gap (and hence the exciton binding energy) are significantly affected by solvents.[1,2]

1 I. Bâldea, *Nanoscale*, 2013, **5**, 9222–9230.
2 I. Bâldea, *Electrochem. Commun.*, 2013, **16**, 19–21.

Gemma Solomon remarked: Very often the molecules that our experimental collaborators are interested in are too big for GW calculations (unless we perform model-GW calculations), yet we know that DFT has significant limitations. What direction do you suggest we take if we want to perform transport calculations on experimentally relevant systems?

Ioan Bâldea replied: For the properties/setups we theorists are interested in, our experimental colleagues are very often unable to provide reliable estimates. Like them, we have to do the best we can. I think that model calculations is the best we can do at present, using model parameters determined from accurate *ab initio* calculations (In view of the examples presented, I would recommend something better than GW.) Once we have reliably estimated model parameters, we can gradually increase the complexity of the model. If at a certain level (say, tight-binding level) theory is at odds with experiment, we should extend the model to account for orbital relaxation (I showed how important it is for molecules used in typical junctions[1]), *e.g.* by allowing orbital energies to depend on the average orbital populations (*i.e.* SCF; single-particle description). If, after doing that, disagreement persists, we can go beyond the single-particle picture and include correlations which adjust (new) model parameters to reproduce, for example, exciton binding energies (again, I showed its importance for typical cases[1]), and so on. At each step, if deviations from experiment remain, we can learn something and attempt further theoretical improvement. Of course, this is just a compromise until reliable (truly≠DFT) *ab initio* transport calculations for sizes of experimental interest become feasible. Otherwise, what can we learn from a transport theory that uses merely mathematical objects as if they were real entities?

1 I. Bâldea, *Faraday Discuss.*, 2014, DOI: c4fd00101j.

Simon Higgins said: Your calculations are carried out on 'bare' molecules rather than upon molecules attached to electrodes. I wonder what is the benefit of carrying out your more sophisticated and computationally-demanding calculations, if things are likely to change drastically when the molecule is connected to the electrodes in a metal–molecule–metal junction?

Ioan Bâldea answered: The messsage conveyed by the results presented in my work is certainly not too optimistic. These results emphasize the need for more sophisticated and computationally demanding calculations than currently done in molecular transport. Can you trust theories claiming agreement with experimental data for molecular junctions on the basis of results obtained by using methods and basis sets that, as shown in my work, turn out to be inadequate even for isolated ("bare") molecules? I believe that this study is useful because it enables us to better realize what are, at present, the limits of molecular transport theory, and what one can expect from theory.

Recent examples demonstrate that cases exist where changes caused by contacting a molecule to electrodes in a junction can be disentangled into contributions with clear physical origin, which can be reliably estimated by using *ab initio* quantum chemical calculations, for example, estimating solvent-driven changes in the molecular orbital alignment[1,2] or estimating differences between orbital alignments in molecular junctions characterized by similar chemical linkages to the electrodes.[3]

Obviously, in general, there is no reason to rule out drastic changes when a molecule is embedded in a junction. Then, in my opinion, the best that theorists can do at present is transport calculations using models, with model parameters determined *via* reliable *ab initio* calculations (I have addressed this issue in my response to Gemma Solomon's second question on my paper). Let me finally mention that resorting to models is the usual current practice for theoretical chemists in cases which are too demanding computationally, for instance, for solving problems of vibronic coupling, which refer to isolated medium-sized molecules. Why should we be more ambitious than, e.g., scientists working on vibronic coupling in dealing with problems that are far more complex (nonequilibrium, infinite systems)? I doubt that approaching problems based on methods which are questionable represents good scientific practice.

1 I. Bâldea, *Nanoscale*, 2013, **5**, 9222–9230.
2 I. Bâldea, *J. Phys. Chem. C*, 2014, **118**, 8676–8684.
3 I. Bâldea, *Nanotechnology*, 2014, **25**, 455202.

Latha Venkataraman opened the discussion of the paper by Ravindra Venkatramani: Conductance measured in a transport experiment usually involves a non-resonant tunneling process for a molecule in the ground state. Charge transfer rates measured in a pump/probe experiment involve charge transfer from an excited state. It is not clear to me why these two measurements should be related.

Ravindra Venkatramani responded: First, I note that the data in Fig. 1 correspond to ground state electron transfer and conductance.[1] More generally however, this study addresses experimental observables (charge transfer (CT) rates/electrochemical rate constants and conductances) which report on the specific property of a molecular bridge to transmit electronic charge. For the case of molecular junction *versus* donor–bridge–acceptor systems that are photoexcited, the source and sink for the electronic charges are different (charge donor/acceptor chemical groups *versus* electrodes), but the medium through which charges propagate is the same, *i.e.* the molecular bridge is identical in the two cases. What will change is the CT barrier, through changes in the so-called "tunneling energy",[2] and this feature is included in the model.

Interestingly, our theoretical study shows that similar types of mechanisms (both resonant and non-resonant) govern the transmission of charges in an electrochemical redox reaction, photoinduced charge transfer in a donor–bridge–acceptor experiment, and the current flowing across a molecular junction. The interplay of different charge transfer mechanisms in the different experimental setups is governed by the same two physical considerations: a) CT barrier heights: the alignment of the source (photoexcited donor or electrode Fermi level) electronic state energy with that of the molecular bridge, and b) bath induced dephasing: the broadening of the bridge electronic states due to interaction with the environment/bath (solvent and molecular vibrational degrees of freedom). Our study shows that if the CT barrier heights and bath-induced dephasing in the two experimental setups are constrained to be the same, then a linear correlation may be obtained between charge transfer rates and molecular conductance. Comparison with actual experimental data indicates non-linearities in the relation between electrochemical rate constants and molecular conductance that can be traced back to differences in CT barrier heights and bath dephasing effects in the two experimental setups.

1 R. Venkatramani, E. Wierzbinski, D. H. Waldeck and D. N. Beratan, *Faraday Discuss.*, 2014, DOI: c4fd00106k.
2 S. S. Skourtis and D. N. Beratan, *Adv. Chem. Phys.*, 1999, 106, 377–452.

Ioan Bâldea queried: I agree with Latha Venkataraman that a relationship between the conductances and charge transfer (CT) rates is problematic due to the different setups (metal–bridge–metal *versus* donor–bridge–acceptor). But what you can do is establish a relationship between conductance G and the electrochemical (EC) CT rate k^0. In your model calculations, the bridge parameters E_B and t_B are intrinsic properties of the isolated molecule (bridge). Therefore, the orbital energies of the isolated molecule are not altered by the coupling to the electrode(s). In real situations, these energies change when the molecule is linked to the electrode(s) (they may also change upon an applied bias, but let us remain with the case of an Ohmic response.) In general, the MO energy shifts for a molecule coupled to two electrodes (transport setup) are larger than for a molecule coupled to one electrode (EC setup); the degree of MO pinning to the Fermi level is different in the two cases. As shown recently,[1] this is the case even for a redox metalloprotein (azurin) with a size of ~3–4 nm and the LUMO located somewhere in the middle: the difference between the equilibrium potential in the EC-STM setup and the redox potential is quite significant. So, because the shift of the relevant molecular orbital (MO) toward the Fermi level (which can be compared to a tunneling barrier) is expected to be larger in a transport setup than in an EC setup, the length attenuation factors β are expected to be different. That is, $\beta_{\sigma} < \beta_{k^0}$, and this immediately yields a sublinear dependence ($m < 1$), like the one visible by inspecting the experimental data shown in Fig. 1 in your paper.[2]

1 I. Bâldea, *J. Phys. Chem. C*, 2013, **117**, 25798–25804.
2 R. Venkatramani, E. Wierzbinski, D. H. Waldeck and D. N. Beratan, *Faraday Discuss.*, 2014, DOI: c4fd00106k.

Ravindra Venkatramani answered: I have previously addressed the issue of comparing electrochemical rate constants, donor–bridge–acceptor charge transfer rates and molecular conductances along with our assumptions therein in response to Prof. Latha Venkataraman's query. With regards to the influence of the CT barrier height on the rate-conductance non-linearity, I agree with Dr Bâldea's assesment. In fact, our calculation in Fig. 6,[1] which includes different charge transport barriers (E_B, and t_B are fixed but E_{DB} and E_{FB} are allowed to vary) for electrochemical rate constants and molecular conductance calculations, shows precisely that the experimental non-linearity can be reproduced when E_{DB} is not equal to E_{FB}. However, we also note that in Fig. 6, the experimental non-linearity is reproduced only when $E_{DB} > E_{FB}$. Dr Bâldea suggests the opposite trend, *i.e.* $E_{FB} > E_{DB}$, based on physical considerations of the two experimental setups. This conflict can be reconciled by considering the effect of bath dephasing. Fig. 5 shows that when dephasing (rate) is not equal to dephasing (conductance), then the rate–conductance relationship is non-linear even when the CT barrier heights are the same ($E_{DB} = E_{FB}$).[1] Thus, the experimental non-linearity may also be reproduced for $E_{FB} > E_{DB}$, providing the dephasing for the charge transfer case is sufficiently less than the dephasing for the conductance. The interplay between bath dephasing and the CT barrier height in the actual experiments of Fig. 1 can only be revealed by a detailed fitting of our model to the experimental data.[1] We are presently working on fitting our model to the experimental data. I would also like to draw attention to our discussion in the conclusion section of the manuscript. The section comments on the physical considerations behind the differences in the CT barrier heights and bath dephasing effects in the rate *vs.* conductance measurements. Electrode effects on the molecular energy levels, which are suggested by Dr Bâldea, are also covered. I thank Dr Bâldea for bringing our attention to his prior study which is relevant to this discussion.

1 R. Venkatramani, E. Wierzbinski, D. H. Waldeck and D. N. Beratan, *Faraday Discuss.*, 2014, DOI: c4fd00106k.

Ioan Bâldea commented: It is certainly very challenging to discriminate between the physical origin of the sublinear dependence *s vs.* k^0 as suggested in my previous comment and that worked out in your paper merely within the realm of theory: you have plenty of parameters, which are hard to estimate. But if/when available, experimental data may provide a clue. If they do not exhibit any temperature dependence, it is very likely that the origin is as I have suggested; otherwise, your dephasing mechanisms are relevant.

Ravindra Venkatramani answered: Our theoretical framework suggests that the sublinear correlation between molecular conductance and the charge transfer rate can arise from differences in the CT barrier heights and/or solvent-induced dephasing rates. Which of these parameters (CT barrier height *vs.* dephasing rate) is dominant for the experimental data in Fig. 1 may only be revealed by a detailed fitting of our model to the experimental data at hand.[1] We are presently working on this. Dr Bâldea makes an excellent suggestion to explore the temperature dependence of the scaling factor (eqn (24) in our manuscript[1]) between the rate and conductance experimentally. While temperature independence of the scaling factor may rule out dephasing as a contributor to the observed rate–conductance non-linearity, temperature dependence does not rule out differences in CT barrier heights. A combination of differences in barrier heights and dephasing effects might still give a temperature dependence for the scaling factor. It would also be interesting to examine this effect theoretically. I thank Dr Bâldea for this suggestion.

1 R. Venkatramani, E. Wierzbinski, D. H. Waldeck and D. N. Beratan, *Faraday Discuss.*, 2014, DOI: c4fd00106k.

Bingqian Xu asked: Single molecular junctions are used to establish the model. As is shown in Fig. 2 in your manuscript,[1] the molecular bridge is defined by tight-binding couplings. If the contact barrier height changes due to the different contact conformations, do you expect different outcomes as shown in Fig. 1,[1] or even different mechanisms?

1 R. Venkatramani, E. Wierzbinski, D. H. Waldeck and D. N. Beratan, *Faraday Discuss.*, 2014, DOI: c4fd00106k.

Ravindra Venkatramani responded: Even within the tight binding model shown in Fig. 2,[1] the effect of different molecule–electrode contact geometries may be captured by the effective coupling parameters (G_L and G_R). Since these parameters effectively broaden the electronic states of the molecular bridge (eqn (17)), they will influence the effective barrier height (and hence the mechanisms of charge transport). Thus, the features of high, medium, and low conductance in Fig. 1, as well as changes in the distance-dependent decay for each of these conductance series, can be reproduced through variation of the coupling parameters (G_L and G_R). We stress that our framework, in general, allows the inclusion of the full electronic structure of the bridge (going beyond the tight binding model used in the present study), as well as more sophisticated distance and angle-dependent electrode–molecule coupling parameters. Thus, the framework can provide a robust analysis of the variation of currents in a molecular junction with different molecule–electrode contact geometries.

1 R. Venkatramani, E. Wierzbinski, D. H. Waldeck and D. N. Beratan, *Faraday Discuss.*, 2014, DOI: c4fd00106k.

Simon Higgins communicated: How does your model compare with the approach of the Kuznetsov–Ulstrup model for describing electron transfer rates in molecules attached to electrode surfaces?

Ravindra Venkatramani communicated in reply: While both models treat the electron transfer using a density of states picture and have many similarities, they differ in some important ways. The framework for computing non-adiabatic electrochemical rate constants in this manuscript includes an explicit description of the molecular bridge which connects the electrode to the redox couple and includes an empirical model for decoherence on the bridge. In general, the electronic structure of the bridge can be described at many levels of sophistication. The explicit description of the bridge allows us to examine different mechanisms for charge transfer between the electrode and the redox couple (resonant, non-resonant and mixed mechanisms), and to relate the charge transfer to the chemical structure of the bridge. In contrast, the Kuznetsov–Ulstrup model assumes a tunnelling mechanism for the charge transfer and only includes the bridge implicitly through an effective tunnelling transmission coefficient.[1] On the other hand, Kuznetsov and Ulstrup have generalized their expressions to include adiabatic charge transfer reactions which are not addressed in our study.

1 J. Zhang, A. M. Kuznetsov, I. G. Medvedev, Q. Chi, T. Albrecht, P. S. Jensen and J. Ulstrup, *Chem. Rev.*, 2008, **108**, 2737–2791.

Ravindra Venkatramani opened the discussion of the paper by Latha Venkataraman: You have presented a handy expression, eqn (2),[1] for the voltage dependence of the electrode–molecule coupling in the molecular orbital basis. Have you tried introducing the broadening functions in the atomic basis (for instance, one could start with the assumption that the electrode contacts only the sulfur atom orbitals)? In this case the broadening matrix elements in eqn (2) can be directly calculated in the molecular orbital basis as a function of the applied voltage (since you already have the electronic structure/eigenvectors as a function of the applied voltage). One could then ask if such a numerically computed broadening matrix element would subscribe to the analytical form predicted in eqn (2). Alternatively, one could ask what form of electrode–atom contacts would give us the analytical form of eqn (2). This may allow a physical interpretation of the voltage scaling parameter ($\nu_{meta/para}$).

1 A. Batra, J. S. Meisner, P. Darancet, Q. Chen, M. L. Steigerwald, C. Nuckolls and L. Venkataraman, *Faraday Discuss.*, 2014, DOI: c4fd00093e.

Latha Venkataraman responded: I believe this is possible, and not very difficult to carry out. There are two issues. Firstly, on the *para* side, the electrode contacts the sulfur. However, on the *meta* side, transport is through ring–electrode coupling. The ring has 6 atoms so we would have to introduce six parameters. The model will then get more complicated than we would like. Alternatively, we could do a full DFT-based calculation to determine what aspect of the molecule controls the scaling parameters.

Ioan Bâldea commented: In your paper,[1] for example in the abstract, you wrote that rectification results from a difference in the voltage dependence of the coupling strength on the through-bond ("*para*") and through-space ("*meta*") side. If one inspects eqn (2), (3) and (4), one can immediately see that rectification (*i.e.* $I(V) \neq \neg I(\neg V)$) exists even if $\Gamma_{meta} = \Gamma_{para}$ and $\nu_{meta} = \nu_{para} \neq 0$. So, I suspect a typo: either the signs in front of ν_{meta} and ν_{para} in eqn (2) and (3) are opposite (one of them should be minus, not plus) or the corresponding ν values given in the caption of Fig. 4 are of opposite signs (ν_{meta} values are positive and ν_{para} values are negative or *vice versa*). Further, you ruled out a rectification resulting from a bias-driven shift of the HOMO energies (Stark effect) and presented Fig. 3(a) as justification.[1] However, if I understand correctly, these are energies of Kohn–Sham "orbitals", and I showed in my work presented at this symposium how inaccurate even estimates for the Kohn–Sham HOMOs are.[2] I checked (please see Fig. 1 and the values in the legend) that using the single-level model and voltage division factors characteristic for other molecular junctions ($\gamma = 0.06$, *cf.* eqn (1) and Table 1, see I. Bâldea[3]) produces a rectification comparable to that shown in Fig. 4(b) in your manuscript.[1]

1 A. Batra, J. S. Meisner, P. Darancet, Q. Chen, M. L. Steigerwald, C. Nuckolls and L. Venkataraman, *Faraday Discuss.*, 2014, DOI: c4fd00093e.
2 I. Bâldea, *Faraday Discuss.*, 2014, DOI: c4fd00101j.
3 I. Bâldea, *Phys. Rev. B*, 2012, **85**, 035442.

Latha Venkataraman answered: Yes, this is a typo. The ν values are not both positive. Indeed, to achieve rectification, we have ν_{para} as positive and ν_{meta} as negative in our model. I have fixed this error in the caption of Fig. 4 in the revised

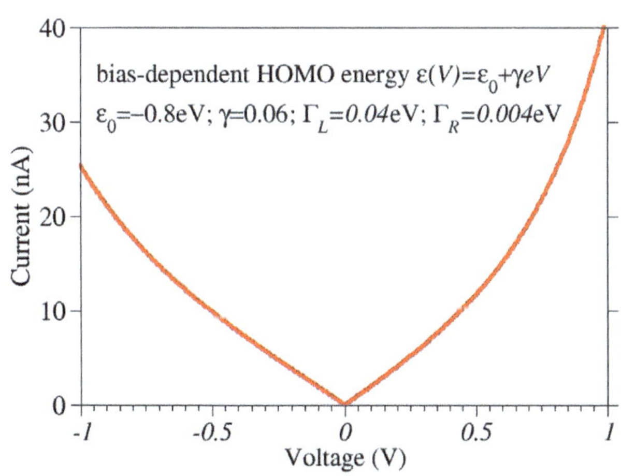

Fig. 1 The rectification computed within a single-level model with Lorentzian transmission, and a typical value of the voltage division factor γ (given in the legend) is comparable to that of Fig. 4(b) of ref. 1.

manuscript. Additionally, yes, DFT HOMOs are not accurate. However, what is shown in Fig. 3A is a general parabolic shape (quadratic dependence) which does not yield any rectification. We need a first order linear dependence to observe rectification. This is what is captured by the ν_{meta} and ν_{para} parameters.

Ioan Bâldea said: Why cannot the rectification observed in your experiments simply be the result of the asymmetric *meta*/*para* conformations, and why should one invoke quantum interference effects? Did you check whether adding side groups to your *meta*/*para* systems has a significant impact on rectification?

Latha Venkataraman responded: An asymmetric coupling does not yield rectification. We have also not tested the dependence of rectification on substituents. This is a good suggestion for future work.

Uli Lemmer enquired: You have shown many ensemble-averaged characteristics. Could you also do time averaging of repetitive measurements on the same molecule? Do the characteristics of one molecule change from one scan to the next?

Latha Venkataraman replied: We can do a few repeated measurements on the same molecule (3–5) but not many more at room temperature. In these cases, we find that some junctions show reproducible characteristics, while others do not.

Simon Higgins asked: Would it be possible to further characterize your rectification mechanism by putting a bulky substituent onto the *meta*-thioether-substituted phenyl ring, such as a *t*-butyl group, to block or at least severely reduce the metal–arene coupling interaction?

Latha Venkataraman replied: This is a good suggestion. We have tested how the conductance varies with substituents on a *para*-terminated system but have not carried out analogous measurements for the *para*/*meta*-coupled system.

Fred Wudl queried: What is your confidence in the measurements? What is the signal to noise? How do the *para*/*para* and *meta*/*meta* results compare? Is there any through-space coupling?

Latha Venkataraman responded: Neither the *para*/*para* nor the *meta*/*meta* junctions should rectify as these are symmetric. We have measured IV characteristics of the *para*/*para* system and shown that there is no rectification.[1] The *meta*/*meta* system does not show a clear conductance signature at low biases therefore we cannot with confidence measure IV curves for these junctions. Again, we have, in past work, demonstrated using atomic force measurements that the *meta*/*meta* system forms junctions but do not conduct.[2]

1 A. Batra, P. Darancet, Q. Chen, J. S. Meisner, J. R. Widawsky, J. B. Neaton, C. Nuckolls and L. Venkataraman, *Nano Lett.*, 2013, **13**, 6233–6237.
2 S. V. Aradhya, J. S. Meisner, M. Krikorian, S. Ahn, R. Parameswaran, M. L. Steigerwald, C. Nuckolls and L. Venkataraman, *Nano Lett.*, 2012, **12**, 1643–1647.

Dmitrii Perepichka commented: Should one expect the same frequency of "up" and "down" rectification events, considering a random orientation of the molecule in the junction? Is this what you have observed in your experiments?

Latha Venkataraman answered: We always find about 50% of the junctions in the up/down configuration. We believe this is because the molecular dipole moment is not strong enough to align the molecule in the field.

Henrik Ottosson said: How much would the observed rectification change if the linkers were changed from methylsulfide to amino groups? Does the observed rectification primarily depend on the strength of the Au–molecule interaction (dative bonding) or could other factors, such as the local polarizability at a particular linker group, also have some importance?

Latha Venkataraman answered: For the system studied in this work, I believe amino groups will yield similar results. We have compared a few other backbones and seen similar rectification ratios for amine *vs.* methylsulfide terminations.

Gemma Solomon remarked: Previously, we proposed a mechanism by which interference effects could be used to induce rectification as the interference features will shift with the applied bias.[1] Why do you conclude that the role of the *meta*-substituted system in this case is simply to induce through-space coupling rather than the interference actually controlling the rectification?

1 D. Q. Andrews, G. C. Solomon, R. P. Van Duyne and M. A. Ratner, *J. Am. Chem. Soc.*, 2008, **130**, 17309–17319.

Latha Venkataraman replied: We have found in these *para*/*meta* linked systems that the conductance decreases significantly with electrode separation.[1] Based on this result, we conclude that conductance in the *para*/*meta* systems is sensitive to the molecule/metal separation. If we were indeed conducting through bonds on both sides, we would expect the conductance to be generally not sensitive to the electrode separation as we see in the *para*/*para* case.

1 J. S. Meisner, S. Ahn, S. V. Aradhya, M. Krikorian, R. Parameswaran, M. Steigerwald, L. Venkataraman and C. Nuckolls, *J. Am. Chem. Soc.*, 2012, **134**, 20440–20445.

Bingqian Xu asked: Compared to low temperature and/or molecular devices with multiple molecules, where the rectification can be at least 1000, most single molecule rectification measurements at room temperature have a rectification ratio of less than 2. Can you comment on this? Do you think low temperature single molecule measurements will improve the rectification ratio?

Latha Venkataraman answered: Most low-temperature measurements are made on a few junctions since these are much harder to carry out, thus the statistical significance of the results is not clear. Also, to the best of my knowledge, very few rectifier systems have been studied both at low temperatures and at room temperature. I cannot, therefore, answer the question of whether low-temperature measurements of the same systems we present would yield higher rectification ratios. *A priori*, I do not think temperature should impact rectification ratios.

Gemma Solomon commented: Interference effects can result in extremely large rectification ratios; however, the part of the molecule responsible for the interference needs to be "protected" from through-space coupling of the kind we see in this instance.

Dmitrii Perepichka commented: You have mentioned that rectification ratios above 5 are not expected in single molecule junctions. At the same time, much higher rectification ratios were reported for a number of molecular monolayer junctions. Do you think this necessarily implies that non-molecular conductance mechanisms (filament artifacts, *etc.*) operate in those cases?

Latha Venkataraman answered: I do not know enough about the monolayer experiments to understand the mechanism of rectification in those systems.

Peter Skabara said: Is there any advantage in shifting the *meta* link further down the chain to note subtle changes in conductance so that you may be able to further probe the true mechanism of through-space conductivity? Some suggested structures are given in Fig. 2 which could be studied by comparing the conductance across the series. Taking the conjugated and *meta* links away from the terminal positions of the molecules may result in a change of conductance if through-space transport from the electrode to the conjugated part of the molecule dominates over intrachain hopping across a *meta* link.

Latha Venkataraman replied: I think this is an interesting suggestion. We have not carried out measurements with such structures; there is a potential for an additional pi-coupling on the meta side which might yield an interesting result.

Ioan Bâldea opened the discussion of the paper by Bingqian Xu: I have a comment on the differences visible in Table 1, where you present results labelled as "Landauer fitting" and "TVS".[1] When you use the first method, you employ the dependence of the dominant molecular orbital on the bias given by eqn (3) in the manuscript of Briechle *et al.*,[2] namely $E_0(V) = E_0 + \gamma_L eV$, where $\gamma_L = (\Gamma_L \neg \Gamma_R)/[2(\Gamma_L + \Gamma_R)]$ is related to the width parameter Γ. Within the second method, the

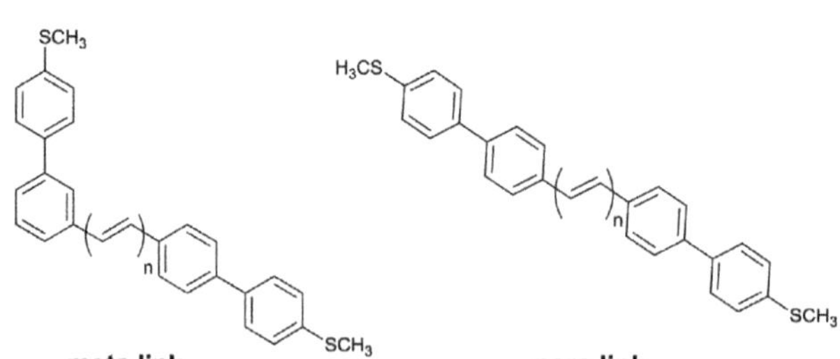

Fig. 2 Suggested structures for shifting the meta and para links away from the ends of the molecule.

dependence looks similar, $E_0(V) = E_0 + \gamma eV$, but is not identical. Here, γ does not depend on Γ, it is an independent parameter that accounts for a different physical fact, namely the asymmetry of the electric potential profile (*cf.*, Bâldea[3]). Eqn (3) in Briechle *et al.*[2] is an *ad hoc* assumption to embody a possible asymmetry of the IV curves (as you indeed measured, *cf.* Fig. 5 (b) and (c) in your manuscript[1]), which is determined by the potential profile and has nothing to do with the difference between the level broadenings $\Gamma_{L,R}$ entering γ_L. This is why the largest difference in ε shown in Table 1 corresponds to the largest IV-asymmetry (last line in Table 1). For nearly symmetric IV curves (the first two lines in Table 1), the difference between ε_{fit} and ε_{TVS} is within the experimental errors (*e.g.* in extracting the transition voltage from the minimum of the Fowler–Nordheim curves), but the first method yields nearly equal $\Gamma_L \approx \Gamma_R$, which is certainly unphysical for an STM setup.[2,4]

1 K. Wang, J. Hamill, J. Zhou, C. Guo and B. Xu, *Faraday Discuss.*, 2014, DOI: c4fd00080c.
2 B. M. Briechle, Y. Kim, P. Ehrenreich, A. Erbe, D. Sysoiev, T. Huhn, U. Groth and E. Scheer, *Beilstein J. Nanotechnol.*, 2012, **3**, 798–808.
3 I. Bâldea, *Phys. Rev. B: Condens. Matter Mater. Phys.*, 2012, **85**, 035442.
4 I. Bâldea, *J. Phys. Chem. C*, 2013, **117**, 25798–25804.

Bingqian Xu responded: Thanks for the detailed comment. First, we agree with your description about the expression of $E_0(V) = E_0 + \gamma eV$.[1,2] The two "γ"s, one in the Landauer formula and the other in TVS, represent two independent parameters that account for different physical meanings. In terms of eqn (3) in Briechle *et al.*, $E_0(V)$ is associated with both Γ_L and Γ_R when studying a specific junction, though overall they contribute to a potential profile that determines the *I–V* behavior. In the paper by Bâldea,[3] high asymmetry was indeed found between the coupling Γ_t of the redox molecule (Azurin) to the tip and the Γ_s of the molecule to the substrate using theoretical calculations. First, there is a major difference in the molecule species studied in this paper and ours. More importantly, theoretical calculations can proclaim that the molecule–tip and molecule–substrate interfaces are different. But this doesn't mean that it is true for the real experimental process. In Scanning Tunneling Microscope Break-junction (STMBJ) experiments, extensive studies have reported that IV curves are symmetric for symmetric junctions (*i.e.* symmetric molecule and symmetric molecule–electrode contact interfaces). These experimental results strongly suggest that the molecule–substrate interface could adjust to its most stable configuration while the tip is stretched, though the substrate looks flat at its initial state. Additionally, this most stable configuration at the molecule–substrate interface is believed to be similar to the molecule–tip interface, indicating a symmetric junction conformation. This is why many previous DFT calculations and MD simulations presumed that the junction layout is symmetric when comparing their results with experimental data extracted from an STM system. To understand the experimental phenomena, our data was just fitted to the standard single-level transport model with no modification in the Landauer fit, and the fitting results gave us a hint to understand the experimental results. We agree that for asymmetric molecule–electrode interfaces, the Γ_s at the two contacts could have a sharp difference, but we still believe that the symmetric IV behavior measured in our system is symmetric for symmetric junction conformations, and thus the broadening factor Γ for the molecule–tip and molecule–substrate interfaces should be symmetric as the fitting results show.

1 B. M. Briechle, Y. Kim, P. Ehrenreich, A. Erbe, D. Sysoiev, T. Huhn, U. Groth and E. Scheer, *Beilstein J. Nanotechnol.*, 2012, **3**, 798–808.
2 I. Bâldea, *Phys. Rev. B: Condens. Matter Mater. Phys.*, 2012, **85**, 035442.
3 I. Bâldea, *J. Phys. Chem. C*, 2013, **117**, 25798–25804.

Gemma Solomon asked: It is tempting to interpret the energy gap that is obtained from the TVS measurements or the Landauer fitting as saying something about a particular orbital of the molecule. However, we know that this is really saying something about an effective single level and that different orbitals may be responsible for transport in different systems (*e.g.* HOMO *vs.* LUMO) or multiple orbitals may be controlling the transport properties at low bias. How do you think we can interpret these values physically (or not) and make the comparison between different molecules? Or should we just see them as indicative parameters?

Bingqian Xu responded: Thanks for the question. We would agree that the way we interpret the results of theoretical fitting, such as TVS and Landauer fitting, is bold and tempting with regard to claiming the change in molecular orbitals. However, in the system we talked about in the paper,[1] these parameters are extracted from the experimental data of junction systems that only differentiate in contact interfaces. Hence the induced change of a specific parameter is predominantly attributed to the asymmetry in the contacts. These parameters offer us a way to understand the experimental results. By fitting the experimental *I–V* to these theories, we obtain a variable factor that could be responsible for the experimental phenomena. For example, the induced increase for ε_{fit} and asymmetric Γ_s, when one of the Au–SH bonds in the Au–S–B–S–Au junction is replaced with an Au–NH_2 bond, illustrates an equivalent effect as the real mechanism proceeds. The real frontier molecular orbital (*e.g.* HOMO/LUMO) is only derivable by theoretical simulations (*e.g.* transmission functions). In this aspect, it is more appropriate and safe to treat the fitting results as indicative parameters. Since this work focuses on the experimental study of molecular junction systems, a thorough simulation of the system is obviously very helpful, so we believe that the fitting results offered a possible explanation for the electronic behavior of the junctions.

1 K. Wang, J. Hamill, J. Zhou, C. Guo and B. Xu, *Faraday Discuss.*, 2014, DOI: c4fd00080c.

Simon Higgins remarked: Turning to the experimental part of your paper, I would like to ask about the histogram in Fig. 2c and the corresponding conductances in Fig. 2d.[1] Could you elaborate on how this histogram is compiled? Do you do the stretch–hold experiment many times and sum the data together? If so, what is the physical reason why the peaks you see in the histogram are so narrow and well-defined that you can confidently de-convolute them in the way indicated in the multi-colour part of the figure?

1 K. Wang, J. Hamill, J. Zhou, C. Guo and B. Xu, *Faraday Discuss.*, 2014, DOI: c4fd00080c.

Bingqian Xu replied: Thank you for the technical question. First of all, yes, we used our homemade Labview computer program to control the SPM tip to do the stretch–hold experiment and simultaneously record the conductance curves. Then about 1000 such curves were used (without pre-selection) to construct the conductance histogram, which resulted in the histogram as shown in blue (Group (1+2+3+4)) in Fig. 2c.[1] This histogram shows 4 sharp peaks instead of a single broad peak (fundamental conductance, $G_1 = 5.2 \times 10^{-5}\ G_0$) obtained by the traditional "continuous stretching" breakjunction method (please see Fig. 5(a) in Xu *et al.*[2]). So, similar to what in electronics people deal with N-bit ADC, we divided the conductance region around the fundamental conductance peak into 5 equal regions from $0–1.5G_1$, with the least significant bit (LSB) of $0.3G_1$. Finally we re-constructed the histogram by grouping according to the finer regions using our home-made computer program, and the results turned out to be 4 very well defined (sharp) conductance peaks (multi-coloured part of Fig. 2c, also see Fig. 7 in Xu *et al.*[2] We are very confident of this method because it does not involve any data pre-selections.

1 K. Wang, J. Hamill, J. Zhou, C. Guo and B. Xu, *Faraday Discuss.*, 2014, DOI: c4fd00080c.
2 J. Zhou, F. Chen and B. Xu, *J. Am. Chem. Soc.*, 2009, **131**, 10439–10446.

Gemma Solomon commented: In Fig. 4 you show an additional barrier when extending the system.[1] Won't there be cases where the barriers that you have for the molecule are also distance-dependent, for example, in the case of charge transfer?

1 K. Wang, J. Hamill, J. Zhou, C. Guo and B. Xu, *Faraday Discuss.*, 2014, DOI: c4fd00080c.

Bingqian Xu responded: First of all, in the system we studied in the paper (metal–linker–molecule–linker–metal; metal: Au or Pt, linker: –SH–, $-NH_2$),[1] the internal chemical bond (C–C bond) of the molecular core has been studied to be much stronger than the molecule–linker (*i.e.* C–S) and metal–linker (*i.e.* Au–S).[2] Therefore, the molecular core can be treated as a rigid body compared with the contact interfaces. When a slight extension of the junction takes place, the change in conductance is believed to predominantly come from the molecule–electrode interface. We suggest that the influence of the distance dependence of the barrier for the molecule in our system is trivial compared with the contact parts. This is why in the modified barrier model we presented in Fig. 4, the major change in barrier length is related to the contact barrier. We agree that there are cases where the barrier for the molecule is distance-dominant, such as for charge transfer, but we still believe that in alkane molecular junctions, the dominating factor is the contact barrier when studying the slight mechanical extension of the junction. It should be safe to state that the modified model in Fig. 4 explains this situation well but may not be a generalized model for all junction systems.

1 K. Wang, J. Hamill, J. Zhou, C. Guo and B. Xu, *Faraday Discuss.*, 2014, DOI: c4fd00080c.
2 B. Xu, X. Xiao and N. J. Tao, *J. Am. Chem. Soc.*, 2003, **125**, 16164–16165.

Gemma Solomon opened the discussion of the paper by Henrik Ottosson: It is not clear to me how aromaticity/antiaromaticity determines whether the switch is "on" or "off". Rather, it seems that the nature of the pathway through the molecule and the presence/absence of saturated centres or cross-conjugation determines this. How should I understand the role of aromaticity/antiaromaticity in this context?

Henrik Ottosson replied: There is no specific connection between the aromaticity/antiaromaticity of a certain isomer and this isomer corresponding to "ON" or "OFF". Instead we use aromaticity/antiaromaticity in the lowest excited states to identify photoreactions that could be used in new photoswitches, *i.e.* we exploit the reversal in the aromaticity/antiaromaticity rules (Baird's rule) for the lowest $\pi\pi^*$ excited states when compared to the electronic ground state (S_0, Hückel's rule) to identify photoisomerizations that either lead to an increase in excited-state aromaticity or to a decrease in excited-state antiaromaticity. For example, in switch **1** we exploit the fact that benzene is antiaromatic in its lowest singlet and triplet excited states (S_1 and T_1) in contrast to the S_0 state where it is aromatic.[1] This antiaromaticity induces a photorearrangement to an *ortho*-quinoid isomer which is found as the central unit in **1-ON**. If (partially) π-conjugated substituents at the central benzene ring are positioned so that they, after the photoisomerization, provide a linearly conjugated path through the molecule, then this isomer pair can function as a molecular conductance switch.

1 H. Löfås, B. O. Jahn, J. Wärnå, R. Emanuelsson, R. Ahuja, A. Grigoriev and H. Ottosson, *Faraday Discuss.*, 2014, DOI: c4fd00084f.

Ravindra Venkatramani asked: It appears that the contrasting aromaticity/anti-aromaticity property of molecules such as benzene in the ground/excited state may be useful to monitor the stacking of aromatic molecules near molecular junctions. Such a situation was presented in Prof. Solomon's theoretical calculations.[1] Upon photo-excitation any strong conductance mediated by stacked geometries would drop dramatically with the current showing photo-switching behavior. Can you comment on the feasibility of this setup?

1 Q. Li and G. C. Solomon, *Faraday Discuss.*, 2014, DOI: c4fd00083h.

Henrik Ottosson replied: Yes, the π-stacking might lead to improved photodimerization, which is well-known for anthracene and which potentially could be used for photoswitching. The anthracene [4+4] photodimerization leads to a dimer where the C atoms in the 9,10-positions of the central rings bind, whereby they are turned into sp^3 hybridized C atoms.[1] This dimer can also be driven back to two separate anthracene monomers photochemically, so it is a photochromic system (where I think the backreaction is a photoreaction that alleviates the S_1-state antiaromaticity of the benzene rings in the dimer). However, I believe that the orientational factor is very important to get such a switch to function because it seems that the orientation which is optimal for photodimerization (structure B0 in the paper by Li and Solomon[2]) is the orientation that displays the strongest interference at the Fermi level (Fig. 4). One would likely have to design the system further, but it should definitely be possible.

1 http://en.wikipedia.org/wiki/Anthracene
2 Q. Li and G. C. Solomon, *Faraday Discuss.*, 2014, DOI: c4fd00083h.

Ioan Bâldea said: The discussion that is touched on in your paper in terms of Hückel and Baird rules has a counterpart in extensive work on tunable *meso*- and nano-scopic systems reported by our group.[1–5] Out of the many aspects that, in my opinion, might be relevant for your study, I would first mention the so-called quantum phase transitions discussed in our papers. To skip details that might not be so important for you, we calculate (we employ a schematic strongly correlated model but perform full CI calculations) the lowest energy of states with given symmetries, spins (singlet, triplet) or other properties (various so-called *e.g* charge density waves (CDW_+, CDW_-) and spin density waves (SDW_+, SDW_-)). You do not have level crossings (points in the parameter space where the total energies of different states become equal, which define the various quantum phase transitions) because you do not have tunable parameters, but you can perform, *e.g.* DFT calculations for "ground" (more properly, lowest energy) singlet and triplet states. Instead of employing highly questionable "energies" of Kohn–Sham "orbitals", you can employ differences between the total energies; DFT poses no problems for ground state calculations, and therefore these energy differences (which are similar to those used in my paper presented at this symposium[6]) do have physical meaning. Secondly, let me mention that the parameter values of our model specific to realistic molecules fall in the SDW region of our phase diagram.[7] Depending on the system (Hückel or anti-Hückel), if the ground state is a SDW-singlet, the lowest excitation is a SDW-triplet or *vice versa*. So, there are similarities to your results; noteworthly, our (CI) calculations exactly include strong correlations. SDW is a state where the spin density is modulated (thence "wave") from site to site (from atom to atom in your case). In this context, it might be interesting to know whether you also have significant spin modulations (differences in local spins) both in the ground state and in the lowest excited state. Again, because you can compute them within "ground" state DFT approaches, these results are meaningful.

1 I. Bâldea, *Phys. Rev. B*, 1999, **60**, 6646.
2 I. Bâldea, H. Köppel and L. S. Cederbaum, *Solid State Commun.*, 2000, **115**, 593–597.
3 I. Bâldea, H. Köppel and L. S. Cederbaum, *Phys. Rev. B*, 2001, **63**, 155308.
4 I. Bâldea, H. Köppel and L. S. Cederbaum, *Eur. Phys. J. B*, 2001, **20**, 289–299.
5 I. Bâldea, H. Köppel and L. S. Cederbaum, *Phys. Rev. B*, 2004, **69**, 075307.
6 I. Bâldea, *Faraday Discuss.*, 2014, DOI: c4fd00101j.
7 I. Bâldea and L. S. Cederbaum, Unusual features in optical absorption and photo-ionisation of quantum-dot nanorings, in *Frontiers in Quantum Systems in Chemistry and Physics*, ed. S. Wilson, P. J. Grout, G. Delgado-Barrio, J. Maruani and P. Piecuch, Springer Science + Business Media B. V., vol. 18, pp. 273–287.

Henrik Ottosson responded: Yes, there are definitely some very interesting analogies between Baird's rule on excited state aromaticity and antiaromaticity and what you have published earlier. Still, I will need to read the papers thoroughly before I can discuss the connection more substantially. At present I would recommend a valence bond theoretical paper by Zilberg and Haas,[1] in which they show that a triplet state [4*n*]annulene, which is Baird-aromatic, can be considered as a $4n\neg2$ Hückel-aromatic ring plus two nonbonding same-spin π-electrons. Also, I think that the paper by Borden and Davidson on disjoint and nondisjoint biradicals[2] might be a good start for further discussions on the singlet and triplet states of Hückel and anti-Hückel rings. Can this be connected to weak correlation and strong correlation, respectively? Did you study charged odd-membered rings such as the cyclopentadienyl cation? If so, what did you find?

1 S. Zilberg and Y. Haas, *J. Phys. Chem. A*, 1998, **102**, 10851–10859.
2 W. T. Borden and E. R. Davidson, *J. Am. Chem. Soc.*, 1977, **99**, 4587–4594.

Ioan Bâldea further commented: Thank you, I know those papers, and they may serve as a basis for further development. In the works of our group mentioned in my previous question, correlations play an essential role both for the phase diagram (ground state) and for excitations. It is possible that, in "natural" molecules, correlations are not so strong as in tunable quantum dot assemblies ("artificial molecules").

I did not systematically study odd-membered rings, but "snapshots" indicate that they considerably differ from those with an even number of sites. For example, they cannot be dimerized; with an odd number of sites, it is impossible to displace one atom in one direction and the next in the opposite direction with the same amount. The distortion is either incommensurate, or is a precursor of a kink–anti-kink pair (there is a short notice in I. Bâldea,[1] ref. 13) as extensively studied within the Su–Schrieffer–Heeger model in polyacetylene, or is non-planar. This has an obvious impact, since the molecular symmetry is affected. Adding charge ("anions") would presumably stabilize a (more) planar geometry in cases where the excess electron is a pi-electron, given the tendency for delocalization.

1 I. Bâldea, *Phys. Rev. B*, 1999, **60**, 6646.

Uli Lemmer asked: Has anybody shown in electrical measurements a clear optical switching of the molecular electronic junction?

Henrik Ottosson replied: Yes, there are several examples. The system which shows the highest switching ratio is the dimethyldihydropyrene switch studied by Royal, Wandlowski and co-workers[1] which is turned from ON to OFF by light and from OFF to ON thermally. This switch has an experimentally measured single-molecule switching ratio which is higher than 10^4. It was also shown to have an excellent reversibility in conductance switching. For additional examples on experimentally investigated optical conductance switching at the single-molecule or molecular assemblies level see references 8–16 in the paper by Royal and Wandlowski.[1] The class of molecules that was studied by Royal, Wandlowski and co-workers are photochromic as they also can be switched between the two isomers by light of different wavelengths (for the unsubstituted dimethyldihydropyrene switch, with wavelengths shorter than 313 nm and longer than 365 nm, respectively). I discuss this switch more later in my reply to the question from Prof. Wudl.

1 D. Roldan, V. Kaliginedi, S. Cobo, V. Kolivoska, C. Bucher, W. Hong, G. Royal and T. Wandlowski, *J. Am. Chem. Soc.*, 2013, **135**, 5974–5977.

Latha Venkataraman commented: We have measured the conductance of an anti-aromatic biphenylene system and found no significant difference between this and the aromatic analog.[1]

1 S. Schneebeli, M. Kamenetska, F. Foss, H. Vazquez, R. Skouta, M. Hybertsen, L. Venkataraman and R. Breslow, *Org. Lett.*, 2010, **12**, 4114–4117.

Henrik Ottosson responded: Our switches do not rely on a difference in conductance between the aromatic and antiaromatic isomers. Instead, the photorearrangements that we exploit for the photoswitching can be viewed as either the alleviation of excited-state antiaromaticity or the gain of excited-state aromaticity. Four of the molecular systems we calculated are designed such that these photorearrangements lead from isomers with low conductance to isomers with higher conductance, whereas one system is designed so that the photoreaction leads from an isomer with higher conductance to one with lower conductance.

Gemma Solomon remarked: The chemical structure does not seem to be a clear predictor for exactly how large the ON/OFF ratio of a switch will be in a DFT calculation. Do you have a sense for when the DFT calculations give surprising results and/or how reliable the chemical structure is?

Henrik Ottosson responded: Actually, I think old-school qualitative theory (MO and VB theory and conformational analysis) is very good at explaining when and when not a certain isomer pair will display a high ON/OFF ratio. In retrospect it is not so surprising that quadricyclane (**3-OFF**), which has two cyclopropane moieties side-by-side, has a high conductance according to DFT calculations because of coupling between the local (bent-bond) C–C orbitals. The **3-OFF** isomer has a seemingly saturated central unit which acts as a π-bonded unit. Similarly, the calculated low conductance of the **4-ON** isomer is not so surprising from a qualitative perspective because its cyclooctatetraene ring is strongly puckered, giving poor π-conjugation and a low conductance. In these two systems I do not think different levels of computations will give extensively different results, but it might be interesting to test (particularly the **3-ON**/**3-OFF** switch) how large are the variations in the computed conductance of **3-OFF** with a central quadricyclane moiety.

Eli Zysman-Colman remarked: Could triphenylene systems of the form Volhardt has studied[1] be useful systems for your molecular switch design?

1 B. Iglesias, A. Cobas, D. Pérez, E. Guitián and K. P. Vollhardt, *Org. Lett.*, 2004, **6**, 3557–3560.

Henrik Ottosson replied: Triphenylene (and biphenylene) derivatives are definitely interesting compounds, however, there needs to be a photoisomerization for the compounds to function as photoswitches. Alternatively, if in a π-stacked arrangement they could transform between a photodimer and two monomers. Biphenylene can undergo photodimerization but it seems to be a rather slow process (3 days of irradiation). With regard to triphenylene I cannot say how they would perform in that context. With regard to biphenylene it can actually be viewed as an "aromatic chameleon" which can adapt some aromatic character in the S_0, S_1 and T_1 states as it can be described as 2 × Hückel-aromatic (two benzene rings) in S_0 and 1 × Baird-aromatic in S_1 and T_1 (one peripheral 12π-electron ring).

Peter Skabara asked: With the introduction of quinoidal structures, what effect does this have on the stability of the molecules? Does this really matter if the molecule is isolated in the junction?

Henrik Ottosson replied: With regard to the first question, isomers with quinoidal structures will be of lower stability for two reasons; they will have a tendency to oligomerize and they will be prone to thermal back-rearrangements as S_0 state aromaticity is regained in that process. Oligomerization will definitely be avoided if the molecules are immobilized in an electrode gap with a sufficient distance between the two photoswitching molecules (or if the junction is single-molecular). However, if the experimental setup involves a solution with photoswitching molecules, then oligomerization will occur upon irradiation and this may complicate the conductance measurements as the solution would consist of a gradually higher concentration of various oligomers. Then, with regard to the intramolecular back-rearrangement, it seems this is possible to hinder by benzannulation as earlier experiments suggest this, and it is also supported by our calculations on the activation energies for the back-rearrangements. So with regard to the second question we think one also needs to be somewhat concerned about the stability of the quinoid structures when the molecules are immobilized in the electrode gaps.

Andrew Mount commented: In this paper the method of switch deactivation, or back reaction, discussed (from ON to OFF) is thermal. However, given the extensive literature about the fluorescence quenching of dyes (aromatic conjugated systems) near metal electrodes, would it not be highly likely that efficient deactivation of the excited state through a similar process would occur in this case? What is the prospect of developing design rules that would allow for this effect, leading to systems with the required molecular conductivity without the promotion of rapid deactivation?

Henrik Ottosson answered: Yes, it is definitely reasonable to expect rapid deactivation of the excited state by the metal. It is difficult to say how easily one can develop such design rules. In the switch of Royal, Wandlowski and co-workers[1] they worked with a large π-conjugated dimethyldihydropyrene connected *via* pyridine anchor groups to Au electrodes. This features π-conjugation that (on paper) stretches from one electrode to the other. The pyridine anchor groups may possibly be rotated somewhat relative to the central dimethyldihydropyrene moiety which would weaken the conjugation leading from one electrode to the other. Yet, Royal, Wandlowski and co-workers selected the pyridine groups as they "enable optimized electron transport through the functional molecular wire."

1 D. Roldan, V. Kaliginedi, S. Cobo, V. Kolivoska, C. Bucher, W. Hong, G. Royal and T. Wandlowski, *J. Am. Chem. Soc.*, 2013, **135**, 5974–5977.

Gemma Solomon stated: There seems to be a fundamental challenge for photoswitches in single-molecule junctions. On the one hand, we need a molecule strongly coupled to two electrodes to yield significant current and a clear "on" state for the switch. On the other hand, the switching element in the molecule needs to be sufficiently isolated from the electrodes to achieve reversible switching without quenching of excited states by the electrodes.

Henrik Ottosson responded: I think the switch reported by Royal, Wandlowski and co-workers[1] is very important in this context because this switch operates photochemically in one direction, from the ON to the OFF state, yet it has pyridine anchor groups linked directly to the central dimethyldihydropyrene moiety, which is the photoswitching element. I think this compound holds a solution on how to design strongly conjugated photoswitches which, in addition, have good contacts to the metal electrodes. It should be very important to investigate the excited state of the ON state (both computationally and experimentally). For example, is the S_1 state only localized to the dimethyldihydropyrene moiety? What is its excited state lifetime? Are the pyridine anchor groups decoupled from the dimethyldihydropyrene moiety by large dihedral angles between the rings?

1 D. Roldan, V. Kaliginedi, S. Cobo, V. Kolivoska, C. Bucher, W. Hong, G. Royal and T. Wandlowski, *J. Am. Chem. Soc.*, 2013, **135**, 5974–5977.

Justin Hodgkiss commented: When an excited state surface induces nuclear rearrangement, it is often the case that this excited state motion couples strongly to high lying vibrational states of the electronic ground state, particularly when the electronic surfaces approach crossing. The end result would be a rapid return to the original ground state *via* vibrational cooling without ever crossing the isomerization barrier.

Do you calculate the full potential energy surfaces along the rearrangement coordinate in order to ensure that they are sufficiently separated to prevent non-productive non-radiative relaxation to the original ground state?

Henrik Ottosson replied: No, in the present study[1] we exclusively investigated the various ON and OFF isomers in their S0 states. For a few of the switches, the transition states on the S0 surfaces for the thermal back-reaction were located. Our present study was aimed at identifying compound pairs that would show high ON/OFF switching ratios in our NEGF calculations. The photoreactions that our different potential switches are based upon have all previously been reported to occur experimentally (yet, in different contexts to optical switching). The confinement into an electrode-electrode gap may change the photoreactivity drastically, however, we believe that this is best examined experimentally. Based on our calculations we find that only candidates 1 and 5 are tentatively interesting, and our next step is now their synthesis and experimental testing. Once we have the experimental results, regardless if they are positive or negative from the perspective of conductance switching, it would definitely be very interesting to carry out calculations on the excited state processes so as to study how the photoreaction progresses and to figure out if it could be modified so as to increase the quantum yield for the photoswitching.

1 H. Löfås, B. O. Jahn, J. Wärnå, R. Emanuelsson, R. Ahuja, A. Grigoriev and H. Ottosson, *Faraday Discuss.*, 2014, DOI: c4fd00084f.

Uli Lemmer remarked: You described conformational switching and indicated that this might be problematic in a real device. Why is this a problem? There are real-world devices (*e.g.* piezo devices) where small mechanical movements continuously occur.

Henrik Ottosson answered: I was misleading if I talked about "real devices" because I was referring to single molecule conductance measurements through a mechanically controllable break-junction setup. What we want to identify are molecular switches where the distances between the two anchor groups (SH groups) do not change significantly along the reaction path. A switching reaction which progresses by a large shortening of the intramolecular distance between the two anchor groups at some intermediate point along the reaction path will not be suitable because these intermediate positions on the potential energy surface will be raised in energy, due to the confinement of the molecule to an electrode–electrode distance which is (much) longer than what is ideal for the intermediate positions. This will affect the efficiency of the switching process in one way or the other when compared to the switching reaction where the molecule is not constrained into an electrode gap.

Gemma Solomon remarked: It is certainly possible to use changes in the length of the molecule to design a photoswitch. In this case, however, it may be only the length of the molecule that determines which state is "on" and which is "off", rather than the chemical structure of the path through the molecule. For this type of switch, we would want to maximise the length change with switching rather than minimise it.

Henrik Ottosson answered: Yes; if it is the length of a flexible conjugated molecule that changes in the photoswitching, it might be the confinement of this molecule in its long isomeric form (state) to an electrode–electrode distance which is too short for optimal conjugation that gives the ON/OFF conductance ratio. For such a switch, the length change between its ON and OFF states should be as large as possible. One can of course also combine different factors that lead to conductance differences between the ON and OFF states. In the switches that we investigated, we had a desire to keep the distance between the anchor groups along the reaction path as constant as possible as we argued that a drastically altered distance, for a molecule confined between two metal electrodes, would impact the profile of the potential energy surface. This would change the reaction rate for the switching (this might be good or it might be bad).

Fred Wudl asked: Have you tried photochromics? It is known that one can "turn on" conjugation by irradiation with a particular wavelength and "turn off conjugation" with a different wavelength. If the turn off wavelength is near IR, it would be the equivalent of heat.

Henrik Ottosson answered: Among the systems that we presently report, we included no (known) photochromic systems as we simply wanted to tailor the activation energies for the thermal back-rearrangements (and thus the thermal stabilities) through DFT transition state calculations. However, several of the optically active molecular conductance switches that already have been reported are based on photochromic compounds such as diarylethenes. One photochrome which I think is particularly interesting in the context of excited-state aromaticity/

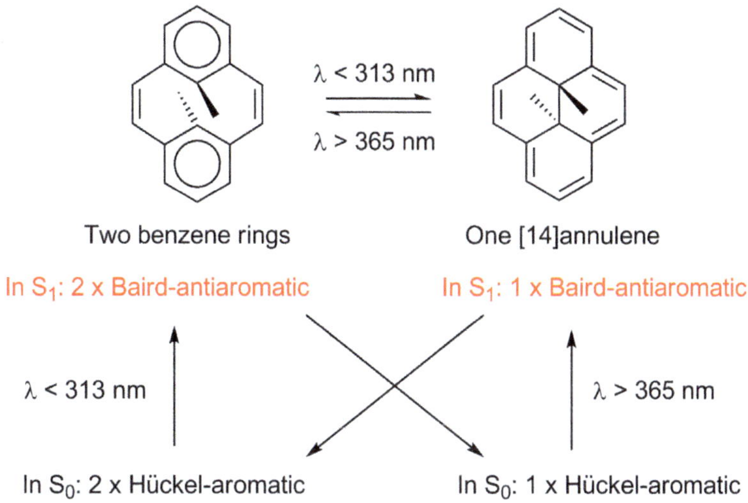

Fig. 3 The dimethyldihydropyrene/cyclophanediene switch is a photochromic system where both isomers rearrange photochemically so as to alleviate S_1-state antiaromaticity, leading to the opposite isomer. Both isomers are S_0-state aromatic and S_1-state anti-aromatic, *i.e.*, they both have "Jekyll and Hyde" character.

antiaromaticity is dimethyldihydropyrene, as this switch goes between a closed form which is a [14]annulene in its perimeter and an open form composed of two essentially isolated benzene rings. Both photoisomerizations can be viewed as relieving S_1-state antiaromaticity. I have described this in Fig. 3. This system was used in the single-molecule conductance switch of Royal, Wandlowski and co-workers,[1] although they used photochemistry in only one direction (from ON to OFF). Expanding on the same principle, one may possibly design photochromic systems where one isomer absorbs in the near-IR region.

1 D. Roldan, V. Kaliginedi, S. Cobo, V. Kolivoska, C. Bucher, W. Hong, G. Royal and T. Wandlowski, *J. Am. Chem. Soc.*, 2013, **135**, 5974–5977.

Faraday Discussions

PAPER

Organic semiconductor distributed feedback laser pixels for lab-on-a-chip applications fabricated by laser-assisted replication

Xin Liu,[ab] Stephan Prinz,†[a] Heino Besser,[cd] Wilhelm Pfleging,[cd] Markus Wissmann,[bd] Christoph Vannahme,‡[ba] Markus Guttmann,[bd] Timo Mappes,§[b] Sebastian Koeber,[be] Christian Koos[be] and Uli Lemmer*[ab]

Received 22nd April 2014, Accepted 3rd June 2014
DOI: 10.1039/c4fd00077c

The integration of organic semiconductor distributed feedback (DFB) laser sources into all-polymer chips is promising for biomedical or chemical analysis. However, the fabrication of DFB corrugations is often expensive and time-consuming. Here, we apply the method of laser-assisted replication using a near-infrared diode laser beam to efficiently fabricate inexpensive poly(methyl methacrylate) (PMMA) chips with spatially localized organic DFB laser pixels. This time-saving fabrication process enables a pre-defined positioning of nanoscale corrugations on the chip and a simultaneous generation of nanoscale gratings for organic edge-emitting laser pixels next to microscale waveguide structures. A single chip of size 30 mm × 30 mm can be processed within 5 min. Laser-assisted replication allows for the subsequent addition of further nanostructures without a negative impact on the existing photonic components. The minimum replication area can be defined as being as small as the diode laser beam focus spot size. To complete the fabrication process, we encapsulate the chip in PMMA using laser transmission welding.

[a]*Light Technology Institute (LTI), Karlsruhe Institute of Technology, 76131 Karlsruhe, Germany. E-mail: uli.lemmer@kit.edu*

[b]*Institute of Microstructure Technology (IMT), Karlsruhe Institute of Technology, 76344 Eggenstein-Leopoldshafen, Germany*

[c]*Institute for Applied Materials - Applied Materials Physics (IAM-AWP), Karlsruhe Institute of Technology, 76344 Eggenstein-Leopoldshafen, Germany*

[d]*Karlsruhe Nano Micro Facility, Hermann-von-Helmholtz-Platz 1, 76344 Eggenstein-Leopoldshafen, Germany*

[e]*Institute of Photonics and Quantum Electronics (IPQ), Karlsruhe Institute of Technology, 76131 Karlsruhe, Germany*

† Currently with TRUMPF Scientific Lasers GmbH + Co. KG, 85774 Unterföhring-München, Germany.

‡ Currently with DTU Nanotech, Technical University of Denmark, 2800 Kgs. Lyngby, Denmark.

§ Currently with Carl Zeiss AG, Corporate Research and Technology, 07745 Jena, Germany.

1 Introduction

Organic semiconductor distributed feedback (DFB) lasers are of particular interest as tunable laser emitters covering the whole visible spectral range.[1–5] Enabled by the efficient conversion in their active material, low threshold organic semiconductor lasers allow for free space and integrated excitation sources in spectroscopic investigations.[6–9] In particular, the integration of miniaturized laser light sources in all-polymer lab-on-a-chip (LOC) platforms is promising for point-of-care systems.[10–14] In order to realize laser pixels, technologies for local definition of the active material, and for the laterally controlled fabrication of DFB structures, have to be developed.

In terms of the gain materials, small molecules such as tris(8-hydroxyquinoline) aluminum (Alq_3) doped with the laser dye 4-(dicyanomethylene)-2-methyl-6-(*p*-dimethylaminostyryl)-4*H*-pyran (DCM) compete with conjugated polymers. The former class of materials can be evaporated locally using shadow masking, and can thus be combined with pre-defined grating structures. On the other hand, conjugated polymers can be processed by inexpensive solution-based techniques such as spin coating, dip-coating and doctor blading. These methods, however, result in an organic semiconductor layer covering the whole substrate. Lateral control can be achieved using printing techniques. A particularly interesting approach is the use of ink-jet printing. Ink-jet printing has already been successfully applied to the fabrication of various organic electronic devices, such as organic transistors,[15,16] organic solar cells[17,18] and organic light-emitting diodes.[19,20] We have recently successfully fabricated organic semiconductor distributed feedback laser pixels by ink-jet printing the active layer from a conjugated polymer solution. Utilizing an optimized mixture of high-boiling and low-boiling solvents for dissolving the polymer, the ink-jet-printed film profile is optimized, thus creating uniformly emitting organic lasers. Such devices show lasing thresholds as low as 76 nJ per pulse and a spectrally homogeneous laser emission over large areas.[21]

The laterally controlled fabrication of DFB gratings for organic semiconductor lasers has only been addressed so far by using nanograting transfer to transfer the gratings onto a homogeneous gain material layer,[22] and a rather complex approach using direct electron beam lithography (EBL) on conjugated polymers.[23] On the other hand, various substrate scale techniques such as laser interference lithography (LIL)[24] and nanoimprint lithography (NIL)[25] are available to achieve the low-cost fabrication of high-quality DFB corrugation nanostructures on an LOC platform. Among them, NIL promises the mass-production of extremely fine structures with feature sizes down to 10 nm.[26] The resolution of NIL can be further improved by the fabrication of a rigid mold with sub-10 nm features.[25] Nowadays, two main subjects of NIL are frequently used in the fabrication of DFB corrugations for organic semiconductor laser applications: thermal nanoimprint lithography (TNIL, also known as hot embossing) and UV-assisted nanoimprint lithography (UV-NIL).[27–30] UV-NIL is an option to fabricate localized DFB corrugations without negative effects on the existing structures, but it is mandatory to use special photoactive materials in the process. This limitation results in a narrow application field in the fabrication of LOC systems. TNIL is advantageous for parallel replication of all structures from a master stamp onto a common polymer substrate. Nevertheless, it has three major drawbacks. First is the difficulty in the fabrication of pre-defined partial structures from a rigid master stamp with structures for different purposes. Secondly, the large-area heating can be detrimental for existing LOC passive photonic elements and microfluidic channels. The third issue is the long heating and cooling time.

These limitations of TNIL can be overcome using a near-infrared high-power diode laser beam.[31,32] In this work, we demonstrate a novel laser-assisted replication to fabricate localized surface-emitting (2nd order) and edge-emitting (1st order) organic distributed DFB laser pixels on a poly(methyl methacrylate) (PMMA) substrate. Our technique allows for fast replication with standard TNIL materials due to the localized heating. Not only is the process time reduced, the localized heating also provides the unique benefit to replicate only parts of the existing structures on a master stamp. Additionally, since only the defined areas are to be heated, it allows for subsequent additive fabrication of DFB corrugations while all the other areas are retained. The minimum localized replication area and the influence of the process parameters on corrugation quality are investigated. A functional LOC is finished after encapsulation *via* laser transmission welding, protecting the device from air and water.[33–35]

2 Fabrication process

In laser-assisted replication, a rigid mold, *e.g.*, a silicon master stamp, is pressed under a polymer substrate, and the contact area is subsequently heated up by a near-infrared laser beam from above, as shown in Fig. 1(a). Due to heat conduction, the temperature of the polymer material, which is in direct contact

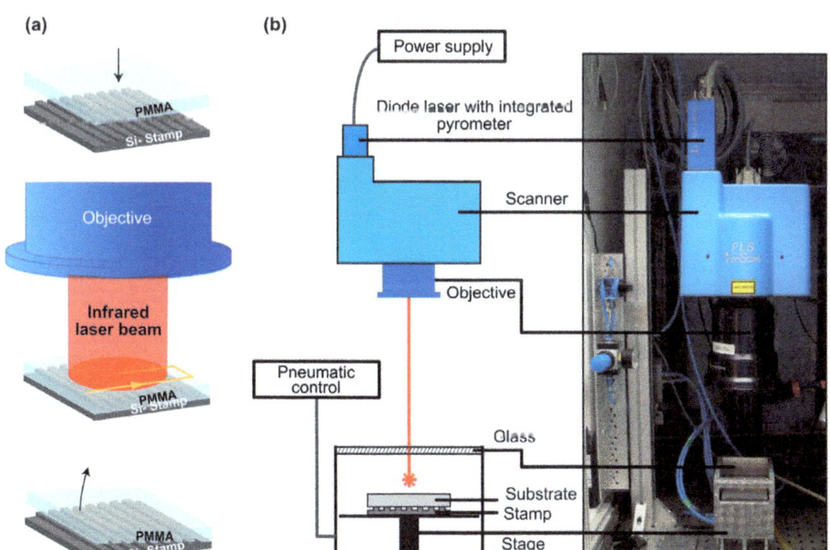

Fig. 1 (a) Scheme of the laser-assisted replication process: positioning of the PMMA substrate on a silicon master stamp (top), execution of the replication (middle), and detachment of the sample (bottom). (b) Schematic illustration (left) and photograph (right) of the applied laser system.

with the master stamp, increases. When the temperature exceeds the polymer glass transition temperature, the nanostructures on the rigid mold are replicated into the polymer substrate. After cooling down, the polymer substrate is detached from the rigid mold, while the nanostructures are preserved.

To perform the experiments, a high-power diode laser system (Fisba Optik, FLS Iron High Power Diode) with an emission wavelength of 940 nm and a maximal power of 50 W is used. As shown in Fig. 1(b), the diode laser and a pyrometer, which provides *in situ* monitoring of the temperature at the laser focus plane, are integrated within the laser scanner unit. The PMMA substrate and the silicon master stamp are fixed in a pneumatic stage, applying a constant pressure. There are two basic operation modes of the system: constant laser power or constant heating temperature. In our case, the temperature regulation is chosen in order to achieve a homogeneous temperature distribution at the contact surface. Utilizing an F-Theta objective lens (Fisba Optik) with a focal length of 163 mm, the size of the elliptical laser focus spot on the work plane is 0.55 mm × 0.7 mm. Attributed to the applied scanner system, the maximum laser processing area can reach 100 mm × 100 mm.

Before replication, a silicon master stamp was fabricated. *Via* electron beam lithography and reactive ion etching (RIE), the DFB corrugations with grating periods ranging from 195 nm to 450 nm, a height of ~120 nm and a duty cycle of 75% were generated. *Via* aligned UV-lithography and subsequent RIE on the same wafer, basins for the DFB lasers with a height of 1.2 μm were completed. A thin anti-adhesion layer was then deposited on the wafer to facilitate the polymer substrate detachment.[13]

For the replication process, a 2 mm-thick PMMA substrate (Evonik, Plexiglas® XT) with a size of 30 mm × 30 mm was positioned on selected corrugations of the silicon master stamp. The following process parameters resulted in the highest-quality replication: a heating temperature of $T = 180$ °C, a laser scan velocity of $v = 40$ mm s^{-1}, a clamping pressure of $p = 0.1$ MPa and a hatch distance (offset between two neighboring laser tracks) of $\Delta x = 300$ μm. To achieve a homogeneous replication, each substrate was treated twice by the laser beam: firstly perpendicular to the grating orientation, as shown in Fig. 1(a), and secondly parallel to the grating orientation. The replication process took about 5 min for the

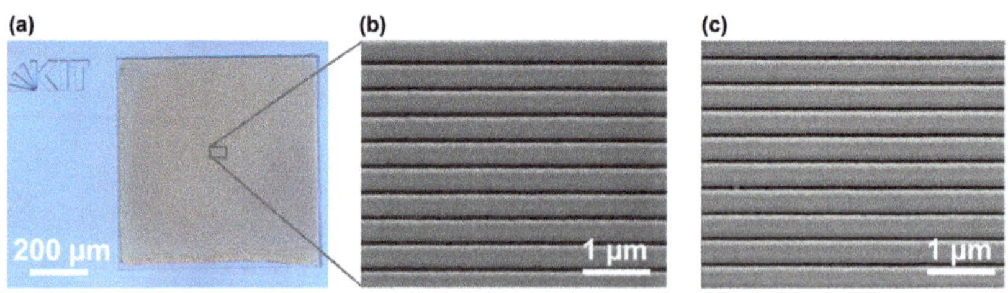

Fig. 2 (a) Microscope image of a one-dimensional PMMA DFB grating with a period of $\Lambda = 400$ nm fabricated by laser-assisted replication. (b) Scanning electron microscope image of the grating structures fabricated by laser-assisted replication. For comparison, (c) shows a scanning electron microscope image of the $\Lambda = 400$ nm grating structures fabricated by thermal nanoimprint lithography (equipment built by Institute of Microstructure Technology-KIT) using the same silicon master stamp as for (b). A temperature of 180 °C and a pressure of 1.85 MPa were applied for 15 min. The total TNIL process lasted for ~45 min.

replication of an area of 40 mm × 40 mm. After replication, the clamping pressure was released. The PMMA substrate was cooled down with a nitrogen spray gun and separated from the silicon master stamp. Fig. 2(a) and (b) show a microscope image and a scanning electron microscope image on PMMA of a grating with a period of $\Lambda = 400$ nm. The grating structures fabricated *via* laser-assisted replication show a high fidelity over the whole molding area. The replicated quality was similar to thermal nanoimprint lithography, as shown in Fig. 2(c). However, the laser-assisted replication process took a much shorter time compared to the 45 min-long TNIL process. Alq_3 doped with the laser dye DCM has been chosen as the active material. It forms a very efficient Förster energy transfer system and exhibits an excellent long-term stability. Evaporating 250 nm Alq_3 doped with 2.8 wt% DCM onto the defined grating fields finally yielded the organic DFB laser pixels. The used stencil shadow mask was fabricated by UV-lithography and nickel electroplating.[14]

3 DFB laser characterization

For optical characterization, the fabricated surface-emitting organic DFB lasers were excited by a diode-pumped, actively Q-switched frequency tripled neodymium:yttrium orthovanadate ($Nd:YVO_4$) laser (Advanced Optical Technology Ltd., AOT-YVO-20QSP) with a wavelength of 355 nm. The pump pulse had a duration of approximately 1 ns at a repetition rate of 1 kHz. The pump pulse energy was adjusted with a variable neutral density filter and measured with a calibrated gallium arsenide phosphide photodiode connected to an oscilloscope (Tektronix, TDS2024C). The sample was kept in a vacuum chamber ($<5 \times 10^{-3}$ Pa) to protect the active material from photooxidation. The chamber can be shifted in the plane perpendicular to the excitation beam using an automated precision stage. This allows for a spectrally and spatially resolved characterization of the samples. A focusing lens was used to adjust the excitation spot size. The emission from the sample was collected using the focusing lens for the pump beam, then directed through a dichroic mirror and coupled into a multimode optical fiber. Further on, the laser spectra were analyzed using a spectrograph (Acton Research Corporation, SpectraPro 300i, variable grating) connected to an intensified charge-coupled device camera (Princeton Research, PiMax 512). The pump spot showed a slightly elliptical shape with a size of 30 μm × 40 μm on the sample.

Fig. 3(a) shows the color-encoded spatially resolved lasing wavelength from a PMMA chip containing 9 laser pixels with individual lateral dimensions of 500 μm × 500 μm and an interspacing of 500 μm. The grating period of the DFB corrugations varies from 370 nm to 450 nm in steps of 10 nm from the outermost left pixel to the outermost right pixel. The chip was probed with a resolution of 100 μm. The laser emission wavelength did not show any deviation on the area of a single laser pixel within the spectral resolution of the setup (approximately 0.5 nm). This indicates a very uniform replication of the nanogratings in PMMA. The corresponding laser emission spectra and the laser threshold values for the TE_0-modes are shown in Fig. 3(b) and (c). The blue-shifted TM_0-mode laser peaks identified with a polariser can be detected at $\lambda = 630.4$ nm and $\lambda = 643.6$ nm. These peaks originate from the lasers with grating periods of $\Lambda = 400$ nm and $\Lambda = 410$ nm. The higher laser threshold from the laser pixel with a grating period

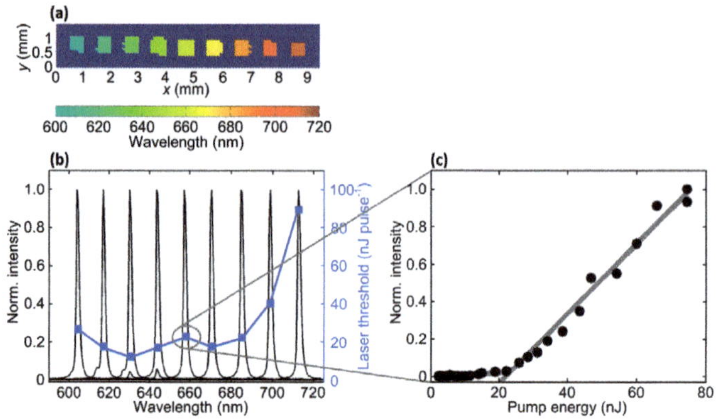

Fig. 3 (a) Color-encoded spatially resolved laser wavelengths of the fabricated organic DFB laser pixels with grating periods from 370 nm (left) to 450 nm (right) in steps of 10 nm. (b) Laser spectra and TE_0-mode laser threshold corresponding to the grating distribution. For some gratings the onset of the TM_0-mode can also be seen as very small peaks. (c) Input–output characteristics of a laser pixel with a grating period of $\Lambda = 410$ nm and a laser emission at $\lambda = 657.5$ nm.

of 450 nm is attributed to the lower gain coefficient at the emission wavelength of 712.5 nm.

For being suitable for LOC spectroscopic applications, the organic semiconductor lasers are usually based on first-order DFB gratings for the emission of light in the chip plane only. By introducing a waveguide onto the chip, the first-order organic laser emission can be efficiently coupled and guided to the analyte sites. PMMA chips with deep ultraviolet (DUV)-induced waveguides have been fabricated previously.[11,13] In this work, we fabricated the active grating and waveguides altogether *via* laser-assisted replication and evaporated Alq_3:DCM on both of them to build a coupled edge-emitting organic laser on chip. This novel configuration allows for a one-step fabrication of the laser and waveguide without mix-and-match pattern-related defects.[36] Besides, the direct connection between the laser and waveguide made from the same active material will facilitate an optimum coupling efficiency.

Different gratings with lateral dimensions of 500 μm, interconnected by 300 μm-wide waveguide structures perpendicular to the grating orientation, were simultaneously fabricated in 1.2 μm-deep basins on the PMMA chip. The grating periods for the first-order organic DFB laser were chosen to be $\Lambda = 195$ nm, $\Lambda = 200$ nm, and $\Lambda = 205$ nm, as marked in Fig. 4(a). A 250 nm-thick layer of Alq_3:DCM was evaporated on the defined grating and waveguide, thus allowing for an optimum light coupling.

The experimental setup for the chip characterization is schematically depicted in Fig. 4(b). The pump laser and the detection system were the same as used for the characterization of the surface-emitting organic lasers, but we changed the detection plane according to the chip orientation. The excitation spot diameter on the sample surface was ~100 μm. In order to limit laser degradation in the ambient atmosphere, we set the excitation repetition rate as low as 100 Hz. Fig. 4(c) shows the laser spectrum of a DFB laser with a grating period of

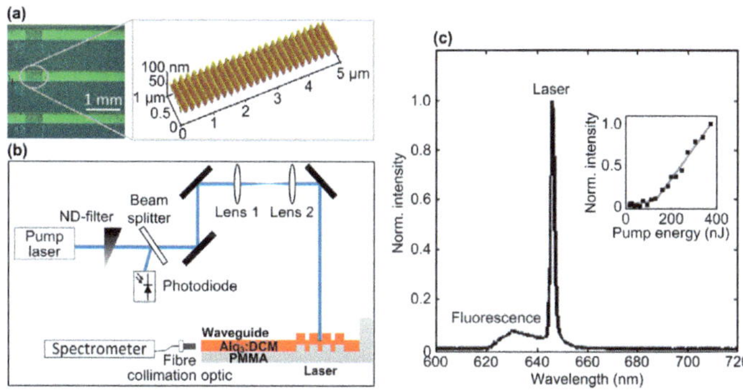

Fig. 4 (a) Microscope image of a coupled organic DFB edge-emitting laser (left) and an atomic force micrograph of the nanograting with a period of Λ = 200 nm (right). (b) Schematic illustration of the optical setup to detect the edge emission from the integrated organic DFB laser. (c) Laser spectrum and input–output curve (inset) taken from the chip facet of a grating of Λ = 200 nm using a multimode optical fiber coupled to a spectrometer.

Λ = 200 nm. The laser emission at λ = 645.5 nm was detected at the chip facet using a multimode optical fiber (Ocean Optics, P400-3-UV-VIS). A stronger background due to the Alq_3:DCM fluorescence than in the case of the surface-emitting lasers is observed. This can be explained by the fact that the fluorescence is coupled into the waveguide with comparable efficiency to the laser radiation. In the case of surface-emitting lasers, the fluorescence is almost not detected as it is emitted in all directions. The corresponding laser threshold was measured to be 145 nJ per pulse. This higher threshold can be explained by the laser being pumped at a larger excitation spot size than for the surface-emitting lasers.

4 Size dependence of laser-assisted replication

The minimum replication area size is determined by the applied laser beam diameter, the used processing time, and the laser-induced heating temperature in the contact zone of the master stamp and polymer. To investigate the spatial resolution limits of the replication process, we fabricated test structures with a tool comprising spatially defined DFB grating areas *via* laser-assisted replication using different heating periods. Fig. 5(a) shows a microscope image of the beginning trail of the replicated structure, which was obtained by laser-induced heating of a single line with a scanning speed of 40 mm s^{-1}, corresponding to a ~17.5 ms-long local heating. The red line indicates a fit to the right end of the replicated structures. The rim is in good accordance with the elliptical laser focus spot size of 0.55 mm × 0.7 mm.

In comparison, applying 2 s-long punctuated heating resulted in an expansion of the heat-affected zone on the silicon stamp and a consequently enlarged replication area. As shown in Fig. 5(b), an extension was measured as ~3.8 times that of the laser focus spot size.

We verified the versatility of the localized replication by the fabrication of pre-selected laser pixels. Fig. 5(c) shows a partial replication from the silicon master stamp with gratings in periods of Λ = 390 nm, Λ = 400 nm, Λ = 410 nm and

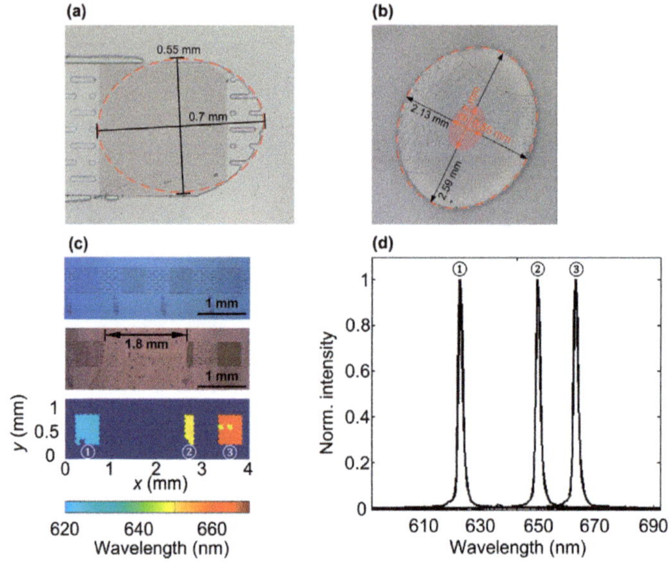

Fig. 5 Microscope images of (a) the beginning of the replication trail obtained by heating a single line with a laser scanning velocity of 40 mm s^{-1}, and (b) the replicated spot after 2 s-long punctuated heating in comparison with the laser focus spot (red). (c) Photograph of the complete set of spatially defined laser gratings (from left to right: Λ = 390 nm, Λ = 400 nm, Λ = 410 nm and Λ = 420 nm) on a silicon master stamp (top). The actual laser grating areas are visible as homogeneous darker grey areas which are separated by auxiliary structures in between. Photographs of the PMMA laser pixels partially replicated on a PMMA chip (middle) and spatially resolved laser wavelengths of the fabricated organic DFB laser pixels (bottom) are also shown. (d) Laser spectra corresponding to the grating areas ①, ② and ③ in (c).

Λ = 420 nm. Each pixel has a dimension of 500 µm × 500 µm with an interspacing of 500 µm. A 2 mm-broad space was excluded from the laser writing path from left to right on the chip. As a result, we observed an unstructured area with a width of 1.8 mm between the replicated structures. The slight difference between the size of the unstructured area and the non-irradiated area is attributed to the heat conduction in the silicon master stamp. The spatially resolved lasing wavelength distribution revealed one completely omitted and one only partially available laser pixel. The latter had a clearly defined border. The corresponding laser pixel spectra are shown in Fig. 5(d).

As previously mentioned, a great advantage of the localized replication is the subsequent fabrication of nanostructures with already existing photonic, fluidic or electronic components, which remain unaffected during the fabrication process. We demonstrated this by replicating the same structures twice on a single 2 mm-thick PMMA substrate and performing two successive replication processes at a distance of 10 mm. Each replicated area was defined to be 5 × 20 mm^2 without overlapping each other, as shown in Fig. 6(a). Fig. 6(b) shows a photograph of the finished laser pixels on the PMMA chip, which can be clearly identified by the interference effect. A 250 nm-thick Alq_3:DCM layer was evaporated on top of the chip to check the laser behavior and determine the replication quality. All the laser pixels showed homogeneous laser emissions throughout the

Fig. 6 (a) Scheme of the first and second laser-assisted replication areas of size 5 mm × 20 mm. (b) Photograph of the subsequently fabricated grating structures 1 and 2. (c) Finished chip after encapsulation with a PMMA lid *via* laser transmission welding. The chip edge appears silver due to the absorption layer of carbon.

whole pixel area with corresponding wavelengths according to their DFB grating periods. Laser pixels featuring the same grating period revealed comparable characteristics, *e.g.*, the laser thresholds of the laser pixels with a grating period of $\Lambda = 420$ nm were measured to be 17.3 nJ per pulse and 14.8 nJ per pulse, both emitting at the same laser peak wavelength of $\lambda = 670$ nm. The existing grating structures were not influenced by the subsequent replication process. This can be widely useful for LOC and other integrated optics applications. The inexpensive fabrication of identical nanostructures without repetitive construction on a master stamp is shown to be possible.

5 Encapsulation *via* laser transmission welding

During operation, organic semiconductors suffer from photoluminescence degradation when exposed to air or water. Therefore, encapsulation is necessary to sustain the long lifetime of the device. Previously, LOCs containing organic semiconductor lasers have been encapsulated in a polymer using thermal bonding,[12,13] which is usually time-consuming and therefore only suitable for parallel processing with large numbers of LOC chips. In addition, the high temperature used for bonding leads to bleaching of the organic dyes. Here, we demonstrate the encapsulation of our fabricated organic laser chip *via* laser transmission welding.[33] For this purpose, the same setup as for laser-assisted replication was used, as shown in Fig. 1. A 2 mm-thick PMMA substrate was prepared as a lid and placed on top of the previously fabricated chip with laser-pixel basins. Since PMMA is transparent to near-infrared radiation, an additional absorption layer of carbon with a thickness of 5–10 nm was deposited under the polymer chip, where the welding process was to be performed (chip edge). Applying the diode laser beam at a wavelength of $\lambda = 940$ nm and a maximum power of 50 W on the chip edge with a 5 mm-wide margin, the absorption layer converted the laser radiation into heat, which melted the two PMMA substrates along the contact area and bound them together (heating temperature $T = 140$ °C, laser scan velocity $v = 40$ mm s^{-1}; clamping pressure $p = 0.4$ MPa, hatch distance $\Delta x = 300$ μm). A photograph of the encapsulated organic laser chip is shown in Fig. 6(c). As expected, the encapsulation process was found not to affect the emission properties of the laser pixels. Laser transmission welding is thus not only a practical tool for chip bonding direct after laser-assisted replication, but also can be a substitute for thermal bonding after the thermal nanoimprint lithography process.

6 Conclusion

In summary, we have successfully applied laser-assisted replication in the fabrication of spatially localized organic semiconductor DFB laser pixels on PMMA chips. We accomplished nine surface-emitting laser pixels based on nanogratings with different periods in lateral dimensions of 500 μm × 500 μm on a PMMA substrate. We also realized the simultaneous fabrication of nanoscale DFB corrugations and neighboring microscale waveguide basins. After evaporation of the same active material, coupled edge-emitting organic lasers were achieved with a combination of the functional waveguides. The minimum fabrication area size was investigated by comparing the replication spot with the laser beam focus size, and the versatility of the spatially localized replication was verified by the fabrication of pre-selected laser pixels. We also proved that using this fabrication method, further nanostructures could be added into the chip platform without a negative influence on the neighboring photonic components. Consequently, laser-assisted replication is an ideal tool for the fast and flexible fabrication of micro- and nanostructures with high quality, and is especially suited for the rapid prototyping of LOC platforms. Finally, we completed the chip fabrication by encapsulation of the polymeric chip using laser transmission welding. This progress in the additive fabrication of gratings can ideally be combined with our recent progress in ink-jet printing the active gain medium of organic semiconductors. By using different laser pixels, this approach allows for the realization of digitally manufactured versatile coherent light sources integrated in photonic microsystems and LOC devices.

Acknowledgements

The authors thank A. Bacher, P.-J. Jakobs and A. Muslija for the silicon master stamp fabrication and P. Abaffy for taking the scanning electron microscope images. This work was supported by the Deutsche Forschungsgemeinschaft and the State of Baden-Württemberg through the DFG-Center for Functional Nanostructures (CFN). The work of X. L. is supported by Carl Zeiss Stiftung and Karlsruhe School of Optics & Photonics (KSOP). We acknowledge support by Deutsche Forschungsgemeinschaft and the Open Access Publishing Fund of Karlsruhe Institute of Technology. Finally, the support for laser processing by the Karlsruhe Nano Micro Facility (KNMF, http://www.knmf.kit.edu), a Helmholtz research infrastructure at the Karlsruhe Institute of Technology is gratefully acknowledged.

References

1 N. Tessler, G. J. Denton and R. H. Friend, Lasing from conjugated-polymer microcavities, *Nature*, 1996, **382**, 695–697.

2 F. Hide, M. A. Diaz-Garcia, B. J. Schwartz, M. R. Andersson, Q. B. Pei and A. J. Heeger, Semiconducting polymers: a new class of solid-state laser materials, *Science*, 1996, **273**, 1833–1836.

3 S. V. Frolov, M. Ozaki, W. Gellermann, Z. V. Vardeny and K. Yoshino, Mirrorless lasing in conducting polymer poly(2,5-dioctyloxy-*p*-phenylenevinylene) films, *Jpn. J. Appl. Phys.*, 1996, **35**, L1371–L1373.

4 I. D. W. Samuel and G. A. Turnbull, Organic semiconductor lasers, *Chem. Rev.*, 2007, **107**, 1272–1295.
5 S. Klinkhammer, X. Liu, K. Huska, Y. Shen, S. Vanderheiden, S. Valouch, C. Vannahme, S. Bräse, T. Mappes and U. Lemmer, Continuously tunable solution-processed organic semiconductor DFB lasers pumped by laser diode, *Opt. Express*, 2012, **20**, 6357–6364.
6 Y. Oki, S. Miyamoto, M. Maeda and N. J. Vasa, Multiwavelength distributed feedback dye laser array and its application to spectroscopy, *Opt. Lett.*, 2002, **27**, 1220–1222.
7 T. Woggon, S. Klinkhammer and U. Lemmer, Compact spectroscopy system based on tunable organic semiconductor lasers, *Appl. Phys. B*, 2010, **99**, 47–51.
8 S. Klinkhammer, T. Woggon, C. Vannahme, T. Mappes and U. Lemmer, Optical spectroscopy with organic semiconductor lasers, *Proc. SPIE*, 2010, **7722**, 77221I.
9 X. Liu, P. Stefanou, B. Wang, T. Woggon, T. Mappes and U. Lemmer, Organic semiconductor distributed feedback (DFB) laser as excitation source in Raman spectroscopy, *Opt. Express*, 2013, **21**, 28941–28947.
10 T. Mappes, C. Vannahme, M. Schelb, U. Lemmer and J. Mohr, Design for optimized coupling of organic semiconductor laser light into polymer waveguides for highly integrated bio-photonic sensors, *Microelectron. Eng.*, 2009, **86**, 1499–1501.
11 C. Vannahme, S. Klinkhammer, A. Kolew, P.-J. Jakobs, M. Guttmann, S. Dehm, U. Lemmer and T. Mappes, Integration of organic semiconductor lasers and single-mode passive waveguides into a PMMA substrate, *Microelectron. Eng.*, 2010, **87**, 693–695.
12 C. Vannahme, M. B. Christiansen, T. Mappes and A. Kristensen, Optofluidic dye laser in a foil, *Opt. Express*, 2010, **18**, 9280–9285.
13 C. Vannahme, S. Klinkhammer, M. B. Christiansen, A. Kolew, A. Kristensen, U. Lemmer and T. Mappes, All-polymer organic semiconductor laser chips: parallel fabrication and encapsulation, *Opt. Express*, 2010, **18**, 24881–24887.
14 C. Vannahme, S. Klinkhammer, U. Lemmer and T. Mappes, Plastic lab-on-a-chip for fluorescence excitation with integrated organic semiconductor lasers, *Opt. Express*, 2011, **19**, 8179–8186.
15 H. Sirringhaus, T. Kawase, R. H. Friend, T. Shimoda, M. Inbasekaran, W. Wu and E. P. Woo, High-resolution inkjet printing of all-polymer transistor circuits, *Science*, 2000, **290**, 2123–2126.
16 H. E. Katz, Recent advances in semiconductor performance and printing processes for organic transistor-based electronics, *Chem. Mater.*, 2004, **16**, 4748–4756.
17 C. N. Hoth, P. Schilinsky, S. A. Choulis and C. J. Brabec, Printing highly efficient organic solar cells, *Nano Lett.*, 2008, **8**, 2806–2813.
18 T. Aernouts, T. Aleksandrov, C. Girotto, J. Genoe and J. Poortmans, Polymer based organic solar cells using ink-jet printed active layers, *Appl. Phys. Lett.*, 2008, **92**, 033306.
19 T. R. Hebner, C. C. Wu, D. Marcy, M. H. Lu and J. C. Sturm, Ink-jet printing of doped polymers for organic light emitting devices, *Appl. Phys. Lett.*, 1998, **72**, 519–521.
20 J. Bharathan and Y. Yang, Polymer electroluminescent devices processed by inkjet printing: I. Polymer light-emitting logo, *Appl. Phys. Lett.*, 1998, **72**, 2660–2662.

21 X. Liu, S. Klinkhammer, K. Sudau, N. Mechau, C. Vannahme, J. Kaschke, T. Mappes, M. Wegener and U. Lemmer, Ink-jet-printed organic semiconductor distributed feedback laser, *Appl. Phys. Express*, 2012, **5**, 072101.

22 X. Liu, S. Klinkhammer, Z. Wang, T. Wienhold, C. Vannahme, P.-J. Jakobs, A. Bacher, A. Muslija, T. Mappes and U. Lemmer, Pump spot size dependent lasing threshold in organic semiconductor DFB lasers fabricated via nanograting transfer, *Opt. Express*, 2013, **21**, 27697–27706.

23 R. Stabile, A. Camposeo and L. Persano, Organic-based distributed feedback lasers by direct electron-beam lithography on conjugated polymers, *Appl. Phys. Lett.*, 2007, **91**, 101110.

24 M. Stroisch, T. Woggon, U. Lemmer, G. Bastian, G. Violakis and S. Pissadakis, Organic semiconductor distributed feedback laser fabricated by direct laser interference ablation, *Opt. Express*, 2007, **15**, 3968–3973.

25 N. R. Hendricks and K. R. Carter, Nanoimprint lithography of polymers, in *Polymer Science: A Comprehensive Reference*, Elsevier, 2012, pp. 251–274.

26 S. Y. Chou, P. R. Krauss and P. J. Renstrom, Nanoimprint lithography, *J. Vac. Sci. Technol., B: Microelectron. Nanometer Struct.–Process., Meas., Phenom.*, 1996, **14**, 4129–4133.

27 D. Pisignano, L. Persano, P. Visconti, R. Cingolani, G. Gigli, G. Barbarella and L. Favaretto, Oligomer-based organic distributed feedback lasers by room-temperature nanoimprint lithography, *Appl. Phys. Lett.*, 2003, **83**, 2545–2547.

28 E. B. Namdas, M. Tong, P. Ledochowitsch, S. R. Mednick, J. D. Yuen, D. Moses and A. J. Heeger, Low thresholds in polymer lasers on conductive substrates by distributed feedback nanoimprinting: progress toward electrically pumped plastic lasers, *Adv. Mater.*, 2009, **21**, 799–802.

29 C. Ge, M. Lu, X. Jian, Y. Tan and B. T. Cunningham, Large-area organic distributed feedback laser fabricated by nanoreplica molding and horizontal dipping, *Opt. Express*, 2010, **18**, 12980–12991.

30 Y. Wang, G. Tsiminis, A. L. Kanibolotsky, P. J. Skabara, I. D. W. Samuel and G. A. Turnbull, Nanoimprinted polymer lasers with threshold below 100 W cm^{-2} using mixed-order distributed feedback resonators, *Opt. Express*, 2013, **21**, 14362–14367.

31 V. Grigaliunas, S. Tamulevicius, R. Tomasiunas, V. Kopustinskas, A. Guobiene and D. Jucius, Laser pulse assisted nanoimprint lithography, *Thin Solid Films*, 2004, **453**, 13–15.

32 V. Grigaliunas, S. Tamulevicius, M. Muehlberger, D. Jucius, A. Guobiene, V. Kopustinskas and A. Gudonyte, Nanoimprint lithography using IR laser irradiation, *Appl. Surf. Sci.*, 2006, **253**, 646–650.

33 W. Pfleging, O. Baldus and A. Baldini, Method for joining plastic workpieces, *EP*, EP20050739878, 27 Nov. 2013.

34 W. Pfleging, R. Kohler, I. Südmeyer and M. Rohde, Laser micro and nano processing of metals, ceramics, and polymers, *in Laser-Assisted Fabrication of Materials*, ed. J. D. Majumdar and I. Manna, Springer, 2013, pp. 319–374.

35 A. Singh, W. Pfleging, M. Beiser and C. K. Malek, Transparent thin thermoplastic biochip by injection-moulding and laser transmission welding, *Microsyst. Technol.*, 2013, **19**, 445–453.

36 X. Cheng and L. Jay Guo, One-step lithography for various size patterns with a hybrid mask-mold, *Microelectron. Eng.*, 2004, **71**, 288–293.

Faraday Discussions

PAPER

Rigid biimidazole ancillary ligands as an avenue to bright deep blue cationic iridium(III) complexes†

Adam F. Henwood,[a] Sloane Evariste,‡[b] Alexandra M. Z. Slawin[a] and Eli Zysman-Colman*[a]

Received 12th May 2014, Accepted 12th June 2014
DOI: 10.1039/c4fd00107a

Herein we report the synthesis and optoelectronic characterisation of three deep blue-emitting cationic iridium complexes, of the form $[Ir(dFppy)_2(N^{\wedge}N)]PF_6$, bearing biimidazole-type N^N ancillary ligands (dFppyH = 2-(2,4-difluorophenyl)pyridine). Complex **1** contains the parent biimidazole, biim, while **2** contains a dimethylated analog, dMebiim, and **3** contains an *ortho*-xylyl-tethered biimidzole, *o*-xylbiim. We explore a strategy of tethering the biimidazole in order to rigidify the complex and increase the photoluminescent quantum yield, culminating in deep blue (λ_{max}: 457 nm in MeOH at 298 K) ligand-centered emission with a very high photoluminescent quantum yield of 68% and microsecond emission lifetime. Density functional theory calculations elucidate the origin of such disparate excited state kinetics across this series, especially in light of virtually identical optoelectronic properties observed for these compounds.

Introduction

For many years Organic Light-Emitting Diodes (OLEDs) have been touted as the technology that will usurp conventional fluorescent tubes as the market's dominant lighting source,[1] owing to their use of environmentally benign,[2] relatively cheap emissive materials[3] and their capacity to achieve external quantum efficiencies of 100%.[4] However, in spite of these desirable features, OLEDs have struggled to attain universal marketability as the solid-state lighting (SSL) technology of choice. The emissive materials employed are incapable of effecting balanced charge injection and mobility, thus necessitating the encapsulation of

[a]EaStCHEM School of Chemistry, University of St Andrews, St Andrews, Fife, KY16 9ST, UK. E-mail: eli.zysman-colman@st-andrews.ac.uk; Fax: +44 (0)1334 463808; Tel: +44 (0)1334 463826

[b]Département de Chimie, Université de Sherbrooke, 2500 Boul. de l'Université, Sherbrooke, QC, J1K 2R1, Canada

† Electronic supplementary information (ESI) available. See DOI: 10.1039/c4fd00107a

‡ Current address: Sciences Chimiques, UMR 6226 CNRS – Université de Rennes 1, Campus de Beaulieu, Rennes Cedex, France 35042. Web: http://www.zysman-colman.com

low work function, air-reactive electrodes within complex multilayer compositions.[4c] Typical fabrication of such sensitive devices thus typically requires vacuum sublimation – a process which is both labor- and cost-intensive, and requires thermally stable, non-ionic materials,[5] which limits the choice of organometallic triplet harvesters that might be used.[6]

A promising alternative lighting technology to OLEDs is Light-Emitting Electrochemical Cells (LEECs). By using charged materials they confer many of the same advantages but they allow for the circumvention of the arduous vacuum sublimation process. Processing is instead carried out by solution printing, using air-stable high work function electrodes in a single- or two-layer device architecture, making large-area artificial illumination a very real possibility.[7] Two classes of emitter materials are typically employed: (1) a mixture of conjugated polymer, ion transport material and inorganic salt such as LiOTf;[8] and (2) an ionic Transition Metal Complex (iTMC).[9] Of the different families of iTMCs, by far the most widely studied and exciting class of emitters for LEECs are heteroleptic cationic iridium(III) complexes, of the form $[Ir(C^\wedge N)_2(N^\wedge N)]^+$, where C^N is a monoanionic cyclometalating bis(chelate) and N^N is a neutral diimine ancillary ligand.[5]

LEECs also present their own design challenges. Issues that still require addressing for iTMC LEECs include slow turn-on times,[10] limited device stability[11] and poor colour quality.[12] In particular, few examples exist of blue-emitting LEECs,[13] which is mainly due to a shortage of deep blue, brightly emitting complexes. Blue emitters are critical both for white light emission and as a component of RGB-based pixels in displays.

Our group has thus devoted serious attention to designing cationic iridium complexes towards obtaining blue emission, combining electron-deficient C^N and electron-rich N^N ancillary ligands, with varying degrees of success.[14] There are now a few reported examples of deep blue emitting cationic iridium complexes in solution (λ_{max} < 470 nm), but significant issues still remain regarding the brightness of these emitters.[14b,15]

In surveying the literature for electron-rich diimine ligand architectures as avenues towards deep blue emitters, we found imidazole-based ligands to be promising candidates. Complexes bearing imidazole ligands have been employed in a diverse set of photophysical applications ranging from bioimaging[16] and sensing[17] to excited state proton-coupled electron transfer (PCET)[18] and solid-state lighting.[19] Of particular interest to us were iridium complexes of the form $[Ir(C^\wedge N)_2(N^\wedge N)]^+$ bearing a 1*H*,1*H*′-2,2′-biimidazole (biim) N^N ligand. As expected, the electron-rich nature of this ligand type has been shown to give a pronounced blue-shift in emission in comparison to the prototypical complex $[Ir(ppy)_2(dtBubpy)]^+$ (where ppyH is 2-phenylpyrdine and d*t*Bubpy is 4,4′-di-*tert*-butyl-2,2′-bipyridine, λ_{max} = 581 nm in MeCN).[14c,20] For instance, Wenger and co-workers[21] reported that $[Ir(tolpy)_2(biim)]^+$, where tolpy is 2-*p*-tolylpyridinato, showed emission maxima at 484 and 514 nm in DCM while Qiu and co-workers[19b] reported a similar complex, $[Ir(ppy)_2(dMebiim)]^+$, where dMebiim is 1,1′-dimethyl-2,2′-biimidazole, that emits at 497 nm in DCM. Most notably, Kim and co-workers[19d] recently showed that combining the biim N^N ligand with an electron-deficient C^N ligand in $[Ir(dFpmpy)_2(biim)]^+$, where dFpmpy is 2-(2′,4′-difluorophenyl)-4-methylpyridine, could achieve a deep blue emission with an emission maxima at 456 and 484 nm in DCM. However, despite these promising

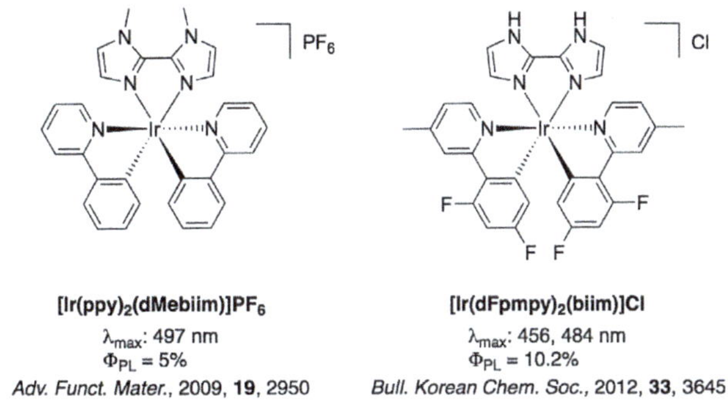

Fig. 1 Literature examples of cationic iridium complexes containing a biim-based N^N ligand.

examples in terms of emission energy, the photoluminescence quantum yield (Φ_{PL}) remains very low (Fig. 1).

The low Φ_{PL} observed in $[Ir(ppy)_2(dMebiim)]PF_6$ was due to increased non-radiative decay kinetics, k_{nr}, where undesired twisting of the dMebiim ligand resulted from the imposed steric strain of the methyl groups. We hypothesized that in adopting a tethering strategy between the two non-coordinating biimidazole nitrogen atoms the torsional strain could be alleviated and the complex rigidified, resulting in a concomitant increase in Φ_{PL}. Four target complexes were identified (Chart 1) to test this hypothesis. Critically, we anticipated the lack of conjugation of the tether groups in these complexes (**3** and **4**) to not adversely affect the optoelectronic properties observed for the parent compounds (**1** and **2**), with the only anticipated effect being an improvement in Φ_{PL}.

Other strategies have been employed for improving Φ_{PL}, and these center around two main approaches: (1) the use of bulky groups to suppress excited state emission quenching;[19a,22] and (2) employing higher-order ter-,[15f] tetra-,[23] or even hexadentate[24] chelates as molecular rigidifiers (chelate effect). In this report, we highlight a third strategy for photoluminescence quantum yield enhancement, whereby restricting the degrees of freedom of a bidentate chelate confers increased molecular rigidity to the complex and results in higher Φ_{PL} for deep-blue emitting cationic iridium complexes. Complex **3**, in particular, has been targeted as an emitter for LEECs.

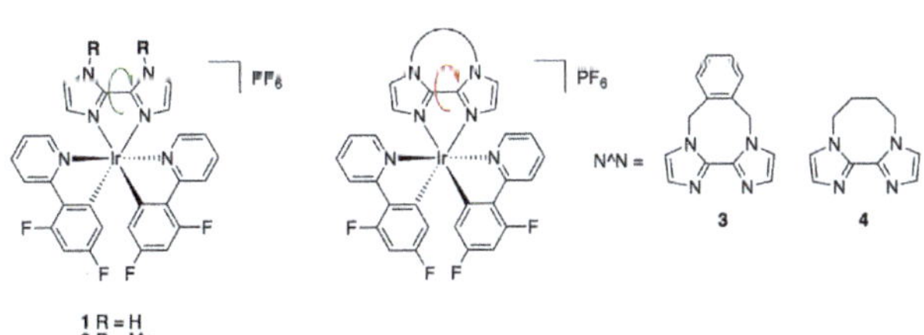

Chart 1 Complexes under investigation in this study. The green arrow connotes a certain degree of rotational flexibility about the bond while the red arrow connotes the opposite.

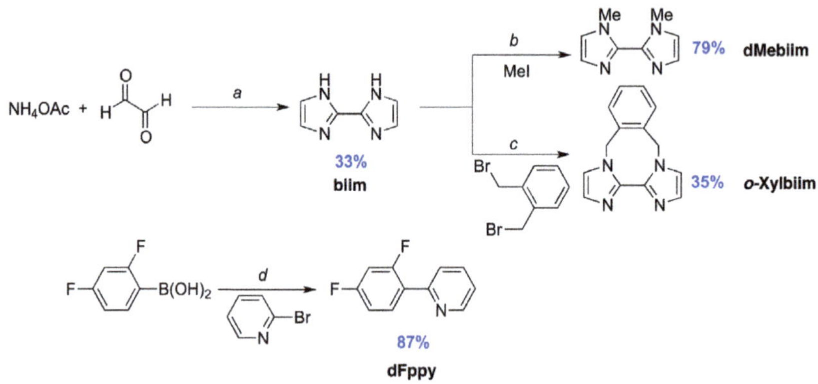

Scheme 1 Synthesis of C^N and N^N ligands. Reagents and conditions: [a] H_2O, 40 °C, 8 h. [b] NaOH (35% w/v), DMF, RT, 12 h. [c] NaOH (35% w/v), MeCN, 82 °C, 12 h. [d] 2.0 equiv. Na_2CO_3, 5 mol% $Pd(PPh_3)_4$, N_2, 1,4-dioxane–H_2O (4 : 1 v/v), 105 °C, 19 h.

Results and discussion

Ligand and complex synthesis

The cyclometalating ligand 2-(2′,4′-difluorophenyl)pyridine, dFppy, was chosen as this is the most common electron-deficient C^N ligand reported and **1** would thus serve as an appropriate benchmark complex. The dFppy ligand was prepared in good yield by a modified method to that reported previously,[25] with the corresponding μ-dichloro-bridged iridium dimer, $[Ir(dFppy)_2Cl]_2$, prepared by the method reported by Nonoyama.[26] 1*H*,1*H*′-2,2′-biimidazole, biim, was prepared in moderate yield by the condensation of glyoxal in the presence of ammonium acetate.[27] The dMebiim and *o*-xylbiim ligands were obtained through alkylation of biim. Alkylation of biim using methyl iodide in the presence of DMF and an aqueous sodium hydroxide base at room temperature[27b] afforded dMebiim in good yield while more forcing conditions were required to obtain *o*-xylbiim (Scheme 1).[28] Despite repeated alkylation attempts, the butylene linked analogue ligand, bubiim, was not able to be isolated.

The iridium complexes were isolated in good yield by cleavage of $[Ir(dFppy)_2Cl]_2$ with the corresponding biimidazole in a refluxing DCM–MeOH solution followed by purification by column chromatography and isolation as the PF_6^- salt by anion metathesis with solid NH_4PF_6 (Scheme 2). The purity and structure of the complexes were established by NMR spectroscopy, HRMS and melting point analyses. The molecular structures of **1** and **3** were determined by single-crystal X-ray structure analysis.

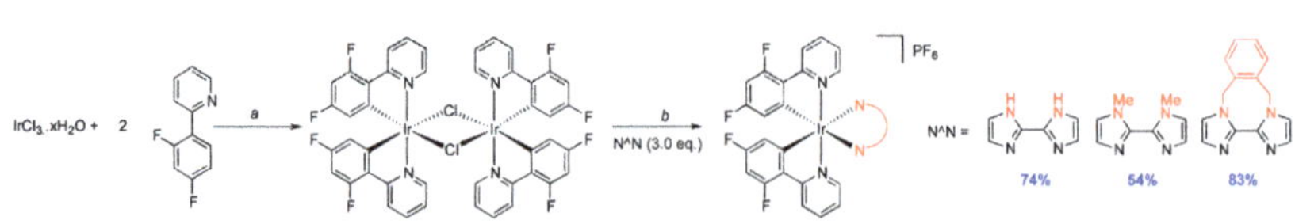

Scheme 2 Synthesis of complexes in this study. Reagents and conditions: [a] 2-EtO-C_2H_4OH–H_2O (4 : 1 v/v), 110 °C, N_2, 19 h. [b] (i) CH_2Cl_2–MeOH (5 : 4 v/v), 55 °C, 19 h, N_2; (ii) excess solid NH_4PF_6.

X-ray and solution state structural elucidation

Crystals of **1** and **3** were grown by vapor diffusion of Et_2O into solutions of MeCN and DCM–MeOH, respectively. The structures have been deposited with the CCDC (deposition numbers CCD 1001467-1001468). The poor quality of the dataset for **3** precludes any detailed analysis of the structural parameters though it does provide convincing proof of connectivity. Both complexes exhibit a distorted octahedral geometry (Fig. 2) with the pyridyl groups in the typical *trans* relationship. In **1** the bite angles of the C^N ligands [80.8(30)° and 81.2(4)°] and biimidazole-based N^N ligand [76.1(2)°] are comparable with those reported for cationic biimidazole complexes[21] and related bis(triazole) complexes,[14a] with the reduced bond angle of the biimidazole attributable to the smaller chelate angle of five-membered ring chelates over six-membered ring chelates.[27a] Similar bond angles to the metal are reported in other biimidazole-to-metal crystal structures such as with iron (biim = 80.39°, *o*-xylbiim = 75.4°)[28] and with rhodium (biim = 79.24).[29] There appears to be some changes in torsion angles between **1** and **3**. In **1** the biim ligand is relatively flat whereas in **3** the $N_{(Ir)}$–C–C–$N_{(Ir)}$ torsion angle is 8°.

The packing in the crystal structures differs between **1** and **3**. In **1** there are strong hydrogen bond interactions between the fluorine atoms on the PF_6^- anion and the N–H biimidazole hydrogen atoms, [H(3)···F (1) 1.96, N(3)···F(1) 2.931(8) Å, N–H···H 170.9°; H(10)···F(5) 1.95, N(10)···F(5) 2.889(9) Å; N(10)–H(10)···F(5) 158.6°]. This type of hydrogen-bonded ion pairing interaction has been previously observed with Wenger's[21] $[Ir(tolpy)_2(biim)]^+$ complex, which crystallized in the presence of the 3,5-dinitrobenzoate anion, as well as with $[Ir(pqx)_2(biim)]Cl$ (pqx = phenylquinoxaline), where the chloride anion hydrogen bonds with the distal NH groups of the biim.[19c]

Solution state NMR spectroscopy suggests that complex **3** exists as at least two sets of diastereomeric atropisomers (*vide infra*). However, in solid form **3** crystallises as a racemate in the $P\bar{1}$ space group. The absence of biimidazole hydrogen

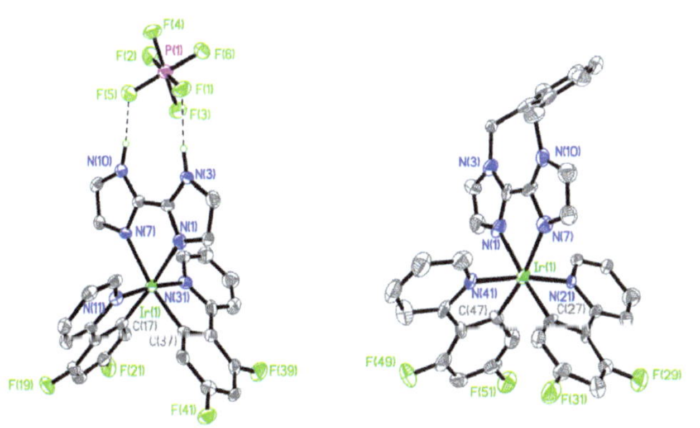

1" **3"**

Fig. 2 Molecular structures of **1** (left) and **3** (right) with 50% probability ellipsoids. The majority of the hydrogen atoms and the counterion in **3** have been omitted for clarity. Selected bond lengths (Å) and angles (°) for **1**: Ir(1)–N(1) 2.158(8), Ir(1)–N(7) 2.173(6), Ir(1)–N(11) 2.027(8), Ir(1)–N(3)1 2.038(8), Ir(1)–C(1)7 2.019(10), Ir(1)–C(3)7 2.033(8), N(1)–Ir(1)–N(7) 76.1(2), N(11)–Ir(1)–C(17) 81.2(4), N(31)–Ir(1)–C(37) 80.8(3).

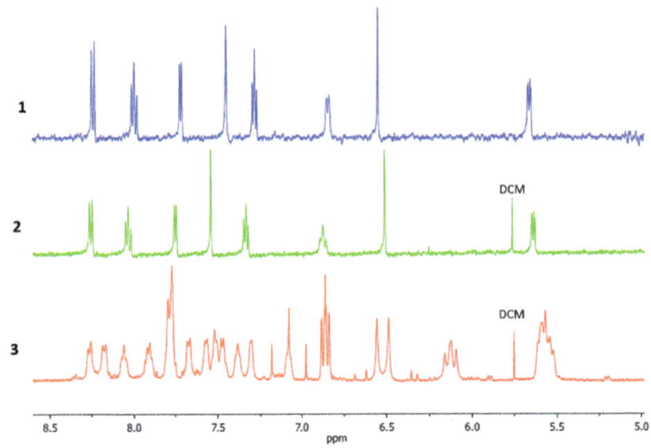

Fig. 3 Stacked plot of ^{1}H NMR spectra of complexes **1**, **2** and **3** at room temperature in DMSO-d_6.

atoms precludes similar interactions to those seen in **1**. There is a weak interaction from a fluorine atom of the dFppy to an *o*-xylyl aryl hydrogen atom [H(50)···F(29) 2.35, C(50)···F(29) 3.29(3) Å; C(50)–H(50)···F(29) 170.2°]. Interactions with the PF_6^- anion in this instance are minimal.

Characterization of complex **3** by solution ^{1}H NMR proved challenging owing to the generally poor solubility in virtually all solvents and its surprisingly complex ^{1}H NMR spectrum, especially when compared with the ^{1}H NMR spectra of complexes **1** and **2** (Fig. 3).

The complexity of this spectrum at room temperature was attributed to the slow fluxional motion of the *o*-xylyl group resulting in the detection of two or more diastereomeric atropisomers at this temperature (Fig. 4). Upon heating, the ^{1}H NMR spectrum simplifies to the expected pattern. For instance, at 318 K, there is an observed coalescence of the doublet at 6.55 ppm. Eyring analysis of this coalescence phenomenon suggests the activation barrier to *o*-xylyl ring flipping is 82.97 kJ mol^{-1}.[30]

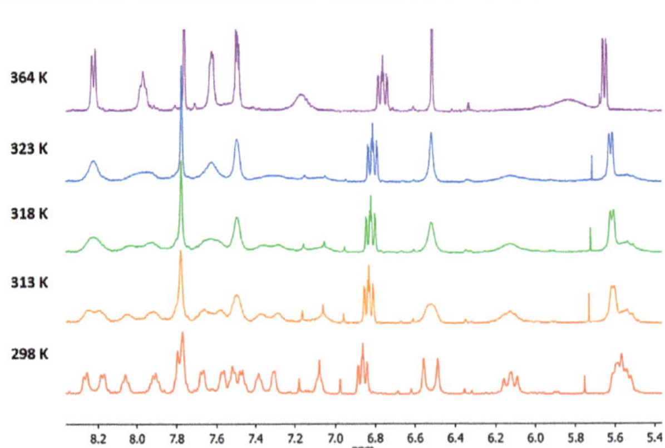

Fig. 4 Variable temperature ^{1}H NMR study of **3** in DMSO-d_6.

Table 1 Electrochemical data and orbital energies for **1–3**[a]

Compound	$E_{1/2}^{ox}$ (V)	ΔE_p (mV)	E_{HOMO}[b] (eV)	E_{LUMO}[c] (eV)	$E_{0,0}$[d] (eV)
1	1.51	76	−6.56	−3.71	2.85
2	1.45	79	−6.50	−3.64	2.86
3	1.44	72	−6.49	−3.59	2.90

[a] All measurements were performed at 50 mV s^{-1} in deaerated MeCN solution using Fc/Fc^+ as an internal standard, and are referenced with respect to SCE ($Fc/Fc^+ = 0.38$ V in MeCN).[31] [b] $E_{HOMO} - [E^{ox}_{pa\ vs.\ Fc/Fc^+} + 5.39]$ eV.[32] [c] $E_{LUMO} = E_{HOMO} + E_{0,0}$ eV. [d] $E_{0,0}$ estimated from the intersection point of the absorption and emission spectra at 298 K in MeOH.

Cyclic voltammetry

The electrochemical behavior of complexes **1–3** was investigated by cyclic voltammetry (CV) in deaerated MeCN solution containing n-NBu_4PF_6 as the supporting electrolyte and using Fc/Fc^+ as an internal standard at 298 K. All potentials are referenced with respect to SCE ($Fc/Fc^+ = 0.38$ V in MeCN)[31] and all reported data were carried out at a scan rate of 50 mV s^{-1}. The HOMO energy levels were determined from the relation $E_{HOMO} - [E^{ox}_{pa\ vs.\ Fc/Fc^+} + 5.39]$ eV,[32] while the lack of a detectable reduction wave in the accessible solvent window necessitated estimating the E_{LUMO} energies from the sum of the E_{HOMO} values and the optical band gap values, $E_{0,0}$, for each complex. $E_{0,0}$ was inferred from the intersection point between the absorption and emission spectra obtained at 298 K in MeOH. Table 1 summarizes the relevant electrochemical data.

The oxidation potentials of **1–3** are expectedly virtually unchanged across the series, with each complex demonstrating a single, quasi-reversible oxidation wave in the region of 1.5 V.

These oxidation potentials are very similar to that previously reported[14c] for $[Ir(dFMeppy)_2(bpy)]PF_6$ (dFMeppy is 2-(2′,4′-difluorophenyl)-4-methylpyridine and bpy is 2,2′-bipyridine), where $E_{1/2}^{ox} = 1.55$ V under similar conditions. The oxidation is thus assigned to the Ir^{III}/Ir^{IV} redox couple with contribution from the C^N ligands. DFT calculations (*vide infra*) corroborate this analysis. Surprisingly, despite virtually identical photophysical properties to **1**, the electrochemical properties of Kim's[19d] $[Ir(dFpmpy)_2(biim)]^+$ complex differ somewhat to our own, with a reported $E_{1/2}^{ox}$ value of 1.59 V *versus* Fc/Fc^+ in DCM, resulting in modestly higher reported E_{HOMO} (−6.25 eV) and E_{LUMO} (−3.56 eV) energies. The HOMO–LUMO energy gap, estimated from the optical gap ($E_{0,0}$), is however similar (2.69 eV) to those in this study.

Solution state photophysical behavior

A comprehensive summary of the relevant photophysical data undertaken in this study is given in Table 2; absorptivity data may be found in the ESI, Table S1.† Fig. 5 shows the normalized absorption and emission spectra for complexes **1–3** at room temperature in MeOH and also shows the normalized 77 K emission spectra in a 1 : 1 MeOH–EtOH glass. The absorption spectra for the complexes are relatively unstructured, typical for iridium–biimidazole complexes,[19b,d] with the intense band at around 250 nm assigned to spin-allowed ligand centered

Table 2 Relevant photophysical data for **1–3**[a]

	λ_{em}[b] (nm)		Φ_{PL}[c] (%)	τ_e		k_r	k_{nr}
	77 K	298 K	298 K	77 K (μs)	298 K (ns)	($\times 10^5$ s^{-1})	($\times 10^5$ s^{-1})
1	453, 486	464, 490	20	3.682	1559	1.28	5.13
2	451, 484	457, 486	2	3.718	91	2.20	107.69
3	450, 483	457, 487	68	3.956	3840	1.77	0.83

[a] 298 K measurements in deaerated MeOH and 77 K measurements in 1 : 1 MeOH–EtOH glass. [b] Principal emission peaks listed. [c] Quinine sulfate used as the reference (Φ_{PL} = 54.6% in 0.5 M H_2SO_4 at 298 K).[33]

(^{1}LC) $^1\pi \rightarrow \pi^*$ transitions. All three complexes also demonstrate a distinct lower energy absorption band at about 370 nm, as well as a small tail into the near UV region. These bands are also present in [Ir(dFpmpy)$_2$(biim)]$^+$ and were attributed by Kim to be comprised of a mix of $^3\pi \rightarrow \pi^*$ and spin-allowed and spin-forbidden metal-to-ligand charge transfer (^{1}MLCT) and (^{3}MLCT) transitions.[19d] Alkylation of the biim ligand leads to a more structured absorption profile, particularly between 250–370 nm. Excitation spectra for **1–3** (Fig. S16–S18†) reproduce these characteristic features.

The structured emission profiles at 77 K and 298 K arise from a ^{3}LC emission (Fig. 5). The absence of any rigidochromic shift in the emission maxima further corroborates the ^{3}LC nature of the emission. At both 298 and 77 K, two high-energy emission maxima are observed at around 455 nm and 484 nm, along with lower vibronic emission peaks tailing out to about 650 nm. The near identical emission spectra across the three complexes verifies our assertion that the electronics across the series are unchanging, with only a slight blue shift in emission arising from incorporating the alkyl groups in place of the parent protons.

Though the Stokes shifts for **1–3** are very small, the presence of iridium and the microsecond emission lifetimes point to a phosphorescence emission. Low temperature emission lifetimes (τ_e) of all three complexes in 1 : 1 MeOH–EtOH glass are similar and are in the range of 3.6–4.0 μs. However, at 298 K while τ_e for **1** and **3** remains in the microsecond regime, the value for **2** drops significantly to 90 ns, indicative of substantial contributions to k_{nr} for this complex at room temperature. The photoluminescence quantum yield, Φ_{PL}, for **1** is 20% and decreases markedly for **2** to only 2%. Gratifyingly, the Φ_{PL} for **3** is a remarkable 68%! These figures are reflected in the excited state kinetics. While the radiative rate constants, k_r, are similar for **1–3** (ranging from 1.3 to 2.2 $\times$ 10^5 s^{-1}), k_{nr} values differ dramatically across the series. Complex **2** has a k_{nr} of 107 $\times$ 10^5 s^{-1}, which is two orders of magnitude larger than that calculated for **1** at 5.1 $\times$ 10^5 s^{-1}. The brightest complex, **3**, has a calculated k_{nr} of 0.8 $\times$ 10^5 s^{-1}, which is six-fold smaller than that of **1**.

Theoretical calculations

A combined density functional theory (DFT) and time-dependent DFT (TDDFT) study was undertaken to rationalize the optoelectronic properties and to verify the hypothesis that tethering in this instance leads to a less strained geometry.[34]

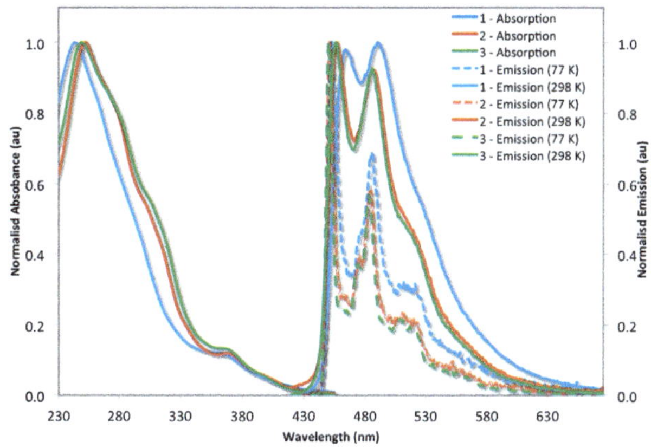

Fig. 5 Normalized absorption spectra of **1–3** in aerated MeOH at 298 K and normalized emission spectra in deaerated MeOH at 298 K and 1 : 1 MeOH–EtOH glass at 77 K.

Complexes **1–3** and butyl tethered analog **4** were modeled by Gaussian 09 (ref. 35) using the following DFT protocol at the B3LYP[36] level of theory with the SBKJC-DVZ[37] basis set for iridium, 6-31G* for heavy atoms directly coordinated to iridium and 3-21G* for all other atoms[37a,38] in the presence of the solvent MeCN.[39]

The geometry of the ground state structures was fully optimized without the imposition of symmetry restrictions. Each complex adopts a pseudo-octahedral geometry. Selected structural parameters for **1–4** are summarized in Table 3. Computed geometries for **1** and **3** generally reproduce those found in the crystal structure though the Ir–$N_{N^\wedge N}$ and Ir–N_{dFppy} bonds for **1** and **3** are predicted to be slightly elongated by around 0.02 Å while there is a slightly more pronounced torsion between the two imidazole fragments in the crystal structure of **3** of 8°. The geometries of the triplet state were optimized using spin-unrestricted DFT calculations at the UB3LYP level. In the triplet state, the Ir–$N_{N^\wedge N}$ is predicted to be slightly elongated while modest bond length contractions are predicted between the iridium center and the dFppy ligands. Structural differences manifest most markedly in the $(N\text{–}C\text{–}C\text{–}N)_{N^\wedge N}$ dihedral angle. There is a large change in the $(N\text{–}C\text{–}C\text{–}N)_{N^\wedge N}$ torsion observed in **2** between the S_0 and T_1 states while in **4**, a larger twist is predicted compared to **1–3**. This results from a conformational

Table 3 Selected calculated average structural parameters for **1–4**[a]

	1		**2**		**3**		**4**	
Complexes	S_0	T_1	S_0	T_1	S_0	T_1	S_0	T_1
Ir–$N_{N^\wedge N}$	2.1966	2.2138	2.1702	2.1902	2.1820	2.1936	2.0188	2.2006
Ir–$N_{C^\wedge N}$	2.0737	2.0650	2.0733	2.0635	2.0736	2.0653	2.0756	2.0668
Ir–$C_{C^\wedge N}$	2.0198	2.0072	2.0228	2.0086	2.0213	2.0087	2.0209	2.0081
$N_{N^\wedge N}$–Ir–$N_{N^\wedge N}$	75.4	75.1	74.3	74.3	74.9	74.5	75.1	74.8
$N_{C^\wedge N}$–Ir–$N_{C^\wedge N}$	80.4	81.2	80.4	81.3	80.4	81.2	80.4	81.1
$(N\text{–}C\text{–}C\text{–}N)_{N^\wedge N}$	0.1	1.6	3.1	11.5	1.4	1.5	8.7	9.2

[a] Bond lengths in Å and bond angles in °. C^N = dFppy.

compensation in this dihedral angle in order to minimize repulsive interactions in the pseudo-gauche conformation of the butyl linker (47.2°). Complex **3** exhibits the most rigid conformation.

Fig. 6 shows a comparison of the relative energies of the five highest energy occupied and five lowest energy unoccupied molecular orbitals (MOs) for **1–3**. The HOMO is localized on both the aryl ring of the C^N ligands and the iridium atom (t_{2g}). The LUMO is also situated on the dFppy ligands though there is now increased contribution from the pyridine moiety. The large HOMO–LUMO gap for all three is calculated to be *ca.* 4.16 eV. The effect of alkylation of the biim ligand does not significantly perturb the energies of the frontier molecular orbitals.

The computations reproduce the principal features in the UV-visible spectra for the complexes (*cf.* Fig. S19–S21†). TDDFT analysis for **1–3** predicts a T_1 state that contains two major contributions: HOMO → LUMO (51%) and HOMO − 2 → LUMO + 1 (22%). The qualitative description of the triplet state that results is predominantly ligand-centered (^{3}LC) on the dFppy ligands with some metal-to-ligand charge transfer (^{3}MLCT) from the iridium t_{2g} orbitals to the dFppy ligands. The spin densities for the T_1 state for **1–4** are shown in Fig. 7 and all show similar topologies with the spin density localized on one of the two dFppy ligands, implying an emission resulting from a ^{3}LC state. This assignment is consistent with the observed structured emission at 298 K in MeOH.

The emission energy was predicted using two different methodologies. The phosphorescence is estimated as the difference between the T_1 and S_0 states in their respective optimized geometries ($E_{0,0}$), which is a good indicator of the $E_{0,0}$ emission measured at 77 K. For **1–3**, an emission at 430 nm was predicted. The λ_{max} at 77 K ranged from 451–453 nm. The adiabatic electronic emission (E_{AE}) is determined from the vertical energy difference between the T_1 and S_0 states at the

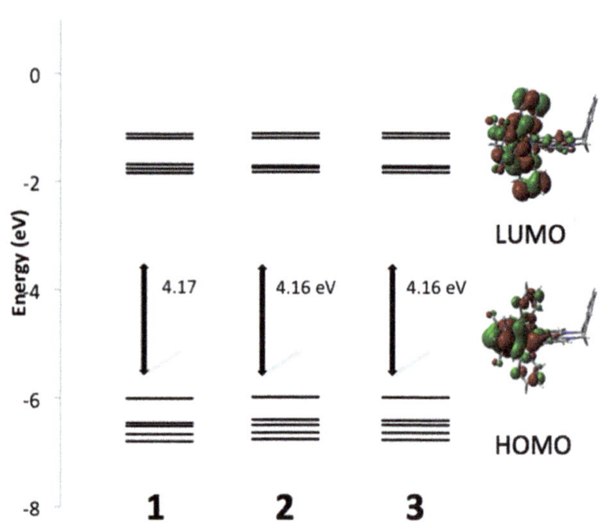

Fig. 6 Calculated energy level scheme for the Kohn–Sham orbitals between HOMO − 4 to LUMO + 4 of **1–3**, and the associated DFT calculated HOMO–LUMO energy gap (in eV). Electron density contour plots for **3** (0.002 e bohr^{-3}). The contour plots for **1** and **2** mirror those of **3**.

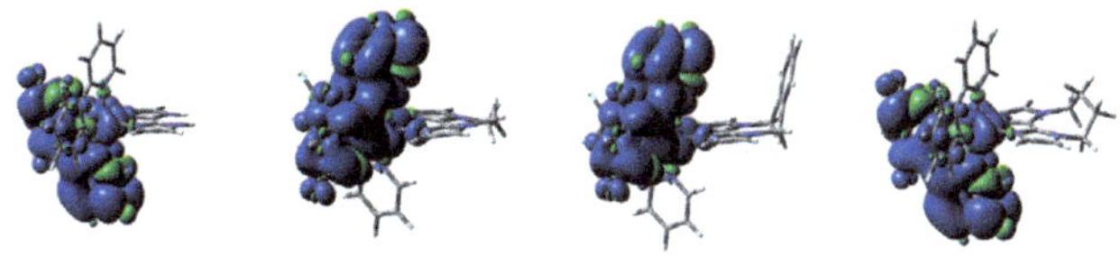

Fig. 7 Calculated spin density contours of the T_1 state for **1–4** (isocontour value of 0.0004 au).

optimized geometry of the T_1 state. For **1–3**, an emission at 487 nm was predicted. The λ_{max} at 298 K ranged from 457–464 nm. The calculations quite accurately reproduce (E_{AE}) the solution state emission observed at 298 K, with relative errors of about 6%.

Conclusions

In summary, three new cationic iridium(III) complexes bearing biimidazole-type ancillary ligands have been reported. By exploring the role of alkylation of the biimidazole, we have dramatically modified the excited state kinetics of these three complexes without any significant changes to the electronics of these deep blue emitters, which all emit in the deep blue region at around 455 nm in a MeOH solution. Crucially, we have reported a dramatic increase in the photoluminescence quantum yield for complex **3** of 68% compared to 2% for **2** – among the very brightest of deep blue (λ_{max} < 470 nm) cationic iridium emitters reported in the literature. Microsecond emission lifetimes in **3** were maintained. However, poor solubility of these compounds has hindered their subsequent device fabrication. Current efforts are underway to make **3** more soluble and thus processable as an emissive layer in a LEEC and results thereof will be reported in due course.

Experimental section

General synthetic procedures

Commercial chemicals were used as supplied. All reactions were performed using standard Schlenk techniques under an inert (N_2) atmosphere with reagent grade solvents. Flash column chromatography was performed using silica gel (Silia-P from Silicycle, 60 Å, 40–63 μm). Analytical thin layer chromatography (TLC) was performed with silica plates with aluminum backings (250 μm with indicator F-254). Compounds were visualized under UV light. ^{1}H, ^{13}C and ^{19}F NMR spectra were recorded on a Bruker Avance spectrometer at 500 MHz, 126 MHz and 471 MHz, respectively. The following abbreviations have been used for multiplicity assignments: "s" for singlet, "d" for doublet, "t" for triplet, "m" for multiplet and "br" for broad. Deuterated chloroform ($CDCl_3$) and deuterated DMSO (DMSO-d_6) were used as the solvents of record. Melting points (Mps) were recorded using open-ended capillaries on an electrothermal melting point apparatus and are uncorrected. High-resolution mass spectra were recorded on a quadrupole time-of-flight (ESI-Q-TOF), model ABSciex 5600 Triple TOF in positive electrospray ionization mode and spectra were recorded using sodium formate solution as a calibrant. The iridium(III) dimer, $[(dFppy)_2Ir(\mu\text{-}Cl)]_2$ was prepared according to the procedure described by Nonoyama.[26]

Ligand syntheses

2-(2,4-Difluorophenyl)pyridine (dFppy). The synthesis of this ligand is by a modified version of a previously reported method.[40] 2,4-Difluorophenylboronic acid (1.1 equiv.), 2-bromopyridine (1.0 equiv.), and sodium carbonate (2.0 equiv.) were added to a Schlenk tube containing a mixture of 1,4-dioxane and distilled water (4 : 1 v/v) to obtain a concentration of 0.15 to 0.20 M. The reaction mixture was degassed *via* three freeze–pump–thaw cycles. Upon warming to room temperature from the third cycle, $Pd(PPh_3)_4$ (5 mol%) was added to the tube under a positive nitrogen pressure and the tube was sealed. The mixture was refluxed for 19 h and then cooled to room temperature. The mixture was poured onto distilled water and extracted multiple times with dichloromethane. The organic fractions were combined, washed with a portion of brine and dried over magnesium sulfate. Filtration and evaporation under reduced pressure gave the crude product (1.45 g). The crude product was purified by flash column chromatography (silica, hexane–ethyl acetate gradient 100 : 0 to 80 : 20) to give 1.31 g of pure compound as a colourless oil. Yield: 87%. R_f: 0.48 (20% EtOAc/hexanes on silica). ^{1}H NMR (500 MHz, $CDCl_3$) δ (ppm): 8.72 (dt, J = 4.5, 1.0 Hz, 1H), 8.03–7.99 (m, 1H), 7.75 (d, J = 4.0 Hz, 2H), 7.28–7.23 (m, 1H), 7.03–6.99 (m, 1H), 6.93–6.90 (m, 1H). ^{13}C NMR (126 MHz, $CDCl_3$) δ (ppm): 164.6, 162.1, 159.5, 152.7, 149.9, 136.6, 132.3, 124.4, 122.6, 112.1, 104.5. ^{19}F {^{1}H} NMR (471 MHz, $CDCl_3$) δ (ppm): −109.3 (d, J = 6.8 Hz, 1F), −113.0 (d, J = 6.8 Hz, 1F). GCMS: (13.6 min) $[M]^+$: 191. The characterisation matches that reported.[40]

***1H*,1′*H*-2,2-Biimidazole (biim).** Synthesis of this ligand was as outlined in the literature.[41] To a mixture of ammonium acetate (2.7 equiv.) in distilled water at 40 °C, 40% aqueous glyoxal solution (1.0 equiv.) was added dropwise over a period of 3 h to give a concentration of 0.01 M. The mixture was allowed to stir for a further 5 h at room temperature. The reaction mixture was filtered and washed multiple times with distilled water and acetone to give 8.31 g of a brown crude product. This material was added to ethylene glycol (0.5 M), heated to 150 °C and treated with decolourising carbon. Filtration saw product precipitate immediately, with further washings with distilled water to maximise product precipitation. The product was filtered and dried to give 2.47 g as a cream white powder. Yield: 33%. R_f: 0.12 (10% MeOH–DCM on silica). Mp: 350–352 °C. Litt: >300 °C.[27a] ^{1}H NMR (500 MHz, DMSO-d_6) δ (ppm): 12.67 (s, 2H), 7.14 (s, 2H), 7.00 (s, 2H). ^{13}C NMR (126 MHz, $CDCl_3$) δ (ppm): 139.8, 128.7, 117.9. The ^{1}H NMR differs from that previously reported but matches that determined by us from a commercial source.[27a]

1,1′-Dimethyl-2,2′-biimidazole (dMebiim). Synthesis of this ligand was as outlined in the literature.[41] 1*H*,1′*H*-biimidazole (1.0 equiv.) was added to a mixture of aqueous sodium hydroxide (5.6 equiv., 35% w/v) in DMF to give a concentration of 0.9 M. This was stirred for 1 h. The mixture turned green and then black over the course of the hour. Methyl iodide (3.0 equiv.) was then added slowly to the reaction mixture. The mixture was left to stir for 19 h at room temperature. The crude reaction mixture was then poured onto distilled water and extracted with chloroform multiple times. The combined organic layers were washed with water and dried over sodium sulfate. Filtration and evaporation under reduced pressure gave the crude product (0.29 g). Purification by flash column chromatography (silica, dichloromethane–ethanol gradient 100 : 0 to 95 : 5) afforded 0.19 g of the product as an off-white solid. Yield: 79%. R_f: 0.25 (10% EtOAc/hexanes on silica). Mp: 117–118 °C. ^{1}H NMR (500 MHz, $CDCl_3$) δ (ppm): 7.11 (s, 2H), 6.96 (s, 2H), 4.04 (s, 6H). ^{13}C NMR (126 MHz, $CDCl_3$) δ (ppm): 128.0, 122.8, 35.5. Characterisation matches that previously reported, although we only detected three ^{13}C resonances.[41]

1,1′-(α,α′-*o*-Xylylene)-2,2-biimidazole (*o*-xylbiim). Synthesis of this ligand was as outlined in the literature.[28] To a solution containing α,α′-dibromo-*o*-xylene (1.0 equiv.) in acetonitrile (0.1 M), 1*H*,1′*H*-biimidazole (1.2 equiv.) was added with stirring followed by an aqueous sodium hydroxide (5.6 equiv., 35% v/w) solution. The temperature was increased to reflux, where after about 10 min a yellow-brown solution formed. The mixture was maintained at reflux overnight, before being cooled to room temperature. After addition of distilled water the mixture was extracted with dichloromethane multiple times. The organic fractions were combined, dried over anhydrous magnesium sulfate and then evaporated to dryness under reduced pressure. The crude product was washed with portions of diethyl ether, affording 0.23 g of the pure compound as an off-white solid. Yield: 35%. Mp: 288–291 °C. Litt: 284–292 °C.[27a] ^{1}H NMR (500 MHz, DMSO-d_6) δ (ppm): 7.47 (d, J = 1.0 Hz, 2H), 7.41–7.47 (m, 4H), 7.11 (d, J = 0.5 Hz, 2H), 4.97 (s, 4H). ^{13}C NMR (126 MHz, $CDCl_3$) δ (ppm): 139.4, 133.9, 130.1, 128.9, 128.7, 122.1, 49.0. Characterisation matches that previously reported.[27a]

General procedure for the synthesis of [(C^N)$_2$Ir(N^N)]PF_6 complexes

To a Schlenk tube containing [Ir(dFppy)$_2$Cl]$_2$ (1.0 equiv.) and N^N ligand (3.0 equiv.) DCM and MeOH (5 : 4 v/v) were added to give a concentration of 0.03 M. The mixture was degassed *via* three freeze–pump–thaw cycles, before backfilling with N_2 upon thawing from the third cycle. The reaction mixture was heated to 55 °C for 19 h. Over the course of the reaction the mixture darkened in colour. The solution was cooled to room temperature and solid NH_4PF_6 (10.0 equiv.) was added and the reaction mixture was left to stir for a further 1 h. The resulting suspension was evaporated to dryness, with the residue then copiously washed with Et_2O and distilled water. This crude product was purified by flash column chromatography (silica, DCM–MeOH gradient 100 : 0 to 95 : 5). Fractions containing the desired complex were combined and solid NH_4PF_6 (10 equiv.) was added. The suspension was stirred at room temperature for 0.5 h. This mixture was then evaporated to dryness, washed vigorously with distilled water and dried to afford the pure material.

Iridium(III) bis[2-(4′,6′-difluorophenyl)-pyridinato-N,C$^{2\prime}$]-*N*,*N*′-(1*H*,1′*H*-2,2′ biimidazole) hexafluorophosphate: [(dFppy)$_2$Ir(biim)](PF_6), 1. Yellow powder (0.094 g). Yield: 74%. Mp: 310–311 °C. ^{1}H {^{19}F} NMR (500 MHz, DMSO-d_6) δ (ppm): 8.24 (d, J = 8.5 Hz, 2H), 8.0 (td, J = 1.5, 7.8 Hz, 2H), 7.72 (d, J = 5.5 Hz, 2H), 7.45 (s, 2H), 7.28 (td, J = 1.0, 7.0 Hz, 2H), 6.84 (dd, J = 2.0, 8.8 Hz, 2H), 6.55 (d, J = 1.0 Hz, 2H), 5.66 (dd, J = 2.5, 8.0 Hz, 2H). ^{19}F {^{1}H} NMR (471 MHz, DMSO-d_6) δ (ppm): −70.09 (d, J = 712.2 Hz, 6F), −107.72 (d, J = 9.42 Hz, 2F), −109.7 (d, J = 9.89 Hz, 2F). HR-MS (ES-Q-TOF): [M − PF_6]$^+$ calculated: ($C_{28}H_{18}N_6F_4Ir$) 707.1158; found: 707.1130.

Iridium(III) bis[2-(4′,6′-difluorophenyl)-pyridinato-N,C$^{2\prime}$]-*N*,*N*′-(1,1′-dimethyl-2,2′-biimidazole) hexafluorophosphate: [(dFppy)$_2$Ir(dMebiim)](PF$_6$), 2. Yellow powder (0.062 g). Yield: 54%. Mp: 325–326 °C. ^{1}H {^{19}F} NMR (500 MHz, 364 K, DMSO-d_6) δ (ppm) 8.25 (d, J = 7.5 Hz, 2H), 8.03 (t, J = 7.8 Hz, 2H), 7.74 (d, J = 6.0 Hz, 2H), 7.53 (d, J = 1.5 Hz, 2H), 7.33 (td, J = 1.0, 6.5 Hz, 2H), 6.87 (t, J = 6.0 Hz, 2H) 6.50 (d, J = 1.5 Hz, 2H), 5.63 (dd, J = 2.5, 8.0 Hz, 2H) 4.22 (s, 6H). ^{19}F {^{1}H} NMR (471 MHz, DMSO-d_6) δ (ppm): −70.14 (d, J = 712.2 Hz, 6F), −107.75 (d, J = 9.9, 2F), −109.77 (d, J = 9.9, 2F). HR-MS (ES-Q-TOF): [M − PF$_6$]$^+$ calculated: ($C_{30}H_{22}N_6F_4Ir$) 735.1471; found: 735.1442.

Iridium(III) bis[2-(4′,6′-difluorophenyl)-pyridinato-N,C$^{2\prime}$]-*N*,*N*′-1,1′-(α,α′-*o*-Xylylene)-2,2-biimidazole hexafluorophosphate: [(dFppy)$_2$Ir(xylbiim)](PF$_6$), 3. Yellow powder (0.062 g). Yield: 83%. Mp: 359–360 °C. ^{1}H {^{19}F} NMR (500 MHz, DMSO-d_6) δ (ppm): 8.22 (d, J = 8.0 Hz, 2H), 7.97 (t, J = 7.5 Hz, 2H), 7.76 (d, J = 1.0 Hz, 2H), 7.69–7.57 (m, 3H), 7.56–7.46 (m, 3H), 7.17 (s, br, 2H), 6.79–6.73 (m, 2H), 6.51 (d, J = 1.0 Hz, 2H), 5.86 (s, br, 4H), 5.65 (dd, J = 2.0, 8.8 Hz, 2H). ^{19}F {^{1}H} NMR (471 MHz, DMSO-d_6) δ (ppm): 70.13 (d, J = 712.2 Hz, 6F), −107.60 (m, 2F), −109.7 (m, 2F). HR-MS (ES-Q-TOF): [M − PF$_6$]$^+$ calculated: ($C_{36}H_{24}N_6F_4Ir$) 809.1628; found: 809.1597.

Photophysical measurements

All samples were prepared in HPLC grade methanol with varying concentrations in the order of μM. Absorption spectra were recorded at RT using a Shimadzu UV-1800 double beam spectrophotometer. Molar absorptivity determination was verified by linear least-squares fit of values obtained from at least three independent solutions at varying concentrations with absorbance ranging from 1.26×10^{-4} to 3.43×10^{-5} M.

The sample solutions for the emission spectra were prepared in HPLC grade MeOH and degassed *via* three freeze–pump–thaw cycles. Steady state emission and excitation spectra and time-resolved emission spectra were recorded at 298 K and 77 K using an Edinburgh Instruments F980. All samples for steady state measurements were excited at 360 nm while samples for time-resolved measurements were excited at 378 nm. Emission quantum yields were determined using the optically dilute method.[42] A stock solution with absorbance of *ca.* 0.5 was prepared and then four dilutions were prepared with dilution factors of 5, 6.6, 10 and 20 to obtain solutions with absorbances of *ca.* 0.1 0.075, 0.05 and 0.025, respectively. The Beer–Lambert law was found to be linear at the concentrations of the solutions. The emission spectra were then measured after the solutions were rigorously degassed *via* three freeze–pump–thaw cycles prior to spectrum acquisition. For each sample, linearity between absorption and emission intensity was verified through linear regression analysis and additional measurements were acquired until the Pearson regression factor (R^2) for the linear fit of the data set surpassed 0.9. Individual relative quantum yield values were calculated for each solution and the values reported represent the slope value. The equation $\Phi_s = \Phi_r(A_r/A_s)(I_s/I_r)(n_s/n_r)_2$ was used to calculate the relative quantum yield of each of the samples, where Φ_r is the absolute quantum yield of the reference, n is the refractive index of the solvent, A is the absorbance at the excitation wavelength, and I is the integrated area under the corrected emission curve. The subscripts s and r refer to the sample and reference, respectively. A solution of quinine sulfate in 0.5 M H_2SO_4 ($\Phi_r = 54.6\%$) was used as the external reference.[33].

Electrochemistry measurements

Cyclic voltammetry (CV) measurements were performed on an Electrochemical Analyzer potentiostat model 600D from CH Instruments. Solutions for cyclic voltammetry were prepared in ACN and degassed with ACN-saturated nitrogen bubbling for about 10 min prior to scanning. Tetra(*n*-butyl)ammoniumhexafluorophosphate ($TBAPF_6$; *ca.* 0.1 M in ACN) was used as the supporting electrolyte. An Ag/Ag^+ electrode (silver wire in a solution of 0.1 M KCl in H_2O) was used as the pseudo-reference electrode; a Pt electrode was used for both the working and counter electrodes. The redox potentials are reported relative to a saturated calomel electrode (SCE) with a ferrocenium/ferrocene (Fc^+/Fc) redox couple as an internal reference (0.38 V *vs.* SCE).[31]

Density functional theory (DFT) calculations

All calculations were performed with the Gaussian 09 (ref. 43) suite. The level of theory for all DFT[34c,44] and TD-DFT[34d–f] calculations was B3LYP; excited-state triplet geometries were calculated using the unrestricted B3LYP method (UB3LYP).[36b,36c,45] The 6-31G* basis set[46] was used for C, H and N directly linked to iridium while the other C, H, N and F atoms where undertaken with 3-21G* basis set,[37a,38a–e] and the VDZ (valence double ζ) with SBKJC effective core potential basis set[37] was used for iridium. The predicted phosphorescence wavelengths were obtained by energy difference between the Triplet and Singlet states at their respective optimized geometries.[47] The energy, oscillator strength and related MO contributions for the 100 lowest singlet–singlet and 5 lowest singlet–triplet excitations were obtained from the TD-DFT/singlets and the TD-DFT/triplets output files, respectively. The calculated absorption spectra were visualized with GaussSum 2.1 (fwhm: 1000 cm^{-1}).[48]

Acknowledgements

EZ-C acknowledges the University of St Andrews for financial support.

References

1 S. Reineke, F. Lindner, G. Schwartz, N. Seidler, K. Walzer, B. Lussem and K. Leo, *Nature*, 2009, **459**, 234.
2 H. Sasabe and J. Kido, *Eur. J. Org. Chem.*, 2013, **2013**, 7653.
3 C. J. Humphreys, *MRS Bull.*, 2008, **33**, 459.
4 (*a*) F. So, J. Kido and P. Burrows, *MRS Bull.*, 2008, 663; (*b*) Y. Tao, C. Yang and J. Qin, *Chem. Soc. Rev.*, 2011, **40**, 2943; (*c*) J. D. Slinker, J. Rivnay, J. S. Moskowitz, J. B. Parker, S. Bernhard, H. D. Abruña and G. G. Malliaras, *J. Mater. Chem.*, 2007, **17**, 2976.
5 R. D. Costa, E. Ortí, H. J. Bolink, F. Monti, G. Accorsi and N. Armaroli, *Angew. Chem., Int. Ed.*, 2012, **51**, 8178.
6 H. Yersin. *Highly Efficient OLEDs with Phosphorescent Materials*, Wiley-VCH, Weinheim, 2008.

7 (*a*) D. A. Bernards, J. D. Slinker, G. G. Malliaras, S. Flores-Torres and H. D. Abruna, *Appl. Phys. Lett.*, 2004, **84**, 4980; (*b*) H.-C. Su, C.-C. Wua, F.-C. Fang and K.-T. Wong, *Appl. Phys. Lett.*, 2006, **89**, 261118.
8 S. Tang, J. Pan, H. A. Buchholz and L. Edman, *J. Am. Chem. Soc.*, 2013, **135**, 3647.
9 S. Bernhard, J. A. Barron, P. L. Houston, H. D. Abruña, J. L. Ruglovsky, X. Gao and G. G. Malliaras, *J. Am. Chem. Soc.*, 2002, **124**, 13624.
10 E. Zysman-Colman, J. D. Slinker, J. B. Parker, G. G. Malliaras and S. Bernhard, *Chem. Mater.*, 2008, **20**, 388.
11 G. E. Schneider, H. J. Bolink, E. C. Constable, C. D. Ertl, C. E. Housecroft, A. Pertegas, J. A. Zampese, A. Kanitz, F. Kessler and S. B. Meier, *Dalton Trans.*, 2014, **43**, 1961.
12 J. D. Slinker, D. A. Bernards, P. L. Houston, H. D. Abruña, S. Bernhard and G. G. Malliaras, *Chem. Commun.*, 2003, (19), 2392.
13 J. M. Fernández-Hernández, S. Ladouceur, Y. Shen, A. Iordache, X. Wang, L. Donato, S. Gallagher-Duval, M. de Anda Villa, J. D. Slinker, L. De Cola and E. Zysman-Colman, *J. Mater. Chem. C*, 2013, **1**, 7440–7452.
14 (*a*) L. Donato, P. Abel and E. Zysman-Colman, *Dalton Trans.*, 2013, **42**, 8402; (*b*) J. M. Fernandez-Hernandez, S. Ladouceur, Y. Shen, A. Iordache, X. Wang, L. Donato, S. Gallagher-Duval, M. de Anda Villa, J. D. Slinker, L. De Cola and E. Zysman-Colman, *J. Mater. Chem. C*, 2013, **1**, 7440; (*c*) S. Ladouceur, D. Fortin and E. Zysman-Colman, *Inorg. Chem.*, 2011, **50**, 11514; (*d*) S. Ladouceur, K. N. Swanick, S. Gallagher-Duval, Z. Ding and E. Zysman-Colman, *Eur. J. Inorg. Chem.*, 2013, **2013**, 5329.
15 (*a*) L. He, L. Duan, J. Qiao, R. Wang, P. Wei, L. Wang and Y. Qiu, *Adv. Funct. Mater.*, 2008, **18**, 2123; (*b*) M. Mydlak, C. Bizzarri, D. Hartmann, W. Sarfert, G. Schmid and L. De Cola, *Adv. Funct. Mater.*, 2010, **20**, 1812; (*c*) F. Kessler, R. D. Costa, D. Di Censo, R. Scopelliti, E. Ortí, H. J. Bolink, S. Meier, W. Sarfert, M. Gratzel, M. K. Nazeeruddin and E. Baranoff, *Dalton Trans.*, 2012, **41**, 180; (*d*) F. De Angelis, S. Fantacci, N. Evans, C. Klein, S. M. Zakeeruddin, J.-E. Moser, K. Kalyanasundaram, H. J. Bolink, M. Gratzel and M. K. Nazeeruddin, *Inorg. Chem.*, 2007, **46**, 5989; (*e*) C.-H. Yang, J. Beltran, V. Lemaur, J. Cornil, D. Hartmann, W. Sarfert, R. Fröhlich, C. Bizzarri and L. De Cola, *Inorg. Chem.*, 2010, **49**, 9891; (*f*) N. Darmawan, C. H. Yang, M. Mauro, M. Raynal, S. Heun, J. Pan, H. Buchholz, P. Braunstein and L. De Cola, *Inorg. Chem.*, 2013, **52**, 10756; (*g*) F. Monti, F. Kessler, M. Delgado, J. Frey, F. Bazzanini, G. Accorsi, N. Armaroli, H. J. Bolink, E. Orti, R. Scopelliti, M. K. Nazeeruddin and E. Baranoff, *Inorg. Chem.*, 2013, **52**, 10292; (*h*) W.-T. Chen, Y.-J. Chen, C.-S. Wu, J.-J. Lin, W.-L. Su, S.-H. Chen and S.-P. Wang, *Inorg. Chim. Acta*, 2013, **408**, 225; (*i*) B. Chen, Y. Li, W. Yang, W. Luo and H. Wu, *Org. Electron.*, 2011, **12**, 766.
16 (*a*) L. Murphy, A. Congreve, L. O. Palsson and J. A. Williams, *Chem. Commun.*, 2010, **46**, 8743; (*b*) X. Wang, J. Jia, Z. Huang, M. Zhou and H. Fei, *Chem.–Eur. J.*, 2011, **17**, 8028.
17 Y. Bao, H. Wang, Q. Li, B. Liu, Q. Li, W. Bai, B. Jin and R. Bai, *Macromolecules*, 2012, **45**, 3394.
18 (*a*) Y. Cui, H.-J. Mo, J.-C. Chen, Y.-L. Niu, Y.-R. Zhong, K.-C. Zheng and B.-H. Ye, *Inorg. Chem.*, 2007, **46**, 6427; (*b*) O. S. Wenger, *Acc. Chem. Res.*, 2013, **46**, 1517; (*c*) O. S. Wenger, *Coord. Chem. Rev.*, 2014, DOI: 10.1016/j.ccr.2014.03.025.

19 (*a*) G. G. Shan, H. B. Li, H. Z. Sun, H. T. Cao, D. X. Zhu and Z. M. Su, *Dalton Trans.*, 2013, **42**, 11056; (*b*) L. He, J. Qiao, L. Duan, G. Dong, D. Zhang, L. Wang and Y. Qiu, *Adv. Funct. Mater.*, 2009, **19**, 2950; (*c*) N. Sengottuvelan, H.-J. Seo, S.-K. Kang and Y.-I. Kim, *Bull. Korean Chem. Soc.*, 2010, **31**, 2309; (*d*) S.-J. Yun, H.-J. Seo, M. Song, S.-H. Jin and Y. I. Kim, *Bull. Korean Chem. Soc.*, 2012, **33**, 3645.

20 J. D. Slinker, A. A. Gorodetsky, M. S. Lowry, J. Wang, S. T. Parker, R. Rohl, S. Bernhard and G. G. Malliaras, *J. Am. Chem. Soc.*, 2004, **126**, 2763.

21 J. C. Freys, G. Bernardinelli and O. S. Wenger, *Chem. Commun.*, 2008, 4267.

22 L. Sun, A. Galan, S. Ladouceur, J. D. Slinker and E. Zysman-Colman, *J. Mater. Chem.*, 2011, **21**, 18083.

23 D. A. Vezzu, J. C. Deaton, J. S. Jones, L. Bartolotti, C. F. Harris, A. P. Marchetti, M. Kondakova, R. D. Pike and S. Huo, *Inorg. Chem.*, 2010, **49**, 5107.

24 G. St-Pierre, S. Ladouceur, D. Fortin and E. Zysman-Colman, *Dalton Trans.*, 2011, **40**, 11726.

25 L.-C. Campeau, S. Rousseaux and K. Fagnou, *J. Am. Chem. Soc.*, 2005, **127**, 18020.

26 M. Nonoyama, *Bull. Chem. Soc. Jpn.*, 1974, **47**, 767.

27 (*a*) R. P. Thummel, V. Goulle and B. Chen, *J. Org. Chem.*, 1989, **54**, 3057; (*b*) J.-C. Xiao and J. N. M. Shreeve, *J. Org. Chem.*, 2005, **70**, 3072.

28 H. V. Phan, P. Chakraborty, M. Chen, Y. M. Calm, K. Kovnir, L. K. Keniley, Jr, J. M. Hoyt, E. S. Knowles, C. Besnard, M. W. Meisel, A. Hauser, C. Achim and M. Shatruk, *Chem.–Eur. J.*, 2012, **18**, 15805.

29 E. Laurila, L. Oresmaa, M. Niskanen, P. Hirva and M. Haukka, *Cryst. Growth Des.*, 2010, **10**, 3775.

30 H. S. Gutowsky and C. H. Holm, *J. Chem. Phys.*, 1956, **25**, 1228.

31 V. V. Pavlishchuk and A. W. Addison, *Inorg. Chim. Acta*, 2000, **298**, 97.

32 C. M. Cardona, W. Li, A. E. Kaifer, D. Stockdale and G. C. Bazan, *Adv. Mater.*, 2011, **23**, 2367.

33 W. H. Melhuish, *J. Phys. Chem.*, 1961, **65**, 229.

34 (*a*) P. Hohenberg and W. Kohn, *Phys. Rev.*, 1964, **136**, B864; (*b*) W. Kohn and L. J. Sham, *Phys. Rev.*, 1965, **140**, A1133; (*c*) in *The Challenge of d and f Electrons*, ed. D. R. Salahub and M. C. Zerner, ACS, Washington, DC, 1989; (*d*) R. E. Stratmann, G. E. Scuseria and M. J. Frisch, *J. Chem. Phys.*, 1998, **109**, 8218; (*e*) R. Bauernschmitt and R. Ahlrichs, *Chem. Phys. Lett.*, 1996, **256**, 454; (*f*) M. E. Casida, C. Jamorski, K. C. Casida and D. R. Salahub, *J. Chem. Phys.*, 1998, **108**, 4439.

35 M. J. Frisch, G. W. Trucks, H. B. Schlegel, G. E. Scuseria, M. A. Robb, J. R. Cheeseman, V. G. Zakrzewski, J. A. Montgomery, R. E. Stratmann, J. C. Burant, S. Dapprich, J. M. Millam, A. D. Daniels, K. N. Kudin, M. C. Strain, O. Farkas, J. Tomasi, V. Barone, M. Cossi, R. Cammi, B. Mennucci, C. Pomelli, C. Adamo, S. Clifford, J. Ochterski, G. A. Peterson, P. Y. Ayala, Q. Cui, K. Morokuma, A. Malik, A. D. Rabuck, K. Raghavachari, J. B. Foresman, J. Cioslowski, J. V. Ortiz, A. G. Baboul, B. B. Stefanov, G. Liu, A. Liashenko, P. Piskorz, I. Komaromi, R. Gomperts, R. L. Martin, M. Challacombe, P. M. W. Gill, B. G. Johnson, W. Chen, M. W. Wong, J. L. Andres, M. Head-Gordon, E. S. Replogle and J. A. Pople*Gaussian 98 (Revision A.6)*, Pittsburgh, PA, 1998.

36 (*a*) A. D. Becke, *J. Chem. Phys.*, 1993, **98**, 5648; (*b*) C. Lee, W. Yang and R. G. Parr, *Phys. Rev. B: Condens. Matter Mater. Phys.*, 1988, **37**, 785; (*c*) B. Miehlich, A. Savin, H. Stoll and H. Preuss, *Chem. Phys. Lett.*, 1989, **157**, 200.
37 (*a*) J. S. Binkley, J. A. Pople and W. J. Hehre, *J. Am. Chem. Soc.*, 1980, **102**, 939; (*b*) W. J. Stevens, W. J. Basch and M. Krauss, *J. Chem. Phys.*, 1984, **81**, 6026; (*c*) W. J. Stevens, M. Krauss, H. Basch and P. G. Jasien, *Can. J. Chem.*, 1992, **70**, 612; (*d*) T. R. Cundari and W. J. Stevens, *J. Chem. Phys.*, 1993, **98**, 5555.
38 (*a*) M. S. Gordon, J. S. Binkley, J. A. Pople, W. J. Pietro and W. J. Hehre, *J. Am. Chem. Soc.*, 1982, **104**, 2797; (*b*) W. J. Pietro, M. M. Francl, W. J. Hehre, D. J. Defrees, J. A. Pople and J. S. Binkley, *J. Am. Chem. Soc.*, 1982, **104**, 5039; (*c*) K. D. Dobbs and W. J. Hehre, *J. Comput. Chem.*, 1986, **7**, 359; (*d*) K. D. Dobbs and W. J. Hehre, *J. Comput. Chem.*, 1987, **8**, 861; (*e*) K. D. Dobbs and W. J. Hehre, *J. Comput. Chem.*, 1987, **8**, 880; (*f*) R. Ditchfield, W. J. Hehre and J. A. Pople, *J. Chem. Phys.*, 1971, **54**, 724; (*g*) W. J. Hehre, R. Ditchfield and J. A. Pople, *J. Chem. Phys.*, 1972, **56**, 2257; (*h*) P. C. Hariharan and J. A. Pople, *Theor. Chim. Acta*, 1973, **28**, 213; (*i*) P. C. Hariharan and J. A. Pople, *Mol. Phys.*, 1974, **27**, 209; (*j*) M. S. Gordon, *Chem. Phys. Lett.*, 1980, **76**, 163.
39 J. Tomasi, B. Mennucci and R. Cammi, *Chem. Rev.*, 2005, **105**, 2999.
40 Y. You and S. Y. Park, *J. Am. Chem. Soc.*, 2005, **127**, 12438.
41 J.-C. Xiao and J. N. M. Shreeve, *J. Org. Chem.*, 2005, **70**, 3072.
42 (*a*) G. A. Crosby and J. N. Demas, *J. Phys.Chem.*, 1971, **75**, 991; (*b*) S. Fery-Forgues and D. Lavabre, *J. Chem. Educ.*, 1999, **76**, 1260.
43 M. J. Frisch, G. W. Trucks, H. B. Schlegel, G. E. Scuseria, M. A. Robb, J. R. Cheeseman, G. Scalmani, V. Barone, B. Mennucci, G. A. Petersson, H. Nakatsuji, M. Caricato, X. Li, H. P. Hratchian, A. F. Izmaylov, J. Bloino, G. Zheng, J. L. Sonnenberg, M. Hada, M. Ehara, K. Toyota, R. Fukuda, J. Hasegawa, M. Ishida, T. Nakajima, Y. Honda, O. Kitao, H. Nakai, T. Vreven, J. A. Montgomery Jr., J. E. Peralta, F. Ogliaro, M. Bearpark, J. J. Heyd, E. Brothers, K. N. Kudin, V. N. Staroverov, R. Kobayashi, J. Normand, K. Raghavachari, A. Rendell, J. C. Burant, S. S. Iyengar, J. Tomasi, M. Cossi, N. Rega, J. M. Millam, M. Klene, J. E. Knox, J. B. Cross, V. Bakken, C. Adamo, J. Jaramillo, R. Gomperts, R. E. Stratmann, O. Yazyev, A. J. Austin, R. Cammi, C. Pomelli, J. W. Ochterski, R. L. Martin, K. Morokuma, V. G. Zakrzewski, G. A. Voth, P. Salvador, J. J. Dannenberg, S. Dapprich, A. D. Daniels, Ö. Farkas, J. B. Foresman, J. V. Ortiz, J. Cioslowski and D. J. Fox, 7.0, Wallingford, CT, 2009.
44 (*a*) P. Hohenberg and W. Kohn, *Phys. Rev. B: Condens. Matter Mater. Phys.*, 1964, **136**, 864; (*b*) W. Kohn and L. Sham, *Phys. Rev. A*, 1965, **140**, 1133; (*c*) R. G. Parr and W. Yang, *Density-functional theory of atoms and molecules*, Oxford Univ. Press, Oxford, 1989.
45 A. D. Becke, *J. Chem. Phys.*, 1993, **98**, 5648.
46 V. A. Rassolov, J. A. Pople, M. A. Ratner and T. L. Windus, *J. Chem. Phys.*, 1998, **109**, 1223.
47 (*a*) S. Ladouceur, D. Fortin and E. Zysman-Colman, *Inorg. Chem.*, 2010, **49**, 5625; (*b*) M. S. Lowry, W. R. Hudson, R. A. Pascal Jr and S. Bernhard, *J. Am. Chem. Soc.*, 2004, **126**, 14129.
48 N. M. O'Boyle, *GaussSum 2.0*, Dublin City University, Dubin Ireland, 2006, available at http://www.gausssum.sf.net.

Faraday Discussions

PAPER

Lasing of Tamm states in highly efficient organic devices based on small-molecule organic semiconductors

R. Brückner, V. G. Lyssenko, S. Hofmann and K. Leo*

Received 5th May 2014, Accepted 4th June 2014
DOI: 10.1039/c4fd00094c

We discuss approaches to increase the light outcoupling efficiency in organic microcavity (MC) lasers and organic light-emitting diodes (OLEDs). We find that the introduction of metals into the cavities leads to additional Tamm-plasmon polariton modes, while the corrugation of metal contacts, such as perforated μ-size holes or a periodic array of metal stripes, leads to 2D confinement of the cavity modes, which in turn reduces the lasing threshold in MCs. Furthermore, we elucidate light loss mechanisms in OLEDs and reveal how external dielectric layers and periodic gratings can be used to enhance outcoupling from the OLED cavity.

Introduction

In recent years, devices based on organic semiconductors have made rapid progress. In order to gain knowledge of the interactions between light and matter, organic microcavities (MCs), *i.e.* an organic light-emitting material sandwiched between two dielectric mirrors (distributed Bragg reflectors, DBRs), are widely studied. The introduction of metal layers is especially of interest to construct an electrically driven organic solid state laser and important progress has already been achieved.[1]

The first organic light-emitting diodes (OLEDs) were demonstrated by Tang and Van Slyke in 1987.[2] Nowadays, OLEDs are widely used in displays for smart phones or for lighting applications. In comparison to common light sources, OLEDs have several advantages: they are flat area emitters offering a diffuse light perception, wide viewing angles, and they can be realized on flexible substrates. For both display and lighting applications, high efficiency is crucial. The efficiencies reached with the first OLEDs were quite low. For optimization, an adequate understanding of the electric, excitonic, and optical processes limiting these efficiencies is required. The reasons for low efficiency are: (i) the inefficient injection and transport of charge carriers which is hindered by space-charge limited currents and contact resistance, (ii) the

Institut für Angewandte Photophysik, TU Dresden, 01069 Dresden, Germany. E-mail: karl.leo@iapp.de

fluorescent emitter system used, *i.e.* all electrically excited triplets are lost, and (iii) the light outcoupling is not optimized for the thin film structure, *i.e.* light is trapped inside the OLED.

Electrical improvements regarding the charge injection from the contacts and the charge carrier transport can be achieved by adding blocking layers and (molecularly) doped transport layers. With the latter, a pin-structure could be realized which has been intensively studied in recent years.[3–7]

Excitons are excited states that are responsible for light emission. They are generated by the recombination of holes and electrons, either forming a singlet or a triplet. Due to spin statistics, about 25% singlets and 75% triplets are generated in the OLED. Their distribution on the different emitter molecules is crucial for the design of efficient OLEDs. The internal quantum yield, *i.e.* the conversion efficiency of charge carriers into photons, in OLEDs comprising only phosphorescent emitters is nowadays almost 100%.

However, due to internal losses (total internal reflection, absorption, and surface plasmon polaritons[8]), about 70 to 80% of the generated photons are trapped inside the OLED thin film structure and the substrate.[9] Hence, optical properties and light outcoupling enhancement methods are the main remaining obstacles to achieve high efficiency[10–14] of OLEDs.

In the following, we cover the theoretical background of lasing in MCs. Furthermore, we investigate the introduction of metal layers and corrugation into the cavities and their influence on the lasing threshold. Finally, we elucidate light loss mechanisms in OLEDs and reveal how external dielectric layers and periodic gratings can be used to enhance the efficiency of top-emitting OLEDs.

Microcavities

In a planar microcavity,[15] photons are confined between DBRs in the cavity layer with thickness, L_c, and refractive index, n_c. The emission of the excited cavity layer propagates in growth *z*-direction with a resonance wave vector, $k_z = \pi/n_cL_c$, and can propagate at an oblique angle, φ, with a longitudinal wave vector, k_x, and a "parabolic-like" dispersion:

$$E(k_x) = \frac{\hbar c}{n_c}\sqrt{\left(\frac{\pi}{L_c}\right)^2 + {k_x}^2} \approx E(0) + \frac{\hbar^2 {k_x}^2}{2m_x}, \tag{1}$$

where $\pi\hbar c/n_cL_c = E(0)$ is the energy of the fundamental mode, $m_x = \pi\hbar n_c/cL_c$ corresponds to an "effective mass" for photons, while $k_x = \omega \sin \varphi/c$ is related to the angle φ of the emitted light with respect to the optical axis. By measuring the angular-resolved reflection, transmission, or emission spectra of the MCs, one can restore the true dispersion curves of exciton–polaritons and their nature. In Fig. 1, angular-resolved TE- and TM-polarized spontaneous emission spectra of the red fluorescent emitter 4-(dicyanomethylene)-2-methyl-6-[*p*-(dimethyl-amino)-styryl]-4*H*-pyran (DCM) doped into the tris(8-hydroxy-quinolinato)-aluminium (Alq_3) matrix, embedded between two DBRs and excited by a cw non-resonant 405 nm laser below the lasing-threshold intensity, are shown. Both the cavity-mode (620 nm) and emission through short-wavelength DBR sidebands (570 nm) show a parabolic-like dispersion according to eqn (1).

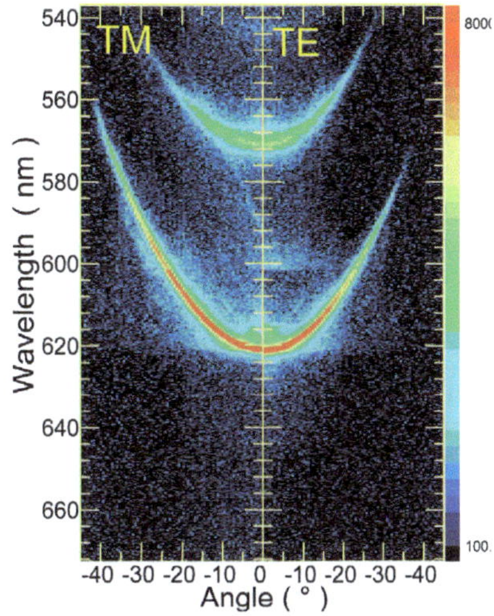

Fig. 1 Experimental angular-resolved TM- and TE-polarized emission spectra of the cavity parabola-like mode at 620 nm (see eqn (1)) and transmitted through DBR side-band at 570 nm.

Purcell enhancement[16]

In 1946, Mills Purcell proposed to control the spontaneous emission (SE) rate of a monochromatic dipole with an emitting wavelength, λ_e, and linewidth, $\Delta\lambda_e$, by using a microcavity (MC) with a single cavity mode to tailor the number of electromagnetic modes to which it is coupled.[17] For the case of the "weak coupling" $\Delta\lambda_e \ll \Delta\lambda_c$, the escape time of SE photons out of the cavity is much shorter than the radiative lifetime and reabsorption is negligible. A simple derivation shows that the SE rate in the microcavity mode is given by the Purcell factor, $F = 3Q\lambda_c^{\ 3}/4\pi^2 n_c V$, where n_c is the refractive index of the medium and V is the effective mode volume. Additionally, the radiative rate Γ_r is strongly enhanced at above-threshold excitations, proportional to the density, n, of coherent photons. Therefore, the ratio $\Gamma_r/(\Gamma_r + \Gamma_{nr})$ can reach ~1.

Rate equations and lasing[18]

Above the lasing threshold, lasing occurs when the round-trip gain is greater than the round-trip loss. The threshold is determined as the pump power at which the mean stimulated and spontaneous emission rates into the lasing mode are equal.[19,20] The rates of spontaneous emission into the lasing and non-lasing modes are described as fractions, β and $1 - \beta$, of the total spontaneous emission rate. The corresponding angular-resolved emission spectra at the above-threshold excitation is presented in Fig. 2. Due to the increased number of excited state molecules, both the gain and the rate of spontaneous emission, Γ_r, into non-lasing modes increased.

The discussion of lasing in MC with a four-level homogeneously broadened gain medium can be carried out using rate equations.[21] The model is defined by the Maxwell–Bloch equations:

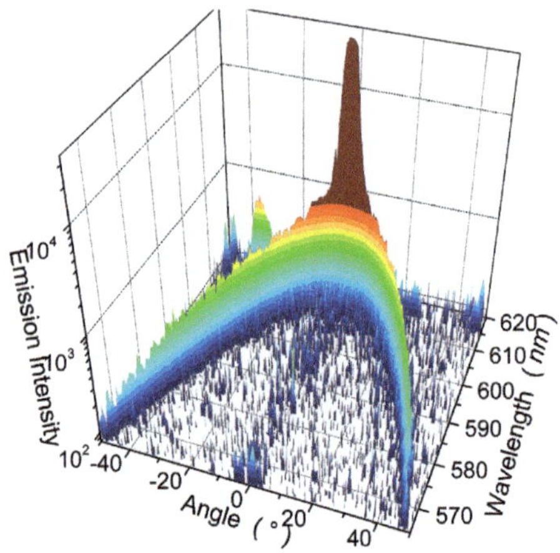

Fig. 2 Experimental angular-resolved emission above the lasing threshold.

$$\dot{\alpha} = -\kappa\alpha + gv, \tag{2a}$$

$$\dot{v} = -\gamma_{\perp}/2v + g\alpha N, \tag{2b}$$

$$\dot{N} = -\gamma_{\|}N + \Gamma + g(\alpha v^* + v\alpha^*), \tag{2c}$$

where α is the complex amplitude of the laser field (in photon number units), N is a number of atoms in the upper level of the lasing transition, v is the gain-medium polarization amplitude summed over all of the atoms, 2κ is the photon decay rate, $\gamma_{\|}$ is the spontaneous emission rate to modes other than the laser mode, $\gamma_{\perp}$ is the full width at half maximum (FWHM) of the homogeneous gain, g is the dipole coupling constant, and Γ is excitation rate.

The polarization $\gamma_{\perp}$ damps on a subpicosecond time scale and may therefore be adiabatically eliminated. As a result, eqn (2) can be replaced by the following equivalent rate equations:

$$\dot{n}/\gamma_{\|} = -\lambda n + \beta n N, \tag{3a}$$

$$\dot{N}/\gamma_{\|} = -N + P - \beta n N, \tag{3b}$$

where $\lambda = 2\kappa/\gamma_{\perp}$, $P = \Gamma/\gamma_{\perp}$, $\beta = 4g^2/\gamma_{\perp}\gamma_{\|}$, and $n = |\alpha|^2$ is the photon number.

Spontaneous emission into the laser mode can be included by the replacement of the photon number n by $n + 1$ in the stimulated emission term βNn. Thus the generalized rate equations are:

$$\dot{n}/\gamma_{\|} = -\lambda n + \beta n N + \beta N, \tag{4a}$$

$$\dot{N}/\gamma_{\|} = -N + P - \beta n N - \beta N, \tag{4b}$$

where λ and P are the cavity decay rate and pumping rate, respectively, measured in units of the spontaneous emission rate; β is the branching ratio which specifies the fraction of spontaneous emission directed into the laser mode. The steady-state solutions of eqn (4) are:

$$n = \frac{2P - \lambda \pm \sqrt{(2P - \lambda)^2 + 4\lambda\beta^2 P}}{2\lambda\beta}, \tag{5a}$$

$$N = P/(1 + \beta n),\ v = 2g\alpha N/\gamma_{\perp}. \tag{5b}$$

Fig. 3 presents the experimentally measured (circles) intensities of the spontaneous and stimulated emission from an organic MC and calculated using eqn (5a). Here, the experimental and calculated results are in remarkable agreement.

In the next step, the temporal development of the emitted laser pulses above the threshold is investigated in an up-conversion experiment, when the 800 nm pulse of the amplifier is split into two laser beams. For one beam, the frequency is doubled and used as the ~400 nm pump beam of the organic MC. The MC emission is guided into another non-linear crystal and is added with the second 800 nm pulses, the gate beam, from the amplifier. Its temporal arrival can be varied by a delay stage in the range of up to ~500 ps. The time resolution of this experiment is given by the gate pulse duration of about 200 fs. Fig. 4(a) presents five time traces of the organic laser for different pump powers above the threshold. At the lowest recorded pump pulse power (0.78 nJ), the maximum output occurs about 150 ps after the arrival of the pump pulse with a pulse width of 60 ps. With increasing power, the MC pulse appears earlier and exhibits a narrower temporal width. After nearly doubling the excitation power to 1.3 nJ, the organic laser shows the highest output 40 ps after pump pulse incidence with a reduced width of 14 ps. The existence of the gain switched state is much shorter than the natural lifetime of the excited state as observed by the spontaneous emission lifetime. An additional requirement for gain switching is a short photon lifetime which is also fulfilled by $\tau_{cav} = 1.5$ ps. In our experiment, a 400 fs pump pulse excites a large number of Alq_3 molecules, which have a comparably long

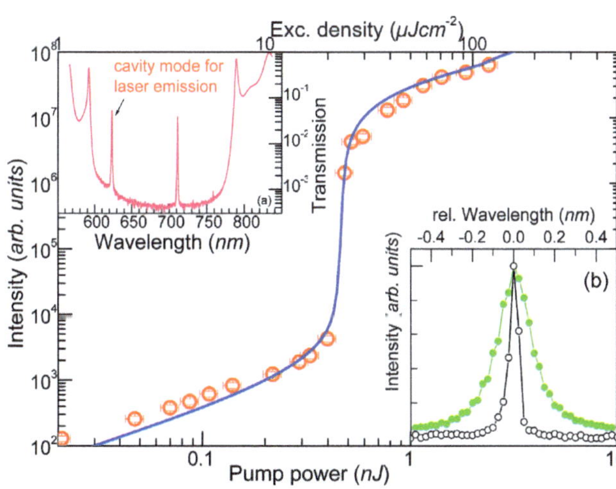

Fig. 3 Experimentally measured (circles) intensities of spontaneous and stimulated emission from organic MC and calculated using eqn (5a). Reprint with permission from ref. 22. Copyright 2005, AIP LLC.[22]

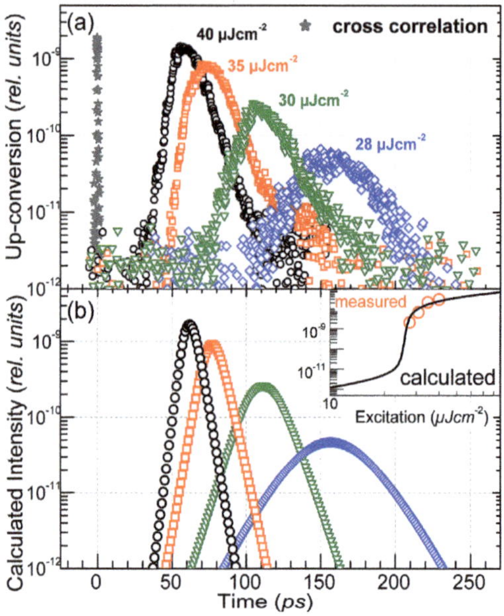

Fig. 4 (a) Time-resolved organic MC emission as measured by an up-conversion technique for pump powers above the threshold. (b) The laser output is modeled by a set of rate equations for the different excitation energies. Reprint with permission from ref. 22. Copyright 2005, AIP LLC.[22]

radiative lifetime of about $\tau_{\mathrm{Alq3}} = 20$ ns. The excitation is non-radiatively transferred to the DCM molecules with a lifetime of $\tau_{\mathrm{DCM}} = 5$ ns. As the up-conversion data show, the laser emission occurs after several picoseconds and thus appears sufficiently earlier than the spontaneous emission. Again, the simulated time-resolved intensities using the rate equations (Fig. 4(b)) are in good agreement with the experimental data.

Metal layers: Tamm-plasmon polaritons

In order to realize optoelectronic devices, electrical contacts are needed. In the case of OLEDs, In_2O_3:Sn (ITO) with typical layer thicknesses of 80–150 nm is most commonly used as a semitransparent electrode. The second way to efficiently contact an organic active layer is the utilization of thin metallic layers, which can easily be deposited by means of thermal evaporation under high vacuum. This fact motivates the study of metallic structures inside organic MCs which may enable electrical contacting. In addition, metals may also serve as broadband high-reflectivity mirrors and act as a heat sink due to their high thermal conductivity.[23]

Similar to electronic Tamm states at the surface of a crystal, optical Tamm states can form at the interface between two photonic structures with overlapping band gaps.[24] The complex interplay between a metal layer and a periodic DBR results in the formation of localized surface states at the interface.

In order to verify this interpretation, the transfer matrix formalism[25,26] is utilized to numerically determine the transmission properties of a half-

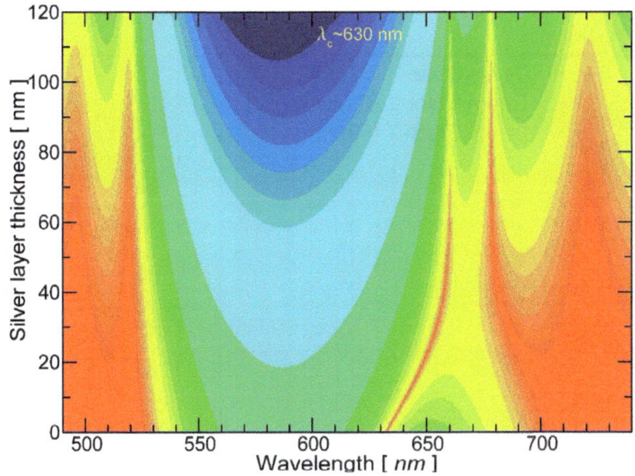

Fig. 5 Calculated transmission spectra of the organic MC with a Ag layer of different thicknesses inserted between the organic cavity layer and TiO_2 top layer of the bottom DBR.

wavelength organic microcavity with a silver layer of increasing thickness (Fig. 5). At a silver thickness of zero, only one cavity resonance (TS1) exists at the design wavelength, $\lambda_c = 632$ nm. Upon increasing the silver thickness up to ~40 nm, we observe a continuous shift of the cavity resonance towards lower energies and the emergence of a new mode (TS2) from the DBR side band shifting to higher energies. Both spectral movements of the modes saturate at a silver thickness of ~40–50 nm, at a minimal spectral distance of 22 nm. In Fig. 6, the simulated spatial distribution of the resonance cavity mode at λ_c ~620 nm demonstrates a substantial enhancement of the electric field in the cavity resulting in an enhancement of the spontaneous emission.

In Fig. 7, the calculated and experimental transmission spectra of unpolarized light are presented. The simulated parameters are selected to obtain the best agreement between experimental results and simulations for two cases, with and without the embedded metallic layer. When the metal is present, both of the Tamm states are visible, with the base eigenenergy of TS1 at 640 nm. This state is therefore supported by the gain of DCM and may preferentially show lasing. With no metal inside the structure, the ordinary cavity state is located at 620 nm.

To experimentally demonstrate and investigate the lasing from the Tamm-plasmon states, nanosecond pulses of 532 nm wavelength are utilized to pump neighboring regions either with or without 40 nm Ag layer. In this experiment, regions with homogeneous metal, 15 μm away from metal free regions, are pumped *via* the focused beam (diameter of 3 μm). In Fig. 8, the angle-resolved emission patterns of the cavity against the excitation energy are presented. In panel (a), both the cavity state and the TS1 appear.

Examining the quality factor, $Q = \lambda_c/\Delta\lambda_c$, of the modes leads to values of ~1400 for the cavity state and about ~650 for TS1.[27] In panel (b), the emission intensity increases by a factor of two to 0.11 nJ. By increasing the excitation energy further, the coherence threshold of the cavity state is reached at 620 nm (panel (c), excitation energy 0.2 nJ). At $k = 0$ of the corresponding parabola, a super-linear

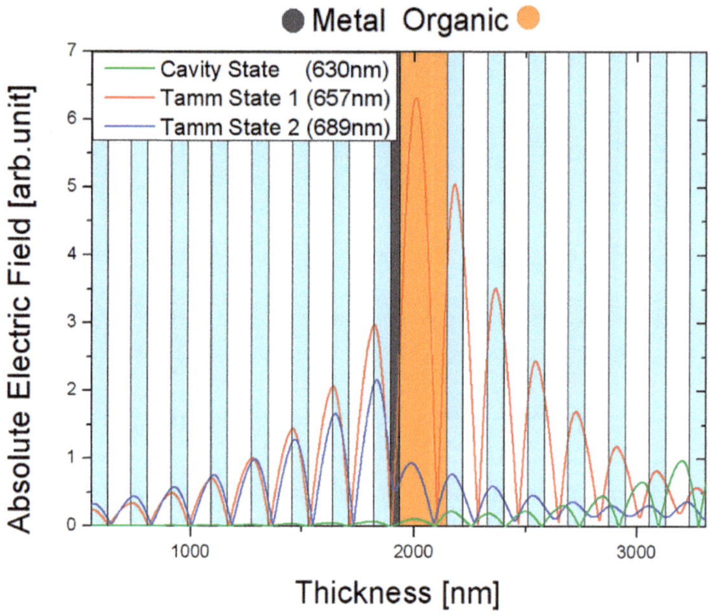

Fig. 6 Spatial distribution of the cavity modes with wavelengths of 630, 657, and 686 nm in MC with an inserted 40 nm silver layer between the left DBR and organic layer.

increase in emission intensity is observed accompanied by a reduction in line-width. A further increase in pump energy leads to the observation of a second non-linear threshold, this time of the TS1 at ~640 nm. A significant reduction in

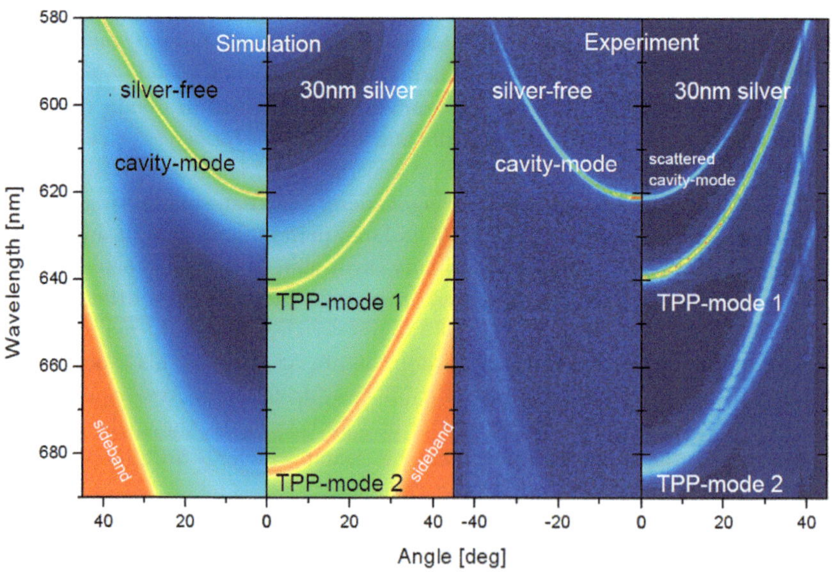

Fig. 7 Calculated (left panel) and experimental (right panel) angular-resolved transmission spectra without (left part of each panel) and with (right part of each panel) inserted 30 nm Ag layer.

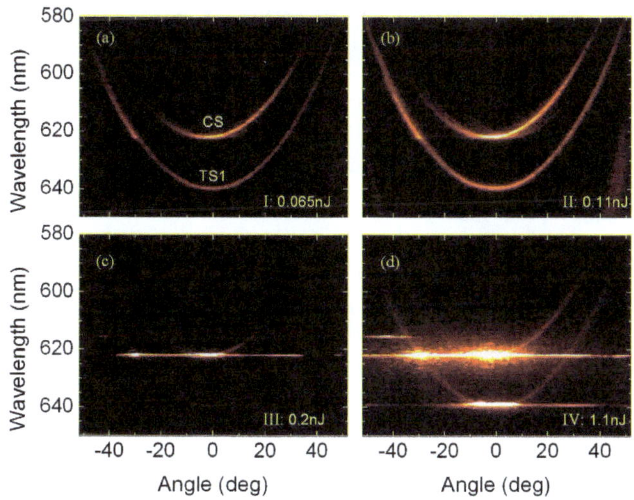

Fig. 8 Angular dispersion of the CS and TS1 against the excitation energy. (a) At low excitation energy, both modes are apparent with different intensities. (b) Increasing the pump energy leads to a linear increase in the spontaneous emission until (c), the threshold of the cavity mode is reached. (d) By increasing the excitation further, the emission of the TS1 mode at $k = 0$ increases super-linearly indicating lasing. Graph taken from ref. 27.[27]

the spectral emission wavelength is further evidence of lasing from a Tamm-plasmon polariton, see panel (d).

Confined modes

Elliptic holes in the metal layer

Light propagating in 3D space is reflected from the interface between two media with different refractive indices, or with different propagation constants for a 2D waveguide. In a planar MC, light can be confined in the *z*-direction between two DBRs, and additionally in elliptic islands in the *xy* space within the thin metal layers.

The far field emission of holes with different diameters (4, 7, and 10 μm) excited exactly in the middle of the hole is shown in Fig. 9(a)–(c). The parabola-shaped dispersion curves with cutoff wavelengths at 655 nm (TS1) and 690 nm (TS2) originate from Tamm states formed due to the incorporated 40 nm silver layer. Each of the states is represented by a set of two parabolas as a result of the splitting between the TE and TM polarized light at nonzero angles.

In addition to the states TS1 and TS2, a parabola-shaped discretized cavity state (CS) is present at 619 nm. The coexistence of Tamm states and the cavity parabola is explained by the lateral extent of the wave functions which may substantially overlap at regions next to the metal edge. The discrete far field pattern of the CS is a typical fingerprint of confined optical systems which is caused by the boundary conditions of the holes in the metal layer.[28]

As expected for good optical confinement, the energy spacing between adjacent discrete levels decreases and the angular linewidth narrows when the hole diameter is increased (Fig. 9(b)).[29,30] By increasing the diameter further, the energy

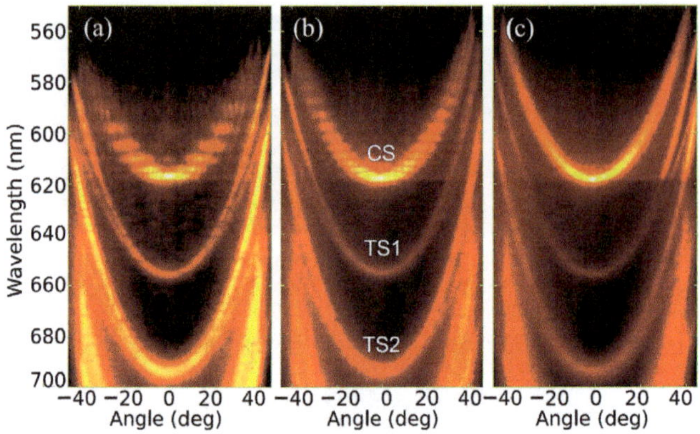

Fig. 9 Angle-resolved photoluminescence spectra of the metal–organic MC excited centrally in three holes of (a) 4 μm, (b) 7 μm, and (c) 10 μm diameter by a 405 nm cw laser beam with a 2 μm-diameter spot. Reprint with permission from ref. 30. Copyright 2005, AIP LLC.[30]

splitting vanishes (Fig. 9(c)) and the parabola narrows until, at a diameter of 15 μm, it becomes indistinguishable from the delocalized cavity mode.

As presented in Fig. 10, the lasing threshold and efficiency of the emission from μ-holes are improved in comparison to surrounded Tamm-plasmon polariton states.

External efficiency control *via* gain spectrum and periodic structuring

Lai *et al.*[31] investigated emission from the array of periodically spaced metal stripes and observed coherent emission from both the zero-ground state and

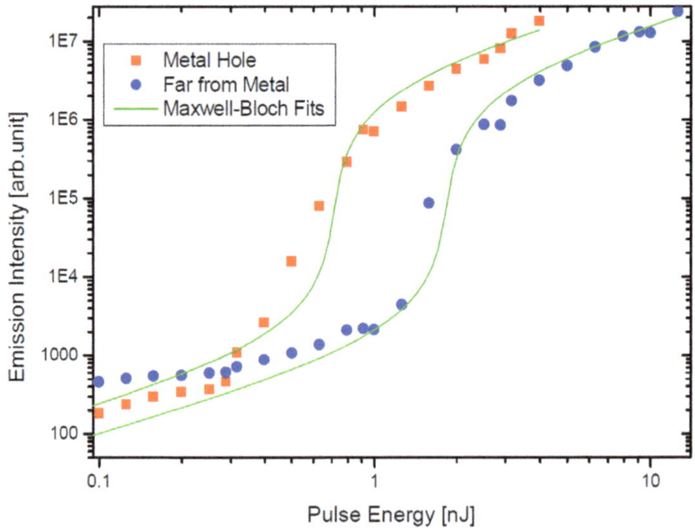

Fig. 10 Cavity state (squares) lasing threshold of a 12 μm hole compared to unstructured spot (circles) confinement improves the lasing threshold by decreasing the cavity decay rate by a factor of 3.

Bloch-like excited π-states. In the following publications,[32,33] it was shown that at higher excitations the extended and confined modes can simultaneously coexist and be manipulated optically, depending on the excitation spot position in periodically modulated inorganic MC structures. These coherent phenomena in photonic boxes and wires have been described theoretically.[34]

Periodic distribution of gain in planar MCs can be created by two or more interfering laser beams with a period of a few microns depending on the angle between the excitation beams. In these cases, coherent emissions have a periodic distribution, resulting in two or more normal and oblique lasing beams.

In Fig. 11(a–d), the coherent multi-beam laser emission from organic planar MCs at above-threshold periodically-modulated excitations is shown for four different angles between two interfering excitation beams. In addition to the spontaneous emission with a parabola-like dispersion (green), substantially strong multi-beam lasing emission peaks (pink) occur near the bottom of the parabola at $\theta = 0°$. The corresponding spatial distributions of the emission patterns are presented in Fig. 11(e–h).

In the case of a periodically rectangular patterned cavity layer of the organic MC with periods of $\sim$7 and 11 μm, coincidence of the extended and confined modes were observed in the far-field emission spectra (Fig. 12).

Organic light-emitting diodes

To achieve high luminous efficacy, the driving voltage of the OLED needs to be low. With the introduction of doped transport layers and exciton blocking layers, pin-OLEDs achieve driving voltages close to the thermodynamic limit. We will thus not discuss this aspect further in this work. In the following, the key factors which determine the external quantum efficiency (EQE) of OLEDs will be described and explained.

External quantum efficiency[35,36]

Regarding the electrical excitation, creation of electron–hole pairs and excitons, their radiative emission, and optical outcoupling processes, the external quantum

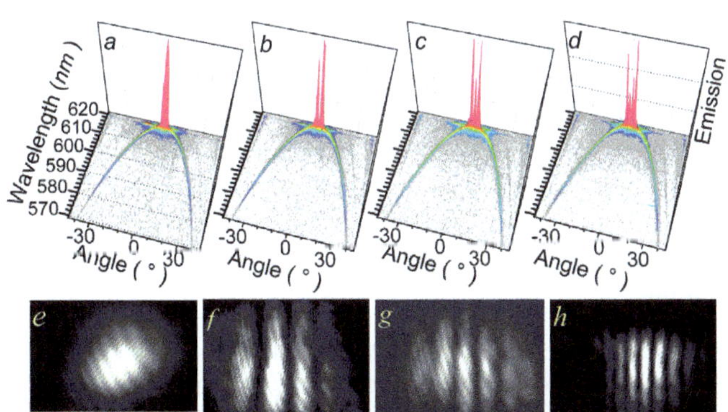

Fig. 11 Coherent multi-beam laser emission from an organic planar MC at above-threshold periodically-modulated excitations with four different angles between two interfering excitation beams.

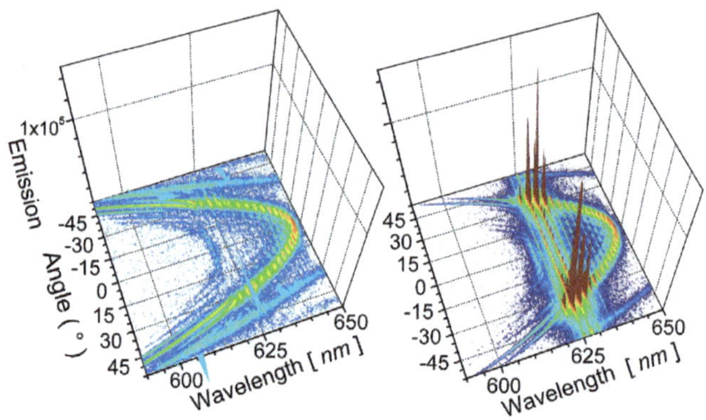

Fig. 12 Experimental angular-resolved emission spectra from an organic MC with a periodically-modulated cavity layer below (left) and above (right) the lasing threshold.

efficiency, η_{EQE}, is defined as the ratio of the number of photons emitted from the device to the number of carriers injected into the molecular layers. It can be split into four terms:

$$\eta_{EQE} = \gamma\eta_{S/T}\eta_r\eta_{out}, \tag{6}$$

where γ is the electrical efficiency (also known as the charge carrier balance factor). It describes the ratio of generated excitons to the number of injected electrons. The ratio of singlets/triplets is described by the exciton spin factor $\eta_{S/T}$, with $\eta_{S/T} = 1$ for phosphorescent emitters, and $\eta_{S/T} = 0.25$ for fluorescent emitters. The effective radiative efficiency, η_r (or internal quantum efficiency of the emitter), describes the number of generated photons to the number of generated excitons:

$$\eta_r = F\Gamma_r/(F\Gamma_r + \Gamma_{nr}). \tag{7}$$

The Purcell[14] factor, F, accounts for the quantum mechanical effect that the radiative efficiency of an emitter depends on its surrounding. In free space $F = 1$, while in OLEDs, F ranges values below and above 1. Γ_r and Γ_{nr} are the radiative and nonradiative decay rates of the emitter, which are strongly dependent on the emission spectrum. Here, the wavelength dependency of the respective efficiencies (*i.e.* the photoluminescence, absorption and gain spectra) needs to be considered. Often, the product of the charge balance, the exciton spin factor, and the efficiency of radiative decay are summarized as the internal quantum efficiency, $\eta_i \leq 1$.

Light outcoupling

The last term in eqn (6) is the outcoupling efficiency, η_{out}. It accounts for the ratio of outcoupled photons to generated photons, *i.e.* the probability that light generated within the device is coupled out through the transparent substrate (Fig. 13). The outcoupling losses are caused by the different indices of refraction of the organic materials, metal, ITO, and the glass substrate. Thus, reflected light will be trapped within the organic and the glass, which leads to absorption in the

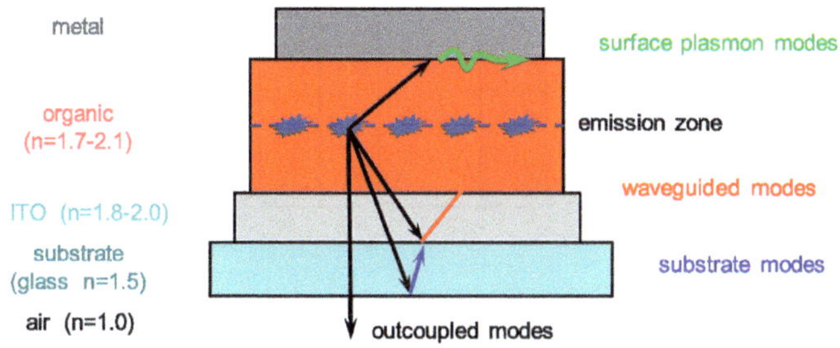

Fig. 13 Light loss mechanisms inside the OLED resulting from total internal reflection. The light modes can be distinguished into outcoupled, wave-guided, and surface plasmon modes. Absorption losses in the ITO and the doped transport layers can be expected to play a minor role. Refraction at the interfaces has been omitted for simplification.

ITO and the metal cathode. As shown in Fig. 13, three different light modes can be distinguished within the device: (i) surface plasmon modes at the organic–metal interface, (ii) wave-guided modes, and (iii) substrate modes. These losses indicate that there is huge potential for efficiency improvement by outcoupling the trapped modes.[37,38]

The outcoupled emission exhibits a strong dependence on the emission layer (EML) position and thickness. Fig. 14 shows the experimentally obtained EQE (dots) in conjunction with the simulated loss mechanisms for a red phosphorescent OLED. The outcoupled emission is characterized by a first maximum, a first minimum, and a second maximum. This can be understood by regarding the resonance condition inside the OLED cavity for 0°: $2L = m\lambda$, with $L = n_c d$ as the optical thickness of the cavity (*i.e.* the cavity length, d, multiplied by the refractive index, n_c) and $m = 1, 2, \ldots$, the resonance order. The cavity condition is a result of

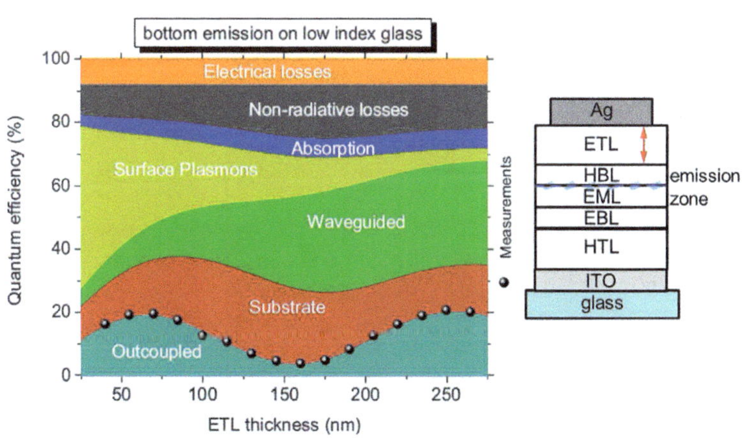

Fig. 14 Quantified loss mechanisms in a red phosphorescent pin-OLED for different ETL thicknesses by Meerheim *et al.*[36] The simulated external quantum efficiency agrees nicely with experimental results (at low current density of 1.51 mA cm^{-2}, configuration as shown on the right side). Reprint with permission from ref. 39. Copyright 2010, American Institute of Physics.[39]

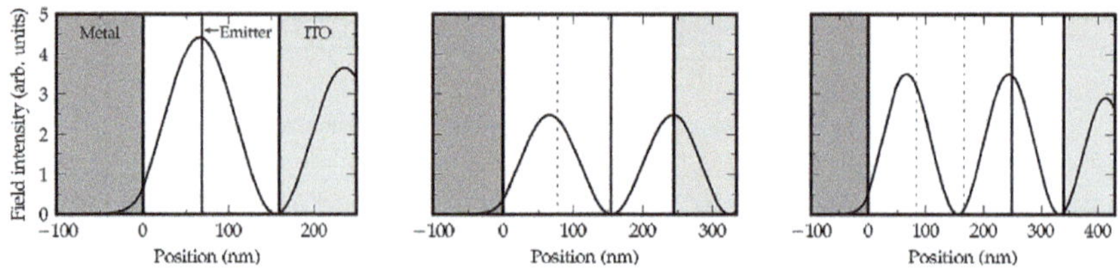

Fig. 15 Electromagnetic field intensities of the waveguide modes for the devices shown in Fig. 14 at the peak wavelength of the emitter (610 nm, 0°) for metal/organic/ITO structures. From left to right: 3/4λ cavity with the emitter located 70 nm from the metallic contact, λ cavity with the emitter at 155 nm and 5/4λ cavity with the emitter at 250 nm. Increasing the ETL thickness and thus the cavity thickness shifts the emitting molecules through a first maximum, a minimum, and a second maximum. Reprint with permission from ref. 9. Copyright 2010, American Institute of Physics.[9]

constructive and destructive interference of the light modes. The calculation of the electromagnetic field confirms this condition. The calculated field intensity for different cavity lengths, L, and emitter positions is shown in Fig. 15.

If the emitting molecules are placed in the maximum of the electro-magnetic field, a large number of radiation modes are available, and therefore a pronounced coupling to the outcoupled modes can be expected. The coupling of the emitting molecules to the electromagnetic field strongly influences the outcoupling efficiency and qualitatively explains the dependence of the outcoupling efficiency in Fig. 14. In contrast, placing the emitters into the node of the field, leads to low outcoupling efficiencies. The influence of the emitter coupling on the field and the Purcell effect on the outcoupling efficiency are often referred to as cavity effects.

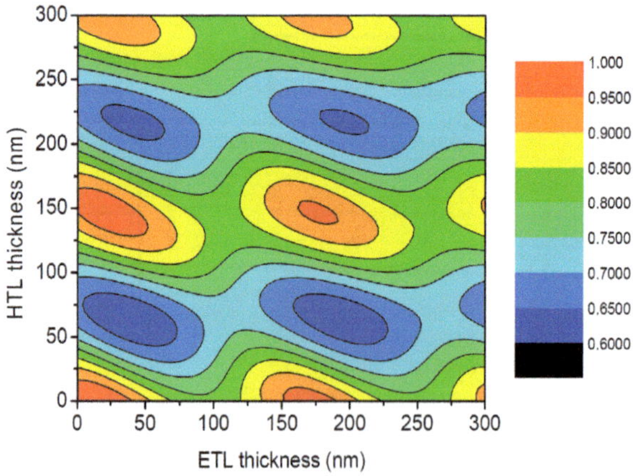

Fig. 16 Normalized light transmission of the top contact including the capping layer at a wavelength of 510 nm as a function of the capping and spacer 1 layer thickness calculated based on the following OLED structure: Ag (80 nm)/spacer 1 (Y nm)/emitter (15 nm)/spacer 2 (35 nm)/Ag (15 nm)/capping layer (X nm). The refractive index of all of organic materials used is assumed to be 1.75.

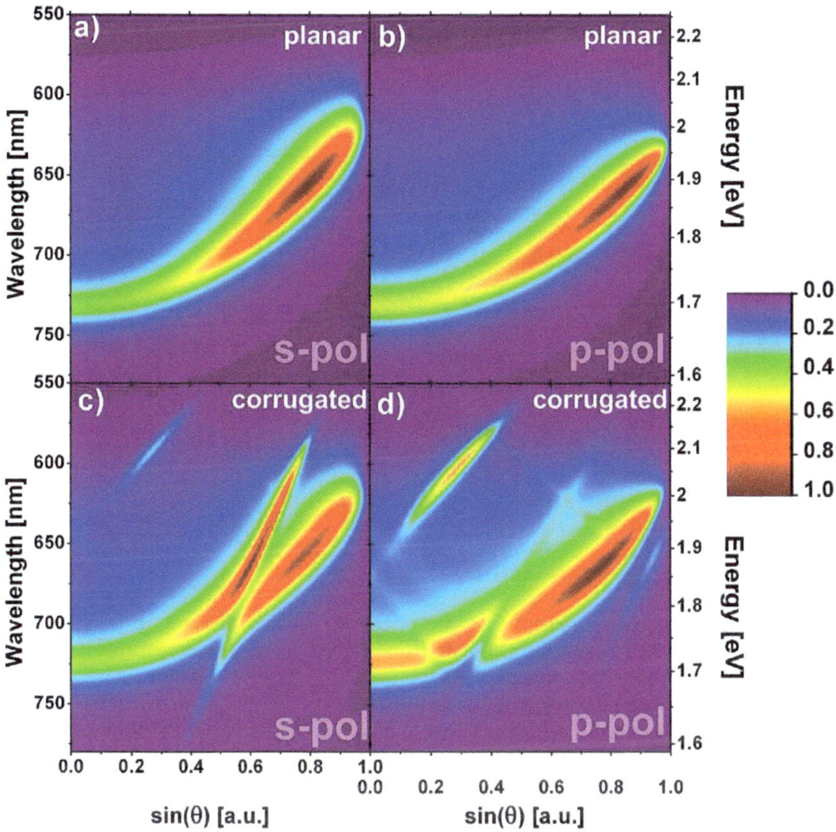

Fig. 17 Measured angular dependent emission spectra for s- and p-polarized light of planar (a and b) and corrugated top-emitting OLED (c and d) operating in the second optical maximum. Bragg scattering at the one dimensional shallow sub-μm grating superimposes sharp spectral features to the resonant cavity emission, resulting from constructive and destructive interference of the cavity emission and Bragg scattered modes. Reproduced from ref. 48. Copyright 2013, International Society for Optics and Photonics.

A large number of methods have been studied to improve light outcoupling, for example, the use of a resonant cavity,[40] the excitation of surface plasmons,[41] the use of periodic structures placed in the optically active layer to introduce Bragg scattering normal to the substrate plane,[42] high refractive index glass in combination with a half-sphere,[43] and many more. Despite the fact that white OLEDs with fluorescent tube efficiencies can be achieved at a lab scale, outcoupling methods for OLED fabrication on an industrial scale are still pending.

Top-emitting OLEDs

Top-emitting OLEDs are characterized by a highly reflective bottom electrode and a semitransparent top electrode, and in contrast to the previously described bottom-emitting OLEDs emit light away from the substrate. Thus, the OLED forms an MC, where the light is reflected back and forth between both electrodes

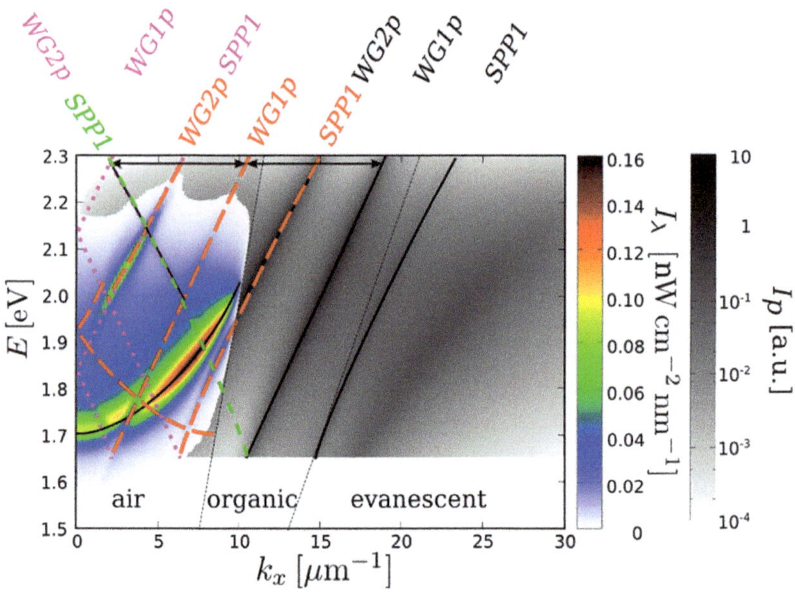

Fig. 18 Measured spectral radiant intensity of the corrugated device in p-polarization (false color scale) along with the simulated power dissipation spectra of the identical planar reference (gray-scale) outside the experimentally accessible region. Black arrows indicate the reciprocal lattice vector and link the modes of the simulated planar MC to the measured modes arising from Bragg scattering (colored lines: scattered surface plasmon (SPP) or wave-guided (WG) modes). Reproduced from ref. 50. Copyright 2013, International Society for Optics and Photonics.

leading to narrowing of the emitted spectrum and inefficient light outcoupling at non-resonant conditions.

Capping layer

To improve light outcoupling and to weaken the microcavity effects, a capping layer of a high-index material is deposited onto the top electrode. This changes the transmittance of the top contact, as known from thin film optics, and improves the EQE of the OLED.[44] In Fig. 16, it is demonstrated that the maximum efficiency is obtained at neither the highest nor the lowest transparency of the top contact, but at a certain value in between, which depends on the interference effects within the device.

In fact the best outcoupling results are obtained when not only the transmittance of the top electrode is optimized by varying the capping layer, but when the optical structure of the whole device is adjusted.

Bragg scattering[45–49]

In Fig. 17(a and b), angular-resolved s- and p-polarized spontaneous emission spectra of a planar top-emitting OLED are shown. As is typical for MCs, the emitted light demonstrates a parabolic angular dependence of the emission spectra. In order to enhance the EQE, a corrugation in the form of a line grating with a period of 750 nm and 40 nm height is introduced by a nano-imprint photoresist layer beneath the bottom electrode.

The corresponding emission spectra are given in Fig. 17(c and d). The Bragg scattering at the 1D grating superimposes sharp spectral features to the cavity-mode. By analyzing the measured emission spectra with the simulated emission for the whole *k* space, these features can be identified as Bragg scattered wave-guided and surface plasmon modes of different orders (Fig. 18).

Moreover, the EQE of the corrugated OLED is enhanced by a factor of 1.16 in comparison to the planar reference.

Conclusions

The external efficiency of light-emitting structures is strongly influenced by their optical design. Our article discusses some approaches to increase the external efficiency of organic microcavities and organic light-emitting devices. In organic microcavities with a high *Q*-factor $\sim$1000, where the emitting layer is embedded by distributed Bragg reflectors, the 1D confinement increases the spontaneous emission due to the Purcell effect. Further optimization of the thicknesses of the metal, active organic and adjusting dielectric layer minimizes absorption of light by the metal, allowing the lasing of the coupled Tamm-plasmon polaritons. High above the threshold excitation, lasing substantially increases the ratio between radiative and non-radiative emission. Corrugation of the metal contacts, such as perforated μ-size holes or a periodic array of metal stripes, leads to 2D confinement of the cavity modes, decreasing the lasing threshold.

Using light refractive, diffractive, or scattering elements in OLEDs, the external quantum efficiency can be strongly enhanced. For top-emitting OLEDs a dielectric organic capping layer and periodically structured line-grating especially increase the efficiency. It is expected that by optimization of the grating period, grating height, and the implementation of two-dimensional gratings, the light out-coupling and thus the EQE can be further improved.

Acknowledgements

We thank Mauro Furno, Qiang Huang, Rico Meerheim, Tobias Schwab, Cornelius Fuchs, Andreas Mischok, and Markas Sudzius. The authors gratefully acknowledge the financial support of this work by the Bundesministerium für Bildung und Forschung (BMBF), Deutsche Forschungsgemeinschaft (also *via* the excellence cluster CfAED), and the European Community.

Notes and references

1 R. Brückner, A. A. Zakhidov, R. Scholz, M. Sudzius, S. I. Hintschich, H. Fröb, V. G. Lyssenko and K. Leo, *Nat. Photonics*, 2012, **6**, 322.
2 C. W. Tang and S. A. Van Slyke, *Appl. Phys. Lett.*, 1987, **51**, 913.
3 J. Blochwitz, M. Pfeiffer, T. Fritz and K. Leo, *Appl. Phys. Lett.*, 1998, **73**, 729.
4 M. Pfeiffer, T. Fritz, J. Blochwitz, A. Nollau, B. Plönnigs, A. Beyer and K. Leo, *Adv. Solid State Phys.*, 1998, **39**, 77.
5 J. Blochwitz, Organic light-emitting diodes with doped charge transport layers, PhD Thesis, TU Dresden, 2001.

6 J. Huang, M. Pfeiffer, A. Werner, J. Blochwitz, K. Leo and S. Liu, *Appl. Phys. Lett.*, 2002, **80**, 139.

7 M. Pfeiffer, K. Leo, X. Zhou, J. S. Huang, M. Hofmann, A. Werner and J. Blochwitz-Nimoth, *Org. Electron.*, 2003, **4**, 89.

8 W. L. Barnes, A. Dereux and T. W. Ebbesen, *Nature*, 2003, **424**, 824.

9 M. Furno, R. Meerheim, S. Hofmann, B. Lüssem and K. Leo, *Phys. Rev. B: Solid State*, 2012, **85**, 115205.

10 H. Benisty, H. D. Neve and C. Weisbuch, *IEEE J. Quantum Electron.*, 1998, **34**, 1632.

11 J. A. E. Wasey and W. L. Barnes, *J. Mod. Opt.*, 2000, **47**, 725.

12 S. Astilean and W. L. Barnes, *Appl. Phys. B: Lasers Opt.*, 2002, **95**, 591.

13 Y. Rag Do, Y.-C. Kim, Y.-W. Song and Y.-H. Lee, *J. Appl. Phys.*, 2004, **96**, 7629.

14 B. C. Krummacher, J. Frischeisen, N. A. Reinke and W. Brütting, *J. Appl. Phys.*, 2008, **104**, 123109.

15 C. Weisbuch, M. Nishioka, A. Ishikawa and Y. Arakawa, *Phys. Rev. Lett.*, 1992, **96**, 3314.

16 J. M. Gérard, B. Sermage, B. Gayral, B. Legrand, E. Costard and V. Thierry-Mieg, *Phys. Rev. Lett.*, 1998, **81**, 1110.

17 E. M. Purcell, *Phys. Rev.*, 1946, **69**, 681.

18 P. R. Rice and H. J. Carmichael, *Phys. Rev. A: At., Mol., Opt. Phys.*, 1994, **50**, 4318.

19 G. Bjork and Y. Yamamoto, *IEEE J. Quantum Electron.*, 1991, **27**, 2386.

20 Y. Yamamoto, S. Machida and G. Bjork, *Opt. Quantum Electron.*, 1992, **24**, S215.

21 H. Yokoyama and S. D. Brorson, *J. Appl. Phys.*, 1988, **66**, 4801.

22 M. Koschorreck, R. Gehlhaar, V. G. Lyssenko, M. Swoboda, M. Hoffmann and K. Leo, *Appl. Phys. Lett.*, 2005, **87**, 181108.

23 C.-Y. Lu, S.-W. Chang, S. L. Chuang, T. D. Germann and D. Bimberg, *Appl. Phys. Lett.*, 2010, **96**, 251101.

24 A. V. Kavokin, I. A. Shelykh and G. Malpuech, *Phys. Rev. B: Solid State*, 2005, **72**, 233102.

25 M. Kaliteevski, I. Iorsh, S. Brand, R. Abram, J. Chamberlain, A. Kavokin and I. Shelykh, *Phys. Rev. B: Solid State*, 2007, **76**, 165415.

26 R. Brückner, M. Sudzius, S. I. Hintscich, H. Fröb, V. G. Lyssenko and K. Leo, *Phys. Rev. B.*, 2011, **83**, 033405.

27 R. Brückner, Coherence and Coupling of Cavity Photons and Tamm Plasmons in Metal-Organic Microcavities, PhD Thesis, TU Dresden, 2013.

28 M. Langner, M. Sudzius, H. Fröb, V. G. Lyssenko and K. Leo, *Appl. Phys. Lett.*, 2009, **95**, 091109.

29 A. Muller, C.-K. Shih, J. Ahn, D. Lu, D. Gazula and D. G. Deppe, *Appl. Phys. Lett.*, 2006, **88**, 031107.

30 C. Reinhardt, R. Brückner, J. Haase, M. Sudzius, S. I. Hintschich, H. Fröb, V. G. Lyssenko and K. Leo, *Appl. Phys. Lett.*, 2012, **100**, 103306.

31 C. W. Lai, N. Y. Kim, S. Utsunomiya, G. Roumpos, H. Deng, M. D. Fraser, T. Byrnes, P. Recher, N. Kumada, T. Fujisawa and Y. Yamamoto, *Nature*, 2007, **450**, 529.

32 E. Wertz, L. Ferrier, D. D. Solnyshkov, R. Johne, D. Sanvitto, A. Lemaître, I. Sagnes, R. Grousson, A. V. Kavokin, P. Senellart, G. Malpuech and J. Bloch, *Nat. Phys.*, 2010, **6**, 860.

33 L. Ferrier, E. Wertz, R. Johne, D. D. Solnyshkov, P. Senellart, I. Sagnes, A. Lemaître, G. Malpuech and J. Bloch, *Phys. Rev. Lett.*, 2011, **106**, 126401.
34 M. Wouters, T. C. H. Liew and V. Savona, *Phys. Rev. B: Solid State*, 2010, **82**, 245315.
35 K. Walzer, B. Maennig, M. Pfeiffer and K. Leo, *Chem. Rev.*, 2007, **107**, 233.
36 R. Meerheim, B. Lüssem and K. Leo, *Proc. IEEE*, 2009, **97**, 1606.
37 L. Xiao, Z. Chen, B. Qu, J. Luo, S. Kong, Q. Gong and J. Kido, *Adv. Mater.*, 2011, **23**, 926.
38 G. He, M. Pfeiffer, K. Leo, M. Hofmann, J. Birnstock, R. Pudzich and J. Salbeck, *Appl. Phys. Lett.*, 2004, **85**, 3911.
39 R. Meerheim, M. Furno, S. Hofmann, B. Lüssem and K. Leo, *Appl. Phys. Lett.*, 2010, **97**, 253305.
40 T. Shiga, H. Fujikawa and Y. Taga, *J. Appl. Phys.*, 2003, **93**, 19.
41 P. Andrew and W. L. Barnes, *Science*, 2004, **5**, 1002.
42 J. Feng, T. Okamoto and S. Kawata, *Appl. Phys. Lett.*, 2005, **87**, 241109.
43 J. M. Lupton, B. J. Matterson, I. D. W. Samuel, M. J. Jory and W. L. Barnes, *Appl. Phys. Lett.*, 2000, **77**, 3340.
44 R. Reineke, F. Lindner, G. Schwartz, N. Seidler, K. Walzer, B. Lüssem and K. Leo, *Nature*, 2009, **459**, 234.
45 Q. Huang, K. Walzer, M. Pfeiffer, K. Leo, M. Hofmann and T. Stübinger, *J. Appl. Phys.*, 2006, **100**, 064507.
46 P. A. Hobson, S. Wedge, J. A. E. Wasey, I. Sage and W. L. Barnes, *Adv. Mater.*, 2002, **14**, 1393.
47 L. H. Smith, J. A. E. Wasey, I. D. W. Samuel and W. L. Barnes, *Adv. Funct. Mater.*, 2005, **15**, 1839.
48 T. Schwab, C. Fuchs, R. Schol, X. Li, F. Xie, W. Choy, K. Leo and M. C. Gather, *Proc. SPIE*, 2013, 882911.
49 T. Schwab, C. Fuchs, R. Scholz, A. Zakhidov, K. Leo and M. C. Gather, *Opt. Express*, 2014, **22**, 7524.
50 C. Fuchs, T. Schwab, T. Roch, S. Eckard, A. Lasagni, S. Hofmann, B. Lüssem, L. Müller-Meskamp, K. Leo, M. C. Gather and R. Scholz, *Opt. Express*, 2013, **21**, 16319.

Faraday Discussions

PAPER

Dynamic amplification of light signals in photorefractive ferroelectric liquid crystalline mixtures

Takeo Sasaki,* Satoshi Kajikawa and Yumiko Naka

Received 14th April 2014, Accepted 4th June 2014
DOI: 10.1039/c4fd00068d

The photorefractive effect in photoconductive ferroelectric liquid crystals that contain photoconductive chiral compounds was investigated. Terthiophene compounds with chiral structures were chosen as the photoconductive chiral compounds, and they were mixed with an achiral smectic C liquid crystal. The mixtures exhibit the ferroelectric chiral smectic C phase. The photorefractivity of the mixtures was investigated by two-beam coupling experiments. It was found that the ferroelectric liquid crystals containing the photoconductive chiral compound exhibit a large gain coefficient of over 1200 cm^{-1} and a fast response time of 1 ms. Real-time dynamic amplification of an optical image signal of over 30 fps using the photorefractive ferroelectric liquid crystal was demonstrated.

1 Introduction

The photorefractive effect in ferroelectric liquid crystals has been attracting significant interest.[1] The photorefractive effect is a phenomenon that allows for the formation of holographic images within a material; it provides the potential to realize dynamic holograms by recording holograms as a change in the refractive index of a medium.[2,3] The characteristic phenomenon of the photorefractive effect is the asymmetric energy exchange in a two-beam coupling that can be used to coherently amplify signal beams; therefore, the photorefractive effect has the potential to be used in a wide range of optical technologies similar to a transistor in electrical circuits. Optically transparent materials that show both photovoltaic and electro-optic effects can exhibit the photorefractive effect.

Several photorefractive materials have been developed, including inorganic ferroelectric photoconductive crystals, organic crystals, photoconductive nonlinear optical organic polymers, amorphous organic photoconductive materials, photoconductive amphiphilic compounds, and photoconductive liquid crystals (LCs).[4–6] Organic materials in particular have attracted significant interest

Department of Chemistry, Faculty of Science, Tokyo University of Science, 1-3 Kagurazaka, Shinjuku-ku, Tokyo 162-8601, Japan. E-mail: sasaki@rs.kagu.tus.ac.jp

in this context since 1994, because they exhibit large photorefractivity and short response times.[6]

The photorefractive effect induces a change in the refractive index by a mechanism that involves both photovoltaic and electro-optic effects (Fig. 1). When two laser beams interfere in an organic photorefractive material, charge generation occurs at the bright positions of the interference fringes. At the same time, an electric field is usually applied to the material. The generated charges diffuse or drift within the material. The mobilities of the positive and negative charges are different in most organic materials, so that a charge separated state is formed. The charge with a higher mobility diffuses over a longer distance than the charge with a lower mobility; therefore, while the low mobility charge stays in the bright areas, the high mobility charge moves to the dark areas. The bright and dark positions of the interference fringes are thus charged with opposite polarities, and an internal electric field (space charge field) is formed in the area between the bright and dark positions. The refractive index of the area between the bright and dark positions is changed through the electro-optic effect. Thus, a refractive index grating (in other words, a hologram) is formed.

A high electric field of 10–50 V μm^{-1} is usually applied to a polymer film, in addition to the internal electric field, to obtain photorefractivity. The polymer material is typically 100 μm thick, so that the voltage necessary to achieve photorefractivity is 1–5 kV, which is comparable to the breakdown voltage of the polymer film. This electric field is necessary to increase the charge generation efficiency. One class of materials that exhibits high photorefractivity is glassy photoconductive polymers doped with high concentrations of D–π–A chromophores, in which the donor and acceptor groups are attached to a π-conjugate system.[6,7] A multiplex hologram has been demonstrated using a photorefractive polymer film during the development of a 3D display,[8,9] and clear 3D images were recorded in the film. However, the slow response (*ca.* 100 ms) and high electric field (30–50 V μm^{-1}) required to activate the photorefractive effect in polymer materials require improvement.

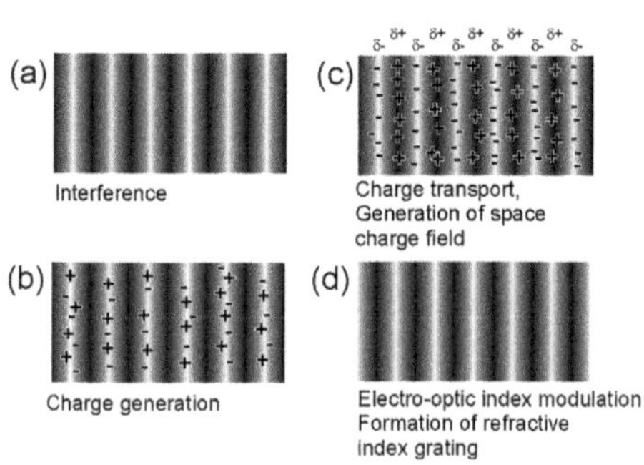

Fig. 1 Schematic illustration of the mechanism of the photorefractive effect. (a) Two laser beams interfere in the photorefractive material; (b) charge generation occurs in the bright areas of the interference fringes; (c) electrons are trapped at trap sites in the bright areas, holes migrate by diffusion or drift in the presence of an external electric field and generate an internal electric field between the bright and dark positions; and (d) the refractive index of the corresponding area is altered by the internal electric field that is generated.

The photorefractive effect in LCs has been investigated previously.[7] LCs are basically liquid, so they can be easily driven by a low electric field. LCs are classified into several groups, the most well known of which are nematic and smectic LCs. Nematic LCs are used in liquid crystal displays (LCDs), whereas smectic LCs are very viscous and are thus seldom utilized in practical applications. Nematic LCs were first used as photorefractive LCs and a large photorefractivity was obtained with the application of an electric field of only a few volts per micrometer.[10] The photorefractive effect has been reported in surface-stabilized ferroelectric liquid crystals (SS-FLCs) doped with a photoconductive compound.[14,15] Ferroelectric liquid crystals (FLCs) belong to a class of smectic LCs that have a layered structure.[16,17] FLCs exhibit a chiral smectic C phase (SmC*) that possesses a helical structure. In order to observe ferroelectricity in the FLCs, they must be formed into thin films.[14] The thickness of the film needs to be a few micrometers. When an FLC is sandwiched between glass plates to form a film a few micrometers thick, the helical structure of the SmC* phase uncoils and a surface-stabilized state (SS-state) is formed in which spontaneous polarization (Ps) appears.

For display applications, the thickness of the film is typically 2 μm. In these thin films, FLC molecules can align in only two directions, which is the SS-state, and the alignment direction of the FLC molecules changes according to the direction of the spontaneous polarization. When an alternating electric field is applied to the SS-FLC, the FLC molecules show a continuous switching motion. The electrical switching response time of FLCs is typically shorter than 1 ms. The direction of the spontaneous polarization is governed by the applied electric field, which gives rise to a change in properties according to the direction of polarization. Thus, when an internal electric field is created in an SS-FLC material, the direction of the spontaneous polarization is changed by the field.

Fig. 2 shows a schematic illustration of the mechanism of the photorefractive effect in FLCs. When laser beams interfere in a mixture of an FLC and a photoconductive compound, charge separation occurs between the bright and dark

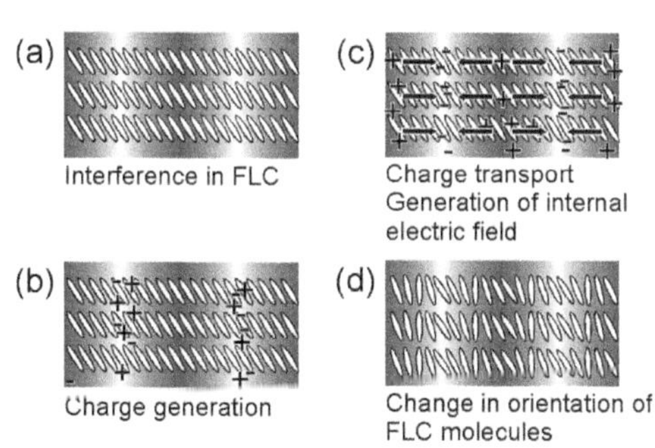

Fig. 2 Schematic illustration of the mechanism of the photorefractive effect in the FLCs. (a) Two laser beams interfere in the SS-state of the FLC–photoconductive compound mixture; (b) charge generation occurs in the bright areas of the interference fringes; (c) electrons are trapped at trap sites in the bright areas, holes migrate by diffusion or drift in the presence of an external electric field to generate an internal electric field between the bright and dark positions; and (d) the orientation of the spontaneous polarization vector (*i.e.*, orientation of mesogens in the FLCs) is altered by the internal electric field.

positions and internal electric fields are produced. The internal electric field alters the direction of the spontaneous polarization in the area between the bright and dark positions of the interference fringes, which induces a periodic change in the orientation of the FLC molecules. This is different from the processes that occur in other organic photorefractive materials, in that the molecular dipole rather than the bulk polarization responds to the internal electric field. The switching of FLC molecules is due to the response of the bulk polarization and is thus extremely fast. We have investigated the photorefractive effect in commercially available FLCs mixed with a photoconductive dopant and that of FLC mixtures comprising liquid crystalline compounds and photoconductive chiral dopants. The FLC mixtures containing photoconductive chiral dopants exhibited a large photorefractive effect. In this study, the composition of the FLC mixture was further optimised and application to optical signal amplification was demonstrated.

1.1 Characteristics of the photorefractive effect

As mentioned earlier, a change in the refractive index *via* the photorefractive effect occurs in the areas between the bright and dark positions of the interference fringes so that the phase of the resulting index grating is shifted from that of the interference fringes. A characteristic of the photorefractive effect is that the phase of the refractive index grating is $\pi/2$-shifted from the interference fringes under conditions where one of the photogenerated charges does not move from the bright position. When the material is photochemically active and is not photorefractive, then a photochemical reaction occurs at the bright areas, and a refractive index grating with the same phase as that of the interference fringes is formed [Fig. 3(a)]. The interfering laser beams are diffracted by this grating; however, when the intensities of the interfering beams are equal, the apparent transmitted intensities of the laser beams do not change, because the diffraction is symmetric. Beam 1 is diffracted in the direction of beam 2 and beam 2 is diffracted in the direction of beam 1. However, if the material is photorefractive, then the phase of the refractive index grating is shifted from that of the interference fringes and this affects the propagation of the two beams. Beam 1 is energetically coupled with beam 2 for two laser beams, and consequently, the apparent transmitted intensity of beam 1 increases while that of beam 2 decreases [Fig. 3(b)]. This phenomenon where one beam is amplified by another beam is termed asymmetric energy exchange in a two-beam coupling experiment.[2,3] The

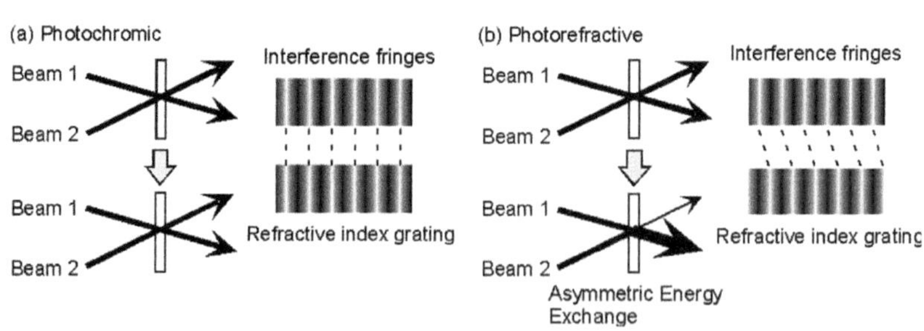

Fig. 3 (a) Photochromic grating and (b) photorefractive grating.

photorefractivity of a material is confirmed by the occurrence of this asymmetric energy exchange, which has the potential for a vast variety of optical applications.

2 Experimental

2.1 Samples

Compared to nematic LCs, FLCs are more crystalline than liquid, so that the preparation of fine FLC films requires sophisticated techniques.[16,17] It is very difficult to obtain a uniformly aligned, defect-free SS-FLC using a single FLC compound. Consequently, mixtures of several LC compounds are usually used to obtain fine SS-FLC films. An FLC mixture is composed of the base LC, which is a mixture of several LC compounds, and a chiral compound. The chiral compound introduces a helical structure to the LC phase through supramolecular interactions. To utilize a FLC as a photorefractive material, photoconductive compounds must be added to the FLC. However, the introduction of such non-LC compounds to the FLC often hinders the formation of a uniformly aligned SS-state. Thus, appropriate design of the photoconductive compounds is crucial.

The structures of the LC compounds, the electron acceptor trinitrofluorenone (TNF) and the photoconductive chiral compounds used in this study are shown in Fig. 4. The photoconductive chiral compounds were synthesized as described in a previous paper.[11] TNF was obtained from Tokyo Kasei Co. The smectic C (SmC) liquid crystal used in this study was a 1 : 1 : 2 mixture of 5-octyl-2-(4-octyloxyphenyl) pyrimidine (8PP8), 2-(4-decyloxyphenyl)-5-octyl pyrimidine (8PP10), and 2-(4-hexyloxy)-5-octyl pyrimidine (8PP6). 8PP8, 8PP10, and 8PP6 were obtained from Wako Chemicals Co., and were purified by recrystallization from a mixture of methanol and ethyl acetate solution. The mixing ratio of 8PP8, 8PP10, and 8PP6 was set to 1 : 1 : 2 because the 1 : 1 : 2 mixture exhibits the SmC phase over the widest temperature range. Hereafter, the 1 : 1 : 2 mixture of 8PP8, 8PP10, and 8PP6 is referred to as the base LC. The concentration of TNF was 0.1 wt%. The base LC, TNF, and photoconductive chiral compound were dissolved in dichloroethane, and the solvent was evaporated. The mixture was then dried in vacuum at room temperature for one week. The samples were subsequently injected into a 10 μm-gap glass cell equipped with 1 cm^2 ITO electrodes and a polyimide

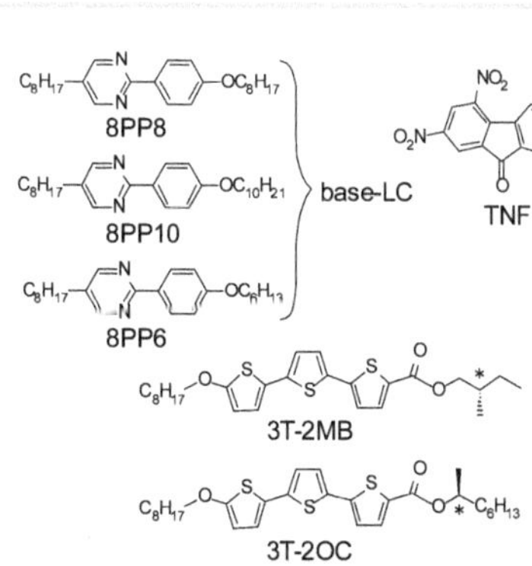

Fig. 4 Structures of the smectic LCs (8PP8, 8PP10 and 8PP6), photoconductive chiral dopants (3T-2MB and 3T-2OC) and the sensitizer TNF.

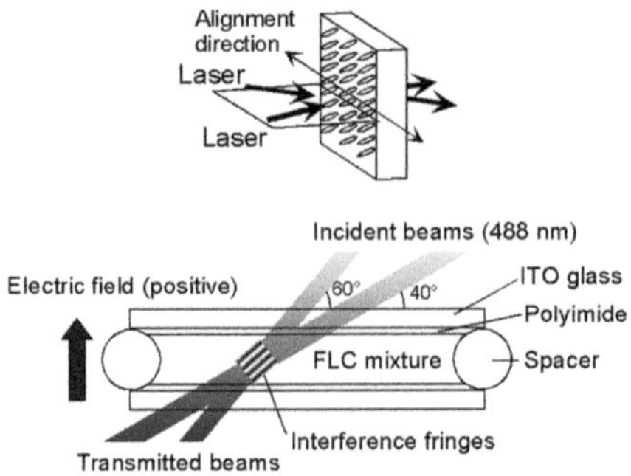

Fig. 5 Laser beam incidence condition and the structure of the LC cell.

alignment layer (LX-1400, Hitachi Chemicals Co.) for the measurements. The samples were heated to the temperature where the FLCs exhibit the isotropic phase. The samples were then gradually cooled to ambient temperature at a rate of 0.1 °C min^{-1}. This process is necessary for preparing uniformly aligned FLC samples.

2.2 Measurements

Phase transition temperatures were determined using differential scanning calorimetry (DSC; DSC822, Mettler) and microscopic observations (FP-80, FP-82, Mettler; BX-50 polarizing microscope, Olympus). Spontaneous polarization (Ps) was measured by the triangular waveform voltage method (10 V_{p-p}, 100 Hz). The photorefractive effect was measured in a two-beam coupling experiment. A linearly polarized beam from a diode-pumped solid state laser (DPSS laser, Spectra Physics, Cyan-PC13689, 488 nm, continuous wave or Oxxious, 473S-50-COL-PP, 473 nm, continuous wave) was divided in two by a beam splitter; the two beams were then interfered in the sample film. A p-polarized beam was used in most of the experiments in this study. The laser intensity was 1 mW cm^{-2} for each beam (1 mm diameter). The orientation of the rubbing direction and the beam incidence plane are shown in Fig. 5. The incident beam angles to the glass plane were 40° and 60°. Each interval of the interference fringe was 1.87 μm. The sample was maintained at 25 °C using a thermo-controller (DB1000, Chino Co.). An electric field in the range of 0 to 10 V μm^{-1} was applied to the sample from a regulated DC power supply (Kenwood DW36-1), and the change in the transmitted beam intensity was monitored by photodiodes (ET-2040, Electro-Optics Technology, Inc.) and recorded by an oscilloscope. The formation time for the refractive index grating in the FLC was measured based on the simplest single-carrier model of photorefractivity,[4,5] in which the gain transient is exponential. The rising signal of the two-beam coupling was fitted by a single exponential function.

3 Results and discussions

3.1 POM observation of SS-FLC samples

The textures of the FLC mixtures in 10 μm-gap cells were observed using polarizing optical microscopy (POM). The alignment of the FLC molecules is dominated not only by the properties of the FLCs but also by the affinity of FLC molecules with the alignment layer (polyimide coating). LC molecules interact with the alignment layer and form a homogeneous, anisotropic film in which the LC molecules are aligned in a specific direction. The fabrication of FLC cells involves the coating of ITO glass with the polyimide, rubbing treatment of the polyimide surface, and precise assembly of these glasses into a gap in a cell of only a few micrometers in width (Fig. 5). All of these processes affect the quality (homogeneity) of the surface-stabilized states, and the appropriate preparation conditions differ from FLC to FLC.

In this study, Hitachi Chemicals LX-1400 polyimide, used industrially in nematic LC displays, was chosen for the alignment layer. The thickness of the polyimide coating was set to 20–30 nm, and the surface of the polyimide was rubbed with a polyester velvet roll using a rubbing machine (RM-300/50, EHC Co. Ltd.) under specific conditions. Fig. 6 and 7 show typical examples of textures observed in 3T-2MB samples and 3T-2OC samples under the polarizing microscope. As the concentration of the photoconductive chiral dopant increased, defects appeared in the texture. The 3T-2MB sample retained the uniformly aligned state with a few defects for 3T-2MB concentrations below 8 wt% (Fig. 6). However, the 3T-2OC sample retained the uniformly aligned state with a few defects for 3T-2OC concentrations below 6 wt% (Fig. 7). The spontaneous polarization in the 3T-2MB samples was smaller than 1 nC cm^{-2}, while that in the 3T-2OC samples was about 5 nC cm^{-2}. The smaller spontaneous polarization (thus, smaller intermolecular interactions) of the 3T-2MB sample might have contributed to the uniformity of the LC phase. It must be highlighted that a striped pattern was observed in the textures of the 3T-2MB samples (Fig. 6). This indicates that the 3T-2MB samples do not form a complete SS-state in the 10 μm-gap cell. The striped pattern denotes the helical structure of the aligned LC molecules in the 10 μm-gap cell. Zig-zag defects, which are typical defects in the SS-state (book shelf structure), are also observed in the texture (Fig. 6). It was considered that the FLC mixtures showed the SS-state at the area close to the glass surface and formed a helical structure around the centre of the thickness of the 10 μm-gap cell.

3.2 Asymmetric energy exchange in FLC mixtures

The photorefractivity of the FLC mixtures was evaluated by a two-beam coupling experiment. Fig. 8 shows a typical example of the asymmetric energy exchange observed in a mixture of the base LC, 3T-2MB, and TNF at 25 °C with the application of an electric field of 1 V μm^{-1}. Interference of the divided beams in the sample resulted in an increased transmittance of one of the beams and decreased transmittance of the other beam. These transmittance characteristics were reversed when the polarity of the applied electric field was reversed. Asymmetric energy exchange was only observed when an electric field was applied, indicating that beam coupling was not caused by a thermal grating. About 40% of the energy of the laser beam L_2 was transferred to the L_1 beam.

In order to calculate the two-beam coupling gain coefficient, the diffraction condition needs to be correctly identified. The two possible diffraction conditions are the Bragg regime and the Raman–Nath regime, distinguished by the dimensionless parameter Q.[2,3]

$$Q = 2\pi\lambda L/n\Lambda^2, \tag{1}$$

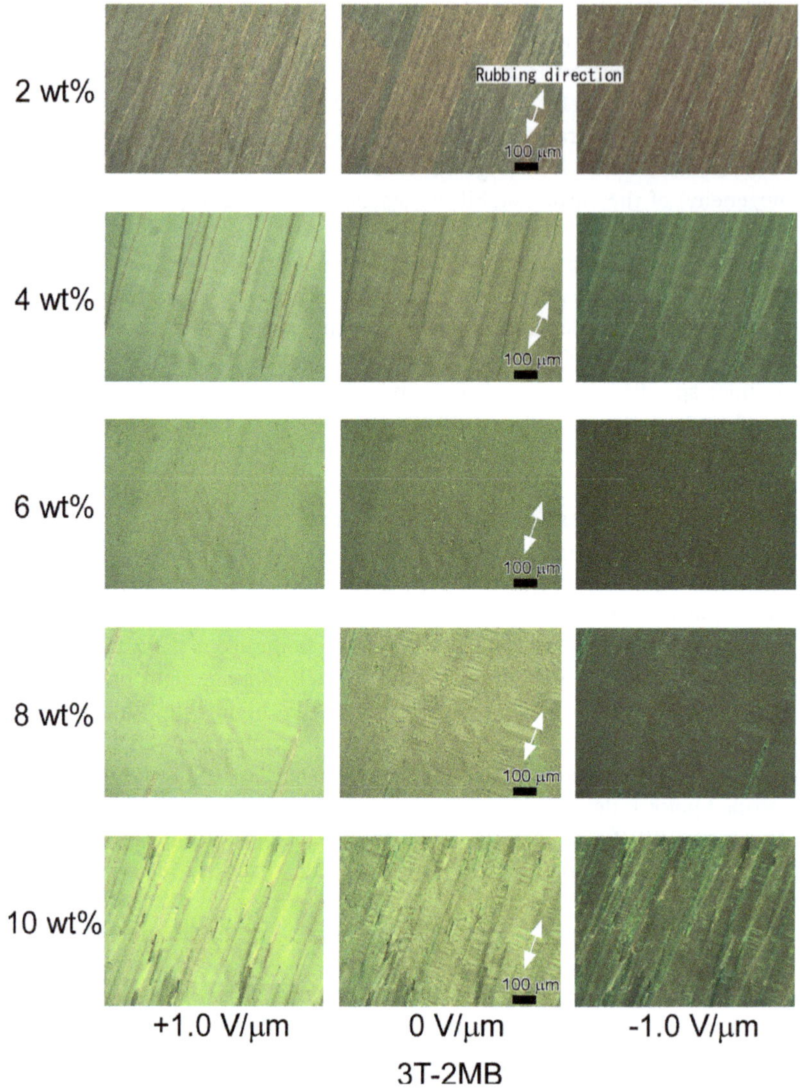

Fig. 6 Textures of FLC mixtures containing 3T-2MB in a 10 μm-gap LC cell observed under a polarizing microscope. The strengths of the external electric field were +1.0 V μm^{-1}, 0 V μm^{-1}, and −1.0 V μm^{-1}. The direction of the applied electric field is shown in Fig. 5. The 3T-2MB concentrations were in the range of 2–10 wt%, and 0.1 wt% of TNF was added.

where λ is the wavelength of the laser, L is the interaction path length, n is the refractive index, and Λ is the grating spacing. The Bragg regime of optical diffraction is defined as $Q > 1$, and excludes multiple scattering to produce only one order of diffraction of light. Conversely, $Q < 1$ is defined as the Raman–Nath regime of optical diffraction, in which many orders of diffraction can be observed. A Q value greater than 10 is usually required to guarantee that diffraction occurs entirely in the Bragg regime. Under the present experimental conditions, Q is

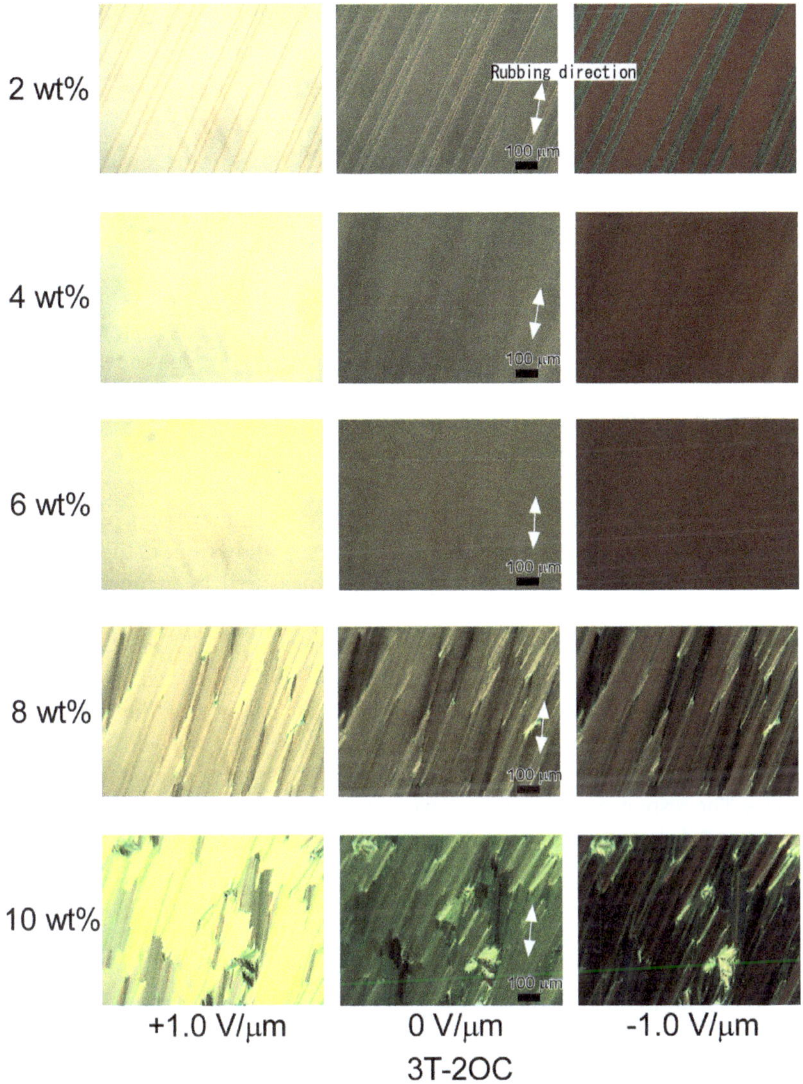

Fig. 7 Textures of FLC mixtures containing 3T-2OC in a 10 μm-gap LC cell observed under a polarizing microscope. The strengths of the external electric field were +1.0 V μm^{-1}, 0 V μm^{-1}, and −1.0 V μm^{-1}. The direction of the applied electric field is shown in Fig. 5. The 3T-2OC concentrations were in the range of 2–10 wt%, and 0.1 wt% of TNF was added.

calculated to be 6.2. Therefore, the diffraction observed in this experiment occurs predominantly, but not entirely, in the Bragg regime, with a small Raman–Nath component. However, because higher-order diffraction was not observed, the two-beam coupling gain coefficient Γ was calculated assuming the Bragg diffraction.[2–6]

$$\Gamma = \frac{1}{D}\ln\left(\frac{gm}{1+m-g}\right), \tag{2}$$

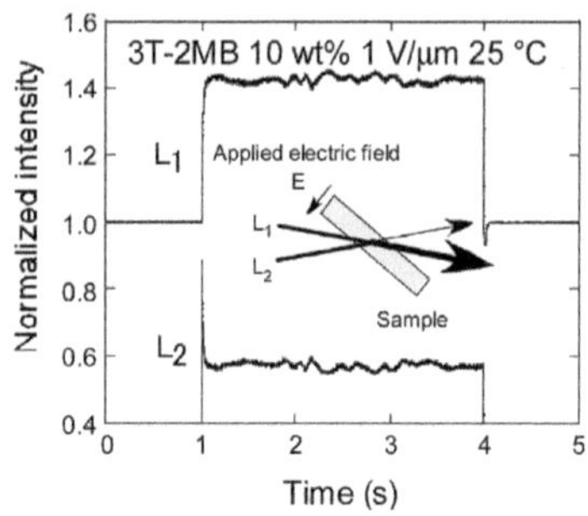

Fig. 8 Typical results of two-beam coupling experiments for a ternary mixture base LC, 3T-2MB, and TNF measured at 25 °C. The pump beam L_2 was incident at 1 s and closed at 4 s.

where $D = L/\cos(\theta)$ is the interaction path for the signal beam (L = sample thickness, θ = propagation angle of the signal beam in the sample), g is the ratio of the signal beam intensities behind the sample with and without a pump beam, and m is the ratio of the beam intensities (pump/signal) in front of the sample.

The gain coefficients of the samples were measured as a function of the applied electric field strength (Fig. 9). The gain coefficient was calculated to be 1200 cm^{-1} in the 10 wt% sample with the application of only 1.5 V μm^{-1}. This gain coefficient is much higher than those of FLCs reported to date (800 cm^{-1}).[1,11–15] It was considered that a higher transparency of the LC mixture contributed to the high gain coefficient. The small electric field required to activate the photorefractive effect in FLCs is a great advantage for photorefractive devices. The response time decreased with increasing electric field strength due to the increased charge separation efficiency. The shortest formation time obtained was 0.9 ms for an external electric field of 1.9 V μm^{-1}. The large gain and fast response are advantageous for realizing optical devices such as real-time image amplifiers and accurate measurement devices.[2,3,18]

3.3 Temperature dependence of the photorefractive effect of FLCs

The temperature dependence of the gain coefficient of the 3T-2MB samples is shown in Fig. 10. Asymmetric energy exchange was observed at temperatures below the SmC*–SmA phase transition temperature. The helical pitch of the 3T-2MB samples observed in the polarizing microscope is plotted as a function of temperature in Fig. 11. Similarly, the helical pitch diverged when the temperature approached the phase transition temperature. In our previous study, the asymmetric energy exchange in an FLC sample was observed only in the temperature range in which the sample exhibits spontaneous polarization. Thus, asymmetric energy exchange was observed only in the temperature range in which the sample exhibits ferroelectric properties (SmC* phase).

3.4 Effect of the concentration of photoconductive chiral dopant

The gain coefficient of the samples with different 3T-2MB concentrations is plotted as a function of the magnitude of the external electric field in Fig. 12. The gain coefficient increased with the strength of the external electric field up to 0.2–0.6 V μm^{-1} and reached a constant value. It is considered that the decrease in the

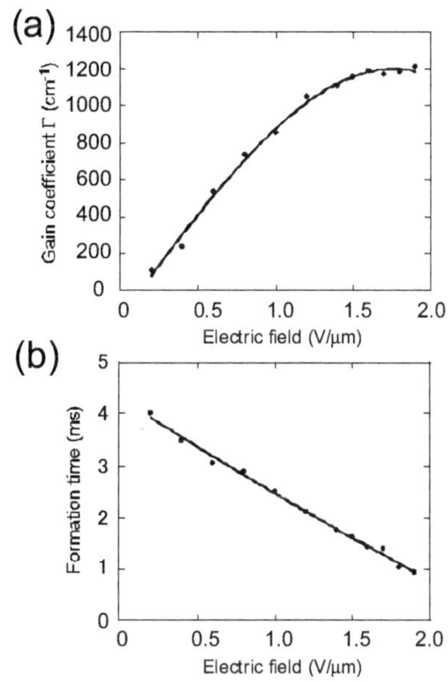

Fig. 9 (a) Electric field dependence of the gain coefficients for mixtures of the base LC, 3T-2MB (10 wt%), and TNF (0.1 wt%) measured at 25 °C. (b) Refractive index grating formation times (response time) for mixtures of the base LC, 3T-2MB (10 wt%), and TNF (0.1 wt%) measured at 25 °C.

gain coefficient at a high external electric field is due to the realignment of the FLC molecules being restricted because the strength of the external electric field exceeded that of the internal electric field. As the concentration of the photoconductive chiral dopants increased, so did the gain coefficient. This may be due to increased density of charge carriers in the FLC medium and an increase in the magnitude of spontaneous polarization (Ps). The magnitude of the gain coefficient was independent of the concentration of TNF. It was considered that conduction based on a hopping mechanism occurred. In this case, TNF acts as just the electron acceptor that introduces electron holes into the photoconductive

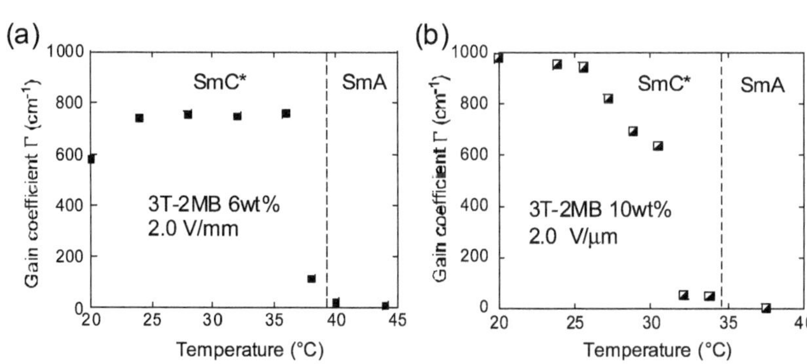

Fig. 10 Temperature dependence of gain coefficients and formation times. The concentration of 3T-2MB was (a) 6 wt% and (b) 10 wt%.

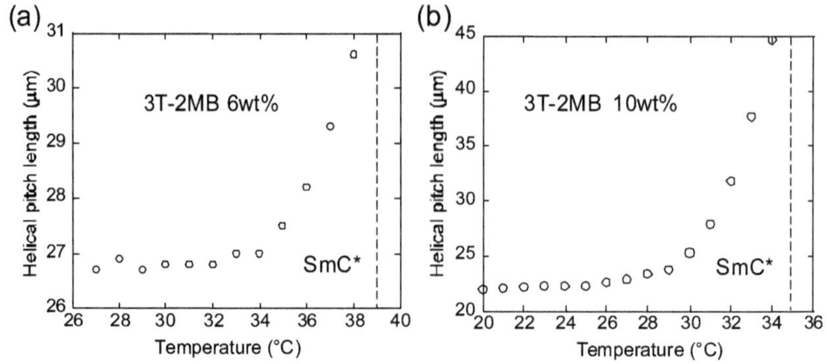

Fig. 11 Temperature dependence of helical pitch lengths. The concentration of 3T-2MB was (a) 6 wt% and (b) 10 wt%.

chiral dopant. Since the molecular weight of the photoconductive chiral dopant is not very different from that of the LC molecules, approximately four photoconductive dopant molecules are dispersed in 96 LC molecules. A cube wherein each side includes 5 LC molecules contains 125 LC molecules. Thus, the average distance between the photoconductive chiral dopant molecules in the LC phase is less than 3 LC molecules. In this case, charge transport based on a hopping mechanism might not be impossible. However, a more detailed understanding of the charge transport mechanism in the dye-doped FLC medium requires further investigation. The gain coefficient was calculated to be 1200 cm^{-1} in the 10 wt% 3T-2MB sample with the application of only 1.6 V μm^{-1}. This gain coefficient is twice as high as that of the 3T-2OC sample.

The response time of the samples is plotted as a function of the magnitude of the external electric field in Fig. 13. The response time decreased with increasing electric field strength due to increased charge separation efficiency. The shortest formation time of 0.93 ms was obtained in the 10 wt% 3T-2MB sample with the application of 2.0 V μm^{-1}. This response time was shorter than that of the 3T-2OC samples. Although the magnitude of the spontaneous polarization in the 3T-2OC samples (5 nC cm^{-2}) is higher than that in the 3T-2MB sample (<1 nC cm^{-2}), the

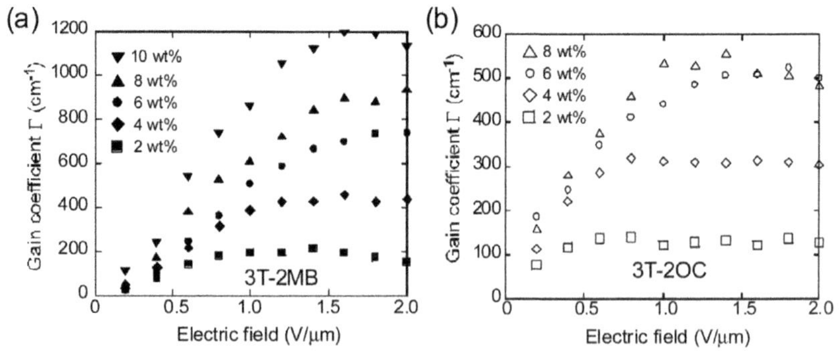

Fig. 12 Electric field dependence of the gain coefficients of the mixtures of the base LC, 3T-2MB, and TNF (0.1 wt%) measured at 25 °C. The concentrations of 3T-2MB were in the range of 2–10 wt%.

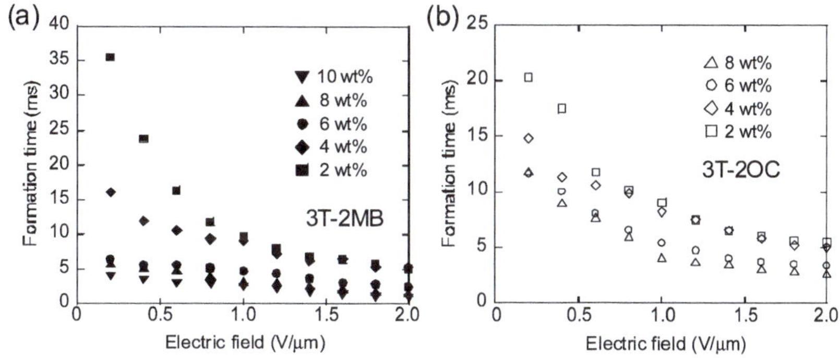

Fig. 13 Refractive index grating formation time (response time) of the mixtures of the base LC, 3T-2MB, and TNF (0.1 wt%) measured at 25 °C. The 3T-2MB concentrations were in the range of 2–10 wt%.

gain coefficient and response speed are higher for the 3T-2MB sample. This indicates that transparency is more important for the photorefractive effect of FLCs than the magnitude of spontaneous polarization.

3.5 Dynamic amplification of the optical image in photorefractive FLCs

Optical image amplification was demonstrated in this work (Fig. 14). A computer-generated image was displayed on a spatial light modulator (SLM) and a 473 nm

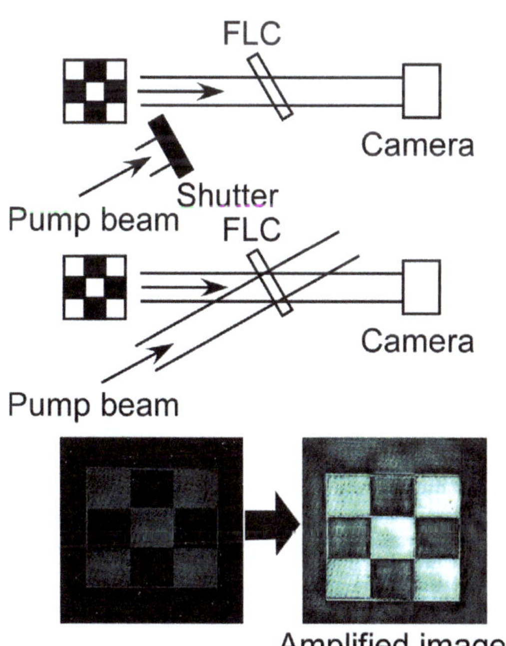

Fig. 14 Optical image amplification experiment. A computer-generated image was displayed on the SLM. The SLM modulated the object beam (473 nm), which was irradiated on the FLC sample and interfered with the pump beam. The image transmitted through the FLC sample (3T-2MB, 10 wt%) was monitored by a CCD camera.

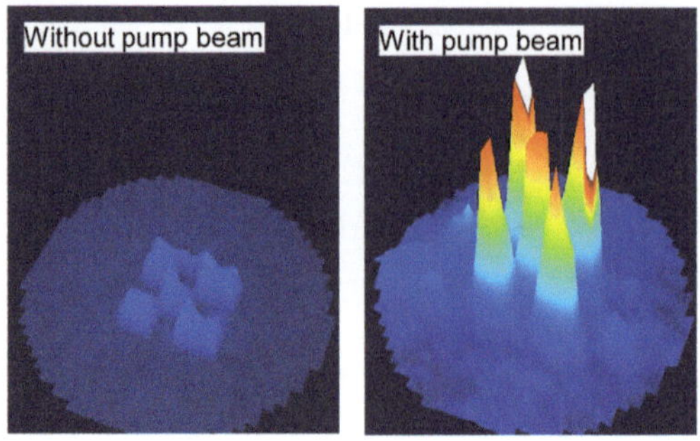

Fig. 15 Signal beam intensities without and with the pump beam.

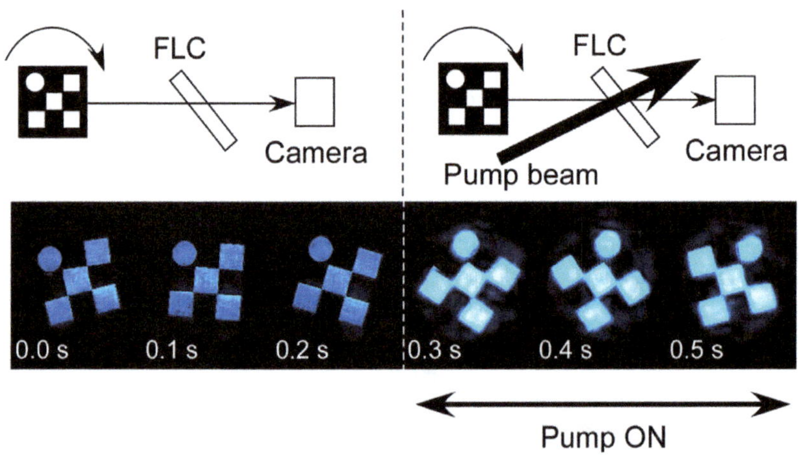

Fig. 16 Optical image amplification experiment. A computer-generated animation was displayed on the SLM. The SLM modulated the object beam (473 nm), which was irradiated on the FLC sample, and interfered with the reference beam. The image transmitted through the FLC sample (3T-2MB, 10 wt%) was monitored by a CCD camera.

DPSS laser beam was irradiated. The laser beam containing the image (signal beam) was transmitted through the FLC sample and monitored by a CCD camera. When the pump beam (a beam divided from the signal beam before SLM) was incident to the FLC sample and interfered with the signal beam, the signal was amplified through asymmetric energy exchange. The crossing angle of the beams at 15° yielded the largest amplification and the best contrast. The intensities of the optical signal beam with and without the pump beam are shown in Fig. 15. The intensity of the signal beam was amplified 6-fold compared to that without the pump beam.

A computer-generated animation was displayed on the SLM. The frame rate was 30 fps. A 473 nm beam was irradiated on the SLM and the reflected beam was incident on the FLC sample. A pump beam interfered with the beam from the SLM in the FLC sample. A laser beam containing the moving image of the animation was amplified by the incident pump beam (Fig. 16). This result shows that the response of the photorefractive FLC was fast enough to amplify the optical image in real time. If a typical photorefractive polymer with a response time of ~100 ms is used in place of the FLC sample, the amplification would not occur. In that case, although a still image can be amplified, the intensity of the image would be weakened to the original magnitude when the image starts to move at a video rate.

4 Conclusion

The photorefractivity of photoconductive FLC mixtures containing photoconductive chiral dopants was investigated. Relatively large gain coefficients higher than 1200 cm^{-1} with a response time (refractive index grating formation time) of 1 ms were obtained under application of only 1.5 V μm^{-1} in the SS-state of a FLC mixture. This response time is sufficiently short to realize real-time dynamic holograms. Because the molecular shape of the terthiophene-type photoconductive chiral dopants is similar to that of the base LC molecules, the miscibility with the base LC is high. FLC mixtures containing the photoconductive chiral dopants exhibited high gain coefficient and first response, making them useful for photorefractive devices. Real-time dynamic amplification of the optical image signal using the photorefractive ferroelectric liquid crystal was demonstrated.

References

1 T. Sasaki, M. Ikegami, T. Abe, D. Miyazaki, S. Kajikawa and Y. Naka, *Appl. Phys. Lett.*, 2013, **102**(063306), 1–4.

2 L. Solymar, J. D. Webb and A. Grunnet-Jepsen, *The Physics and Applications of Photorefractive Materials*, Oxford, New York, 1996.

3 P. Yeh, *Introduction to Photorefractive Nonlinear Optics*, John Wiley, New York, 1993.

4 W. E. Moerner and S. M. Silence, *Chem. Rev.*, 1994, **94**, 127–155.

5 B. Kippelen and N. Peyghambarian, *Advances in Polymer Science, Polymers for Photonics Applications II*, Springer, pp. 87–156, 2002.

6 O. Ostroverkhova and W. E. Moerner, *Chem. Rev.*, 2004, **104**, 3267–3314.

7 T. Sasaki, *Polym. J.*, 2005, **37**, 797–812.

8 S. Tay, P. A. Blanche, R. Voorakaranam, A. V. Tunc, W. Lin, S. Rokutanda, T. Gu, D. Flores, P. Wang, G. Li, P. Hilarie, J. Thomas, R. A. Norwood, M. Yamamoto and N. Peyghambarian, *Nature*, 2008, **451**, 694–698.

9 P. A. Blanche, A. Bablumian, R. Voorakaranam, C. Christenson, W. Lin, T. Gu, D. Flores, P. Wang, W. Y. Hsieh, M. Kathaperumal, B. Rachwal, O. Siddiqui, J. Thomas, R. A. Norwood, M. Yamamoto and N. Peyghambarian, *Nature*, 2010, **468**, 80–83.

10 H. Ono and N. Kawatsuki, *Opt. Lett.*, 1997, 22, 1144–1146.

11 T. Sasaki, D. Miyazaki, K. Akaike, M. Ikegami and Y. Naka, *J. Mater. Chem.*, 2011, **21**, 8678–8686.

12 G. P. Wiederrecht, B. A. Yoon and M. R. Wasielewski, *Adv. Mater.*, 2000, **12**, 1533–1536.

13 M. Talarico, R. Termine, P. Prus, G. Barberio, D. Pucci, M. Ghedini and A. Goelemme, *Mol. Cryst. Liq. Cryst.*, 2005, **429**, 65–76.

14 M. Talarico and A. Goelemme, *Nat. Mater.*, 2006, **5**, 185–188.

15 T. Sasaki, *Chem. Rec.*, 2006, **6**, 43–51.

16 K. Skarp and A. A. Handschy, *Mol. Cryst. Liq. Cryst.*, 1988, **165**, 439–509.
17 P. Oswald and P. Pieranski, *Smectic and Columnar Liquid Crystals*, Taylor & Francis, New York, 2006.
18 N. Koukourakis, T. Abdelwahab, M. Y. Li, H. Höpfner, Y. W. Lai, E. Darakis, C. Brenner, N. C. Gerhardt and M. R. Hofmann, *Opt. Express*, 2011, **19**, 22004–22023.

Faraday Discussions

PAPER

Organic semiconducting single crystals as solid-state sensors for ionizing radiation

Beatrice Fraboni,*[a] Andrea Ciavatti,[a] Laura Basiricò[a] and Alessandro Fraleoni-Morgera[bc]

Received 9th May 2014, Accepted 24th June 2014
DOI: 10.1039/c4fd00102h

So far, organic semiconductors have been mainly proposed as detectors for ionizing radiation in the indirect conversion approach, *i.e.* as scintillators, which convert ionizing radiation into visible photons, or as photodiodes, which detect visible photons coming from a scintillator and convert them into an electrical signal. The direct conversion of ionizing radiation into an electrical signal within the same device is a more effective process than indirect conversion, since it improves the signal-to-noise ratio and it reduces the device response time. We report here the use of Organic Semiconducting Single Crystals (OSSCs) as intrinsic direct ionizing radiation detectors, thanks to their stability, good transport properties and large interaction volume. Ionizing radiation X-ray detectors, based on low-cost solution-grown OSSCs, are here shown to operate at room temperature, providing a stable linear response with increasing dose rate in the ambient atmosphere and in high radiation environments.

1 Introduction

The detection of ionization radiation (*e.g.* X-rays, electrons and alpha particles) is a constantly growing area of research thanks to its vast and numerous application fields, which span from astrophysics to nuclear power plants, to industrial and civil security and to medical imaging and diagnostics.

High energy photons (X- and gamma-rays) can be detected with two different categories of functional materials: scintillators and semiconductors. In both cases, the interaction with a high energy photon first induces primary excitations and ionization processes (ions and electrons) which, at a second stage, interact within the volume of the detection material and produce a majority of secondary excitations (electron–hole pairs), within a picosecond timeframe. The by-products of both the primary and secondary excitations are electron–hole pairs (excitons) that can be transduced into an output signal following different pathways in

[a]*Dipartimento di Fisica e Astronomia, Università di Bologna, viale Berti Pichat 6/2, Bologna, Italy. E-mail: beatrice.fraboni@unibo.it; Fax: +39 051 2095113; Tel: +39 051 2095806*
[b]*Dipartimento di Ingegneria e Architettura, Università di Trieste – V. Valerio 10, Italy, 34100 Trieste, Italy*
[c]*Elettra Sincrotrone Trieste S.C.p.A, Strada Statale 14, Km 163.5, Basovizza (TS), Italy*

semiconductor detectors and in scintillators, described in more detail in the following.[1]

In a scintillator, the excitons transfer their energy to luminescent centers which are often intentionally introduced. These centers release the energy radiatively, and the resulting photons, typically in the visible wavelength range, escape the scintillator and are collected by a coupled photo-multiplier tube (PMT) or a photodiode to obtain an electrical signal associated to the incident radiation beam.

In a semiconductor detector (*e.g.* CdTe, SiC), an electric field is applied to dissociate the electron–hole pairs and to sweep the electrons and holes to the positive and negative electrodes, respectively. The resulting photocurrent is directly recorded as the output electrical signal associated to the high energy radiation particles. The direct conversion of ionizing radiation into an electrical signal within the same material, and thus within one single device, is a more effective process than indirect conversion, since it improves the signal-to-noise ratio and it reduces the device response time.

The material requirements for the two different detection mechanisms share some similarities: high stopping power to maximize the absorption efficiency of the incident radiation, high purity to minimize exciton trapping, and good uniformity to reduce scattering and good transparency, possibly coupled to the ability to grow the material into a large size to increase the interaction volume. For semiconductors, a high and balanced carrier mobility and a low intrinsic carrier density are essential to obtain a high sensitivity and a low background current. On the other hand, scintillators must have an efficient cascade energy transition series to achieve a high light emission yield.[2]

Among the novel materials which could push the boundaries of radiation detection further, organic π-conjugated small molecules and polymers are interesting emerging candidates, thanks to their potential to realize large area and flexible optoelectronic devices[3] by cost effective printing methods,[4] which is due to their processing from solution at room temperature. Moreover, by controlling the molecular structure and the π-electron conjugated chains, their luminescence and optoelectronic properties can be easily tuned. In fact, it is known that organic semiconductors can work as effective photodetectors, *via* the photoinduced generation of excitons which, within the material and in suitable conditions (presence of a local electron donor–electron acceptor interface usually formed at the interface between two different organic semiconducting materials), are able to split into free charge carriers that can be collected at the electrodes.[8]

For this reason, organic semiconductors are now seen as an attractive alternative to inorganic compounds for photon detection in the UV to the near infrared (NIR) region where, besides the vast field of organic photovoltaic devices,[5] light pulse detectors for optical communications[6] and digital flat-panel arrays for imaging applications[7] have been proposed in the literature.[8]

In the UV-NIR range, very interesting values for the figures of merit such as the photo-conversion efficiency, speed and minimum detectable signal level have been reported,[8] and even though the simultaneous attainment of all of these relevant parameters has been demonstrated in only a limited number of papers, real applications are within reach for this technology, where the best reported photo-responsivities outperform amorphous silicon-based devices. Such achievements have been obtained by overcoming a few major bottlenecks for organic electronic devices, such as current injection from contacts to reduce leakage currents, charge collection efficiency and the stability during exposure to air.[8]

Organic materials could also be viable candidates for the detection of higher energy photons (X- and gamma-rays), and they were first suggested for radiation detection in the early 1980s.[9] However, the interest in these materials was mostly focused on their scintillating properties,[1] possibly because the material requirements are particularly stringent in the case of direct detectors, as evidenced by the few reports present in the literature on radiation detection based on organic thin films,[10–14] where stability, reproducibility and the attainment of good sensitivity (likely limited by the small interaction volume) are still open issues.

In this report we will discuss the very interesting and promising radiation detecting properties of a particular class of semiconducting organic materials: Organic Semiconducting Single Crystals (OSSCs).

We have recently reported how solution-grown OSSCs are ideal candidates to directly detect X-ray radiation (*i.e.* to directly convert an X-ray photon into an electron), thanks to their particular properties (which will be discussed in detail in the following paragraph), coupled to all the above mentioned advantages offered by organic conjugated materials.[15] Moreover, being based on carbon, their low effective atomic number is similar to the average human tissue-equivalent Z value and makes them ideal candidates for radiotherapy and medical applications, which would benefit greatly from the improved accuracy of tissue-equivalent dosimeters. In fact, there are currently no low-cost, large-area detectors with tissue-equivalence response available.

It is noteworthy that, unlike organic photodiodes, OSSCs show direct and intrinsic charge photogeneration and collection processes with no need for the heterogeneous donor–acceptor material interface, which is usually employed to improve the collection efficiency in organic photodiodes and solar cells.

We will assess here the reproducible process of the conversion of X-ray photons into an electrical signal within OSSCs, and how this process allows the fabrication of stable radiation detectors which provide a linear response to an increasing X-ray dose. The observed performance indicates that OSSCs are very promising candidates for a novel generation of low-cost, real-time and room temperature-operated X-ray detectors.

2 Semiconductor-based direct detectors

The ability to detect high energy radiation such as X-rays, gamma-rays, and other uncharged and charged particles with solid-state semiconductor detectors has improved dramatically in the last 20 years thanks to two main factors: (i) the enormous advances in semiconductor science and technology, and (ii) the strong and increasing demand for high-performing solid-state radiation detectors.

High purity silicon and germanium were the first materials to be used as solid-state detectors, and are still widely employed thanks to their extremely good energy resolution (below 0.2%) which, however, can only be achieved at cryogenic temperatures. This prompted the development of novel compound semiconductors such as CdTe, SiC and CdZnTe which can offer excellent performance at room temperature, superior in a few aspects to Ge.

Nonetheless, the difficulty to grow large-size, high-quality crystals of these II–VI compound materials at a low cost is limiting their application in very high-tech and specific detectors, *e.g.* in satellites and as pioneering medical diagnostic tools. A non-negligible further drawback of these materials is their limited availability, and often their toxicity. These limitations have prompted the need to find alternative novel semiconducting materials.

The main requirements for a good solid state semiconductor detector are common to all semiconductors and are briefly detailed below:[1,16]

(a) High resistivity ($>10^9$ Ω cm) and low leakage current. Low leakage currents when an electric field is applied during operation are critical for low noise operation. The necessary high resistivity is achieved by using larger band gap materials (>1.5 eV) with low intrinsic carrier concentrations.

(b) A small enough band gap so that the electron–hole ionization energy is small (<5 eV). This ensures that the number of electron–hole pairs created is reasonably large and results in a higher signal to noise ratio.

(c) High atomic number (Z) and/or a large interaction volume for efficient radiation–atomic interactions. The cross-section for photoelectric absorption in a material of atomic number Z varies as Z^n, where $4 < n < 5$. For high-sensitivity and efficiency, large detector volumes are required to ensure that as many incident photons as possible have the opportunity to interact in the detector volume. A further related requirement is that the detector material must have a high density, although this is essentially guaranteed simply by the fact that a solid material is employed for the detector material in contrast to gas-based detectors.

(d) High intrinsic $\mu\tau$ product. The carrier drift length is given by $\mu\tau E$, where μ is the carrier mobility, τ is the carrier lifetime and E is the applied electric field. Charge collection is determined by which fraction of photo-generated electrons and holes effectively traverses the detector and reaches the electrodes.

(e) High-purity, homogeneous, defect-free materials, to ensure good charge transport properties, low leakage currents, and no conductive short circuits between the detector contacts.

(f) Electrodes which produce no defects, impurities or barriers to the charge collection process and which can be used effectively to apply a uniform electric field across the device. This requirement is also related to the need to avoid material polarization effects which may affect the time response of the detector.

(g) Surfaces should be highly resistive and stable over time to prevent increases in the surface leakage currents over the lifetime of the detector.

It is obvious that not all of the above requirements can be easily met by a single material, but the dramatic advancements in the organic semiconductor research field recently have stimulated studies on the potential application of organic semiconductors as solid-state detectors.

The first studies performed featured the neutron irradiation of polyacetylene and polythiophene films, which induced an increase in the film conductivity, linear with the irradiation dose but irreversible.[17] To date, only a few examples of simple, direct detectors based on organic semiconductors have been reported, and they all refer to thin films based on organic semiconducting polymers[11,12,14] and small molecules.[13] When compared to silicon-based detectors, conjugated polymers, despite their lower carrier mobility and inferior radiation tolerance, exhibit the advantage that large areas can be covered and possibly nanostructured *via* wet-processing, thus increasing the detector's active surface at a much lower cost.[18]

The already mentioned issue of the degradation of semiconducting polymers represents a practical problem for the fabrication of effective semiconducting polymer-based intrinsic, direct detectors. In fact, many of the above cited detectors base their operability on the measurement of the resistivity (conductivity) of the polymeric semiconductor, which increases (decreases) upon device exposure to the ionizing radiation, due to material degradation. This means that the above described devices are not able to perform for prolonged periods, neither to be repeatedly used with reproducible performance, hence resulting in detectors with a very short operative lifetime and, in the best possible case, as disposable devices.

A first step towards the solution of these problems was made recently, when semiconducting polymer-based drop-cast films with a thickness of approximately 10–20 μm (hence a rather thick film, to maximize X-ray attenuation) showed linearity up to dose rates of 60 mGy s^{-1}.[13,14] The sensitivity of these devices reached values of 100–400 nC mGy^{-1} cm^{-3}, which are comparable to silicon devices. The time response of the X-ray photocurrent measured for the thin film devices was less than 150 ms, and encouragingly, no sign of radiation damage was observed for doses in excess of 10 Gy, although no information on the reproducibility of the results was reported.[13,14]

Though the authors of these reports do not always mention it explicitly, it has to be noted that in these devices the polymeric semiconductor thin films are always coupled to metallic electrodes or substrates, which are exposed to the ionizing radiation together with the organic layer. The importance and the role of the metallic electrodes and/or the substrate in the performance of these devices has been stressed in several dedicated works,[6,7,13] since they act as the primary X-ray photo-conversion layers, producing secondary electrons which are injected into the organic thin film and thus produce the electrical signal output.

We have recently reported how solution-grown OSSCs are ideal candidates to directly detect X-ray radiation, as their particular properties overcome most of the major limitations inherent to organic materials discussed to date.[15] Their long-range molecular packing order and lack of grain boundaries impart unique transport properties to all OSSCs such as high carrier mobilities (up to 40 cm^2 V^{-1} s^{-1} (ref. 19)), transport anisotropy and long exciton diffusion lengths (up to 8 μm (ref. 20), as opposed to the few nanometers in organic thin films and blends used in photovoltaic and photodetecting devices[8]). These features, coupled with their high resistivity and low dark currents due to the relatively large band gap, are enhanced, in the case of solution-grown OSSCs, by the possibility of tuning the crystal dimensions up to mm^3, as will be discussed in more detail in the next paragraph. Moreover, a notable unique property of OSSCs is that they can efficiently and intrinsically photo-convert X-ray photons into an electrical signal, without the need for the intervention of extra metal/substrate layers.[15]

3 Solution-grown organic semiconducting single crystals

The best-performing OSSCs are usually grown by chemical vapour deposition (CVD), which involves the use of multi-zone heated tubes, in which a vapour of molecules is carried by a convenient inert gas onto a cold wall. However, due to the nature of this growth process, it is difficult to grow very large crystals.[21] Crystal growth from the melt could represent an interesting alternative to CVD; nonetheless, although in some cases crystallization from the melt of OSSCs proved to be successful,[22] stability problems (especially due to enhanced photo-oxidation rates at temperatures approaching the melting temperature) often limit the possibility of taking advantage of this crystallization technique for organic materials.

These problems may be overcome using growth from solution. This approach has a high versatility, with the capability to deliver very large (up to several cm^3) and pure crystals, and with low energy requirements (*i.e.* no need for dramatic heating, cooling or vacuum) and ease of implementation. This latter point is worthy of particular attention, since it implies an extremely low cost, especially when large-area detectors are envisaged.

Solution growth has also been used to produce organic (not necessarily semiconducting) single crystals used as scintillators in ionizing radiation detectors,[23] evidencing that this technique delivers crystals which are able to usefully interact with ionizing radiation. In addition, solution-grown OSSCs have repeatedly demonstrated good electronic quality.[24] The possibility of growing OSSCs from solution offers not only exciting opportunities for performing fundamental studies, but, even more interestingly, paves the way for the practical application of these materials in electronic devices. In fact, it has been recently shown that it is possible to fabricate high-quality OSSCs *via* inkjet printing[25] or shear-aided crystallization.[26]

We will report here on two different types of OSSCs which we have grown from solution: 4-hydroxycyanobenzene (4HCB) and 1,5-dinitronaphthalene (DNN), as shown in Fig. 1.

These two molecules have been chosen because they generate crystals with different geometries: platelets in the case of 4HCB and needles for DNN. Both types of crystals can be easily grown from widely available solvents/non-solvents (*i.e.* ethylic ether and/or petroleum ether for 4HCB[27] and chloroform or benzonitrile for DNN), and their size can be controlled by varying parameters such as the solvent/non-solvent volume ratio or the organic molecule concentration in the

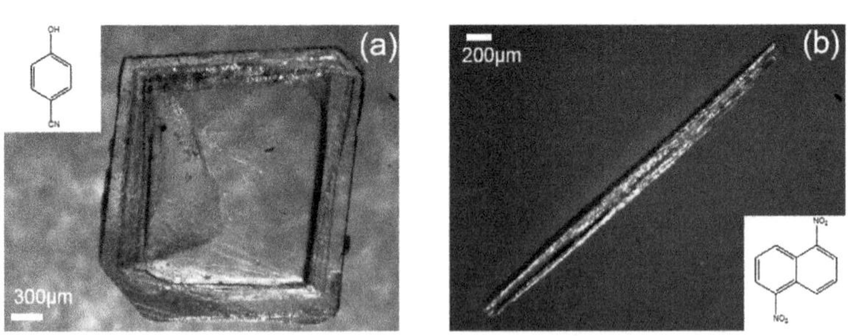

Fig. 1 Optical microscopy images of (a) 4-hydroxycyanobenzene (4HCB) and (b) 1,5-dinitronaphthalene (DNN) organic single crystals. The images clearly show the different shapes of the crystals, *i.e.* the platelet shape of the 4HCB crystal and the needle-like shape of the DNN crystal. In the insets the chemical structures of the two molecules are reported.

starting solutions, reaching sizes of tens of mm^3. This translates into a radiation interaction volume which is much larger than those typically accessible with polymer films (which are a few tens of micrometers thick at most).[27] The as-obtained single crystals are very robust to physical manipulation (they can be easily moved and positioned on substrates, electrodes arrays, sample holders, *etc.*) and to environmental conditions (air, light, room temperature).

It is known that the low symmetry of organic molecules leads to anisotropically packed crystal structures, where the direction of the strongest π-orbital overlap usually coincides with the direction of the highest carrier mobility, thus also affecting the crystal transport properties.[19,21,28] On one hand this asymmetry introduces difficulties for clearly understanding the transport behaviour of the organic crystal, but on the other hand it offers the possibility to investigate the correlation between the three-dimensional molecular stacking order of OSSCs and their known anisotropic electronic transport properties. In particular, in recent years, different mobilities have been measured along the three dimensions for micro or nanometer-sized organic single crystals, either vapour deposited[19,21,28] or solution-grown.[24,29–31]

A complete characterization of several key electronic transport parameters of OSSCs can be carried out by means of Space Charge Limited Current (SCLC)[21,29,31,32] and by Field Effect Transistor (FET) analyses.[29,30]

We have recently reported on the electrical characterization of solution-grown, macroscopic 4HCB single crystals, evidencing their good electronic transport properties and their clear anisotropic behaviour along the three crystallographic axes,[29,31] and we will here focus our attention on the results obtained for DNN single crystals. Due to their needle-like shape, there is only one crystallographic direction which can be electrically accessed, and which coincides with the higher π-orbital overlap direction (visible by examining the DNN crystal structure[33]). The good electronic transport properties of DNN single crystals have been assessed by SCLC analysis, and a typical SCLC curve of a DNN crystal is shown in Fig. 2, together with a sketch of the electrical contact configuration. The intrinsic carrier mobility of the material is reported in Table 1, together with the average carrier mobility determined for the 4HCB crystals, also from SCLC analysis, along the three crystallographic directions.

Indeed, as shown in Table 1, the transport properties of 4HCB crystals reflect their anisotropic packing, and a clear difference in mobility values is found along

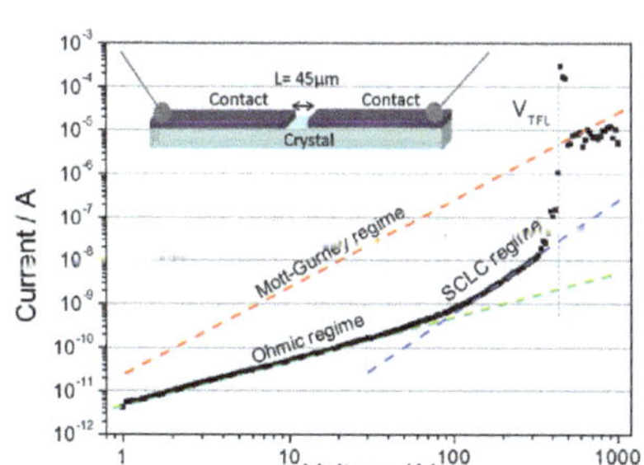

Fig. 2 Current *vs.* voltage curve, in the double-logarithmic scale, of a DNN crystal within the space-charge conduction regime. The inset shows the schematic layout of the electrical ohmic contacts on a DNN crystal.

Table 1 Carrier mobility values reported in the literature[29–31] for a 4HCB crystal along the three crystallographic axes, *a*, *b* and *c*, in comparison with the mobility value measured for a DNN crystal along its axis of growth (only one axis as it is needle-shaped)

Crystal	μ_{SCLC} (cm^2 V^{-1} s^{-1})
4HCB *a* axis	$(1.0 \pm 0.5) \times 10^{-1}$
4HCB *b* axis	$(4 \pm 2) \times 10^{-2}$
4HCB *c* axis	$(2.0 \pm 0.5) \times 10^{-5}$
DNN	$(2.2 \pm 0.8) \times 10^{-3}$

the three axes. Charge carrier mobilities of 0.1 cm^2 V^{-1} s^{-1} can be easily and routinely obtained along the *a* axis of the 4HCB solution-grown crystals, with very reproducible and stable results.[31] While the anisotropic electrical response of OSSCs is extremely interesting from the point of view of fundamental studies, it poses the problem of properly aligning the crystals and the electrodes in order to select and control the appropriate carrier transport properties needed for specific device applications. Even if we have demonstrated how linearly polarized infrared analyses allow the fast, easy and reliable determination of the crystallographic orientation of OSSCs,[34,35] it is important to investigate the X-ray photo-response of single crystals where the electronic transport axis is clearly identifiable. For this reason, needle-like crystals, such as DNN, could be of practical interest.

4 X-ray photo-response of OSSCs

After having assessed that solution-grown 4HCB and DNN OSSCs have appropriate volume sizes and electronic transport properties to grant the collection of photo-generated charge carriers, we performed a complete characterization of the electrical photo-response of the organic crystals under a molybdenum (Mo) tube X-ray broad spectrum with an accelerating voltage of 35 kV and a dose rate of 2.5–140 mGy s^{-1}, All of the reported measurements were made in air at room temperature.

Fig. 3 reports for the first time the typical current–voltage and current–time curves measured for a DNN crystal exposed to an X-ray beam from a Mo tube (35 keV, 140 mG s^{-1}), with two Ag electrodes evaporated at a distance of 45 μm, in the same configuration as reported in the inset of Fig. 2. The strong modification induced in the current, measured as a function of the applied voltage when the DNN crystal is exposed to the X-ray beam, due to the photo-induced generation of charge carriers, is visible in Fig. 3a. It is noteworthy that this sharp response was obtained with applied voltages as low as 2 V, a very small value for radiation detectors. Moreover, the "off" current recorded after the exposure to X-rays does not show significant hysteresis or degradation effects, and its value always lies below 500 pA, again a very good low value for radiation detection applications. A further positive feature of this device is that the current *vs.* time curve recorded under an on–off switching X-ray beam (Fig. 3b) clearly shows how repetitive exposures to X-rays do not alter the sensor photo-response in terms of $\Delta I = I_{on} - I_{off}$ and the baseline current shift. Fig. 3c shows the virtually null effect of UV-Vis irradiation on the same crystal. It is hence evident that the physical processes generating the X-ray-induced photo-response do not have much in common with those responsible for the UV-IR detection in organic photodiodes or phototransistors.[8]

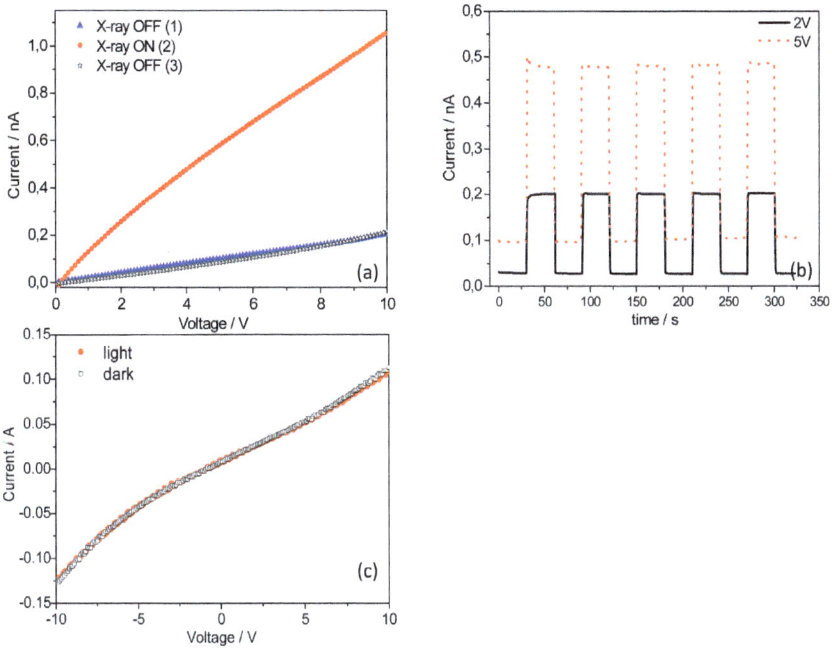

Fig. 3 (a) Current–voltage curves of a DNN crystal with Ag contacts in the dark (blue solid triangles), under X-ray irradiation at 117 mGy s^{-1} (red circles), and again in the dark after X-ray irradiation (black empty stars). (b) Current–time curves of the DNN-based device recorded switching the X-ray beam on–off for two different applied bias voltages, *i.e.* 2 V and 5 V. (c) Current–voltage curve of the DNN-based device under UV-Vis irradiation (red solid circles) and in the dark (black open squares); the two curves overlap perfectly.

Several DNN crystals were tested under X-ray dose rates ranging from 20 to 120 mGy s^{-1}, at different bias voltages. The induced photocurrent values, $\Delta I = I_{on} - I_{off}$, are reported in Fig. 4a and b and demonstrate that the crystal response for an increasing X-ray dose rate is linear for each tested bias voltage, assessing that it is possible to effectively drive an OSSC detector even at a few bias volts. The detector sensitivity, defined as $S = \Delta I/D$, where D is the radiation dose rate, was evaluated to be about 6 nC mGy^{-1} at 10 V.

Similar results have been obtained for platelet-shaped 4HCB crystals, which showed a linear response to X-ray irradiation for bias values up to 500 V with X-ray photoinduced current variations ($\Delta I = I_{on} - I_{off}$) of up to ten nA.[15] The radiation hardness of OSSCs, *i.e.* their tolerance to high doses of ionizing radiation, is surprisingly high. Fig. 5 shows the current *vs.* time curve of a 4HCB crystal, with an on–off switching 35 keV X-ray beam, before and after having received a total X-ray dose of about 2 kGy. The detector performance of the same 4HCB crystal was measured again after 1 month of storage in the dark and its photo-response, again for an on–off switching 35 keV X-ray beam, is notably not altered. It is noteworthy that the $\Delta I = I_{on} - I_{off}$ value is quite stable both after strong irradiation and after prolonged aging, even if the baseline current obviously shifts a little, which is likely due to bulk radiation damage and/or degradation effects which could be located either within the crystal or at the electrodes.

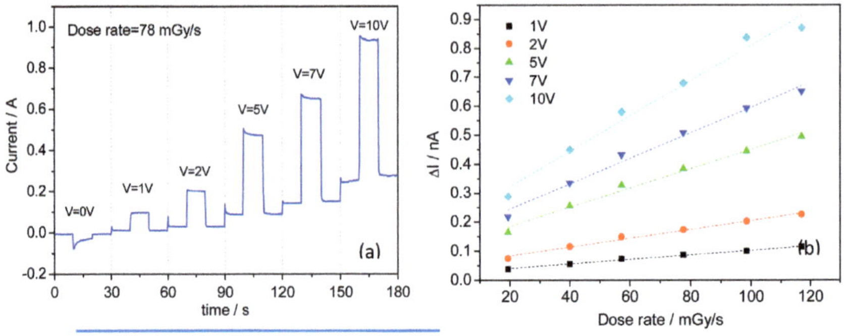

Bias (V)	Sensitivity (nC/Gy)
1	0.8
2	1.5
5	3.4
7	4.4
10	6.1

(c)

Fig. 4 (a) Current–time curve of the DNN-based device recorded by switching on–off the X-ray beam (dose rate of 78 mGy s^{-1}) and sweeping the bias voltage after each cycle. (b) X-ray-induced current variation $\Delta I = I_{on} - I_{off}$ for increasing dose rates at different bias voltages. (c) Detector sensitivity values for different bias voltages, calculated from a linear fit of the plots reported in (b).

To investigate the role and the effects of the molecular anisotropic packing in the X-ray photo-response of OSSCs, it is useful to compare the responses of the DNN molecule, that crystallizes in a needle-like shape, and of the platelet-shaped crystals obtained with 4HCB (Fig. 1). While DNN crystals allow us to sample only one crystal axis, the axis parallel to the strongest π-stacking and faster crystal growth rate, 4HCB crystals allow us to investigate their photo-response along all three crystallographic directions (inset of Fig. 5). As can be inferred from the

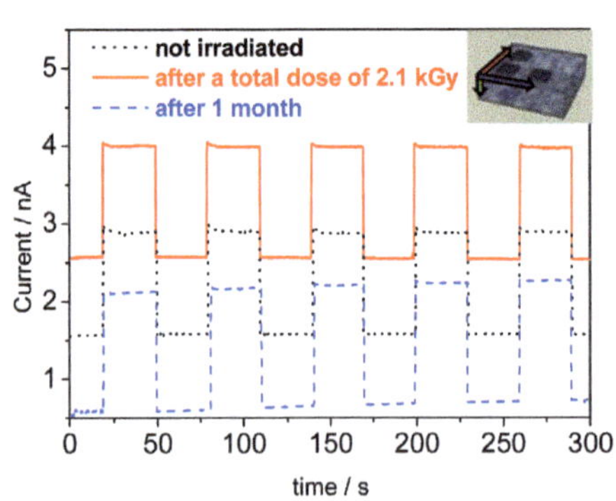

Fig. 5 Current–time curves recorded along the c axis (see inset) under an on–off switching X-ray beam (35 keV @ 150 mGy s^{-1}) before (black dotted line) and after (red solid line) a total dose of 2.1 kGy; the blue dashed line indicates the same measure repeated after one month.

measured mobility values, the tighter π-stacking axes are the two planar axes (labelled *a* and *b*) which possess good mobility, while the vertical axis *c* is characterized by a poorer π-stacking degree and by a much smaller mobility (Table 1).[29–31]

Quite surprisingly, a direct comparison between the X-ray-induced photocurrent of a 4HCB crystal in the vertical collection geometry (small electrode on the top and on the bottom of the crystal) and in the planar collection geometry (two small electrodes on the top crystal surface), in identical experimental conditions (*i.e.* the same electrode distance of 200 μm and the same X-ray irradiation conditions) shows that the X-ray sensitivity is smaller along the planar axes (7 *vs.* 24 nC Gy^{-1} at 10 V) despite the fact that the charge carrier mobility is about 3 order of magnitude higher than along the vertical axis (Fig. 6). Moreover, the sensitivity is comparable along the main π-stacking axis for both DNN and 4HCB crystals (approx. 7 nC Gy^{-1} @10 V), which is accessed in both types of crystal with a planar electrode geometry.

These results (*i.e.*, the crystal axes characterized by the strongest π-stacking, that generally coincide with the best current transport axes, show a poorer X-ray photo-response) suggest that the planar electrode configuration is less performing regardless of the crystal shape (either needle-like or platelet). Three main causes can be hypothesised for this behaviour.

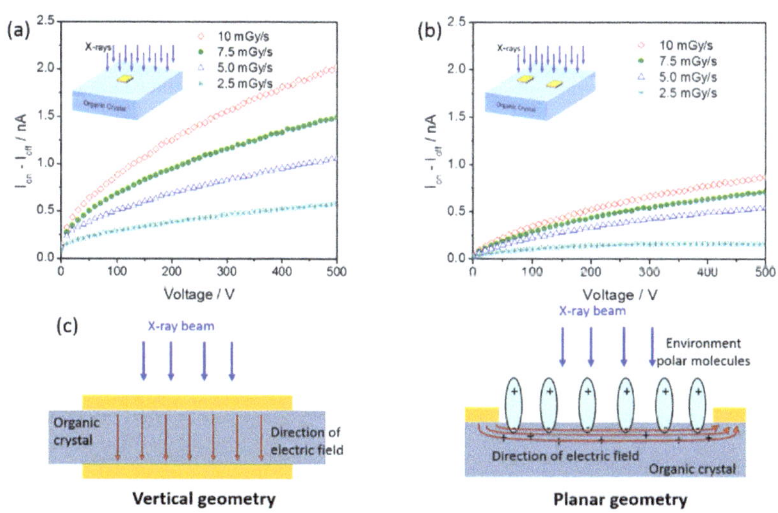

(d)

Bias (V)	Sensitivity (nC/Gy)		
	@10V	@50V	@100V
Planar	7	18	28
Vertical	24	49	70

Fig. 6 X-ray-induced current variation of a 4HCB crystal in the vertical geometry (a) and planar geometry (b) for different X-ray doses. (c) Sketch of the electric field distribution in the vertical and planar geometries. The hypothetical effect of polar environmental molecules at the crystal surface in the planar geometry is also shown. (d) Sensitivity values recorded for a 4HCB-based detector in the planar and vertical geometries for three different bias voltages, *i.e.* 10 V, 50 V and 100 V.

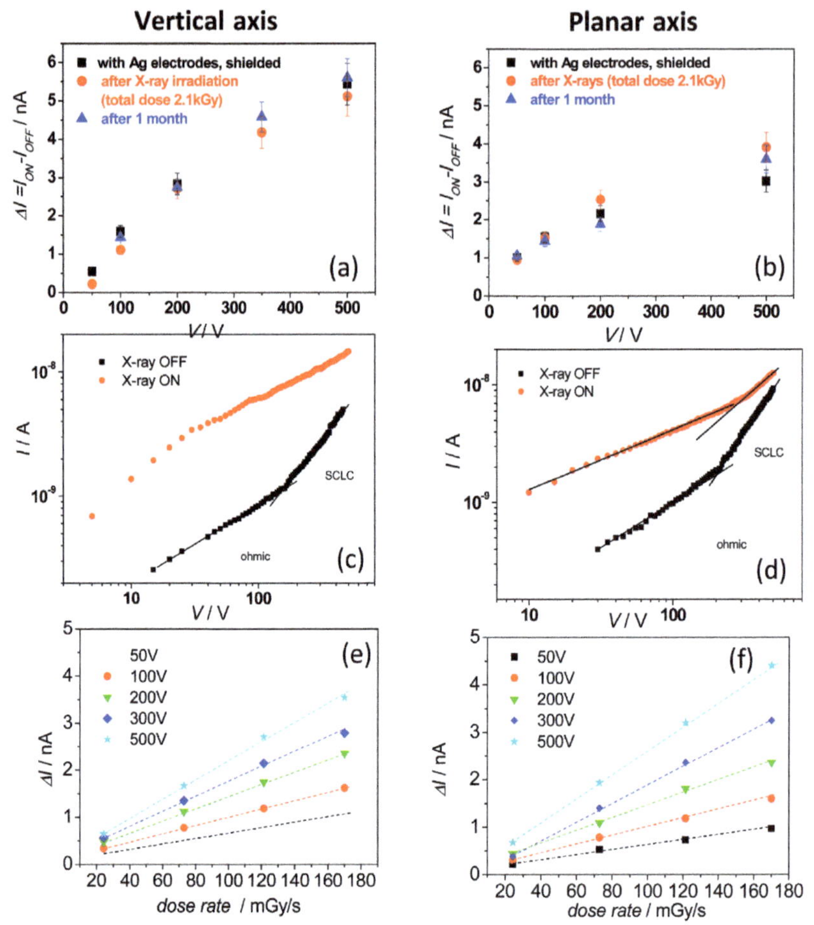

Fig. 7 X-ray-induced photocurrent response of a 4HCB-based detector as a function of the bias voltage, collected along the vertical (a) and the planar (b) crystal axes, before (black squares) and after (red circles) a total dose of 2.1 kGy of X-ray irradiation and after one month (blue triangles). (c and d) Current–voltage curves for the two axes in the dark (black squares) and under X-ray irradiation (dose rate of 170 mGy s^{-1}). The lack of the ohmic–SCLC transition under irradiation is shown. (e and f) X-ray-induced current variation $\Delta I = I_{on} - I_{off}$ for increasing the dose rates at different bias voltages along the two axes.

(i) The better electrical transport properties along crystal axes with strong π-stacking may be a limiting factor with respect to the crystal's sensitivity to X-rays, due to the higher off currents.

(ii) The electric field distribution in the vertical geometry is much more effective in collecting the photogenerated carriers that in the planar geometry (see Fig. 6c). In the vertical geometry the whole electrode area can actively collect the induced charge carriers, while only a thin region around the electrode edge is effective in the planar geometry, even if the exciton diffusion length in OSSCs has been recently assessed to be about 8 μm.[20]

(iii) The high polarizability of the π-electrons at the crystal surface, possibly further enhanced by the direct exposure to X-rays that are known to cause the

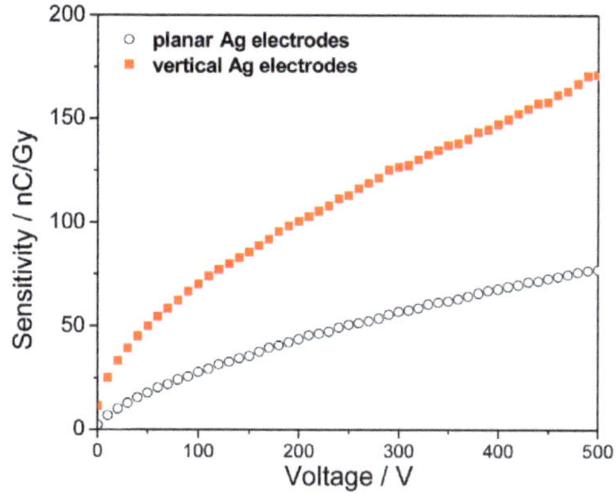

Fig. 8 Comparison between the sensitivity values at different bias voltages for a 4HCB-based detector under 35 keV X-ray irradiation in the planar (black circles) and vertical (red squares) electrode configuration.

ionization of organic molecules, may result in interactions with polar environment molecules (water *etc.*), that when adsorbed at the surface can affect the carrier density distribution or trapping states in the first monolayer below the surface, as sketched in Fig. 6c (an effect which is analogous to the operating principle of organic field effect transistors).

Fig. 7a and b give a better insight into the different ΔI ($=I_{on} - I_{off}$) response along the vertical and planar axes. The ΔI clearly remains smaller along the planar axes even after strong irradiation or a month-long aging. These results suggest that either the crystal surface plays a minor role in these measurements or its interaction with the environment and atmosphere always induces the same effects even after the strong radiation-hardness and aging tests which could have significantly altered the surface chemistry, its charge state or its conductivity. The current–voltage curves measured on the two axes of the same crystal in the dark (*i.e.* when the X-ray beam is off) (Fig. 7c and d) follow the SCLC behaviour, typical of high resistivity semiconductors.[32] However, it is noteworthy that under X-ray exposure the electrical response of our crystals along the vertical direction remains purely ohmic, while along the planar axes it follows the same SCLC behaviour found in the dark, even at higher voltages. This suggests that the electrically active traps which control the SCLC transport in the dark along the vertical axis play no major role when the crystal is probed under X-rays. From a wider perspective, it is remarkable that the strong 3D anisotropy in the carrier mobility does not affect the resulting X-ray photo-response and sensitivity, and that OSSCs can be reliably used as detectors along all three crystallographic directions, irrespective of the mobility value in the dark, as is verified by the good linear response to increasing the X-ray dose rate along both the planar and vertical axes (Fig. 7e and f). However, the device electrode geometry appears to play a major role in the performance of the radiation detector, with the vertical sandwich geometry being much higher performing than the planar geometry (as clearly shown in Fig. 8, which shows the X-ray sensitivity for the planar and vertical axes of a 4HCB crystal under a 35 keV X-ray beam). Further research is ongoing to better clarify this issue.[36]

5 Conclusions and outlook

Organic materials can be a novel and smart alternative to traditional inorganic semiconducting solid-state radiation detectors due to their low cost, wide availability and the possibility of achieving novel functionalities such as flexibility or transparency. In particular, OSSCs are better performing than organic thin films thanks to the following particular properties: (a) low dark current due to the wide effective band gap; (b) high degree of packing order; (c) long exciton diffusion length; (d) good carrier mobility; (e) possibility to tune their volume up to mm^3; (f) greater stability in air.

Solution-grown OSSCs based on different molecules can perform well as solid state radiation detectors, as we have shown for crystals based on the molecules 4HCB and DNN. The crystal shape and geometry does not affect the detector performance, since both needle-like (DNN) and platelet crystals (4HCB) have been successfully assessed as X-ray detectors. Even if the anisotropic packing of the molecule affects the electronic transport properties of the crystal, inducing a carrier mobility which can vary up to 4 orders of magnitude along the three crystal axes, it is noteworthy that all axes can be used for the effective detection of ionizing radiation, paving the way to unprecedented radiation detector architectures. Interestingly, in 4HCB crystals, the largest sensitivity is obtained in the vertical geometry, even if in this configuration the electrodes do not connect the crystal axis with the best molecular π-stacking and carrier mobility. Further work is ongoing to better understand this point and the photo-physical processes generating the X-ray-induced charge carriers, together with their transport and collection mechanisms within the organic semiconducting single crystal.

Acknowledgements

The authors acknowledge the financial support from the European Community under the FP7-ICT Project "i-FLEXIS" (GA n. 611070).

References

1 G. Knoll, *Radiation Detection and Measurement*, Wiley, 4th edn 2011.
2 Q. Chen, T. Hajagos and Q. Pei, *Annu. Rep. Prog. Chem., Sect. C: Phys. Chem.*, 2011, **107**, 298.
3 H. Dong, H. Zhu, Q. Meng, X. Gong and W. Hu, *Chem. Soc. Rev.*, 2012, **41**, 1754.
4 G. C. Schmidt, M. Bellmann, H. Kempa, M. Hambsch, K. Reuter, M. Stanel and A. C. Hübler, *MRS Online Proc. Libr.*, 2011, 1285.
5 G. Dennler, M. C. Scharber and C. J. Brabec, *Adv. Mater.*, 2009, **21**, 1323; S. Gunes, H. Neugebauer and N. S. Sariciftci, *Chem. Rev.*, 2007, **107**, 1324.
6 T. Agostinelli, M. Caironi, D. Natali, *et al.*, *J. Appl. Phys.*, 2008, **104**, 114508.
7 T. N. Ng, W. S. Wong, M. L. Chabinyc, S. Sambandan and R. A. Street, *Appl. Phys. Lett.*, 2008, **92**, 213303; P. E. Keivanidis, N. C. Greenham, H. Sirringhaus, *et al.*, *Appl. Phys. Lett.*, 2008, **92**, 023304.
8 K.-J. Baeg, M. Binda, D. Natali, M. Caironi and Y.-Y. Noh, *Adv. Mater.*, 2013, **25**, 4267; X. Gong, M. H. Tong, Y. J. Xia, W. Z. Cai, J. S. Moon, Y. Cao, G. Yu, C. L. Shieh, B. Nilsson and A. J. Heeger, *Science*, 2009, **325**, 1665.

9 K. Yoshino, S. Hayashi and Y. Inuishi, *Jpn. J. Appl. Phys., Part 2*, 1982, **21**, L569; K. Yoshino, S. Hayashi, G. Ishii and Y. Inuishi, *Solid State Commun.*, 1983, **46**, 405.
10 S. Graham, R. Friend, S. Fung and S. Moratti, *Synth. Met.*, 1987, **84**, 903.
11 P. Beckerle and H. Ströbele, *Nucl. Instrum. Methods Phys. Res., Sect. A*, 2000, **449**, 302.
12 F. Boroumand, M. Zhu, A. Dalton, J. Keddie, P. Sellin and J. Gutierrez, *Appl. Phys. Lett.*, 2007, **91**, 33509.
13 A. Intaniwet, J. Keddie, M. Shkunov and P. Sellin, *Org. Electron.*, 2011, **12**, 1903; A. Intaniwet, C. A. Mills, M. Shkunov, H. Thiem, J. L. Keddie and P. J. Sellin, *J. Appl. Phys.*, 2009, **106**, 064513.
14 A. Intaniwet, C. Mills, M. Shkunov, P. J. Sellin and J. L. Keddie, *Nanotechnology*, 2012, **23**, 235502.
15 B. Fraboni, A. Ciavatti, F. Merlo, L. Pasquini, A. Cavallini, A. Quaranta, A. Bonfiglio and A. Fraleoni-Morgera, *Adv. Mater.*, 2012, **24**, 2289.
16 T. E. Schlesinger and R. B. James, *Semiconductors for Room Temperature Nuclear Detector Applications*, Academic Press, San Diego, 1995.
17 M. A. Butler, D. S. Ginley and J. W. Bryson, *J. Electrochem. Soc.*, 1984, **131**, C327.
18 A. Fraleoni-Morgera, G. Palma and J. Plaisier, *RCS Adv.*, 2013, **3**, 15664; Ph. Leclère, M. Surina, P. Brocorensa, M. Cavallini, F. Biscarini and R. Lazzaroni, *Mater. Sci. Eng., R*, 2013, **55**, 16.
19 V. Podzorov, E. Menard, A. Borissov, V. Kiryukhin, J. A. Rogers and M. E. Gershenson, *Phys. Rev. Lett.*, 2004, **93**, 086602.
20 H. Najafov, B. Lee, Q. Zhou, L. C. Feldman and V. Podzorov, *Nat. Mater.*, 2009, **10**, 938.
21 N. Karl, *J. Cryst. Growth*, 1990, **99**, 1009; B. de Boer, *et al.*, *Phys. Status Solidi A*, 2004, **201**, 1302.
22 K. Sankaranarayanan and P. Ramasamy, *J. Cryst. Growth*, 2006, **292**, 445.
23 N. P. Zaitseva, *et al.*, *LLNL-JRNL-414904*, 2009, 1; G. Hull, *et al.*, *IEEE Trans. Nucl. Sci.*, 2009, **56**, 899.
24 S. C. B. Mannsfeld, A. Sharei, S. Liu, M. E. Roberts, I. McCulloch, M. Heeney and Z. Bao, *Adv. Mater.*, 2008, **20**, 4044; S. K. Park, T. N. Jackson, J. E. Anthony and D. A. Mourey, *Appl. Phys. Lett.*, 2007, **91**, 063514; Q. Tang, *et al.*, *Adv. Mater.*, 2008, **20**, 2947.
25 H. Minemawari, T. Yamada, H. Matsui, J. Tsutsumi, S. Haas, R. Chiba, R. Kumai and T. Hasegawa, *Nature*, 2011, **475**, 364.
26 Y. Diao, B. C.-K. Tee, G. Giri, J. Xu, Do H. Kim, H. A. Becerril, R. M. Stoltenberg, T. H. Lee, G. Xue, S. C. B. Mannsfeld and Z. Bao, *Nat. Mater.*, 2013, **12**, 665.
27 A. Fraleoni-Morgera, L. Benevoli and B. Fraboni, *J. Cryst. Growth*, 2010, **312**, 3466.
28 C. Reese and Z. Bao, *Mater. Today*, 2007, **10**, 20; J. Lee, S. Roth and Y. Park, *Appl. Phys. Lett.*, 2006, **88**, 25216.
29 B. Fraboni, C. Femoni, I. Mencarelli, L. Setti, R. Di Pietro, A. Cavallini and A. Fraleoni-Morgera, *Adv. Mater.*, 2009, **21**, 1835.
30 B. Fraboni, R. DiPietro, A. Castaldini, A. Cavallini, A. Fraleoni-Morgera, L. Setti, I. Mencarelli and C. Femoni, *Org. Electron.*, 2008, **9**, 974.
31 B. Fraboni, A. Fraleoni-Morgera and A. Cavallini, *Org. Electron.*, 2010, **11**, 10.
32 D. Braga, N. Battaglini, A. Yassar, G. Horowitz, M. Campione, A. Sassella and A. Borghesi, *Phys. Rev. B: Condens. Matter Mater. Phys.*, 2008, **77**, 115205;

M. Lampert and P. Mark, *Current Injection in Solids*, Academic Press, New York, 1970.

33 J. Trotter, *Acta Crystallogr.*, 1960, **13**, 95.

34 E. Capria, L. Benevoli, A. Perucchi, B. Fraboni, M. Tessarolo, S. Lupi and A. Fraleoni-Morgera, *J. Phys. Chem. A*, 2013, **117**, 6781.

35 A. Fraleoni-Morgera, M. Tessarolo, A. Perucchi, L. Baldassarre, S. Lupi and B. Fraboni, *J. Phys. Chem. C*, 2012, **116**, 2563.

36 A. Ciavatti, E. Capria, A. Fraleoni-Morgera, G. Tromaba, D. Dreossi, P. Sellin, B. Fraboni2014, submitted for publication.

Faraday Discussions

DISCUSSIONS

Photonics: general discussion

Simon Higgins, Luisa Torsi, Guy Whitworth, Nicolas Laurand, Shuyu Zhang, Ifor Samuel, Bernd Ebenhoch, Fred Wudl, Gitti Frey, Audrius Bucinskas, Jesuraj Inigo, Dmitrii Perepichka, Uli Lemmer, Peter Skabara, Henrik Ottosson, Donal Bradley, Iain McCulloch, Robert Brückner, Eli Zysman-Colman, Takeo Sasaki and Beatrice Fraboni

DOI: 10.1039/c4fd90047b

Nicolas Laurand opened the discussion of the paper by Uli Lemmer: What is the smallest 'rigidified' feature size achievable in principle with the laser-assisted replication technique and what sets the resolution limit? If the laser beam profile does not have a top-hat shape, could it be possible then to obtain a feature size smaller than the effective beam size?

Uli Lemmer replied: The feature sizes are basically given by the feature sizes of the tool which is used in the laser assisted replication process. We have recently shown (manuscript submitted) that the feature size can be as small as 10 nm. This is the feature size that is interesting, *e.g.* for plasmonic elements. As far as the overall size of the replicated area is concerned, I doubt that something smaller than the effective beam size is possible as heat conduction always tends to work in the opposite direction, *i.e.* the replicated area will be larger than the illuminated area.

Donal Bradley commented: There are three questions that I would like to ask: (a) How stable are your lasers in operation? (b) What do think the prospects are for the electrical pumping of organic lasers? (c) What are your thoughts on achieving CW lasing?

Uli Lemmer answered: (a): We had encapsulated Alq3 : DCM lasers which showed lasing for more than 10^7 pulses on one single spot which I consider to be enough for any sensing application, *e.g.* Raman spectroscopy as discussed in ref. 1. The mechanisms of degradation in organic lasers are as complicated as in OLEDs. There is certainly a pump intensity dependence, photooxidation and crosslinking plays a role and it depends strongly on the materials. Most likely, nonlinear annihilation phenomena are important. (b): I am convinced that sooner or later electrical lasing will be demonstrated. The key issue is the right combination of low threshold emitter materials with good transport materials and a good photonic design. With all the progress on high mobility materials, the time for further attempts on achieving electrically induced lasing is there. One of the fundamental problems is of course related to your question (c) and this is the question on how to manage or deactivate long living species (*i.e.* triplet exciton and polarons/charges).

1 X. Liu, P. Stefanou, B. Wang, T. Woggon, T. Mappes and U. Lemmer, *Opt. Express*, 2013, **21**, 28941–28947.

Ifor Samuel asked: For the stability data you just quoted, what was the excitation density?

Uli Lemmer responded: For typical stability measurements we start with a certain multiple of the laser threshold and then record the laser output as a function of time. In ref. 1 for example we have started with eleven times the laser threshold. We are also currently performing experiments with UV-illumination not inducing lasing but just photodegradation in the sample in order to elucidate the degradation processes.

1 C. Vannahme, S. Klinkhammer, M. B. Christiansen, A. Kolew, A. Kristensen, U. Lemmer and T. Mappes, *Opt. Express*, 2010, **18**, 24881–24887.

Donal Bradley said: For encapsulation do you use a simple epoxy or something more complicated?

Uli Lemmer replied: We use different but rather conventional encapsulation methods. The typical encapsulation of the lasers in the lab is based on a curable UV adhesive. For the lasers used in lab-on-chip devices, we have employed laser welding[1] and thermal bonding[2] of a polymer lid as a method for encapsulation. In the commercial VISOLAS laser system, a specifically designed package comprising a glass lid and a metal frame is used.

1 X. Liu, S. Prinz, H. Besser, W. Pfleging, M. Wissamann, C. Wannahme, M. Guttman, T. Mappes, S. Koeber, C. Koos and U. Lemmer, *Faraday Discuss.*, 2014, DOI:10.1039/c4fd00077c.
2 C. Vannahme, S. Klinkhammer, M. B. Christiansen, A. Kolew, A. Kristensen, U. Lemmer and T. Mappes, *Opt. Express*, 2010, **18**, 24881–24887.

Ifor Samuel asked: What have you done to try to elucidate the mechanism of degradation?

Uli Lemmer responded: We are currently intensifying our work to address this issue. One of the simplest measurements is to study the stability of the laser output for different pumping conditions in the presence/absence of air or with/without encapsulation. We are also conducting experiments using UV-illumination and we are performing post-mortem analysis by carrying out IR-spectroscopy and TOF-analysis of the degraded laser samples. Reproducibility is a challenge in all of these experiments.

Donal Bradley remarked: We also need to consider the effect of the electrodes; they cause a lot of additional problems.

Uli Lemmer answered: That is certainly true. There is, however, very nice work *e.g.* by the Braunschweig group[1] in which a very low TE2 waveguide mode attenuation of less than 10 cm^{-1} has been demonstrated. Although the electrodes are a challenge I consider the induced absorption by charges and the management of long living species a bigger challenge for the realization of an electrically pumped laser.

1 J. Chen, Z. Li, S. Yue and Q. Gong, *Appl. Phys. Lett.*, 2010, **97**, 041113.

Fred Wudl asked: Regarding the inhomogenous layer, is that because you're using a polymer and these have problems with polydispersity, *etc.*? What do you think will happen if you use a small molecule or oligomer, would it give you a more homogenous stability?

Uli Lemmer answered: This is a very interesting and somehow open question. In general the stability of the best polymer based lasers in our experiments is an order of magnitude smaller than the stability of the best small molecule evaporated lasers. Polydispersity and most likely a lower purity of the polymer as compared to small molecules might lead to degradation. We have not done, however, a systematic comparison of oligomers with corresponding polymers. Furthermore, most of the experiments were done with Alq3 : DCM which seems to be quite unique and remarkably stable even compared to other small molecules.

Donal Bradley asked: Small molecule systems are not necessarily homogeneous. Alq3 has two isomers that have distinct photophysical properties and DCM has a tendency to aggregate. Either or both of these factors could lead to significant heterogeneity within a blend film.

Fred Wudl commented: Small molecules exhibit polymorphism, which is indeed a problem, to solve this problem one needs annealing, either heating or solvent anealing.

Peter Skabara remarked: What are the challenges of scaling (large area and volume), using the laser assisted replication technique?

Uli Lemmer answered: The laser assisted replication was actually done using a commercial laser system with a commercial laser scanner optics. I am very optimistic that a mass fabrication could be implemented in a straightforward manner. We are now already much faster than conventional replication and we could replicate on the order of 10 full wafers per hour. The most time consuming and expensive step is the design and fabrication of the replication tool, *i.e.* the patterned Si wafer.

Ifor Samuel remarked: How does the performance of lasers using gratings made by the laser-assisted method you showed compare with lasers using gratings made by other methods such as nanoimprint lithography?

Uli Lemmer responded: Albeit we have not done a systematic study so far, it looks like the laser thresholds are in the same range as for the other replication processes (conventional hot embossing and UV-embossing).

Donal Bradley opened the discussion of the paper by Eli Zysman-Colman: In terms of use for lighting/displays, it would be useful for the attendees at the meeting if you would describe the limitations of light emitting electrochemical cell (LEECs) compared to the organic light emitting diode (OLED) structures that are used commercially in mobile phones and televisions.

Eli Zysman-Colman replied: Currently, the principal limitations of LEECs are the following: They are not as stable as OLEDs. The most stable light emitting electrochemical cell (LEEC) reported in the literature has lifetimes in the device on the order of a couple of thousand hours. Compared to commercial OLEDs with stabilities an order of magnitude longer, this remains an issue. Bright red emitters remain elusive. Most importantly, no deep-blue LEEC exists and those that are blue-ish do not last for longer than a couple of minutes. This is the key challenge facing LEEC development. If these issues can be solved then the lower driving voltages, the simplified architecture, the ability to solution process devices make LEECs very attractive. They would cost less to produce and to run and large-area lighting and display panels would be available.

Donal Bradley said: Is resolution an issue in terms of ion diffusion and does this affect pixel size?

Eli Zysman-Colman responded: I am not aware of any issues relating to diffusion of ions and how that may affect pixel size. LEECs are normally thinner devices than OLEDs due to the smaller number of layers required.

Dmitrii Perepichka commented: Have you attempted to deprotonate such iridium imidazole complexes, to afford neutral cyclometallated iridium imidazolate derivatives?

Eli Zysman-Colman responded: We have not as of yet tried this on these systems, though we have looked at this concept in the context of an unrelated project. It is a very good suggestion. I think we will explore it.

Simon Higgins asked: If the idea of the o-xylyl bridge was to rigidify your biimidazole ligand, did you consider using a shorter bridge? For example, could you not do the ring-closing reaction using 1,2-dibromoethane or 1,3-dibromopropane?

Eli Zysman-Colman replied: We have not yet. We started with the designs reported in the paper and got a hit for a bright emitter. I think that so long as the tether doesn't force a distortion of the conformation of the biimidazole then it is worth trying. Good suggestion.

Audrius Bucinskas said: What are the perspectives of iridium in large scale applications? Wouldn't it be too expensive because iridium is one of the rarest elements in the Earth's crust?

Eli Zysman-Colman replied: Iridium though expensive represents a minor cost component to the cost of the overall device, if I extrapolate from the cost breakdown of an OLED. That being said, there is a limited amount of iridium. However, let's try and get an optimized LEEC with deep blue emission up and running first and then worry about the issues of sourcing at a later point.

Ifor Samuel asked: You showed a set of materials with very different photoluminescence quantum yields. Can you learn from the molecules that had poor efficiency? Can they inform the design of more efficient emitters?

Eli Zysman-Colman answered: I think from the suite of studies that we have undertaken that there are certain design features one would desire in these emitters. These certainly include very rigid ligand systems. I also think replacing the pyridine moiety on the C^N ligand with something a lot more electron rich serves to destabilize the Ir e_g^* orbitals and so inhibit non-radiative metal-centred emission. Adding sterically bulky moieties about the complex leads to increased photoluminescent quantum yield (PLQY) as well. These are three design features that we try and incorporate into our ligand design at present.

Peter Skabara asked: The reduction processes for the compounds can't be observed within the electrochemical window of the solvent, so this implies instability of the radical anions. From the data presented there should be a reduction wave at about −1.3 to −1.4 V. How does this impact on the stability of the LEEC device? Also, the determination of the LUMO energy by the method presented should be considered carefully. Ref. 1 nicely highlights the differences in determining energy gaps by optical/electrochemical methods.

1 J.-L. Bredas, *Mater. Horiz.*, 2014, **1**, 17–19.

Eli Zysman-Colman replied: Of course we do not as of yet have LEEC data using these complexes but hope to soon. I agree with you that caution must be taken when estimating LUMO energies from the UV-Vis spectra as Bredas so eloquently mentioned in the Materials Horizons paper to which you allude. Aside from the computations, UV-Vis data was all we had at our disposal.

Ifor Samuel said: How do the solution and solid state quantum yields of your materials compare?

Eli Zysman-Colman responded: We are in the process of measuring these. Based on the work by Bryce and co-workers[1] on related FIrpic emitters, I am hopeful to see an improved PLQY in the thin films.

1 V. N. Kozhevnikov, Y. Zheng, M. Clough, H. A. Al-Attar, G. C. Griffiths, K. Abdullah, S. Raisys, V. Jankus, M. R. Bryce and A. P. Monkman, *Chem. Mater.*, 2013, **25**, 2352–2358.

Ifor Samuel remarked: Deep blue phosphorescence is not just a problem for light-emitting electrochemical cells, but also for OLEDs.

Ifor Samuel commented: What are the CIE co-ordinates of your blue materials?

Eli Zysman-Colman answered: We typically only determine CIE co-ordinates of our emitters once solid state PL data has been acquired, which we are doing at present.

Dmitrii Perepichka commented: Based on your experience, how good is TD-DFT in predicting the triplet states energies of iridium-based emitters?

Eli Zysman-Colman responded: Using the methodology we commonly employ, we get quite good agreement with TD-DFT UV-Vis absorption spectra and experiment. TD-DFT is not good at predicting emission energy but can serve to support the assignment of the nature of the triplet. Rather optimization of the triplet state using DFT and then taking the difference in energy between this value and a single point energy calculation of the ground state at the triplet state geometry will give you a good predictor of RT emission, assuming a solvent model is used.

Jesuraj Inigo said: This paper discusses deep blue emission. However, all of the materials shown here exhibit strong emission at 480 nm. This may lead to the wrong calculation or reporting of efficiency numbers since photopic filters are more sensitive at 480 nm than that of deep blue. This emission also has to be blocked by external filters if these materials were to be used in display or lighting applications. Hence, is this emission good or bad? Is there any way to avoid this?

Eli Zysman-Colman answered: To the best of our knowledge the data reported represent accurate determinations of the PLQY of these complexes. The same experimental protocol has been used in this report as well as many previous reports from my group. The sensitivity to which you refer is systematic and we are only comparing complexes that emit in the exact same region. Thus, whatever systematic error may exist is accounted for in the relative comparison between the emission spectra of the three complexes. I think more important than the actual number is the recognition that complex 3 in the paper is quite a bit brighter than complexes 1 and 2 and so has been targeted for further investigation.

Fred Wudl remarked: Regarding the rigidity of the ligands, I can imagine benzobiimidazole being more rigid, have you looked at that? It is very easy to make.

Eli Zysman-Colman replied: We have not looked at this ligand, primarily due to the fact that the increased conjugation in this ligand system will result in an unwanted red-shift in the emission. Now bis(benzimidazole) has been investigated as a ligand but not the tethered analogue. It is a good idea that I think we will pursue.

Henrik Ottosson communicated: This is a comment which is not directly applicable to the results presented in the paper by Eli Zysman-Colman and co-workers,[1] however, it connects emission of cyclometalated transition metal complexes to Baird's rule on triplet state aromaticity and antiaromaticity (the topic I talked about in Glasgow)[2]. In a paper by Yersin, Thompson and co-workers, they report on counter-intuitive blue-shifted emission from one of their Pt(II) complexes, and they explain this through changes in aromaticity in one of their ligands in the excited state (see p 12407, left column at the bottom).[3] The ligands in the present study would not display such changes, however, it might be interesting for future ligand design.

1 A. F. Henwood, S. Evariste, A. M. Z. Slawin and E. Zysman-Colman, *Faraday Discuss.*, 2014, DOI:10.1039/c4fd00107a.
2 H. Löfås, B. O. Jahn, J. Wärnå, R. Emanuelsson, R. Ahuja, A. Grigoriev and H. Ottosson, *Faraday Discuss.*, 2014, DOI:10.1039/c4fd00084f.
3 A. Bossi, A. F. Rausch, M. J. Leitl, R. Czerwieniec, M. T. Whited, P. I. Djurovich, H. Yersin and M. E. Thompson, *Inorg. Chem.*, 2013, **52**, 12403–12415.

Shuyu Zhang opened the discussion of the paper by Karl Leo†: What is the difference between a surface plasmon and a Tamm plasmon?

Robert Brückner answered: Originally, electronic Tamm states were first described by I. Tamm in 1932.[1] He described the case of an electron wavefunction inside a periodic crystal. The periodicity is terminated at the surface (considered as defect). Solving Schrödinger's equation at the surface assuming real values for the k-vector would lead to imaginary eigenenergies. This is not physical. However, when for the calculated imaginary k-vectors and p-type wavefunctions are assumed, Schrödinger's equation yields eigenenergies which are real. As a result, a confined state localizes at the surface decaying evanescently both in and out of the crystal surface (constant potential).

In photonics similar states have been described. In photonic crystals, for example in distributed Bragg reflectors (DBRs), the periodicity can also be perturbed by defects such as cavity layers.[2] This corresponds to the situation where two DBRs are separated by a certain distance, the thickness of the cavity layer. Consequently, a state localizes which can be referred to as the optical Tamm state (or even cavity mode). The condition is that both structures (the top and the bottom DBR) exhibit overlapping bandgaps as $r_L r_R = 1$, with r_i as the reflection coefficients of both structures.

Furthermore, in a configuration where a material with a negative dielectric function below the plasma frequency (for example metals) is attached to a DBR with a characteristic photonic bandgap, a state is confined which is then called a Tamm plasmon polariton (TPP).[3]

In the case of our structure which comprises a metal layer inside an organic microcavity, we observe several states which are hybridized (photonic cavity mode with Tamm plasmon polaritons).[4,5] The confinement of a surface plasmon polariton (TPP) is on the one hand also due to the negative dielectric constant of metal, but, on the other hand due to the total internal reflection into dielectric media. In this context, characteristics of TTPs and differences to surface plasom polaritons (SPPs) are:

(a) The confinement of a TPP is due to the negative dielectric function of the metal and due to the photonic bandgap of the photonic crystal; (b) TPPs are localized inside the DBR stop bands; (c) the dispersion $\omega(k)$ is parabolic in shape; (d) in contrast to SPPs, Tamm states can be excited in both polarizations, in transverse-electric (TE) and transverse-magnetic (TM); (e) the splitting between the TE and TM polarized branches is proposed to scale quadratically with the in-

† Karl Leo's paper was presented by Robert Brückner, *TU Dresden, Germany*.

plane k vector; (f) the in-plane wave vector of TPPs is close to zero (inside the light cone) and they can therefore be excited *via* photons without a grating; (g) the eigenenergy of TPPs depends on the thicknesses of both the metal layer itself and the adjacent dielectric layers.

1 I. E. Tamm, *Phys. Z. Sowjetunion*, 1932, **1**, 733.
2 A. Kavokin, I. Shelykh and G. Maluech, *Appl. Phys. Lett.*, 2005, **87**, 261105.
3 M. Kaliteevski, I. Iorsh, S. Brand, R. A. Abram, J. M. Chamberlain, A. V. Kavokin and I. A. Shelykh, *Phys. Rev. B*, 2007, **76**, 165415.
4 R. Brückner, M. Sudzius, S. I. Hintschich, H. Fröb, V. H. Lyssenko and K. Leo, *Phys. Rev. B*, 2011, **83**, 033405.
5 R. Brückner, M. Sudzius, S. I. Hintschich, H. Fröb, V. G. Lyssenko, M. A. Kaliteevski, I. Iorsh, R. A. Abram, A. V. Kavokin and K. Leo, *Appl. Phys. Lett.*, 2012, **100**, 062101.

Shuyu Zhang commented: The silver layer inserted in the DBR microcavity caused the Tamm plasmons and such geometry presents a promising route towards electrically pumping organic lasers. So have you tried using two layers of silver on each side of active layer? What phenomena have you observed?

Robert Brückner responded: Thank you very much for this question. Inserting a second metal layer has been tried but is not trivial. It is difficult to control a good metal layer growth on both the DBRs and on top of organic layers which are itself deposited on a DBR and electrode structure. This is especially true for thin silver layers. Penetration of silver into rough DBR layers or diffussion into organic layers lowers the quality factors drastically.

Furthermore, the implementation of a second electrode requires a full redesign of the whole structure because of an altered mode distribution resulting in additional introduced free parameters. Simulations can only be carried out applying numerical methods. For instance a further localization of modes and further hybridizations are observed (triplets).

We are currently working on this question – look for our upcoming publications.

Uli Lemmer asked: How important is the quality/crystallinity of the silver films in your samples. Do you see effects on the Q-factor of the cavity?

Robert Brückner answered: We have chosen silver for several reasons. First, it is a very good conductor and easily available. A problem of thin silver contacts is that silver tends to grow in small islands, especially for very thin layers of only a few nanometers. This effect can be minimized by choosing high deposition rates (5–7 A s^{-1}). The gradient in thickness is produced by the sharp edge of a mask within a very short distance of only 100 μm. This enables us to produce closed layers of different thicknesses. The estimate of the Q-factor in these samples is only possible if the spot size of the excitation laser is very small (with a diameter of 2 μm) compared to the change in thickness due to the gradient.

Furthermore, our group has lots of experience in preparing electrical contacts. To enhance the quality of the silver layers, thin barrier and seed layers of aluminum, gold or molybdenum oxide are deposited prior to the silver contacts. In our sample designs, we produce high quality silver layers and consequently high Q-factors of about 1000. Note, that the Q-factor is not only a measure of the quality of the layers but also of losses, for example due to scattering or absorption.

Donal Bradley commented: It would be interesting to hear why you chose to use Alq3–DCM, or more generally an organic exciton system. Is there a reason that you didn't use inorganic semiconductor films or colloidal particles in your structures?

Robert Brückner responded: A strong arguments for the use of Alq3 in our experiments is the good stability during the preparation (sustains the top DBR deposition) and the experiment (more stable than most polymeric systems). We are generally interested in performing complex spectroscopic experiments and need therefore a stable (gain) emitter system like Alq3–DCM even though we don't achieve the lowest possible threshold values. Furthermore, we can perform our experiments under ambient conditions at room temperature and in air. The samples are only poorly encapsulated by the top DBR from which we know it is porous and hygroscopic. In an experiment to study the water content in DBRs, we continuously measured the transmission spectrum of a complete DBR while the vacuum chamber, in which the DBR was prepared, is purged with humid air. As a result, water fills the pores and the whole spectrum is shifted by 30–40 nm into the red spectral region. This effect can be exploited for designing humidity sensors.

Another motivation of our work and to use an organic excitonic system is to study the possibility of an electrically pumped organic solid state laser. In our approach, we investigate and optimize various sample designs. In the future these designs can be combined with optimized gain materials.

The use of inorganic emitter systems with our technology is not trivial since sharp emission lines provided by many inorganic emitters require a controlled and precise way of layer deposition to match the design. This is very challenging for an optical stack comprising more than 40 layers.

Nicolas Laurand remarked: Referring to your work on mode confinement with metal, micro-size holes:

Do you see the metal micro-hole confined structure as a possible pathway towards electrically-driven organic lasers? Could you comment on this cavity configuration and on the associated challenges of carrier injection into the hole region?

Robert Brückner replied: Thank you for this very interesting question. Such a hole configuration yields several advantages regarding the optics. (a) The geometry is favorable for the confinement of light. The symmetry allows for very easy description of emerging modes.[1] (b) The modes are well confined inside the hole, so that losses due to the metal are minimized. (c) Higher order modes can be selectively excited in order to gain control of the emission wavelength angle, and polarization. (d) The technique of hydrophobic spreading allows the creation of holes in arbitrary sizes in micrometer dimensions. A confinement in these dimensions has proven useful and optimal. (e) Coupling of the modes living in the metal regions (to Tamm states) is possible even though they exhibit different eigenenergies depending on the metal thickness.

From the electrical point of view, a circular electrode would enhance the charge carrier injection density in the center of the hole (where the overlap with the optical mode is maximum) and the carrier density would be lower at the edges of the hole at the same time. However, a second electrode needs to be implemented in any way. There are several obstacles to overcome. (a) The electrode should be placed accordingly on top of the organic gain medium covering the hole. (b) The electrode should contact only at regions of the hole, not outside of it where there is a homogeneous bottom electrode (perforated by holes).

To realize such a configuration, photolithography has to be applied on top of organic materials. This is currently bring investigated and to some extent established in our institute. However, a perfect orientation and placing on the micron scale is essential.

1 C. Reinhardt, R. Brückner, J. Haase, M. Sudzius, S. I. Hintschich, H. Fröb, V. G. Lyssenko and K. Leo, *Appl. Phys. Lett.*, 2012, **100**, 103306.

Uli Lemmer commented: There is a recent paper by Rainer Mahrt *et al.*[1] in which room temperature Bose–Einstein condensation is claimed. Wouldn't you expect to observe similar effects in your samples?

1 J. D. Plumhof, T. Stöferle, L. Mai, U. Scherf and R. F. Mahrt, *Nature Materials*, 2014, **13**, 247–252.

Robert Brückner replied: This is an interesting point. The emitter used is Alq3 : DCM which is known as a weakly coupled system. In our case, we can describe our experiments in terms of laser theory by applying a simple rate equation model. Through this we are able to model both time-resolved measurements (*e.g.* from up-conversion experiments) and entire input-output characteristics. In contrast to systems showing Bose–Einstein condensate (BEC), we pump our system strongly to achieve population inversion, meaning that we are strongly out of thermodynamic equilibrium. In the following, I want to add a collection of statements found in the literature:

H. Deng *et al.*[1] – The polariton condensate is not in thermal equilibrium with its host lattice, and the phase transition is accompanied by emission of coherent light, which is analogous to an atom laser created out of an atom BEC. For this reason, we call the dynamic condensation of polaritons – a "polariton laser".

P. Littlewood[2] – The light emitted from the condensate is nearly coherent, which makes the whole device behave like a special kind of laser. In microcavities containing some organic molecules, large Rabi splittings are tantalizing for a room-temperature device. These objects are a new kind of low-threshold laser.

L.V. Butov[3] – Waves in nature were divided into electromagnetic waves, such as the photon, and matter waves, such as the electron. A coherent state of electromagnetic waves is known as a laser; a coherent state of matter waves is termed a Bose–Einstein condensate…

D. Snoke[4] – Rather than haggling over the exact meaning of the term Bose–Einstein condensation, we can adopt a more general concept – the spontaneous emergence of coherence.

L.V. Butov and A.V. Kavokin *et al.*[5] – Several experimental groups have demonstrated that exciton–polaritons–quasiparticles made from a mix of light and matter – can form a coherent state (that is, a condensate in momentum space) in a semiconductor microcavity.[6–9] However, there is little agreement in the community regarding the nature and associated terminology of this condensate: is it a Bose–Einstein condensate (BEC), a laser, or something else?

There is, however, a clear distinction between a BEC and a polariton laser: thermal equilibrium is an essential property of BEC, but it is not required for a laser.

The physics of a polariton condensate, although different from that of a BEC and superfluidity, is nevertheless rich and interesting.

1 H. Deng, G. Weihs, D. Snoke, J. Bloch and Y. Yamamoto, *PNAS*, 2003, **100**, 15318.
2 P. Littlewood, *Science*, 2007, **316**, 989.
3 L. V. Butov, *Nature*, 2007, **447**, 540.
4 D. Snoke, *Nature*, 2006, **443**, 403.
5 L. V. Butov and A.V. Kavokin, *Nat. Photonics*, 2012, **6**, 2.
6 H. Deng, G. Weihs, C. Santori, J. Bloch and Y. Yamamoto, *Science*, 2002, **298**, 199–202.
7 J. Kasprzak, M. Richard, S. Kundermann, A. Baas, P. Jeambrun, J. M. J. Keeling, F. M. Marchetti, M. H. Szymańska, R. André, J. L. Staehli, V. Savona, P. B. Littlewood, B. Deveaud and L. S. Dang, *Nature*, 2006, **443**, 409–414.
8 S. Christopoulos, G. Baldassarri, H. von Högersthal, A. J. D. Grundy, P. G. Lagoudakis, A. V. Kavokin, J. J. Baumberg, G. Christmann, R. Butté, E. Feltin, J.-F. Carlin and N. Grandjean, *Phys. Rev. Lett.*, 2007, **98**, 126405.
9 A. P. D. Love, D. N. Krizhanovskii, D. M. Whittaker, R. Bouchekioua, D. Sanvitto, S. Al Rizeiqi, R. Bradley, M. S. Skolnick, P. R. Eastham, R. André and L. S. Dang, *Phys. Rev. Lett.*, 2008, **101**, 067404.

Ifor Samuel remarked: In the femtosecond up-conversion measurements in the paper, why is the peak output of the laser up to 100 ps after the excitation pulse when the cavity lifetime is only 1.5 ps?

Robert Brückner answered: Thank you for this question. The delay of the peak output after excitation is mainly determined by the transient times in the system. In our case, the Alq3 molecules have been pumped using femtosecond pulses of 400 nm. The Förster transfer (FRET) which is the dominating energy transfer from the Alq3 molecules to the DCM (doped by 2wt%) molecules is regarded as fast, in our case (treated as a fit parameter) 19 ps. Furthermore, the life time of the emitting states plays an important role, which consists of a radiative and a nonradiative part (also a fit parameter). The coupling and the non-linearity of the systems explain the temporal behavior, please also see M. Koschorrek *et al.*[1]

1 M. Koschorrek, R. Cehlhaar, V. G. Lyssenko, M. Swobado, M. Hoffmann and K. Leo, *Appl. Phys. Lett.*, 2005, **87**, 181108.

Ifor Samuel asked: Can you comment on the relative merits of the microcavity geometry you use compared with lateral (for example distributed feedback (DFB)) geometries for organic lasers?

Robert Brückner responded: This is a very interesting point. Both systems, vertical-cavity surface-emitting lasers (VCSELs) and DFB devices, have advantages and disadvantages. We rely on the versatility and the easy to produce microcavity structure. We produce our planar devices usually under high vacuum, starting from the DBR layers to the organic medium, everything is evaporated and the layers are therefore of high quality. The use of the Alq3 : DCM system which is sublimated twice before deposition enables very stable laser devices even without encapsulation, at room temperature. Another advantage is the easy way to combine the longitudinal feedback given by the resonator itself with lateral structuring, for example by shadow mask evaporation or photolithography. By this we can study very interesting effects known from solid state physics such as Bloch oscillations with photons. A clear drawback of VCSEL structures is the usually high threshold density needed to achieve inversion. Compared to DFB devices, where the extent of the lateral mode (in plane) is large and can exceed hundreds of μms, the mode extent in VSCELs is rather small limited by the reflection of the mirrors and the overall absorption the mode encounters. This is usually not more than a few tens of microns. Consequently, the pump energy needs to be stored into small volumes leading to high excitation densities and possible damage. This is also the reason why it is at the moment not possible to pump a VCSEL by a high-power inorganic LED.

Uli Lemmer asked: Do you observe any damage or loss in quantum efficiency in your DCM molecules upon evaporation of the dielectric mirror layers?

Robert Brückner answered: This is an interesting question. Interestingly, our samples do not degrade while the top DBR is deposited. Neither scattered electrons nor the UV- or X-rays which are present due to electron evaporation harm the Alq3 : DCM to a high degree. The stability of Alq3 in the preparation as well as in the experiment is crucial for us.

Guy Whitworth addressed Karl Leo and Robert Brückner: Can you expand on the physical meaning of , ν, the materials polarization amplitude in your rate equation (2b)?

Robert Brückner answered: In a laser, the coherent electromagnetic field $E_{(\mathrm{r,t})} = A_{(\mathrm{r,t})}\exp^{(-i\omega t)}$ propagates through the active (inverted) medium characterized by the complex permittivity $\varepsilon(\omega) = (n' + in'')^2 = 1 + X(\omega)$, sandwiched between two dielectric mirrors with reflectivities R. This electric field E induces a dielectric polarization density $P(\omega) = X(\omega)E(\omega)$. Their evolutions are described by coupled Maxwell–Bloch equations (1). For simplicity, the amplitude of the electromagnetic field α and the amplitude of the macroscopic polarization ν are normalized, *i.e.* measured in units of photon number and number N of excited molecules (dipoles). Two equations (1a and 1b) describe the evolution of the polariton (as quanta of the polarization). The last equation (1c) describes the decay of the population (inversion) *versus* external pump Γ. In the next steps, the polarization ν is adiabatically eliminated since the damping is very fast. For details please see, P. Rice and H. Carmichael.[1]

1 P. R. Rice and H. J. Carmichael, *Phys. Rev. A*, 1994, **50**, 4318.

Ifor Samuel commented: How does the approach of using a Tamm plasmon polariton you showed compare with alternatives like using structures designed so that the contacts are at nodes of the electric field distribution of the waveguide mode?

Robert Brückner responded: The approach is rather similar. The interesting thing on Tamm plasmon polaritons is that there are additional states occurring because of the coupling of the optical mode to the metal. These modes also exhibit a high degree of localization and little losses.

Iain McCulloch opened the discussion of the paper by Takeo Sasaki: What is the reason that a 1 : 1 mixture of liquid crystals is required and not just a single liquid crystal (one or the other) as C8 and C10 are as transparent as each other?

Takeo Sasaki answered: The photorefractive effect is one of the hologram generating phenomena. Thus, transparency of the photorefractive material is crucial. The ferroelectric liquid crystal has a highly ordered phase structure so that it is more like crystal than a liquid. Formation of a large mono-domain (single crystal) of a ferroelectric liquid crystal is very difficult. It has been known that a large mono-domain can be formed by using blends of liquid crystalline compounds. Actually, a mixture of tens of liquid crystalline compounds are used in the practical display application.

The mixture of liquid crystalline compounds used in this study is not the best one. It is crucial to develop a highly transparent and highly photoconductive ferroelectric liquid crystalline mixture.

Donal Bradley said: Is the electro-optic effect that you use all to do with the material polarisability or is there a component due to molecular orientation?

Takeo Sasaki replied: The electro-optic effect of ferroelectric liquid crystals is based on the response of the spontaneous polarization. The spontaneous polarization is not a polarization of a liquid crystalline molecule but a bulk polarization of the liquid crystalline phase. Thus, the response of the ferroelectric liquid crystals to an applied electric field is very fast (a few tens of micro-seconds to a few hundreds of micro-seconds). In ferroelectric liquid crystals, the direction of the liquid crystalline molecules changes according to the change in the direction of the spontaneous polarization. When laser beams interfere in a mixture of ferroelectric liquid crystals and a photoconductive compounds, the direction of the spontaneous polarization is changed by the internal electric field formed between bright and dark positions of the interference fringe. This leads to a periodic change in the orientations of liquid crystalline molecules. Thus, an orientational grating is formed at the interference fringe.

Donal Bradley said: Is it possible to have an LC based system that responds fast enough to generate video-rate images?

Takeo Sasaki answered: The response time of the photorefractive ferroelectric liquid crystal blend is 930 micro-seconds. The response is fast enough to generate video-rate images. The example of the video-rate amplification of an optical image by a photorefractive ferroelectric liquid crystal can be seen at the web site (http://www.rs.kagu.tus.ac.jp/photoref/Amplification_1_WMV.wmv).

Ifor Samuel commented: You mentioned a millisecond response time, but I thought a feature of ferroelectric liquid crystals (FLCs) was that they had micro-second switching times. What is the fastest switching that could be achieved?

Takeo Sasaki replied: The FLC responds to an electric field very quickly in a time scale of a few tens of micro-seconds. However, the response time of photorefractive effect in FLC is 900 micro-seconds. It is considered that the determining step of the photorefractive effect in FLC is the formation of the space-charge field. A photorefractive FLC blend with a high charge mobility at the bright positions of the interference fringe and high insulation at the dark positions can achieve the fast response. It is expected that a photorefractive FLC with a response time of sub-hundred micro-seconds will be obtained by the development of an adequate photoconductive compound.

Uli Lemmer asked: How much energy can be transferred from the pump beam to the signal beam? Can this be described in absolute or relative numbers (*e.g.*, x% of the pump beam energy is transferred)?

Takeo Sasaki answered: The gain coefficient of a photorefractive FLC blend in the early days (a commercially available FLC mixed with a bulky photoconductive compound) was 15 cm^{-1}. Since then blends of FLC mixtures that show high miscibility with dopants and photoconductive compounds that have the rod like structure have been developed and the gain coefficient of 1200 cm^{-1} was obtained. In the 1200 cm^{-1} FLC blend, more than 40% of the energy of one of the interfering beams transferred to another beam within a 10 micro-meter thick FLC medium. The important elements to obtain a high gain coefficient in an FLC blend are high transparency and high charge mobility. The combination of the FLC mixtures and the photoconductive compounds is almost infinite. It is considered that transfer of energy from one of the interfering beams to another beam can be 70 to 80% which means a gain coefficient over 2000 cm^{-1} can be expected.

Ifor Samuel asked: You measure a very high gain coefficient. What is the maximum actual gain that you can obtain?

Takeo Sasaki replied: Please see the response to the above question asked by Uli Lemmer.

Ifor Samuel said: You mentioned an intensity gain of 40%. What is the maximum it could be?

Takeo Sasaki replied: Please see the response to the above question asked by Uli Lemmer.

Donal Bradley said: This is a comment following a question concerning the difference in beam intensities.

One has to form an interference grating and that is likely to limit, in practice, the range of relative intensities that can be used for the weak and strong beams.

Ifor Samuel remarked: How does the photorefractive amplification you showed compare with other possible approaches for optical amplification such as using parametric amplification or stimulated emission?

Takeo Sasaki replied: The amplification of light beam by the photorefractive effect is based on the creation of a hologram in the photorefractive medium. Only the interfering beams can create a hologram. That means only the beams that have a specific wave length, specific polarizing plane and specific phase is amplified. The photorefractive amplifier distinguishes the specific light beam. One can select a beam to be amplified. This amplification is called coherent amplification. That makes the photorefractive amplifier different from other optical amplification devices. We expect that the photorefractive amplification can be utilized in devices that require recognition of specific light signal and amplification such as an object recognition system for safe driving, detection of a flying object and optical tomography.

Donal Bradley commented: This is a comment following a question concerning the geometry of devices.

The photorefractive effect gives a local amplification process, so there should not be too many restrictions on the format required.

Ifor Samuel asked: What is the operating temperature range for the liquid crystals in your amplifier?

Takeo Sasaki answered: The operating temperature of the FLC blend used in this study is 25 to 40 °C, however, the operating temperature can be adjusted to the required range by the choice of FLC mixture.

Uli Lemmer remarked: What is the optimum electric field in your devices? As high as possible as long as no breakthrough is observed? Do you have to make any compromises?

Takeo Sasaki replied: The operating electric field for the photorefractive effect is a few volts per micro-meter in polymer materials and liquid crystals. Also, the internal electric field (space-charge field) formed in a photorefractive material is estimated to be a few volts per micro-meter. The maximum thickness of the FLC film is 10 micro-meters so that the electric field required to obtain a photorefractive effect is a few volts. That is one of the advantageous features of the photorefractive FLCs.

Simon Higgins opened the discussion of the paper by Beatrice Fraboni: I notice that all of your molecules are composed of first-row elements. Have you considered using molecules containing heavier elements, *e.g.* iodoarenes are quite readily available and can be reasonably stable. Wouldn't this be an advantage for 'stopping' X-Rays?

Beatrice Fraboni answered: Indeed, we also think it would be an advantage and we are presently growing new single crystals based on molecules with heavier elements (iodine, Br and Si) with semiconducting electrical properties.

Ifor Samuel asked: How does the quantum efficiency of your device compare with inorganic detectors? Presumably several photoelectrons could potentially be made per absorbed X-ray photon. Also what is the detectivity – something that has been very good in organic photodiodes even if their quantum efficiency was unremarkable?

Beatrice Fraboni replied: The X-ray attenuation of a 500 μm thick crystal is about 4%, much lower than for inorganic detectors (50% for silicon and more than 60% for a-Se detectors) of the same thickness. Nonetheless the devices show good X-ray responses and sensitivities up to 150 nC/Gy, similar to amorphous silicon based devices. A preliminary evaluation of quantum efficiency is reported in a published work[1] and further detailed evaluations are under investigation.

1 B. Fraboni, A. Ciavatti, F. Merlo, L. Pasquini, A. Cavallini, A. Quaranta, A. Bonfiglio and A. Fraleoni-Morgera, *Adv. Mater.*, 2012, **24**, 2289–2293.

Dmitrii Perepichka commented: Regarding the previous question on lower X-ray sensitivity of organic matter and the possibility to include heavy elements in the semiconducting crystal. I should say that sensitivity is often not the main concern in X-ray detectors. Thus, in biomedical applications, the similarity of the energy response between organic semiconductors and living tissue is potentially a very big advantage.

Dmitrii Perepichka asked: In your measurements, how focused was the radiation field? Were the metal electrodes also exposed to radiation? In our own experiments in similar photoresistor device configuration, we have seen a comparable X-ray photoresponse, whether conjugated or non-conjugated polymer films are used. I thus wonder in your case how important is the semiconducting property/charge mobility of the crystal for the X-ray detection?

Beatrice Fraboni responded: We have carried out analyses with different X-ray beam diameters (dimensions varying from 2 mm down to a few tens of microns in a synchrotron facility) specifically to exposes either the whole device (active channel and electrodes) or small portions of the active channel only with shielded electrodes, respectively. The goal was to assess the contribution to the collected photocurrent induced by the interaction of the X-ray beam with the electrodes by selectively removing their contribution. We observed a stronger induced photocurrent when the electrodes were exposed to the X-ray beam, possibly induced by secondary photo-electrons generated from the metal into the crystals. However, this contribution was always smaller than the contribution from the bare crystal, as assessed by using low Z electrodes (made with conductive polymer) and plastic substrates.[1]

1 B. Fraboni, A. Ciavatti, F. Merlo, L. Pasquini, A. Cavallini, A. Quaranta, A. Bonfiglo and A. Fraleoni-Morgera, *Adv. Mat.*, 2012, **24**, 2289–2293.

Uli Lemmer commented: You have shown prospects for ink jet printed crystals on foils. This is very exciting. But wouldn't you lose the advantage of mechanical flexibility of your sensor layer?

Beatrice Fraboni answered: The flexibility could be preserved as we have shown in the present Faraday Discussion manuscript how single crystals down to 10 microns thick can effectively detect X-rays.[1] It has been reported in the literature how thin single crystals can be bendable[2] and we are presently investigating the thickness limit for detecting applications.

1 B. Fraboni, A. Ciavatti, L. Basiricò and A. Fraleoni-Morgera, *Faraday Discuss.*, 2014, DOI: 10.1039/c4fd00102h
2 A. L. Briseno, S. C. B. Mannsfeld, M. M. Ling, S. Liu, R. J. Tseng, C. Reese, M. E. Roberts, Y. Yang, F. Wudl and Z. Bao, *Nature*, 2006, **444**, 913–917.

Uli Lemmer asked: You highlighted the potential low costs of an organic sensor as compared to inorganic systems. Does this hold true when it comes to a comparison with amorphous inorganic detector materials like a-Si or a-Se? I would expect that the growth of a good organic crystal would somehow take time and thus increase costs.

Beatrice Fraboni replied: Indeed, growing self-standing organic crystals can take considerable amounts of time, even up to three to four weeks. On the other hand, the mention of low cost is referred not to these laboratory prototypes, rather to potentially manufacturable devices in which the crystals are obtained deposing the starting solution with different techniques, like inkjet printing, for example. In these cases, that we are actually exploring, the foreseen cost is definitely low, in terms of both direct costs (materials, equipment) and time needed for the growth (down to a few minutes).

Donal Bradley remarked: Nobody wants to work with amorphous selenium

Gitti Frey remarked: Have you considered the hybrid perovskites currently investigated in organic photovoltaics (solution processes, crystalline and Pb-included) for this process?

Beatrice Fraboni responded: No, we have not, but it would indeed, be very interesting to study such materials.

Bernd Ebenhoch asked: Have you varied the X-Ray energy and is there potential to use these detectors for X-Ray spectroscopy?

Beatrice Fraboni responded: We have carried out analyses with Synchrotron monochromatic X-ray beams in the range 10–25 keV to assess the spectroscopic

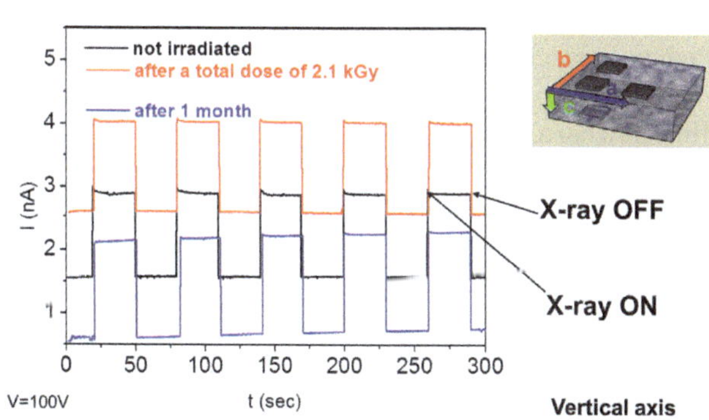

Fig. 1 Repeated on-off switching of the X-ray beam along the vertical direction (axis *c*) of a 4HCB crystal, before exposure to X-ray irradiation (black line), after exposure to a total dose of 2.1 kGy of X-rays at 35k eV (red line) and after a month of storage in air and in the dark (blue line). (B. Fraboni *et al. Adv. Mater*, 2012, **24**, 2289)

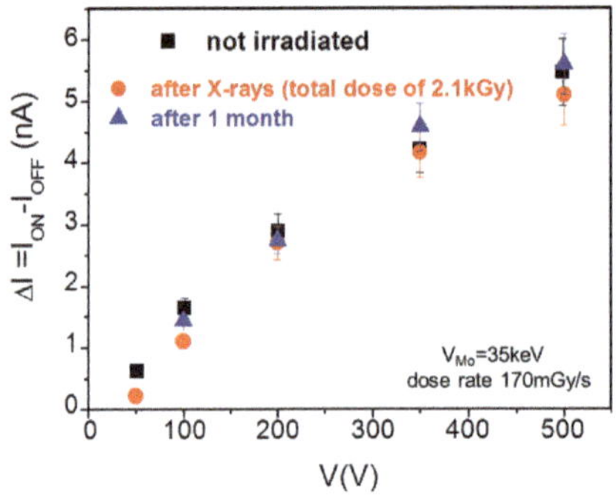

Fig. 2 X-ray induced current variation $\Delta I = I_{on} - I_{off}$ reported for different bias voltages applied to 4HCB crystals before (black squares) and after irradiation with a total delivered X-ray dose of 2.1 kGy (red circles). The response of crystals after an aging period of 1 month (blue triangles) is also reported. (B. Fraboni *et al. Adv. Mater*, 2012, **24**, 2289)

potential of organic single crystals, and we have obtained promising results. More details will be provided in a manuscript that is in preparation

Donal Bradley said: Is it possible that the difference in mobility that you see is a result of the charges being generated from core levels? Might one think of the charges as being akin to those that generate a 'treeing' phenomenon in dielectric breakdown processes?

Beatrice Fraboni responded: We still do not have a definite description of these photo-physical processes, and we are currently in the process of better understanding it. It is indeed likely that the initial ionization caused by the X-rays results in free charges, some of which may directly go into the conduction band while some others may start a cascade effect. We do not have direct experimental evidence of these hypotheses yet.

Dmitrii Perepichka commented: Why do you use the difference between "on" and "off" current (not their ratio, as done more commonly), as a radiation detection signal?

Beatrice Fraboni responded: We tend to discuss and compare the $\Delta I = I_{on} - I_{off}$ signal behaviour instead of the $\Delta I/I$ as the latter one is strongly dependent on the I_{off}, the baseline dark current of the crystal. As we are comparing the performance of crystals grown with different novel molecules, they have quite different dark current values.

Fred Wudl said: Have devices been left to "sit" after being used; is performance still the same afterwards?

Beatrice Fraboni replied: We have assessed how the response of the organic semiconducting single crystals varies with time, up to a few months. The most noticeable results are that despite the observed shift in the dark current baseline, the absolute value of the induced photocurrent ($\Delta I = I_{on} - I_{off}$) remains stable with time. The two figures below (Fig. 1 and 2)[1] show how the dynamic response of the device is still fast, reproducible and reliable after a heavy total dose of X-rays and after 1 month of storage at ambient conditions.

1 B. Fraboni, A. Ciavatti, F. Merlo, L. Pasquini, A. Cavallini, A. Quaranta, A. Bonfiglio and A. Fraleoni-Morgera, *Adv. Mater.*, 2012, **24**, 2289.

Ifor Samuel commented: You showed some drift of baseline, which could to some extent be corrected in the electronics of a system. This correction should be straightforward when measuring a sudden change but more difficult when trying to measure a small change over a long time (for example total X-ray dose received in a month). Which applications would be most suitable?

Beatrice Fraboni responded: The baseline shift can be reduced by lowering the applied bias, thus we are working on the development of lower voltage operating devices. Moreover, the devices are suitable as real-time monitoring radiation detectors instead of accumulation dosimeters. As the X-ray absorption and interaction processes within organic materials are similar to those within human tissues, organic detectors are ideal candidates for medical diagnostic and imaging applications.

Luisa Torsi said: Considering the current decrease you see with time of exposition to X-rays, have you envisaged any operational condition to restore pristine conditions? It is very interesting to see that the current change (ΔI) stays constant irrespective of a background current drift. We have seen very similar effects with an organic semiconductor in a thin-film transistor configuration exposed to different analytes for chemical sensing. To restore pristine condition we use a reverse gate bias pulse that has been shown to effectively work.[1]

1 L. Torsi and A. Dodabalapur, *Anal. Chem.*, 2005, 77, 380 A–387 A.

Beatrice Fraboni responded: The observed drift in the baseline current is due to a combined effect of aging and bias stress, that could depend both on the applied bias and on the nature of the molecule itself. The device is presently a photoconductor, the addition of an organic semiconductor–dielectric interface, like in a transistor configuration could change the shown effects and its behaviour will have to be studied in more depth

Luisa Torsi remarked: As to contact resistance contribution to detection is concerned a good approach to quantify this contribution is to use the 'transfer line method'.[1]

1 L. Torsi, F. Marinelli, M. Daniela Angione, A. Del'Aquila, N. Cioffi, E. De Giglio and L. Sabbatini, *Org. Electron.*, 2009, **10**, 233–239.

Beatrice Fraboni answered: The tested electrodes provide good ohmic contacts with the crystal, and the contact resistance is negligible in such high resistance photoconductors. The electrodes contribution to X-ray signal has been investigated in a previous work.[1]

1 B. Fraboni, A. Ciavatti, F. Merlo, L. Pasquini, A. Cavallini, A. Quaranta, A. Bonfiglo and A. Fraleoni-Morgera, *Adv. Mat.*, 2012, **24**, 2289–2293.

Faraday Discussions

PAPER

Parameter free calculation of the subgap density of states in poly(3-hexylthiophene)†

Jarvist M. Frost,[ab] James Kirkpatrick,[ac] Thomas Kirchartz[ad] and Jenny Nelson*[a]

Received 21st July 2014, Accepted 23rd July 2014
DOI: 10.1039/c4fd00153b

We investigate the influence of intra-chain and inter-chain interactions on the sub-gap density of states in a conjugated polymer using a combination of atomistic molecular dynamics simulation of polymer film structure and tight-binding calculation of electronic energy levels. For disordered assemblies of poly-3-hexylthiophene we find that the tail of the density of hole states is approximately exponential with a characteristic energy of 37 meV, which is similar to experimental values. This tail of states arises mainly from variations in the electronic coupling between neighbouring monomers, and is only slightly influenced by interchain coupling. Thus, knowledge of the disorder in torsion between neighbouring monomers is sufficient to estimate the density of states for the polymer. However, the intrachain torsional disorder is determined largely by the packing of the chains rather than the torsional potential alone. We propose the combination of methods as a tool to design higher mobility conjugated polymers.

Organic semiconductors are promising materials for electronic and optoelectronic applications such as transistors,[1] light emitting diodes[2] and solar cells.[3–7] Among the key parameters of importance for these applications are the mobility of charge carriers and, in bipolar devices, the charge carrier lifetime. In most organic semiconductors charge transport is controlled by the molecular nature of the material, such that charges are localised on individual molecules or conjugated segments and move between weakly coupled sites *via* a hopping process. The variability in molecular conformation and packing which naturally arises in these soft materials then gives rise to variability in the energy of the states on

[a]*Department of Physics and Centre for Plastic Electronics, Imperial College London, London SW7 2AZ, UK. E-mail: jenny.nelson@imperial.ac.uk*
[b]*Department of Chemistry, University of Bath, Claverton Down, Bath BA2 7AY, UK*
[c]*Department of Physics, Oxford University, Parks Road, Oxford OX1 3LB, UK*
[d]*IEK-5 Photovoltiacs, Forschungzentrum Juelich, 52425 Juelich, Germany*

† Electronic supplementary information (ESI) available. See DOI: 10.1039/c4fd00153b

which charge carriers reside (so-called 'energetic disorder'), and variability in the separation and relative orientation of conjugated units (so-called 'configurational disorder'), both of which tend to slow down transport by trapping charges.[8,9] Localised states in the electrical gap (of organic semiconductors) are also expected to increase the rate of non-radiative electron–hole recombination in bipolar devices such as solar cells.[10] Energetic disorder can be described in terms of a density of (localised) states (DoS). In principle, both the charge carrier mobility and lifetime, and their charge density dependence, are functions of the DoS.[11] In order to optimise and diagnose device performance and to develop materials with improved electrical properties, it is essential to develop an improved understanding of the origin of localised states in organic semiconductors.

A variety of experimental studies of organic semiconductors have indicated the presence of a tail of states that extends into the electrical gap. Most commonly the density N of localised states in disordered organic semiconductors is approximated by either a Gaussian distribution[9] or an exponential tail.[12–14] In the case of an exponential tail of electron states, the density of states N in energy interval dE is given by:

$$N(E)\mathrm{d}E = N_0 \exp\left(\frac{E - E_\mathrm{c}}{E_\mathrm{ch}}\right)\mathrm{d}E \tag{1}$$

where E is the energy, E_c is a reference energy at which electrons are free to move, which we will refer to as the mobility edge, N_0 is the density of states at E_c, and E_ch is the characteristic energy of the tail. This characteristic energy is the key parameter controlling the influence of localised states on device performance, such that higher values of E_ch lead to larger concentrations of deep-lying states that can trap charges, slow down transport and facilitate recombination.[15,16]

These tails of states have been studied with a wide range of techniques, in unipolar devices such as transistor structures,[17–19] and especially in bipolar devices such as OLEDs and solar cells. Evidence for a roughly exponential tail of localised states in disordered polymer:fullerene blends used for organic solar cells comes from measurements of the absorption edge using quantum efficiency or spectral photoconductivity methods,[10,20,21] transient photocurrent measurements,[11,22–24] transient absorption measurements,[25,26] charge extraction measurements[27–30] and space charge limited current measurements.[31] The largest body of experimental evidence exists for the blend poly (3-hexylthiophene) (P3HT):1-(3-methoxycarbonyl)propyl-1-phenyl-[6,6]-methano fullerene (PCBM). For that system, transient absorption measurements indicate tail states with E_ch in the range of 35 meV to 70 meV,[32–34] transient photocurrent measurements also indicate an E_ch in the range of 35 meV to 65 meV (ref. 11) and sub-gap quantum efficiency measurements suggest an E_ch between 37 meV (at 300 K) and 32 meV (at 200 K).[20] Although the value of E_ch varies between measurements and with different processing regimes,[32,33] the data converge on a consensus of approximately exponential behaviour with E_ch in the range of 35 to 65 meV. Temperature-dependent quantum efficiency and transient photocurrent measurements further suggest that the tail of states is temperature dependent.[20]

Localised intra-gap (or trap) states in conjugated polymers may arise from chemical or physical defects. Possible chemical defects include in-chain defects such as end groups, sp^3 configured carbon atoms, or oxidised species as well as leaving groups or small molecular weight components remaining from the synthesis. Physical defects may result from variations in the conformation of the conjugated backbone and in intermolecular interactions. Whilst chemical impurities can largely be eliminated by the purification and fractionation of polymers and the choice of end groups, disorder in conformation and packing cannot be avoided in an organic semiconductor film. Amongst the configurational parameters that can influence the energies of electron and hole states in conjugated polymers, the torsion angle θ between consecutive units along the conjugated backbone is one of the most significant. For any compound, the value of θ in the geometrically relaxed structure results from competition between the dispersion of pi-electron states through consecutive units, which tends to planarise the chain, and steric effects which tend to deplanarise it. The potential energy profile for rotation of the units around this energetic minimum tends to be relatively soft, presenting a barrier of the order of 100 meV to chain rotation,[35,36] and thus allowing substantial thermal fluctuations in θ at room temperature. Note that such fluctuations might be expected to make the DoS temperature-dependent. In a solid polymer film at room temperature, torsion angles are further disrupted by steric effects in molecular packing and by chain folding. The net disorder in θ results in disorder in the extent of conjugated segments and in the energy of the electronic states.[36] As well as this intrachain coupling, as controlled by θ, some contribution to the spatial extent and energy of the electronic states may arise from electronic coupling between conjugated units in neighbouring chains. Such interchain coupling may be expected, especially in crystalline materials with a strong degree of pi stacking.

Whilst the concept that torsional disorder and packing disorder can give rise to disorder in electronic states energies is well known, attempts to derive the DoS for realistic molecular assemblies are limited to date. A number of studies have reported distributions of states for assemblies of small molecules such as fullerenes, calculated using quantum chemical methods,[37,38] but these methods are limited to relatively small assemblies. The problem is more challenging for polymers where the effect of torsion on strong intrachain coupling and the effect of the position on weak intrachain coupling must be included. Most studies of P3HT to date have addressed ordered systems. Cheung and co-workers have studied the variation in torsion distribution that results from dynamic chain fluctuations in crystalline P3HT and relate this to charge transport.[37] Rivnay considered the effect of variations in chain spacing, as deduced from X-ray diffraction studies, on tail states in P3HT, assuming a Gaussian distribution of separations.[39] More recently Poelking and co-workers derived a density of states for disordered crystalline P3HT of different regioregularity resulting from fluctuations in the intermolecular spacings.[40] Whilst understanding of the DoS in the crystalline fraction of P3HT is certainly useful, charge transport in devices is likely to be rate-limited by trapping in gap states within the larger, non crystalline fraction of the polymer film.

In this paper we will consider the relative importance of both inter- and intrachain contributions to the density of states in highly disordered films of P3HT. We use atomistic molecular dynamics (MD) to model the morphology of an assembly of regioregular thiophene oligomers of length 16 (corresponding to a low molecular weight of *ca.* 3 kDa). Including both inter- and intra-chain coupling, we find that the tail of the DoS at room temperature is approximately exponential with a characteristic energy of approximately 37 meV, similar to the lower range of experimental values reported.[20,21] More importantly, we are also able to resolve the contributions of inter- and intramolecular coupling to the density of states, finding that this deep tail results mainly from intra-chain coupling and thus results mainly from torsional disorder of the polymer backbone rather than positional or packing disorder. We also find that this deep tail is temperature-dependent. Finally we show that models that rely on a Gaussian distribution of interchain separations are likely to invoke unphysically close minimum separations. We argue that one route to materials with lower energetic disorder is to design stiffer polymers with less torsional freedom.

Methods

The empirical force field used for molecular dynamics was derived from the parameters developed by Moreno *et al.*,[35] supplemented with Optimized Potentials for Liquid Simulations (OPLS)[41] as implemented in the GROMACS[42] molecular dynamics package. The starting geometry of 40 independent extended (3-hexylthiophene) hexadecamers was packed loosely as a bundle into a 10 nm × 10 nm × 10 nm box using PACKMOL.[43] For the room temperature assemblies, steepest descent energy minimisation with the empirical force field was followed by an equilibration molecular dynamics run of 1 ns, with initial atomic velocities drawn from a 300 K Maxwell distribution and using the leapfrog integrator with a Berendsen thermostat and barostat at 300 K and 1 atm respectively. This resulted in the compression of the simulation volume from 1000 nm^3 to 158 nm^3 (density of 1.108 g cm^{-3}). Molecular dynamics was continued for a further 50 ns, with simulation coordinates captured from snapshots taken every 10 ps. Densities of

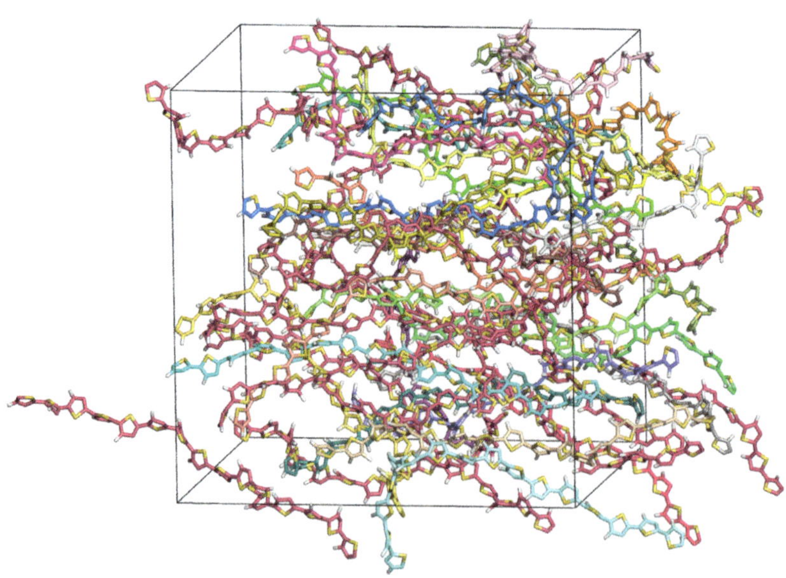

Fig. 1 Sampled assembly of 3HT hexadecamers from molecular dynamics at 300 K, in the periodic simulation cell (black outline), with the molecules reconstructed to be continuous with their centre of mass in the first periodic copy, and the sidechains removed for clarity.

states were computed from over 3000 such snapshots. A typical snapshot is shown in Fig. 1.

The role of inter- and intramolecular transfer integrals is assessed by looking at three different cases: (a) *intermolecular only*, where each oligomer is considered as a single charge transport unit with a given energy. All intermolecular couplings are computed and the resulting tight-binding Hamiltonian is solved, to deduce the distribution of eigenvalues, and hence the density of states. (b) *Intramolecular only*, where each oligomer is decomposed into a set of monomers coupled only to its immediate neighbours *via* a θ dependent transfer integral. A tight-binding Hamiltonian is solved for each oligomer and the set of results collected to give the average density of states, and (c) *inter and intramolecular*, where each monomer in the whole simulation box contributes to the basis set for the tight binding Hamiltonian. All inter- and intramolecular couplings are computed between all monomers, resulting in a large Hamiltonian which is diagonalised, yielding the density of states.

Once an MD trajectory has been computed, we use the VOTCA charge transport toolkit (votca-ctp)[44,45] to map conjugated segments onto the oligomers simulated with molecular dynamics. The inter-chain transfer integrals are computed using molecular orbital overlap[46] from semi-empirical INDO/S Highest Occupied Molecular Orbitals. In the case of the inter-chain only computation (case (a)), orbitals for relaxed hexadecamers are mapped onto the oligomers generated by MD and used to compute the transfer integrals between different oligomers. In the case of the inter- and intra-chain coupling (case (c)), orbitals for relaxed monomers are mapped on to the coordinates of each monomer in the MD-generated assembly, and coupling between monomers located on different oligomers are computed using VOTCA.

Intra-chain inter-monomer electronic coupling terms, needed for both cases (b) and (c), cannot be computed from molecular orbital overlap. In these cases the coupling is assumed to depend only on the torsion angle θ between neighbouring monomers according to the form

$$J_{\text{intra}} = J_0 \cos(\theta). \tag{2}$$

We compute J_0 by calculating the HOMO for planar thiophene oligomers of different number N of repeat units using DFT with a B3LYP functional and a double split 6-31G* basis set. We then fit the N-dependent trend in HOMO energy to the form expected from a tight-binding calculation for a chain of identical units coupled by J_0. This procedure yields a value of 0.8 eV for J_0.

Once all the electronic couplings are computed, we set up the Hückel Hamiltonian:

$$H = |i\rangle\langle i|\varepsilon_i + \sum_j \left(J_{i,j}|i\rangle\langle j| + t_{i,j}^{*}|j\rangle\langle i|\right) \tag{3}$$

where the diagonal element ε_i is set to a single value E_0. Here we set $E_0 = -5.0$ eV, in accordance with the approximate ionisation potential for P3HT of 5.0 eV.[47] We find the eigenvalues of this Hamiltonian using standard GNU Scientific Library solvers and then repeat the process for at least 3000 snapshots of the equilibrated oligomer assembly. The eigenvalues from all snapshots are stored and compiled into distributions representing the DoS functions for the different cases studied.

To repeat the process to study the electron rather than the hole DoS, we would replace the J_0 extracted from the length dependence of the HOMO, as calculated by DFT, with the value extracted from the oligomer length dependence of the LUMO.

Results and discussion

In Fig. 2, the resulting DoS for the three cases for assemblies generated at 300 K are compared. The DoS for case (a) including inter-chain couplings only (open circles) is very sharp, and an exponential fit to the tail reveals a characteristic energy of only 3.6 meV. In contrast, when intra-chain couplings only are considered (open triangles) or when both inter- and intra-chain couplings are included (open squares), a much broader tail in the DoS is obtained. The deep part of this tail can be fit to an exponential tail, with a characteristic energy of 35 meV for the case of only intra-chain couplings (case b) and around 37 meV for case (c), with both types of coupling. These values are in good agreement with the values of E_{ch} obtained from subgap quantum efficiency measurements on P3HT:PCBM solar cells[20,21] and with the lower values obtained from transient photocurrent and transient absorption studies. However, our results show that, in contrast to some previous studies,[20,39] interchain interactions alone cannot explain the breadth of the density of localised hole states in P3HT. The interchain interactions do affect the tail of the DoS slightly, possibly because tail states will

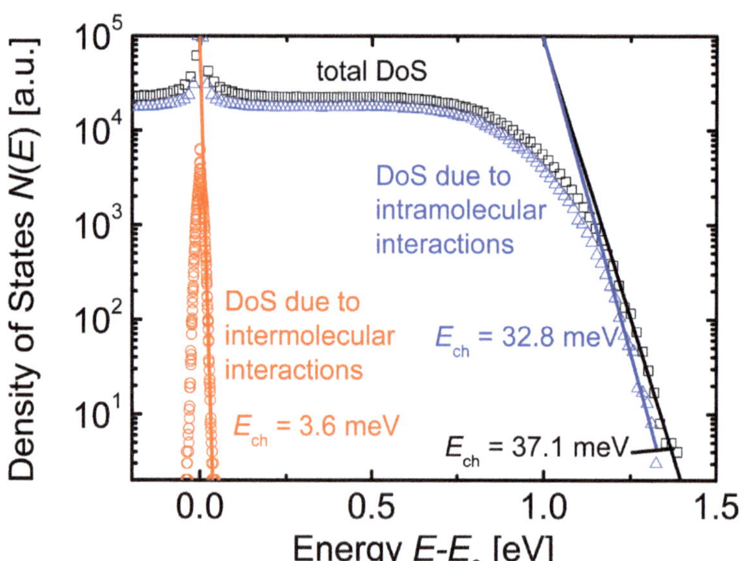

Fig. 2 DoS for molecular assemblies equilibrated at a temperature of 300 K calculated for the cases of intra-oligomer coupling only (triangles), inter-monomer coupling only (circles) and both intra- and inter-monomer coupling (squares). Straight lines represent exponential fits to the tail of states. Energy is measured from the energy of the thiophene HOMO used in the calculation ($E_0 = -5.0$ eV).

tend to result from planarised segments and interchain couplings will tend to be strongest between neighbouring, cofacial planar segments.

Fig. 3 shows the DoS obtained from both intra and intermolecular coupling for assemblies generated at different temperatures. We would expect the range of torsion angles present in the ensemble to broaden with increasing temperature, as increasing kT increases the probability of configurations further from the energetic minimum. The range of torsional angles does indeed broaden (see Fig. S1(a), ESI†) and, accordingly, the DoS becomes broader with increasing temperature. However, the value of E_{ch} obtained from an exponential fit to the deep tail of the DoS is relatively invariant with temperature. It should be noted that the tail is not perfectly exponential and a fit to the DoS higher in the tail would yield values that slowly increase with temperature. The effect of temperature on the interchain-only DoS is also very weak (Fig. S2, ESI†).

The close agreement between the DoS obtained in cases (b) and (c), when the coupling between monomers on different chains are excluded and included, suggests that the DoS for the disordered polymer film may be explained solely by the distribution of inter-monomer torsion angles. This distribution will be influenced both by the size of thermally enabled torsional fluctuations and by chain distortion resulting from the dense packing of the chains and their mutual repulsion. To investigate the extent to which the thermal fluctuations alone control the DoS, we implemented a simple, linear tight binding scheme whereby the torsion angle between neighbouring monomers was drawn from a thermally generated distribution of angles $f(\theta)$.[48] We obtain $f(\theta)$ by mapping a Boltzmann

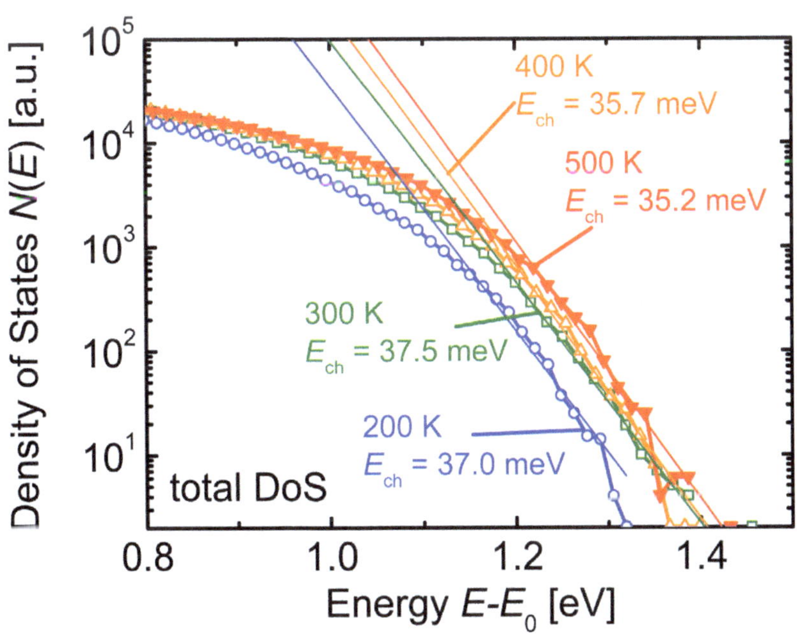

Fig. 3 Calculated DoS due to both intra- and inter-monomer coupling, as a function of temperature. The straight lines represent exponential fits to the tail of states. The slope of the deep part of the tail is insensitive to temperature. Energy is measured from the energy of the thiophene HOMO used in the calculation ($E_0 = -5.0$ eV).

distribution of torsional potential energies $g(U(\theta)) = A\exp(-U(\theta)/kT)$ onto θ, where we have approximated $U(\theta)$ by $U_0\sin^2(\theta)$, with $U_0 = 0.126$ eV, fitted to the published torsional potential.[35] The angular distributions generated this way are broader than those extracted from the MD and are more strongly temperature-dependent (Fig. S1(a) and (b), ESI†). We then populate the first-diagonal elements of the Hamiltonian with coupling terms corresponding to the selected values of θ using eqn (2), and directly calculate the eigenvalue spectrum with a computationally efficient (linear in time and memory) Sturm sequence method.[49]

The resulting temperature-dependent eigenvalue distributions (Fig. 4) show that the tails of states generated by this method are relatively sharp compared to those obtained from the MD-generated torsional configurations, and are too narrow to explain the experimental DoS. This tells us that, whilst the DoS in the disordered film may be dominated by intramolecular torsional variations rather than intermolecular coupling, these torsional variations are controlled by factors other than the torsional potential between conjugated rings. In the MD-generated assembly, the distribution of θ is influenced also by the interactions between side chains, repulsive forces at close atomic separations that may constrain the molecule into an thermodynamically unfavourable conformation (see Fig. S3, ESI†), and by face-on interactions with neighbouring conjugated segments, that will tend to planarise the chain. The atomistic MD exercise is necessary to capture these phenomena and so explain the breadth of the observed DoS. A comparison of the DoS tails for the simple thermodynamic model and the MD model suggests that the occurrence of long conjugated segments that act as traps is more common in the realistic MD case.

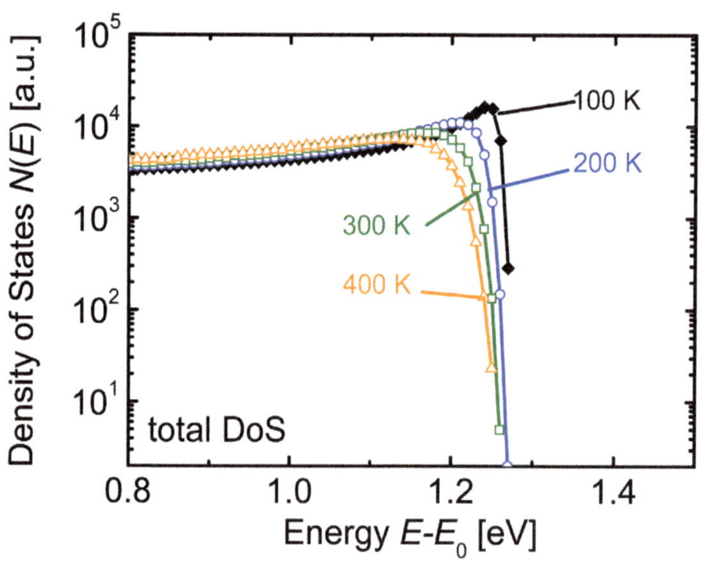

Fig. 4 DoS resulting from the linear chain thermodynamic model for chains of 10^6 monomers at different temperatures. Although the DoS is broad, tails are much sharper than from the torsional distributions generated by MD, meaning that long conjugated segments are more likely. The 300 K DoS is calculated for a large chain of 10^8 monomers for comparison, showing that the DoS tail shape is insensitive to sample size over this range.

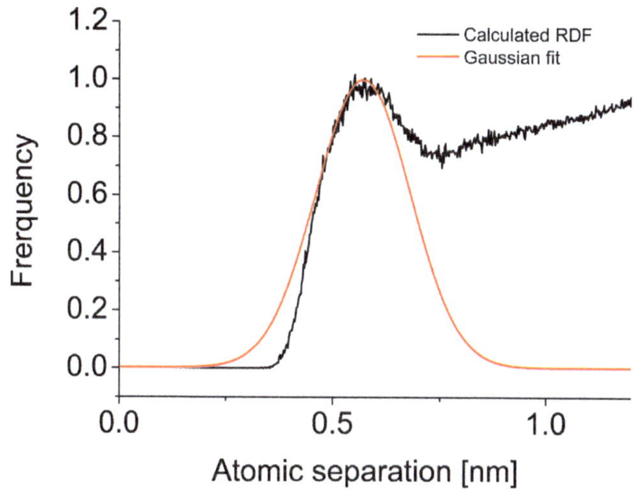

Fig. 5 Radial distribution function for the closest approach of monomers located on different oligomers (black line) and a Gaussian fit to the first peak of the function. The Gaussian function is a poor fit at very small separations.

Finally, we use our findings to address the validity of the Gaussian disorder model for intermolecular distances. Following early approaches to including disorder in electronic coupling into models of charge transport in organic semiconductors[9] several groups have invoked a Gaussian distribution in inter-plane distances to explain electronic phenomena.[20,39] Fig. 5 shows the probability distribution function of finding an intermolecular neighbour at a certain distance, as calculated by molecular dynamics. Also shown is a fit to the tail of this probability density function with a Gaussian distribution. The Gaussian model appears to be a good approximation, except for at short distances (less than 0.4 nm) where the van der Waals repulsion becomes strong and the atoms must not overlap. In the atomistic MD simulation, small inter-chain distances are much less likely than in a Gaussian disorder model, meaning that large inter-chain integrals are also less likely and, therefore, fewer sub-band gap states are expected to result from intermolecular interactions. This observation can resolve the apparent contradiction between our findings and those of previous studies.[20,39]

Conclusions

Using a combination of atomistic molecular dynamics and a tight-binding calculation of hole state energies, we have found that the tail of the density of hole states in P3HT follows an exponential form with a characteristic energy of around 35–40 meV. This value agrees well with values observed experimentally using sub-gap quantum efficiency measurements, and it lies at the lower end of the range of values observed using transient absorption and photocurrent measurements. The breadth of the tail of states arises mainly from intramolecular electronic couplings, which are generally much stronger than interchain couplings. The tail states in this picture thus result from extended planar segments. The key conclusion is that intrachain torsional disorder can account for a significant fraction of the disorder observed experimentally whilst disorder in interchain coupling, at least in the non-crystalline parts of the polymer film, cannot. Deeper tail state distributions often observed in practice for P3HT may be influenced by chemical defects, polydispersity, nanocrystallinity or other factors not included in our model.

The finding that the distribution of tail states is dominated by intra-chain coupling elements and the simple $\cos(\theta)$ torsional dependence of such elements means that, in general, the density of tail states for a conjugated polymer could be estimated from a representative distribution of torsion angles *via* a simple tri-diagonal tight binding Hamiltonian. The generation of such a representative ensemble of torsional angles still requires some realistic—and therefore time consuming—simulation of the polymer chain assembly. We have shown here for the case of 3HT oligomers that the disorder in intrachain torsion responsible for the distribution of tail states results from the way in which the chains are packed in the solid film, and cannot be predicted solely from knowledge of the torsional potential barrier in the conjugated backbone. This reflects the fact that the molecular microstructure within polymer films is normally far from equilibrium.

Regarding practical routes to narrow the distribution of tail states in conjugated polymers, this work suggests that a central focus should be on designing stiffer polymers with higher potential barriers to oppose torsional disorder. However, to make the most of such planarising structures, process routes should allow the chains to find configurations where chain conformation more closely reflects the built-in potential. Improving the organisation and crystallinity of a polymer is also likely to affect the density of states, but mainly as a consequence of the increase in intra-chain order.

Acknowledgements

T. K. acknowledges support by an Imperial College Junior Research Fellowship. JN acknowledges the Royal Society for a Wolfson Merit Award, and the Engineering and Physical Sciences Research Council for support *via* grants EP/K029843/1, EP/G031088/1, EP/J017361/1.

References

1 R. Noriega, J. Rivnay, K. Vandewal, F. P. V. Koch, N. Stingelin, P. Smith, M. F. Toney and A. Salleo, *Nat. Mater.*, 2013, **12**, 1038–1044.
2 S. Reineke, F. Lindner, G. Schwartz, N. Seidler, K. Walzer, B. Lussem and K. Leo, *Nature*, 2009, **459**, 234–U116.
3 L. T. Dou, J. B. You, J. Yang, C. C. Chen, Y. J. He, S. Murase, T. Moriarty, K. Emery, G. Li and Y. Yang, *Nat. Photonics*, 2012, **6**, 180–185.
4 M. Graetzel, R. A. J. Janssen, D. B. Mitzi and E. H. Sargent, *Nature*, 2012, **488**, 304–312.
5 Z. C. He, C. M. Zhong, S. J. Su, M. Xu, H. B. Wu and Y. Cao, *Nat. Photonics*, 2012, **6**, 591–595.
6 G. Li, R. Zhu and Y. Yang, *Nat. Photonics*, 2012, **6**, 153–161.
7 J. Nelson, *Mater. Today*, 2011, **14**, 462–470.
8 A. Salleo, *Mater. Today*, 2007, **10**, 38–45.
9 H. Bässler, *Phys. Status Solidi B*, 1993, **175**, 15–56.

10 R. A. Street, A. Krakaris and S. R. Cowan, *Adv. Funct. Mater.*, 2012, **22**, 4608–4619.
11 R. A. Street, *Phys. Rev. B: Condens. Matter Mater. Phys.*, 2011, **84**, 075208.
12 G. D. Cody, T. Tiedje, B. Abeles, B. Brooks and Y. Goldstein, *Phys. Rev. Lett.*, 1981, **47**, 1480–1483.
13 M. H. Cohen, H. Fritzsche and S. R. Ovshinsky, *Phys. Rev. Lett.*, 1969, **22**, 1065–1068.
14 T. Tiedje, J. M. Cebulka, D. L. Morel and B. Abeles, *Phys. Rev. Lett.*, 1981, **46**, 1425–1428.
15 C. Vanberkel, M. J. Powell, A. R. Franklin and I. D. French, *J. Appl. Phys.*, 1993, **73**, 5264–5268.
16 T. Kirchartz and J. Nelson, *Phys. Rev. B: Condens. Matter Mater. Phys.*, 2012, **86**, 165201.
17 I. N. Hulea, H. B. Brom, A. J. Houtepen, D. Vanmaekelbergh, J. J. Kelly and E. A. Meulenkamp, *Phys. Rev. Lett.*, 2004, **93**, 166601.
18 C. Tanase, E. J. Meijer, P. W. M. Blom and D. M. de Leeuw, *Phys. Rev. Lett.*, 2003, **91**, 216601.
19 A. R. Völkel, R. A. Street and D. Knipp, *Phys. Rev. B: Condens. Matter Mater. Phys.*, 2002, **66**, 195336.
20 R. A. Street, K. W. Song, J. E. Northrup and S. Cowan, *Phys. Rev. B: Condens. Matter Mater. Phys.*, 2011, **83**, 165207.
21 W. Gong, M. A. Faist, N. J. Ekins-Daukes, Z. Xu, D. D. C. Bradley, J. Nelson and T. Kirchartz, *Phys. Rev. B: Condens. Matter Mater. Phys.*, 2012, **86**, 024201.
22 N. Christ, S. W. Kettlitz, S. Zuefle, S. Valouch and U. Lemmer, *Phys. Rev. B: Condens. Matter Mater. Phys.*, 2011, **83**, 195211.
23 R. C. I. MacKenzie, C. G. Shuttle, M. L. Chabinyc and J. Nelson, *Adv. Energy Mater.*, 2012, **2**, 662–669.
24 C. G. Shuttle, N. D. Treat, J. D. Douglas, J. M. J. Frechet and M. L. Chabinyc, *Adv. Energy Mater.*, 2012, **2**, 111–119.
25 J. Nelson, *Phys. Rev. B: Condens. Matter Mater. Phys.*, 2003, 155209.
26 M. Tachiya and K. Seki, *Phys. Rev. B: Condens. Matter Mater. Phys.*, 2010, **82**, 085201.
27 A. Foertig, J. Rauh, V. Dyakonov and C. Deibel, *Phys. Rev. B: Condens. Matter Mater. Phys.*, 2012, **86**, 115302.
28 T. Kirchartz, B. E. Pieters, J. Kirkpatrick, U. Rau and J. Nelson, *Phys. Rev. B: Condens. Matter Mater. Phys.*, 2011, **83**, 115209.
29 R. C. I. MacKenzie, T. Kirchartz, G. F. A. Dibb and J. Nelson, *J. Phys. Chem. C*, 2011, **115**, 9806–9813.
30 C. G. Shuttle, A. Maurano, R. Hamilton, B. O'Regan, J. C. de Mello and J. R. Durrant, *Appl. Phys. Lett.*, 2008, **93**, 183501.
31 Z. M. Beiley, E. T. Hoke, R. Noriega, J. Dacuna, G. F. Burkhard, J. A. Bartelt, A. Salleo, M. F. Toney and M. D. McGehee, *Adv. Energy Mater.*, 2011, **1**, 954–962.
32 P. E. Keivanidis, T. M. Clarke, S. Lilliu, T. Agostinelli, J. E. Macdonald, J. R. Durrant, D. D. C. Bradley and J. Nelson, *J. Phys. Chem. Lett.*, 2010, **1**, 734–738.
33 Y. Kim, S. Cook, S. M. Tuladhar, S. A. Choulis, J. Nelson, J. R. Durrant, D. D. C. Bradley, M. Giles, I. McCulloch, C.-S. Ha and M. Ree, *Nat. Mater.*, 2006, **5**, 197–203.

34 C. G. Shuttle, B. O'Regan, A. M. Ballantyne, J. Nelson, D. D. C. Bradley and J. R. Durrant, *Phys. Rev. B: Condens. Matter Mater. Phys.*, 2008, **78**, 113201.
35 M. Moreno, M. Casalegno, G. Raos, S. V. Meille and R. Po, *J. Phys. Chem. B*, 2010, **114**, 1591–1602.
36 J. L. Bredas, G. B. Street, B. Themans and J. M. Andre, *J. Chem. Phys.*, 1985, **83**, 1323–1329.
37 D. L. Cheung and A. Troisi, *J. Phys. Chem. C*, 2010, **114**, 20479–20488.
38 B. M. Savoie, A. Rao, A. A. Bakulin, S. Gelinas, B. Movaghar, R. H. Friend, T. J. Marks and M. A. Ratner, *J. Am. Chem. Soc.*, 2014, **136**, 2876–2884.
39 J. Rivnay, R. Noriega, J. E. Northrup, R. J. Kline, M. F. Toney and A. Salleo, *Phys. Rev. B: Condens. Matter Mater. Phys.*, 2011, **83**, 121306.
40 C. Poelking and D. Andrienko, *Macromolecules*, 2013, **46**, 8941–8956.
41 W. L. Jorgensen, J. D. Madura and C. J. Swenson, *J. Am. Chem. Soc.*, 1984, **106**, 6638–6646.
42 D. Van der Spoel, E. Lindahl, B. Hess, G. Groenhof, A. E. Mark and H. J. C. Berendsen, *J. Comput. Chem.*, 2005, **26**, 1701–1718.
43 L. Martinez, R. Andrade, E. G. Birgin and J. M. Martinez, *J. Comput. Chem.*, 2009, **30**, 2157–2164.
44 V. Ruehle, C. Junghans, A. Lukyanov, K. Kremer and D. Andrienko, *J. Chem. Theory Comput.*, 2009, **5**, 3211–3223.
45 V. Ruehle, A. Lukyanov, F. May, M. Schrader, T. Vehoff, J. Kirkpatrick, B. r. Baumeier and D. Andrienko, *J. Chem. Theory Comput.*, 2011, **7**, 3335–3345.
46 J. Kirkpatrick, *Int. J. Quantum Chem.*, 2008, **108**, 51–56.
47 Sigma Aldrich Catalogue, http://www.sigmaaldrich.com/catalog/product/aldrich/698997?lang=en&region=GB.
48 J. M. Frost, GitHub - jarvist/LongSnakeMoan, http://dx.doi.org/10.5281/zenodo.11007.
49 J. T. Albrecht, C. P. Chan and A. Edelman, *Foundations of Computational Mathematics*, 2009, **9**, 461–483.

Faraday Discussions

PAPER

In situ formation of organic–inorganic hybrid nanostructures for photovoltaic applications

Sebastian Wood,[a] Oliver Garnett,[b] Nurlan Tokmoldin,[b] Wing C. Tsoi,[a] Saif A. Haque*[b] and Ji-Seon Kim*[a]

Received 29th June 2014, Accepted 21st July 2014

DOI: 10.1039/c4fd00141a

The performance of hybrid (organic–inorganic) photovoltaic devices is critically dependent on the thin film morphology. This work studies the film formation process using the *in situ* thermal decomposition of a soluble precursor to form a well-distributed network of CdS nanoparticles within a poly(3-hexylthiophene) (P3HT) polymer matrix. Resonant Raman spectroscopy is used to probe the formation of the inorganic nanoparticles and the corresponding changes in the molecular order of the polymer. We find that the CdS precursor decomposes rapidly upon heating to 160 °C, but that this has a disruptive effect on the P3HT. The extent of this disruption can be controlled by adjusting the annealing temperature, and nanowire aggregates of P3HT are found to have increased susceptibility. Atomic force microscopy reveals that at high temperatures (>200 °C), cracks form in the film, resulting in a 'plateau'-like microstructure. In order to retain the preferable 'granular' microstructure and to control the molecular disruption, low decomposition temperatures are needed. This work identifies a particular problem for optimising the hybrid thin film morphology and shows how it can be partially overcome.

Introduction

Hybrid (organic–inorganic) thin film photovoltaic devices seek to exploit the complementary attractive processability and optoelectronic properties of both organic and inorganic semiconductor materials.[1] One promising approach to hybrid photoactive layers makes use of a soluble precursor for the inorganic material, which can be blended with an organic semiconductor in solution prior to deposition, and is then thermally decomposed to produce a three-dimensional inorganic semiconducting network within an organic thin film.[2,3] One particular

[a]*Department of Physics and Centre for Plastic Electronics, Imperial College London, London, SW7 2AZ, UK. E-mail: ji-seon.kim@imperial.ac.uk*

[b]*Department of Chemistry and Centre for Plastic Electronics, Imperial College London, London, SW7 2AY, UK. E-mail: s.a.haque@imperial.ac.uk*

advantage of this approach is that it obviates the need for solubilising ligands, which can compromise device performance.[4] The cadmium xanthate precursor has been designed to thermally decompose controllably at temperatures of approximately 150 °C to form cadmium sulphide (CdS) nanoparticles along with volatile side-products which are not expected to be retained in the polymer film. Moreover, solar cells comprising poly(3-hexylthiophene) (P3HT) and CdS produced in this way offer an improved exciton dissociation probability and higher overall device efficiency than comparable devices using simple blends of P3HT with CdS quantum dots.[4]

The performance of hybrid photovoltaic devices depends strongly on the nanoscale morphology of the thin film, ideally seeking a large interfacial area between continuous domains of organic and inorganic components. The morphology of experimental devices is found to be strongly dependent on the film deposition and processing conditions.[5] In the case studied here, the formation of the CdS aggregate network shows a strong dependence on the thermal conditions of the decomposition process, and the weight ratio of the two components has been found to affect the P3HT morphology.[6,7] Here, we investigate the evolution of the hybrid film morphology during thermal annealing using *in situ* resonant Raman spectroscopy to simultaneously monitor the formation of CdS nanoparticles and changes in the molecular order of the P3HT matrix. We find that the chemical decomposition of a cadmium xanthate precursor occurs very rapidly, reaching completion within 2 minutes of heating to 160 °C (previously reported as the optimum temperature for producing efficient solar cells),[6] while the P3HT shows a corresponding disruption in its molecular order, which counteracts the increase in molecular order observed for thermal annealing of neat P3HT. Considering temperatures from 140 to 240 °C, we find that the lowest temperature is preferable for maintaining highly ordered P3HT. Furthermore, we test the preformation of P3HT nanowires in solution as a means to enhance the polymer molecular order, and find that the nanowire morphology shows increased susceptibility to disruption during the thermal decomposition of the CdS precursor. The increased disruption of the P3HT molecular order correlates with the formation of an increasingly intimate P3HT/CdS mixed phase, and ultimately leads to the development of a 'plateau'-like film morphology with a high density of undesirable 'cracks' leading to a low device efficiency.

Experimental methods

P3HT was obtained from Sigma-Aldrich with regioregularity 90.0%, $M_n = 22.3$ kg mol^{-1} and PDI = 1.9. A regioregularity of 90–93% appears to be the optimum for well-controlled nanowire formation since lower regioregularity P3HT does not form ordered aggregates but higher regioregularity P3HT aggregation is difficult to control.[8] Nanowire aggregates were prepared by temperature-controlled precipitation from xylene solution using a method described previously.[9] The cadmium xanthate ($Cd(S_2COEt)_2(C_5H_5N)_2$) precursor solution was prepared as described elsewhere in a dichlorobenzene solution.[3] Solutions were blended to give a 50 : 50 solvent system and an 80 : 20 CdS : P3HT ratio by weight (after thermal annealing).

Devices were prepared in an inverted structure:[7] ITO/TiO_2/CdS/P3HT:CdS/PEDOT:PSS/Au. TiO_2 was deposited by spin coating (4000 rpm) from a precursor solution (0.1 M titanium isopropoxide and acetylacetone in ethanol) followed by heating to 450 °C for 1 hour.[10] A CdS interface layer was deposited from the precursor solution (4000 rpm), followed by annealing at 160 °C for 30 minutes. The active layer was then deposited on top (1000 rpm) before adding a PEDOT:PSS (Ossila AI 4083) layer (4000 rpm). Gold electrodes (100 nm) were then added by thermal evaporation at 5×10^{-6} mbar. Samples for the spectroscopic and microscopic studies were prepared on glass substrates.

UV-visible absorption spectroscopy was performed using a Shimadzu UV-2550 spectrophotometer. Resonant Raman spectra were obtained with a Renishaw inVia Raman spectrometer coupled through a Leica DM2500M microscope. Emission at 457 nm from an argon ion laser was used as the excitation source, with a beam power of 100 μW and a 30 second exposure time. All Raman spectra were measured in a nitrogen environment and the laser excitation was defocussed to a 10 μm spot in order to minimise photodegradation of the sample. *In situ* Raman measurements during thermal annealing were performed with the sample in a Linkam THMS600 hot–cold cell. Atomic force microscopy was carried out using a Park NX10 instrument in the non-contact mode. The photovoltaic device performance was measured using a ScienceTech SS150W solar simulator with IR (Water Filter) and AM 1.5 (ScienceTech) filters, and a Keithley 2400 source measure unit.

Results and discussion

In situ measurements of hybrid film formation

The absorption spectra of the P3HT:CdS blend film before and after thermal annealing are shown in Fig. 1 alongside the spectra for the neat P3HT and CdS films. The absorption is measured from a transmittance measurement which does not distinguish scattering effects from true absorption, resulting in a sub-energy gap 'absorption' tail at long wavelengths (650–700 nm), which we attribute to light scattering in the neat P3HT film. CdS has an absorption onset at around 500 nm, whereas P3HT absorbs strongly between 450 and 650 nm. As a result the

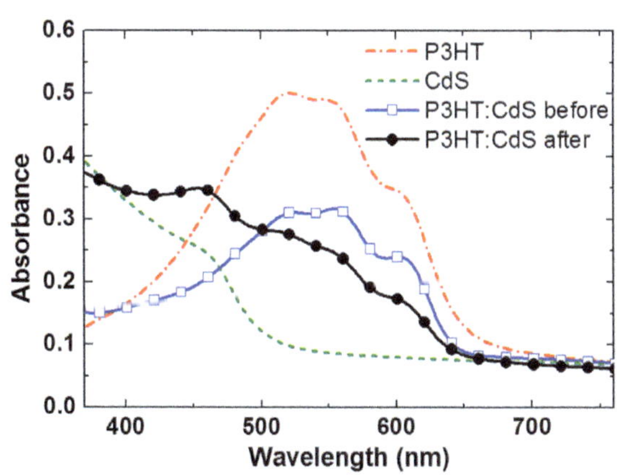

Fig. 1 Absorption spectra of thin films of P3HT and CdS, and the P3HT:CdS precursor blend films before and after thermal decomposition.

separate contributions from the two materials are readily distinguishable in the blend films. Before thermal annealing the absorption spectrum of the blend shows no contribution from CdS, but following thermal annealing there is strong absorption below 450 nm, showing that the CdS phase has formed. There is also a significant reduction in the strength of the P3HT absorption between 500 and 650 nm, suggesting a disruption of the polymer molecules, which we confirm below with Raman spectroscopy. The ratio of absorbances for the first and second absorption peaks (A_{0-0}/A_{0-1}) is a useful measure of intramolecular order, relating strongly to charge mobility.[11] For the P3HT:CdS precursor blend film, this ratio decreases from 0.75 to 0.70 $\pm$ 0.01 during thermal annealing, corresponding with a reduction in the degree of intramolecular order. These conformational changes and their impact on device performance are considered in more detail below.

Resonant Raman spectroscopy is a sensitive technique for elucidating the morphology of conjugated polymer samples, and is particularly useful as an *in situ* structural probe for morphological changes in thin film devices since it is fast, non-destructive, and compatible with other optical, electrical and physical characterisation techniques.[9,12–18]

Fig. 2 shows the Raman spectra measured for the P3HT:CdS blend film before, during and after the thermal annealing at 160 °C. The 457 nm excitation laser is used to match the absorption bands of both materials, resulting in comparable resonant enhancements, so that the Raman scattering from each component can be measured simultaneously. This is shown in Fig. 2a where the main CdS peak (303 cm^{-1}) is clearly visible alongside the main P3HT peaks (1350–1600 cm^{-1}). The development of the CdS Raman peak is shown more clearly in Fig. 2b. The spectrum measured at room temperature before the thermal annealing shows no contribution from the CdS component, but when the heat is applied, the main 303 cm^{-1} peak appears along with a weaker overtone at 605 cm^{-1}. These features appear within two minutes of annealing at 160 °C and remain unchanged up to 30 minutes, suggesting that the chemical decomposition is already complete within two minutes. However, upon quenching to room temperature, the CdS Raman peaks show a strong increase in intensity, though there is no change in their relative intensities. Shiang *et al.* associate this effect with an increase in the lifetime of the resonantly excited electronic state at lower temperature with no change in the size of the CdS nanocrystal domains.[19] The same measurement on a neat film of the CdS precursor showed the same trend, indicating that the presence of P3HT does not significantly affect the CdS thermal decomposition process. The development of the CdS nanoparticles is represented in Fig. 3a where the 303 cm^{-1} peak intensity is plotted over time.

Fig. 2a also shows a clear increase in the fluorescence background in the range 1000–1800 cm^{-1} for spectra measured at 160 °C compared with those measured at room temperature. This increased P3HT fluorescence at high temperatures is indicative of a reduction of intermolecular interactions in the P3HT, but a more detailed analysis can be made from the shapes of the Raman peaks shown in Fig. 2c. There are two main P3HT Raman modes centred around 1380 and 1450 cm^{-1}, usually described as intra-ring C–C and C=C symmetric backbone stretches respectively. The position, width (FWHM) and relative intensities of these peaks are known to be sensitive to the degree of molecular order in P3HT.[14] Here the fluorescence background has been subtracted and the spectra are normalised for ease of comparison. During annealing (2 minutes and 30 minutes at

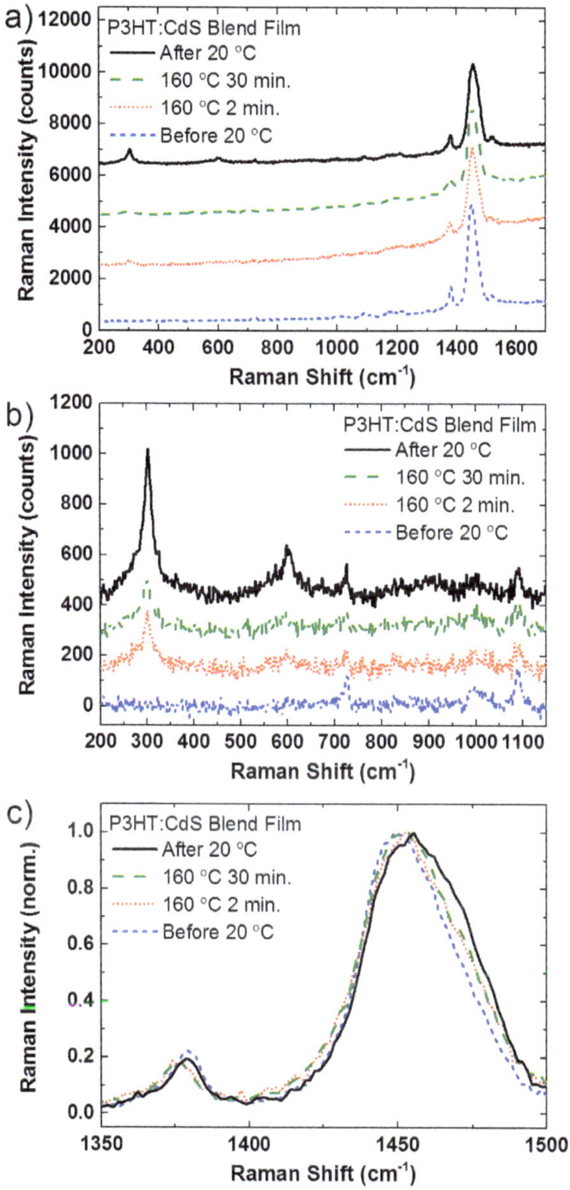

Fig. 2 Resonant Raman spectra of the P3HT:CdS precursor blend films before, during (at 2 and 30 minutes) and after (bottom to top) thermal annealing at 160 °C, followed by quenching to 20 °C; (a) raw spectra separated by vertical offset, (b) main CdS Raman peaks after background correction separated by vertical offsets, and (c) main P3HT Raman peaks normalised after background correction.

160 °C), the measured spectra show a small shift of both the C–C and C=C peaks towards lower frequency and a broadening of the C=C peak towards higher energy (FWHM increases from 37 to 42 ± 1 cm^{-1}). We attribute the increase in peak width (which is preserved in the measurements after cooling to 20 °C) to a broadening of the distribution of molecular conformations in the sample.

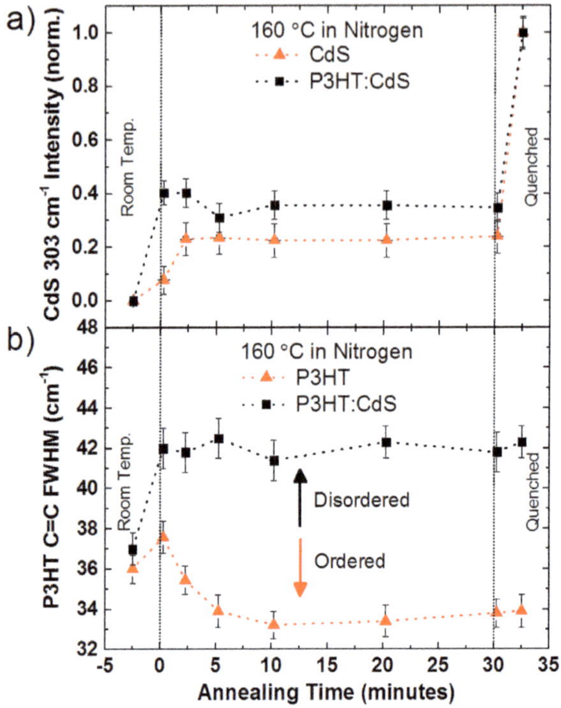

Fig. 3 (a) Peak intensity of the main CdS Raman peak at 303 cm^{-1} as a function of annealing time for the neat CdS precursor and the P3HT:CdS precursor blend films, normalised to the final intensity after quenching to room temperature. (b) FWHM of the main P3HT Raman peak around 1450 cm^{-1} as a function of the annealing time for the neat P3HT and P3HT:CdS precursor blend films. Arrows indicate the change in the overall degree of P3HT molecular order.

Comparing the spectra measured at 20 °C before and after annealing, the broadening is clearly due to an increase in the Raman scattering intensity on the high energy side of the C=C peak (1450–1500 cm^{-1}), which indicates a decrease in the molecular order of the P3HT in the blend.[14] The small peak shift towards lower frequency (which is not preserved after cooling) is assigned to anharmonicity of the vibrational modes.[20] At room temperature the polymer molecules are mostly in the ground vibrational state, but at elevated temperatures a higher proportion will be in vibrationally excited states. The energy spacings of all the vibrational levels of a particular normal mode are equal in harmonic potentials but deviations from harmonicity lead to reduced spacings for higher-lying vibrational transitions. Hence, at elevated temperatures the vibrational modes tend to shift towards lower energies in anharmonic potentials.

The disruption of P3HT molecular order observed here for the P3HT:CdS blend is the opposite trend to that observed for neat P3HT—the change in the C=C peak FWHM against annealing time is shown in Fig. 3b for both of these samples. Since the changes in peak width occur only on the high energy part of the peak, this provides a good measure of the overall degree of P3HT molecular order. We find that the neat P3HT film shows increasing order (FWHM reduces from 36 to 33 ± 1 cm^{-1}) during the thermal annealing, reaching completion in

5–10 minutes. By contrast, the P3HT:CdS blend shows a rapid disruption of polymer ordering (increasing from 37 to 42 ± 1 cm^{-1}) within the first 2 minutes and shows no further change. This effect corresponds with the thermal decomposition of the CdS precursor which is observed simultaneously. These results give clear evidence that the thermal decomposition process, by which the CdS nanoparticles form, causes a simultaneous disruption of the molecular order in the P3HT polymer matrix. The disordered phase of P3HT has low hole mobility and blue-shifted absorption, which means that it is typically detrimental to solar cell performance. Therefore, this molecular disruption during film formation needs to be controlled and reduced. Below, we consider the use of different thermal annealing temperatures and the formation of P3HT nanowires as possible routes to minimise the disruptive interaction.

Decomposition temperature and nanowire P3HT

The cadmium xanthate precursor is designed to decompose at around 150 °C, so we consider temperatures over the range 140–240 °C and their impact on the film morphology. In each case we observe no difference in the rapid formation of the CdS component, but the effects on the P3HT Raman spectra are compared in Fig. 4a. In this case, as the temperature increases, the main C=C peak broadens and then shifts to higher energy, while the C–C peak shifts to slightly lower energy and the relative intensity is reduced. These changes are typical characteristics of P3HT being more disordered, and hence demonstrate clearly that the P3HT molecular disordering effect caused by thermal decomposition of the CdS precursor is more pronounced at higher temperatures, with the maximum peak shift observed at the highest temperature, 240 °C.

We also consider the preparation of P3HT nanowire aggregates in solution, which has previously been reported as a technique for counteracting morphological disruption in blend films, and so it was hypothesised that P3HT nanowires would also either improve or maintain the molecular order of P3HT in blends with CdS upon thermal decomposition.[9,21–24] The distinctive nanowire P3HT morphology can be seen as bright streaks in the atomic force microscopy (AFM) height map of the nwP3HT:CdS blend film before thermal annealing (Fig. 5). Fig. 4b compares the P3HT Raman peaks of the nwP3HT:CdS films formed at different temperatures, which show the same disordering effect as that observed without the nanowires. In this case the disruption also increases with temperature but reaches a maximum at 160 °C, showing no further increase up to 240 °C. This suggests that the nanowires have increased susceptibility to the molecular disorder induced by the CdS nanoparticle formation process.

Molecular order is known to be a critical parameter for charge carrier mobility and other optoelectronic properties. There is some evidence that disordered polymer is active in photocurrent generation but the poor hole mobility is associated with high levels of bimolecular recombination, so disordered P3HT has typically been identified as having a detrimental effect on solar cell performance.[9,11,21,24–30] We therefore suggest that low temperature thermal decomposition is preferable for producing an optimised morphology.

The precise nature of the mechanism by which the thermal decomposition of the cadmium xanthate to produce CdS nanoparticles disrupts the molecular conformation of the P3HT matrix is not clear. Some evidence has been reported

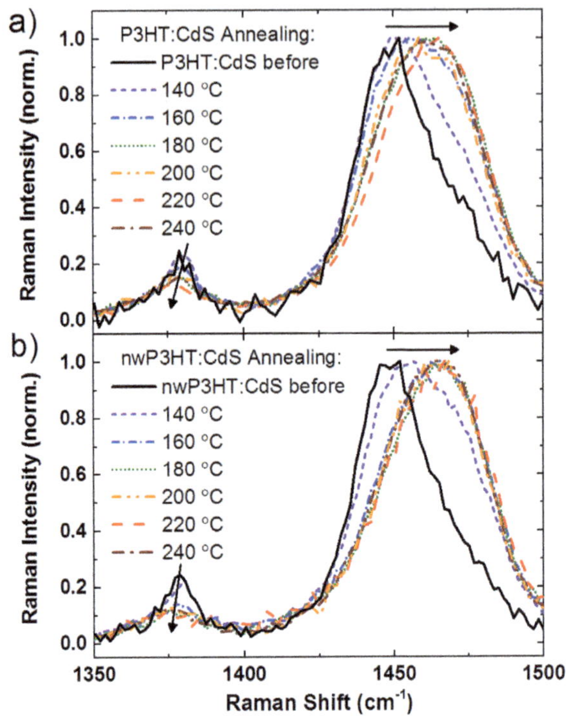

Fig. 4 The main P3HT resonant Raman peak measured at 20 °C after background correction and normalisation, comparing (a) P3HT:CdS blend films before and after annealing at temperatures from 140 to 240 °C, and (b) nwP3HT:CdS blend films before and after annealing at temperatures from 140 to 240 °C.

which suggests that the polymer acts as a capping agent on the CdS nanoparticles.[3] Such an interaction between the CdS nanoparticle and the P3HT matrix might be expected to result in disruption of the P3HT molecular order as the polymer chains conform to the CdS surface, and would also lead to increased solubility of the inorganic nanoparticles in the polymer matrix. Alternatively, the decomposition process itself may cause localised damage as the volatile products are evolved and pass through the polymer film. In either case, some amount of polymer disorder would be intrinsic to this method of forming a hybrid blend film and so gives cause to consider alternative chemical routes.

The increased disruption caused at higher temperatures suggests that the P3HT is more susceptible to damage under these conditions, as might be expected since the polymer chains will show increased mobility. The increased susceptibility of the P3HT nanowires to disruption suggests that the aggregated polymer structures provide a preferential site for decomposition of the CdS precursor molecules. The linear features in the AFM images (Fig. 5) would also be consistent with some kind of templating effect of the nanowires on CdS nanoparticle formation.

The morphological changes probed by resonant Raman spectroscopy relate directly to conformational changes on the molecular scale, whereas AFM images provide a complementary probe for the physical structure of materials

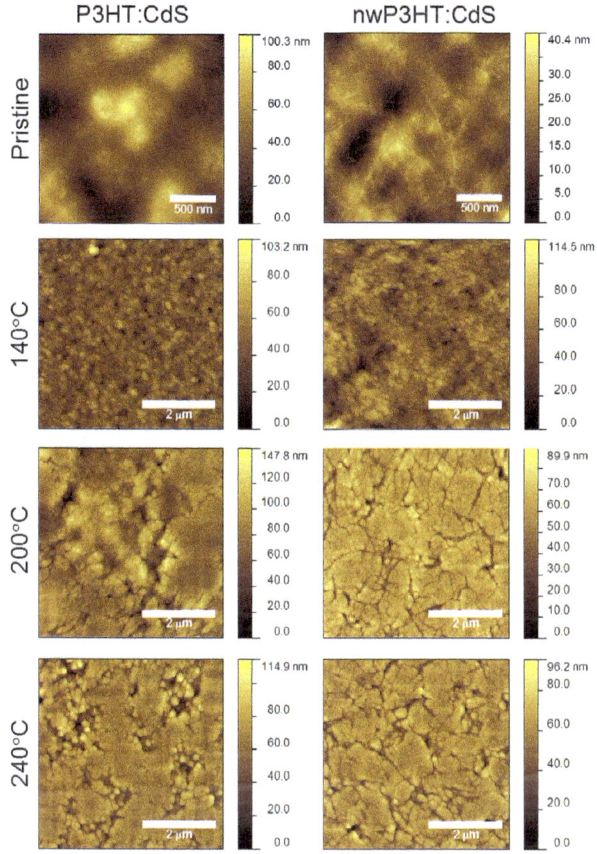

Fig. 5 AFM height images of the P3HT:CdS (left) and nwP3HT:CdS (right) blend films, before and after thermal annealing at 140, 200 and 240 °C (top to bottom).

at the nanometre scale. The images shown in Fig. 5 for the P3HT:CdS and nwP3HT:CdS precursor films before and after annealing at different temperatures show clear morphological differences. After annealing at 140 °C, both films show similar 'granular' surface morphologies with mean grain diameters of 150 ± 60 nm and 160 ± 60 nm for P3HT:CdS and nwP3HT:CdS respectively. In the opposite extreme case, after annealing at 240 °C the morphology is 'plateau'-like, comprising large flat regions (400 ± 300 nm across) with deep localised 'cracks' and holes (40 ± 20 nm deep) between them. The depth of these 'cracks' is likely to be underestimated by this measurement due to the dimensions of the AFM tip, so they could extend throughout the thickness of the film. Both the P3HT:CdS and nwP3HT:CdS films annealed at 240 °C show similar morphological features, though the density of 'cracks' appears higher for nwP3HT:CdS and they tend to be long and more linear. For the intermediate case of films annealed at 200 °C, the nwP3HT:CdS film clearly displays the 'plateau' morphology and looks very similar to the nwP3HT:CdS film annealed at 240 °C. By contrast, the P3HT:CdS film appears to show a transitional state between the 'granular' and 'plateau' morphologies.

Photovoltaic device performance

Photovoltaic devices were prepared using the P3HT:CdS and nwP3HT:CdS blend films as active layers in order to relate the thin film morphology with the solar cell performance. Fig. 6 shows the measured power conversion efficiencies (PCE) for blend films formed at different temperatures. The device efficiencies measured here are comparable with, but a little lower than, other reports based on this same film formation approach (up to 2.1%).[3,7] The overall PCEs for the P3HT:CdS and nwP3HT:CdS devices are similar (1.2%) in the lowest temperature case (140 °C), and in both cases show a sharp drop (to around 0.7%) as the temperature increases; however this drop occurs at a much lower temperature for the nwP3HT:CdS (140–160 °C) than for the P3HT:CdS (220–240 °C).

Comparing these PCEs with the morphological analyses above, there is a good correlation between the sharp drop in device efficiency and the formation of the 'plateau' microstructure, which is suspected to result in shunt connections through the 'cracks' across the active layer. The highest efficiencies correspond with the 'granular' morphology observed for film formation at 140 °C, and the poor performance corresponds with the 'plateau' morphology seen for the 240 °C case. It was noted above that the formation of the 'plateau' microstructure occurs at lower temperatures (140–200 °C) for the nwP3HT:CdS blend, and this correlates

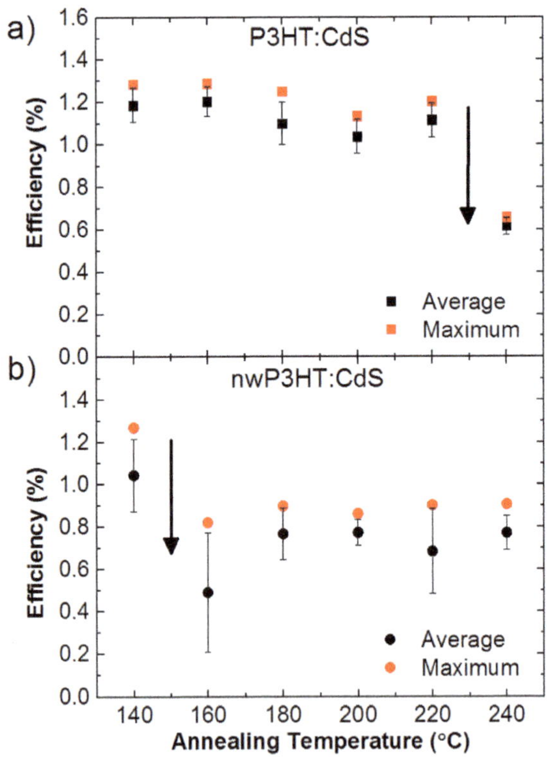

Fig. 6 Solar cell power conversion efficiency plotted against annealing temperature for the (a) P3HT:CdS and (b) nwP3HT:CdS devices. Arrows mark sharp drops in performance with increasing temperature.

well with the sharp drop in efficiency which is measured within the same range (140–160 °C). Aside from this drop in performance which we relate to the microstructure of the film, we observe minimal variation in device efficiency as a function of the annealing temperature. This suggests that the different degrees of molecular disorder in the polymer measured over this temperature range do not affect the overall solar cell efficiency. We do note, however, that all of the samples show a strong component of disordered polymer and so it is quite probable that the device efficiency could be improved if the molecular disruption could be fully overcome.

Conclusions

The results presented here reveal the morphological evolution of a P3HT:CdS blend film formed by the thermal decomposition of a CdS precursor. *In situ* Raman spectroscopy reveals that the chemical decomposition is complete within two minutes of applying heat (160 °C) to the sample, and that this process causes disruption to the molecular order of the P3HT polymer matrix, attributed to interactions between the CdS nanoparticles and the polymer. It is also found that higher temperature processing (up to 240 °C) leads to an increased magnitude of morphological disruption to the P3HT. The formation of P3HT nanowires as a means of mitigating against this disruption was found to be counter-productive, instead resulting in increased susceptibility to disruption at lower temperatures (140–160 °C). AFM images reveal two distinct film microstructures described as 'granular' and 'plateau'. The 'plateau' structure is identified as a compromised morphology comprising an intimate blend of CdS nanoparticles within a highly disordered P3HT matrix, with a high density of 'cracks'. The optimum device performance is achieved by minimising the thermal decomposition temperature to produce the 'granular' microstructure. This also minimises the disruption of the polymer molecular order, though some level of disorder is intrinsic to hybrid blend films formed in this way, which we identify as the key challenge to be overcome for this route to hybrid film formation to attain state of the art efficiencies.[24] We suggest that alternative chemical routes should be explored in order to achieve this.

Acknowledgements

The authors gratefully acknowledge funding provided by the EPSRC Centre for Doctoral Training (EP/G037515/1), SCALLOPS Project (EP/J500021/1), SUPERGEN project (EP/G031088/1), and the UK–India (EP/H040218/2) programme. S. Haque acknowledges support from the Royal Society though an award of a University Research Fellowship.

References

1 M. Wright and A. Uddin, *Sol. Energy Mater. Sol. Cells*, 2012, **107**, 87–111.
2 S. D. Oosterhout, M. M. Wienk, S. S. van Bavel, R. Thiedmann, L. J. A. Koster, J. Gilot, J. Loos, V. Schmidt and R. A. J. Janssen, *Nat. Mater.*, 2009, **8**, 818–824.
3 H. C. Leventis, S. P. King, A. Sudlow, M. S. Hill, K. C. Molloy and S. A. Haque, *Nano Lett.*, 2010, **10**, 1253–1258.

4 L. X. Reynolds, T. Lutz, S. Dowland, A. MacLachlan, S. King and S. A. Haque, *Nanoscale*, 2012, **4**, 1561–1564.
5 N. Bansal, L. X. Reynolds, A. MacLachlan, T. Lutz, R. S. Ashraf, W. Zhang, C. B. Nielsen, I. McCulloch, D. G. Rebois, T. Kirchartz, M. S. Hill, K. C. Molloy, J. Nelson and S. A. Haque, *Sci. Rep.*, 2013, **3**, 1531.
6 S. Dowland, T. Lutz, A. Ward, S. P. King, A. Sudlow, M. S. Hill, K. C. Molloy and S. A. Haque, *Adv. Mater.*, 2011, **23**, 2739–2744.
7 S. A. Dowland, L. X. Reynolds, A. MacLachlan, U. B. Cappel and S. A. Haque, *J. Mater. Chem. A*, 2013, **1**, 13896.
8 J. S. Kim, J. H. Lee, J. H. Park, C. Shim, M. Sim and K. Cho, *Adv. Funct. Mater.*, 2011, **21**, 480–486.
9 S. Wood, J. S. Kim, D. T. James, W. C. Tsoi, C. E. Murphy and J.-S. Kim, *J. Chem. Phys.*, 2013, **139**, 64901.
10 L. Kavan and M. Grätzel, *Electrochim. Acta*, 1995, **40**, 643–652.
11 J. Clark, C. Silva, R. Friend and F. Spano, *Phys. Rev. Lett.*, 2007, **98**, 206406.
12 W. C. Tsoi, W. Zhang, J. Razzell Hollis, M. Suh, M. Heeney, I. McCulloch and J.-S. Kim, *Appl. Phys. Lett.*, 2013, **102**, 173302.
13 S. Wood, J. B. Franklin, P. N. Stavrinou, M. A. Mclachlan and J.-S. Kim, *Appl. Phys. Lett.*, 2013, **103**, 153304.
14 W. C. Tsoi, D. T. James, J. S. Kim, P. G. Nicholson, C. E. Murphy, D. D. C. Bradley, J. Nelson and J.-S. Kim, *J. Am. Chem. Soc.*, 2011, **133**, 9834–9843.
15 J. P. Schmidtke, J.-S. Kim, J. Gierschner, C. Silva and R. H. Friend, *Phys. Rev. Lett.*, 2007, **99**, 167401.
16 J.-S. Kim, P. K. H. Ho, C. E. Murphy, N. Baynes and R. H. Friend, *Adv. Mater.*, 2002, **14**, 206–209.
17 J.-S. Kim, P. K. H. Ho, C. E. Murphy, A. J. A. B. Seeley, I. Grizzi, J. H. Burroughes and R. H. Friend, *Chem. Phys. Lett.*, 2004, **386**, 2–7.
18 D. Li, N. J. Borys and J. M. Lupton, *Appl. Phys. Lett.*, 2012, **100**, 141907.
19 J. J. Shiang, S. H. Risbud and A. P. Alivisatos, *J. Chem. Phys.*, 1993, **98**, 8432.
20 H. Liem, P. Etchegoin, K. S. Whitehead and D. D. C. Bradley, *J. Appl. Phys.*, 2002, **92**, 1154.
21 X. Yang, J. Loos, S. C. Veenstra, W. J. H. Verhees, M. M. Wienk, J. M. Kroon, M. A. J. Michels and R. A. J. Janssen, *Nano Lett.*, 2005, **5**, 579–583.
22 J.-H. Kim, J. H. Park, J. H. Lee, J. S. Kim, M. Sim, C. Shim and K. Cho, *J. Mater. Chem.*, 2010, **20**, 7398–7405.
23 T. Salim, S. Sun, L. H. Wong, L. Xi, Y. L. Foo and Y. M. Lam, *J. Phys. Chem. C*, 2010, **114**, 9459–9468.
24 S. Ren, L.-Y. Chang, S.-K. Lim, J. Zhao, M. Smith, N. Zhao, V. Bulović, M. Bawendi and S. Gradečak, *Nano Lett.*, 2011, **11**, 3998–4002.
25 L. Qiu, W. H. Lee, X. Wang, J. S. Kim, J. A. Lim, D. Kwak, S. Lee and K. Cho, *Adv. Mater.*, 2009, **21**, 1349–1353.
26 H. Sirringhaus, P. J. Brown, R. H. Friend, M. M. Nielsen, K. Bechgaard, B. M. W. Langeveld-Voss, A. J. H. Spiering, R. A. J. Janssen, E. W. Meijer, P. Herwig and D. M. de Leeuw, *Nature*, 1999, **401**, 685–688.
27 P. Westacott, J. R. Tumbleston, S. Shoaee, S. Fearn, J. H. Bannock, J. B. Gilchrist, S. Heutz, J. de Mello, M. Heeney, H. Ade, J. Durrant, D. S. McPhail and N. Stingelin, *Energy Environ. Sci.*, 2013, **6**, 2756.
28 S. D. Dimitrov and J. R. Durrant, *Chem. Mater.*, 2014, **26**, 616–630.

29 B. A. Collins, E. Gann, L. Guignard, X. He, C. R. McNeill and H. Ade, *J. Phys. Chem. Lett.*, 2010, **1**, 3160–3166.

30 J. A. Bartelt, Z. M. Beiley, E. T. Hoke, W. R. Mateker, J. D. Douglas, B. A. Collins, J. R. Tumbleston, K. R. Graham, A. Amassian, H. Ade, J. M. J. Fréchet, M. F. Toney and M. D. McGehee, *Adv. Energy Mater.*, 2013, **3**, 364–374.

Faraday Discussions

PAPER

Computational investigation of hole mobilities in organic semiconductors: comparison of single crystal structures and surface adsorbed clusters

Rose A. Krawczuk,*[a] Steven Tierney,[b] William Mitchell[b] and Joseph J. W. McDouall[a]

Received 26th April 2014, Accepted 29th May 2014
DOI: 10.1039/c4fd00082j

We report hole mobilities obtained computationally based on both single crystal geometries and those obtained from crystal fragments optimised on a model surface. Such computational estimates can differ considerably from experimentally measured thin film mobilities. One source of this discrepancy is due to a difference in the morphology of the thin film compared with that of the crystal. Here, predictions of thin film hole mobilities based on optimised structures are given. A model surface is used to provide an inert geometric platform for the formation of an organic monolayer. The model is tested on pentacene and TIPS-pentacene for which experimental information of the surface morphology exists. The model has also been applied to four previously uninvestigated structures. Two of the compounds studied had fairly low predicted mobilities in their single crystal structures, which were vastly improved post-optimisation. This is in accord with experiment.

1 Introduction

Owing to their potential for low-cost manufacture, thin films of molecular organic materials have attracted interest in technologies such as organic field-effect transistors (OFETs), light-emitting diodes (OLEDs) and photovoltaics (OPVs). Molecular acenes such as pentacene exhibit relatively high field-effect mobilities, a measure of the conductivity, which has prompted the derivatisation of acene cores in order to optimise these properties.

Herein, particular emphasis is placed on OFETs, for which a wide range of differently configured devices can be created, each affecting the mobility in a specific way. Most research in this area focusses on bottom-gate bottom-contact devices due to ease of production,[1] the layout of which is shown in Fig. 1. Many

[a]School of Chemistry, The University of Manchester, Manchester, M13 9PL, UK. E-mail: rose.krawczuk@postgrad.manchester.ac.uk
[b]Merck Chemicals Ltd., Chilworth Technical Centre, University Parkway, Southampton, SO16 7QD, UK

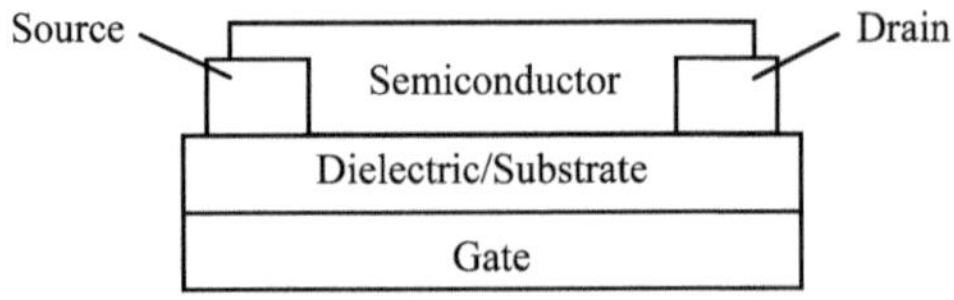

Fig. 1 Diagram of a bottom-gate bottom-contact thin film transistor.

methods also exist for the production of thin films of molecular materials. Each method of production (*e.g.* physical vapour deposition, spin-coating) subtly influences the morphology of the thin film and consequently the mobility. Crystalline and polycrystalline thin films tend to have higher mobilities than their amorphous counterparts. Charge transport in polycrystalline films is hampered by boundaries between crystal grains. By changing certain conditions during deposition, for example optimising the substrate temperature, one can alter these types of defects, thus significantly changing the mobility of a device.

On a molecular level, the underlying packing of the molecules at the surface is also critical to the conductivity. Most charge carriers reside at the semiconductor–dielectric interface, therefore it is the first few layers of a material which define its mobility. It is the balance between molecule–molecule and molecule–substrate interactions that determine the overall structure of the thin film.[2] For instance, pentacene lies with its face down on metal surfaces due to strong molecule–substrate interactions, whereas it tends to arrange itself perpendicularly to inert surfaces, for which molecule–substrate interactions are much weaker.

Since molecular organic materials generally contain some degree of anisotropy in charge transport, the orientation of the surface layers is very important. In order to achieve good charge transport, the molecules need to be orientated in such a way that their face-to-face interactions within the layers are perpendicular to the surface. It was this realisation that led John Anthony's group[3] to produce functionalised pentacene molecules, whose bulky substituents promote these types of contacts in the crystal structure. This is in contrast with pentacene molecules, which crystallise in a herringbone motif; both structures are shown in Fig. 2.

Morphology information is available for pentacene and TIPS-pentacene in their thin film phases on SiO_2. During the early stages of vacuum deposition,

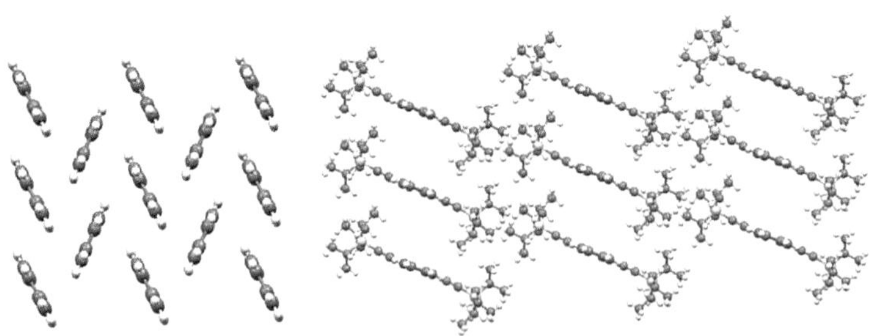

Fig. 2 On the left, the herringbone packing motif for pentacene. On the right, the 'brick-wall' packing motif for TIPS-pentacene.

single molecules of pentacene arrange themselves face down on the SiO_2 surface before they begin to form clusters.[4,5] Once a cluster of pentacene molecules reaches a critical size, it is energetically more favourable for the molecules to align their long axes perpendicular to the surface. In the single crystal there is a tilt in the angle in which one pentacene unit aligns itself with another of 25°, which is 11° for the thin film phase.[6] Other studies have found that the first layer arranges itself completely upright on the surface.[7] This means that the unit cell for the bulk crystal differs from that of the thin film. It is important to note, however, that the general structural arrangement of the molecular units in their single crystal is qualitatively similar to that in the thin film. In particular, the thin film phase of TIPS-pentacene is known to coincide almost exactly with the single crystal structure.[8]

Many computational investigations focus on the dynamics of pentacene on silica surfaces, for instance surface diffusion, molecule–substrate binding energies, defect states and thin film growth. The growth of pentacene on SiO_2 has been widely studied under various deposition conditions. An investigation into the nucleation of pentacene on SiO_2 surfaces found that the second layer nucleates when the first layer is around 80% complete, hence rationalising the observed island growth.[9] This suggests that the modelling of a monolayer on a surface could provide a good first approximation to the structure.

On a molecular scale, one study used molecular dynamics (MD) to probe the structural properties of the monolayers of pentacene on SiO_2 in order to establish the energetic stability of competing phases.[10] In a similar investigation, MD was also used to provide a model for pentacene on a SiO_2 surface. The results were found to be in agreement with previous theoretical work but not experimental grazing incidence X-ray diffraction (GIXD) data. They found tilt angles for pentacene molecules on the surface of between 11.0–15.0°. The paper states that features of the SiO_2 surface, for instance the surface roughness, can have quite an impact on the molecular structure and hence the transport at the interface.[11]

Generally, the surface of the SiO_2 dielectric is functionalised in order to improve the morphology and hence the mobility of the organic semiconductor. These coatings tend to provide a better surface for wetting of the molecular semiconductor to promote layer-by-layer growth.[12] Self-assembled monolayers (SAMs) are often used to produce more uniform interfaces with reduced surface roughness compared to SiO_2,[13] which has been reported to reduce the mobility of pentacene OFETs.[14] Attempts have been made to draw a correlation between the size of the pentacene grains and the mobility, but this is still unclear.[15] It has also been pointed out that a reduction in charge trapping at the SiO_2 semiconductor interface may be responsible for the higher mobilities on SAMs.[16] Thus there are many factors which may lead to the higher mobilities found when the dielectric surface is chemically treated.

Mobility calculations

Device-to-device reproducibility of hole mobilities tends to be poor due to the multitude of variables associated with production which affect their properties. Computationally derived mobilities can provide useful comparisons between different molecular materials. This means that mobilities for different crystals can be benchmarked against one another ignoring complications such as impurities, which can differ from one crystal to the next.

Conduction in molecular organic semiconductors proceeds by a charge hopping regime, for which Marcus's semiclassical theory of electron transfer provides a description. As depicted in Fig. 3, an electron hop is a tunnelling event that moves from the potential energy surface of the reactants, $M^+ + M$ in our case, to that of the products, $M + M^+$.

The rate of charge transfer relies upon both the electronic coupling between the states and the reorganisation energy. The rate of charge transfer, W, within Marcus theory is given by,

$$W = \frac{V^2}{\hbar}\left(\frac{\pi}{\lambda k_B T}\right)^{\frac{1}{2}} e^{\frac{-\lambda}{4k_B T}} \tag{1}$$

where V is the electronic coupling and λ is the reorganisation energy. The electronic coupling represents the strength of interaction between the initial and final electronic states and is given by the coupling matrix element, V_{IJ}.

$$V_{IJ} = \langle \Psi_I | \hat{H} | \Psi_J \rangle \tag{2}$$

where Ψ_I is the electronic wavefunction for the reactant state and Ψ_J is the electronic wavefunction for the product state. V_{IJ} is dependent upon the distance between the molecules and their relative orientations. It is shown in ref. 17 that V_{IJ} is largest when two molecules interact face-to-face and falls off as one of the molecules rotates around its long axis into a face-to-edge configuration.

The internal reorganisation energy, λ_{int}, is the energy required for the reactants in their equilibrium geometry to reorganise to the geometry of the products, plus the energy for the products undergoing the analogous process. The rate of electron transfer is exponentially dependent upon the reorganisation energy; hence small reorganisation energies are critical for high rates of transfer.

In a molecular organic crystal, each molecule is surrounded by a number of others, which vary in their relative distances and orientations. There is a finite probability that the central molecule will exchange an electron with any of the surrounding molecules. Hence, there are charge transfer rates associated with

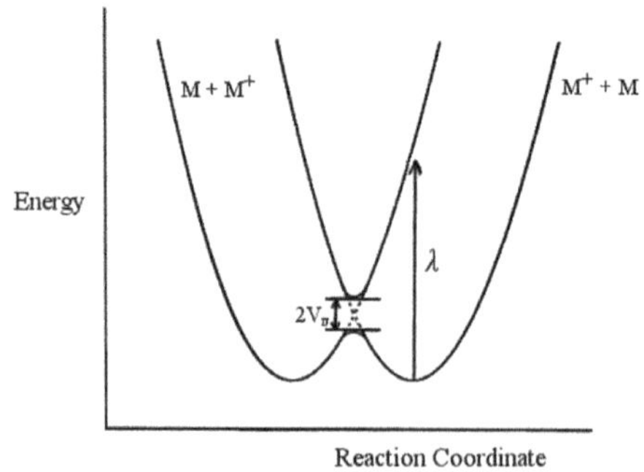

Fig. 3 The potential energy curves for $M + M^+$ and $M^+ + M$ undergoing a self-exchange reaction. When the coupling matrix element, V_{IJ}, is much smaller than the reorganisation energy, transfer of an electron proceeds by a 'hopping' mechanism.

each molecular path. These rates and probabilities are incorporated into the diffusion constant, D, which is then used to evaluate the mobility, μ, *via* the Einstein relation.

$$\mu = \frac{e}{k_B T} D \tag{3}$$

There are two different types of charge carrier: holes and electrons. For hole [electron] transport to occur, an electron must be removed [injected] at the electrode, thus the ionisation energy [electron affinity] for the material must match the Fermi level of the electrode. During hole [electron] transport, the electron is passed between the successive HOMO [LUMO] levels of neighbouring molecules, resulting in the formation of cations [anions] during the transfer. For most organic semiconductors, the hole mobility exceeds the electron mobility by orders of magnitude.[18]

Computational studies of hole mobilities are often performed using single crystal geometries, producing results that are often in agreement with experimentally measured values obtained from single crystal devices. Complications arise, however, when this value is compared with the thin film mobility. Since it is heavily dependent upon the molecular geometry, morphological differences between the single crystal and the thin film may result in large changes in the mobility. The present investigation attempts to simulate the morphology of a molecular organic material in a monolayer in order to obtain a better estimate for the hole mobility of a thin film.

Surface structure

The sole purpose of the surface in this investigation was to provide a geometric platform for formation of the organic layer. An inert surface was therefore chosen, which was unable to chemically bond to the adsorbent. The final calculations to obtain the mobility did not consider any coupling with the surface.

Whilst satisfying these requirements, it was also sensible to choose a surface for which there were data in the literature. There are many experimental and computational studies that use amorphous silica as a dielectric layer in OFETs. Additionally, it is representative of a surface before any modifications have taken place.

The structure of amorphous silica is very complicated; it has no long range order and due to the disordered distribution of ring sizes, exhibits some degree of surface roughness. For simplicity, surface roughness has been explicitly avoided in this study. Local order is provided by four-coordinate Si atoms, bridged by O. A number of silicon-oxide compounds share these short-range properties, whilst preserving long-range order. It is shown in ref. 19 that a surface adapted from the sanidine feldspar mineral can be used as a simplified model of amorphous silica. The mineral is a crystalline lattice consisting of Al, Si and O atoms, with extra cations to balance the overall negative charge. All Al atoms were converted to Si atoms and the lattice stripped of its cations. The resultant structure provides the necessary short-range order and comprises a distribution of ring sizes, with 4-, 6- and 8- member silicons. It is also important to note that silanol groups have not been taken into account in the design of this surface. Since it met all basic requirements, this structure was incorporated into the study and is shown in Fig. 4. The surface possesses a negative electrostatic potential as shown in Fig. 5.

Computational details

The primary structures investigated in this study are four functionalised benzo[*k*]tetraphene molecules, **3**, **4**, **5** and **6**, see Fig. 6. These molecules were synthesised by Merck and both thin film and single-crystal devices were produced.[20] Two other molecules were chosen for the study, pentacene (**1**) and TIPS-pentacene (**2**), principally as points for correlation of known morphologies and assessment of our procedures.

Quantum chemical calculations were carried out in order to investigate the morphologies of monolayers of molecular organic materials. Optimisations were performed using both MOPAC and MOZYME,[21] at the semiempirical PM7 level, which provides reliable intermolecular interactions energies for the S22 dataset. The effects of the deposition conditions on film formation have been excluded. Hole mobilities have been calculated for the experimental single crystal structure, and also for optimised crystal fragments and monolayers at the B3LYP/6-31G(d,p) level of theory.[22]

In order to carry out the hole mobility calculation, a molecule was selected and all of its nearest neighbours were identified. Electronic couplings and reorganisation energies were evaluated for each dimer pair. The rates of electron transfer for each path were calculated, using eqn (1). The diffusion constant was obtained as,

$$D = \frac{1}{2n}\sum_{i=1}^{m} r^2 P_i W_i \tag{4}$$

where n is the number of dimensions, which was set to 3, m is the number of nearest neighbours, r is the distance between the centroids for each molecule in

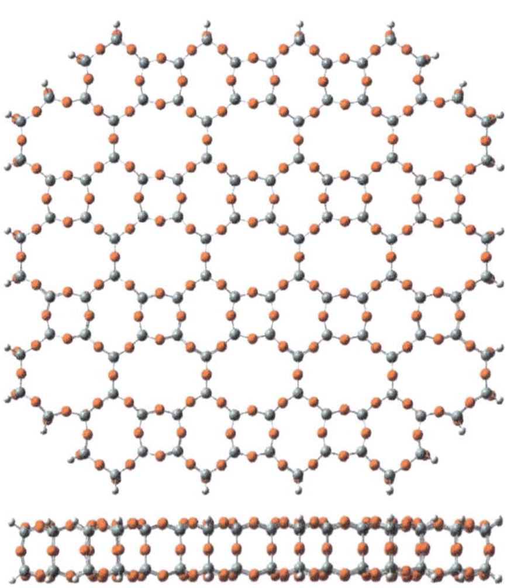

Fig. 4 A surface slab of crystalline all-Si sanidine feldspar. Top: the view of the slab from above. Bottom: the view of the slab from the side, showing the two layers of Si atoms. This surface is incorporated into the study.

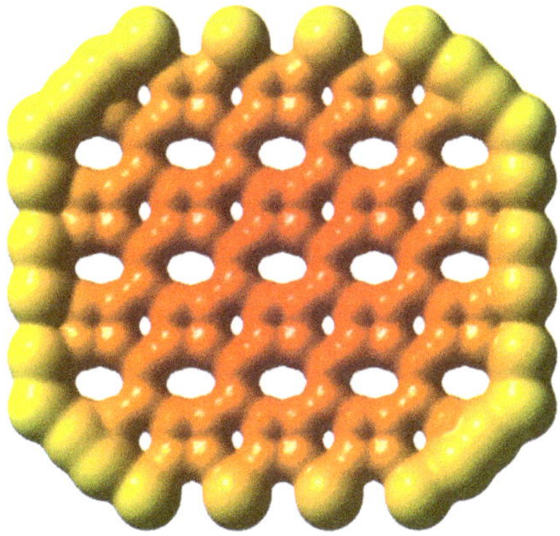

Fig. 5 An ESP map of the surface used.

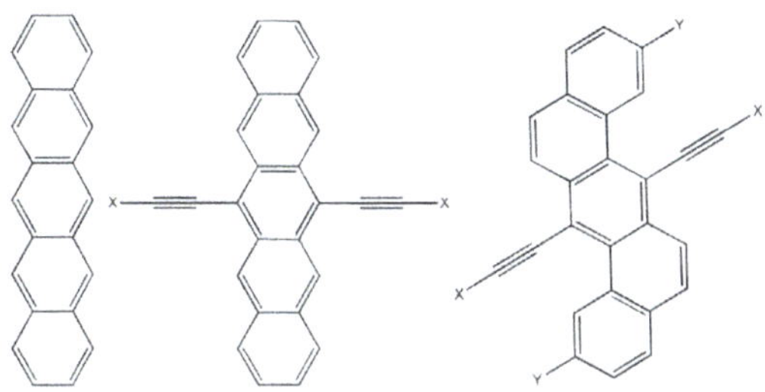

1 2 X = Si(*i*-Pr)$_3$ 3, 4, 5, 6

Compound Number	X	Y
3	$SiMe_3$	F
4	$SiMe_3$	H
5	$SiEt_3$	H
6	Si(*i*-Pr)$_3$	H

Fig. 6 The molecules used in this study.

the dimer, P_i is the probability that a certain 'hop' will take place and W_i is the rate of charge transfer for each possible 'hop'. The probability is given by eqn (5).

$$P_i = \frac{W_i}{\sum_i W_i} \tag{5}$$

For the reorganisation energy, the energies for the neutral and the cation molecules in their equilibrium geometries were calculated, along with the energies for the neutral molecule in the cation geometry and *vice versa* for the cation molecule. The overall internal reorganisation energy was evaluated using eqn (6) (see Fig. 7).

$$\lambda_{\text{int}} = \lambda_1 + \lambda_2 \tag{6a}$$

$$\lambda_1 = (E^* - E) \quad \lambda_2 = \left(E_+^* - E_+\right) \tag{6b}$$

A program was written and interfaced to Gaussian 09 (ref. 23) to evaluate the electronic couplings. This was based on the site-energy correction method, which takes into account the differences in the site energies of two molecules when they

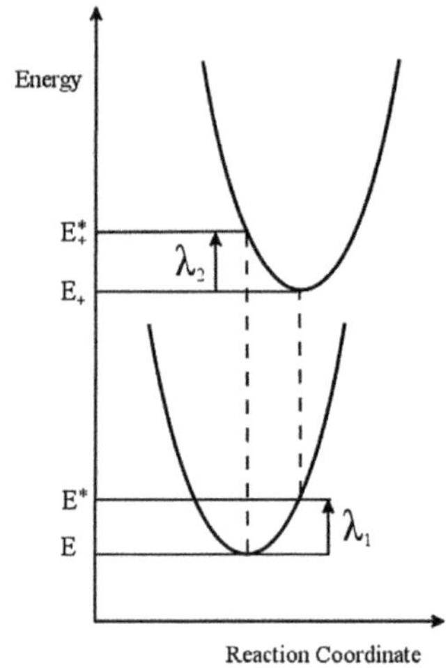

Fig. 7 Pictorial representation of the reorganisation energy calculation.

are not in the same environment.[17] The calculation assumes the HOMO and HOMO−1 levels of the dimer can be obtained by mixing of the HOMO levels of the monomers. In order to obtain the electronic coupling, the dimer Fock matrix was evaluated and transformed into the monomer basis according to eqn (7).

$$\tilde{\mathbf{F}} = \tilde{\mathbf{S}}\tilde{\mathbf{C}}\mathbf{E}\tilde{\mathbf{C}}^{\mathrm{T}}\tilde{\mathbf{S}} \tag{7}$$

where, **E** is the diagonal matrix of orbital energies for the dimer, $\tilde{\mathbf{S}}$ is the overlap matrix and $\tilde{\mathbf{C}}$ is the matrix of coefficients. These are block matrices formed from the **S** and **C** matrices of the monomers, as shown in eqn (8) and (9).

$$\tilde{\mathbf{C}} = \begin{pmatrix} \mathbf{C}_1 & 0 \\ 0 & \mathbf{C}_2 \end{pmatrix} \tag{8}$$

$$\tilde{\mathbf{S}} = \begin{pmatrix} \mathbf{S}_1 & 0 \\ 0 & \mathbf{S}_2 \end{pmatrix} \tag{9}$$

The matrix $\tilde{\mathbf{F}}$ now contains the site energies and the electronic couplings.

$$\tilde{\mathbf{F}} = \begin{pmatrix} \varepsilon_1 & \mathbf{V}_{12} \\ \mathbf{V}_{12} & \varepsilon_2 \end{pmatrix} \tag{10}$$

$\tilde{\mathbf{F}}$ is then symmetrically orthogonalised before the electronic couplings are extracted.[17]

2 Results and discussion

Testing the model

Monomer binding energies. The binding energies for attaching monomers of **1** and **2** over a range of locations on the surface, along each molecule's three molecular axes as depicted in Fig. 8, are shown in Table 1. For **1**, the binding energy increases in the order end < edge < face down on the surface.

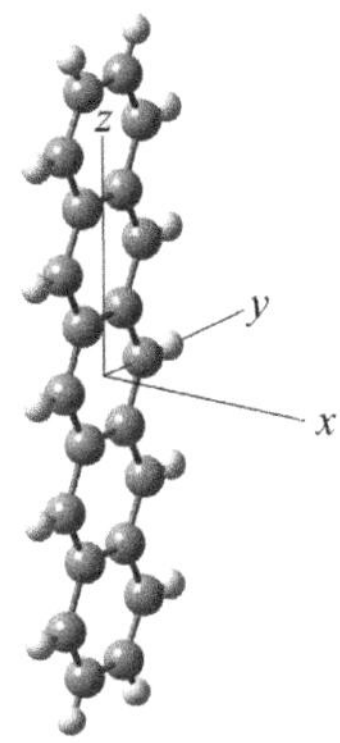

Fig. 8 Three molecular axes of pentacene.

Table 1 The average binding energies for monomers **1** and **2** across different sites on the surface

Molecule	Average binding energy/kJ mol^{-1}		
	End	Face	Edge
1	−15	−69	−36
2	−17	−95	−25

For **2**, the difference between the end- and edge-down configurations is reduced, indicating that the greater the area of contact of the molecule with the surface, the higher the binding energy. The end- and edge-down binding energies are comparable in magnitude to two molecules interacting end-to-end or edge-to-edge. The face-down affinities are quite high, approaching the value for the binding energy of two molecules face-to-face.

As discussed previously, single molecules of **1** arrange themselves face-down on the surface of SiO_2 before they reach a critical cluster size. At this point they reorientate their long axes to become perpendicular with the surface. The reason for migration from a face-down to an end-down configuration can be explained in terms of binding energies. Two systems were taken; the first of which probed the evolution of the binding energy of a face-down monomer, **a**, as more molecules were added to the surface in the same orientation (Fig. 9a). The second probed an analogous system for an end-down monomer, **b** (Fig. 9b).

Fig. 10 shows the change in the binding energy of **a** compared to **b**. As before, one can see that the binding energy for one molecule on a surface is greater for **a** than for **b**. As a second molecule is added, the binding energy for **b** surpasses that of **a**, since it is able to form a strongly bound face-to-face dimer. Once a third molecule has been added, the binding energy for **b** is significantly greater than the energy for **a**. This suggests that only two or three pentacene molecules must be interacting with one another before it is more favourable for them to stack in an upright configuration in the model used.

Orientation of crystalline clusters on the surface. Clusters of molecules were optimised on a silica surface in order to probe the morphology of a monolayer.

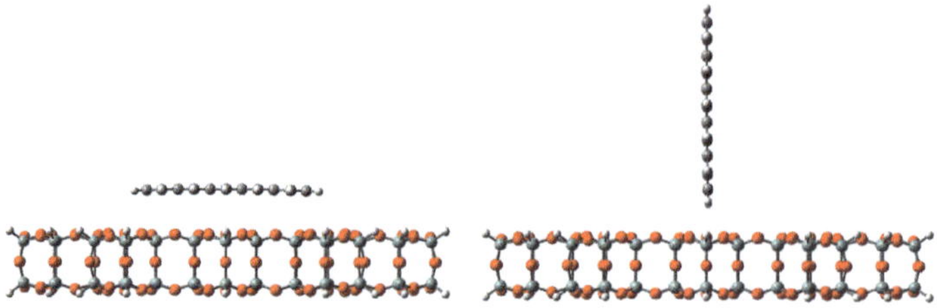

Fig. 9 The face down (left) and end down (right) systems; the pentacene monomers in these systems are **a** (left) and **b** (right).

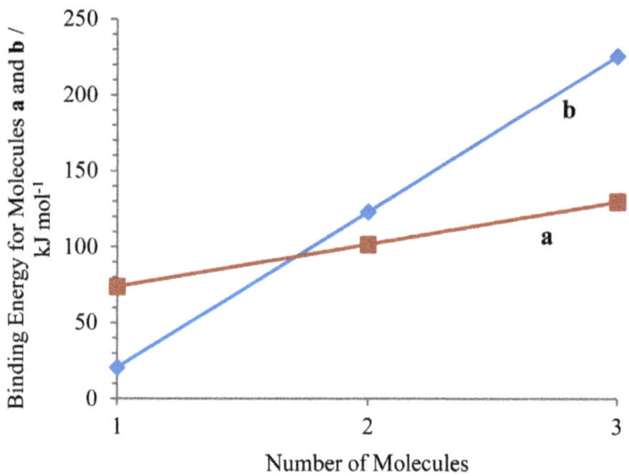

Fig. 10 Binding energies for monomers **a** and **b** (see Fig. 9) as more molecules are added to the surface.

The initial geometry was taken from the single crystal structure. In a pentacene monolayer, the (001) plane is in contact with the surface (Fig. 11). Taking a molecule on one side of a plane and calculating the binding energy to its nearest neighbour molecules on the other side, quantifies the interaction energy. This represents the loss of binding energy when the molecule is removed from the bulk onto the surface. This is akin to a molecular surface energy. Table 2 details the binding energy that would be missing for pentacene and TIPS-pentacene molecules if the crystal were 'cut' along planes which leave the ends, edges or faces of the molecules exposed.

The energy lost when the end of molecule **1** is left exposed is 26 kJ mol^{-1} compared to 153 kJ mol^{-1} for an edge; an almost five-fold increase. For **2**, the exposure of an edge is much more favourable than either of the other orientations. This data agrees and rationalises the known morphologies of **1** and **2** on the surface of amorphous silica.

Benzo[*k*]tetraphenes

The initial geometries on the surface were chosen by the procedure outlined above for pentacene. Table 3 below gives the planes identified and their binding energies.

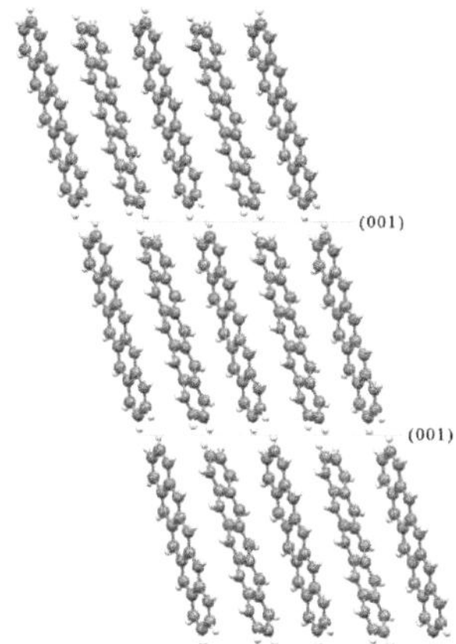

Fig. 11 The (001) plane of crystalline pentacene.

Table 2 The molecular surface energies along low energy planes for **1** and **2**

	Molecular surface energy/kJ mol^{-1}		
Molecule	End	Face	Edge
1	26	214	153
2	209	276	60

Table 3 The molecular surface energies along low energy planes for **3**, **4**, **5** and **6**

Structure	Plane	Molecular surface energy/kJ mol^{-1}
3	(001)	33
	(0$\bar{2}$1)	122
4	(001)	33
	(0$\bar{2}$1)	124
5	(010)	57
	(100)	119
6	(100)	142

For molecules **3**, **4** and to a slightly lesser extent, **5**, one plane of particularly low energy dominates in the crystal. For **6**, the lowest energy plane was found to be considerably larger in energy than any of the others in the set. In general, the bulky side groups tended to be exposed at the lowest energy planes.

Optimisations

To model the morphology of a monolayer, a crystal fragment was placed on the surface, comprising of a central molecule and its nearest neighbours. The system was allowed to relax and the hole mobility for the resultant structure was calculated. In addition, hole mobilities of optimised crystal fragments were calculated.

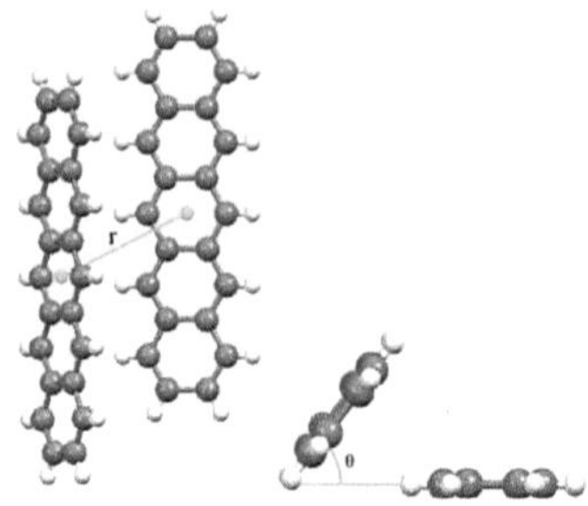

Fig. 12 The centroid–centroid distance, r, between the central molecule and one of its nearest neighbours (left) and the angle between their planes, θ, (right).

These provided a computational comparison for the surface optimisations. Unless stated otherwise, the crystal fragments consisted of a central molecule and its nearest neighbours.

Crystal fragment optimisations are dependent upon the number of molecules in the system. Molecules on the periphery, which are not fully saturated, have higher degrees of freedom than those in the centre. This means that smaller systems are poorer representations of the single crystal structure. However, computational limitations preclude the modelling of very large systems. For smaller molecules such as pentacene, with inherently fewer degrees of freedom than the functionalised cores, some analysis has been provided to indicate the differences between systems of varying size. A comparison for molecule **4** is given in order to investigate the effects of treating larger fragments on the geometries of the functionalised cores.

Hole mobilities can be rationalised by geometry changes from the crystal to the optimised structures. The parameters measured to quantify the geometries are the centroid–centroid distances between the central molecule and each of its nearest neighbours, and the angles between their planes (Fig. 12). In each case, the averages in the differences of these parameters from the experimental to optimised cases have been evaluated to give an idea of the overall movement of a system.

Fragment optimisation. Three differently sized systems were optimised for **1**, depicted in Fig. 13. For system **A**, only the nearest neighbours were present; for system **B**, a few further stabilising molecules were inserted; and for system **C**, another molecular shell around the central molecules was included.

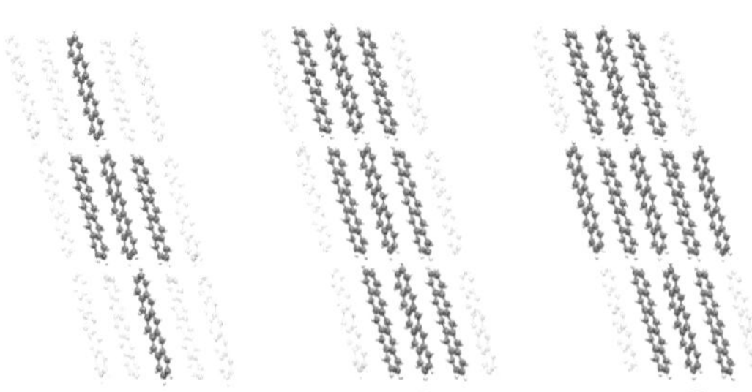

Fig. 13 Systems of varying sizes for fragment optimisation.

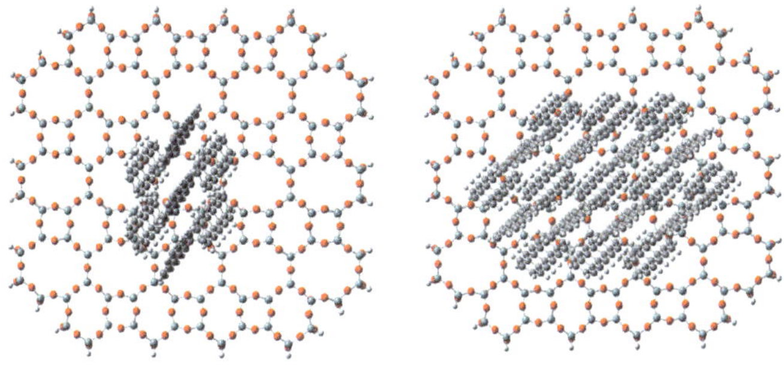

Fig. 14 Two pentacene monolayers; on the left is the smaller system containing only the nearest neighbours, **D**, and on the right is the larger system with an additional molecular shell around the central units, **E**.

As the number of molecules in the system increases, there is a corresponding increase in the hole mobility. Comparing the values, we find that the mobility for the experimental crystal, **A** is approximately half, **B** is comparable to and **C** surpasses the mobility of the single crystal. The centroid–centroid distances differ from the experimental value on average between 0.4–0.6 Å and the angles change on average between 3.2–4.6°. This suggests that whilst the distances between the molecules vary during the optimisation, their angles remain roughly the same. This behaviour is shown distinctly in **C**, in which some of the molecules have moved considerably closer to the central molecule therefore increasing their electronic couplings and hence, the overall mobility.

Monolayer optimisation. The monolayer optimisations comprise a central molecule and its nearest neighbours on the surface, shown in Fig. 14 for system **D**. To investigate the effect of the presence of more molecules in the optimisation, a calculation was run in which an additional molecular shell was added, shown in Fig. 14 as system **E**.

For **D**, the mobility is approximately the same as the single crystal value. In **E**, a similar behaviour to **C** is observed as the molecules are compressed with little change in their angles. As a result, the hole mobility for **E** is much higher than for **D**; 6.8 $cm^2 V^{-1} s^{-1}$. The hole mobilities for **C** and **E**, however, differ by approximately 3.6 $cm^2 V^{-1} s^{-1}$. In the monolayer system, the electronic coupling value for one of the paths has increased. It is suspected that this behaviour has been effected by local variations in the structure of the surface, for instance whether a molecule resides above one of the middle of rings or over a group of atoms.

Hole mobilities

Table 4 details the calculated mobilities for structures **1–6** along with their experimentally measured thin film mobilities, where available. The experimentally measured single crystal mobilities were found to be poor. **5** and **6** formed amorphous thin films in which no field-effect mobility was observed. For **2**, the predicted mobility for the experimental structure is 0.50 $cm^2 V^{-1} s^{-1}$. This value is dominated by two equivalent hopping paths with electronic coupling values of approximately 0.036 eV. In the optimised fragment, these couplings have reduced

Table 4 Hole mobilities for all monomers in their single crystals, optimised fragments and monolayers, and their experimentally measured thin film mobilities

Molecule	Reorganisation energies/kJ mol^{-1}	Hole mobility/cm^2 V^{-1} s^{-1} Single crystal	Optimised fragment		Optimised monolayer		Experimental thin film
1	9.19	2.16	**A**	0.83	**D**	1.91	>5 (ref. 24)
			B	1.82			
			C	3.20	**E**	6.85	
2	14.18	0.50		0.05		0.75	0.4 (ref. 6)
3	22.05	0.16		2.84		2.38	1.3
4	20.74	0.39	Small	3.36	Small	3.01	1.1
			Large	1.87	Large	3.17	
5	22.31	0.20		0.86		0.47	N/A
6	19.69	0.44		0.12		0.77	N/A

to approximately 0.007 eV. The major contributing factor is an increase in the displacement of the molecules' long axes. It is precisely the change in these coupling values, which has reduced the mobility for the optimised fragment by an order of magnitude.

For the optimised monolayer of **2**, the corresponding paths have electronic coupling values between 0.032 and 0.051 eV, which is similar to those in the experimental crystal structure. Since the rest of the electronic coupling values are larger for the optimised monolayer than the experimental crystal, the hole mobility is ~0.25 cm^2 V^{-1} s^{-1} greater.

In their experimental crystal structures, **3** and **4** share the same packing motif, differing mainly in their centroid–centroid distances. For both cases, optimisation of the crystal fragment has led to a significant improvement in the hole mobility, due to large changes from the experimental structure. However, the centroid–centroid distances have reduced to a greater extent in the structure of **4** than in **3**, which is reflected in its higher hole mobility. This may be due to repulsion between F-atoms on neighbouring molecules in system **3**. The optimised fragment and monolayer share similar geometries, accounting for minimal differences in the mobility.

A further, larger fragment was taken for **4**, in order to investigate the effect of system size on the relaxation of the structures for the functionalised molecules. When more molecules are included in the optimisation, the hole mobility almost halves. Crucially, some of the centroid–centroid distances increase with respect to the experimental structure, reducing the electronic coupling values and consequently, the overall mobility. Interestingly, for the monolayer optimisation of **4**, the mobility changes little on increasing the number of molecules in the system. The geometries however, are quite different.

For structure **5**, there is a four-fold increase in the mobility for the optimised fragment compared with the experimental crystal, due to significant differences in geometry. In fact, some of the changes during the optimisation would be unphysical in the actual crystal environment. For the optimised monolayer of **5**, the structure is again significantly different from the experimental crystal, with some larger and some smaller centroid–centroid distances. Overall, more of the hopping paths have electronic coupling values that are larger in the optimised monolayer than the crystal, which has led to a two-fold increase in the hole mobility. The mobilities for both the fragment and monolayer optimisations are not as high as the corresponding mobilities for structures **3** and **4**. The larger substituents in structure **5** preclude shorter distances between molecules due to steric hindrance.

The single crystal packing motif for **6** differs from the other structures in this series, a change that may be caused by the larger size of the substituent. A decrease in the hole mobility compared with the single crystal has been found for optimisation of the crystal fragment. This is likely due to changes in the relative orientations of the molecules. Following optimisation of the monolayer, the hole mobility is almost double that of the single crystal. The intermolecular distance in the starting geometry is significantly greater than in the optimised structure. The average change in the centroid–centroid distances for the molecules is 2.92 Å. In addition, the angles between molecular planes are altered on average by 42.1°. Overall, this optimisation represents a case in which there is such significant reorganisation of the structure that it no longer resembles the initial geometry. This may be related to the likelihood of formation of an amorphous film.

The hole mobilities for molecules **3** and **4**, which have similar structures, have significantly increased post-optimisation. On the other hand, the mobilities for molecules **5** and **6** have changed more modestly. This indicates a correlation between the size of the substituent and the ability for the structure to reorganise in such a way that increases the mobility.

3 Conclusions

It has been shown in this investigation that optimisation of a structure on a surface can significantly affect the predicted hole mobility. Structures **3** and **4** have relatively low predicted mobility values in the single crystal phase compared with those for the optimised monolayers. This is in accord with experimental observation. The calculations using different sized fragments suggest that more molecules must be incorporated to obtain accurate hole mobilities.

As a further refinement of the model, different surface structures should be investigated. This would enable the study of any surface modifications on the mobility.

Acknowledgements

R. A. K. thanks The University of Manchester Alumni Fund for the award of a Research Impact Scholarship and Merck Chemicals Ltd. for financial support.

References

1 B. Kumar, B. K. Kaushik and Y. S. Negi, *Polym. Rev.*, 2014, **54**, 33.
2 D. Choudhary, P. Clancy, R. Shetty and F. Escobedo, *Adv. Funct. Mater.*, 2006, **16**, 1768.
3 J. E. Anthony, J. S. Brooks, D. L. Eaton and S. R. Parkin, *J. Am. Chem. Soc.*, 2001, **123**, 9482.
4 S. Chen and J. Ma, *J. Chem. Phys.*, 2012, **137**, 074708.
5 C. Kim and D. Jeon, *Ultramicroscopy*, 2008, **108**, 1050.

6 S. E. Fritz, S. M. Martin, C. D. Frisbie, M. D. Ward and M. F. Toney, *J. Am. Chem. Soc.*, 2004, **126**, 4084.
7 S. C. B. Mannsfeld, A. Virkar, C. Reese, M. F. Toney and Z. Bao, *Adv. Mater.*, 2009, **21**, 2294.
8 S. C. B. Mannsfeld, M. L. Tang and Z. Bao, *Adv. Mater.*, 2011, **23**, 127.
9 F.-J. Meyer zu Heringdorf, M. C. Ruter and B. M. Tromp, *Nature*, 2001, **412**, 517.
10 R. G. Della Valle, E. Venuti, A. Brillante and A. Girlando, *ChemPhysChem*, 2009, **10**, 1783.
11 L. Viani, C. Risko, M. F. Toney, D. W. Breiby and J.-L. Brédas, *ACS Nano*, 2014, **8**, 690.
12 P. Clancy, *Chem. Mater.*, 2011, **23**, 522.
13 J. Veres, S. Ogier, G. Lloyd and D. de Leeuw, *Chem. Mater.*, 2004, **16**, 4543.
14 D. Knipp, R. A. Street, A. Volkel and J. Ho, *J. Appl. Phys.*, 2003, **93**, 347.
15 S. A. DiBenedetto, A. Facchetti, M. A. Ratner and T. J. Marks, *Adv. Mater.*, 2009, **21**, 1407.
16 J. E. McDermott, M. McDowell, I. G. Hill, J. Hwang, A. Kahn, S. L. Bernasek and J. Schwartz, *J. Phys. Chem. A*, 2007, **111**, 12333.
17 E. F. Valeev, V. Coropceanu, D. A. da Silva Filho, S. Salman and J.-L. Brédas, *J. Am. Chem. Soc.*, 2006, **128**, 9882.
18 H. T. Nicolai, M. Kuik, G. A. H. Wetzelaer, B. de Boer, C. Campbell, C. Risko, J. L. Brédas and P. W. M. Blom, *Nat. Mater.*, 2012, **11**, 882.
19 A. Rimola, B. Civalleri and P. Ugliengo, *Phys. Chem. Chem. Phys.*, 2010, **12**, 6357.
20 N. Blouin, M. D'Lavari, W. Mitchell, S. Tierney, C. Wang, WO. Pat., WO2012076092A1, 2012.
21 (*a*) MOPAC2012, J. J. P. Stewart, Stewart Computational Chemistry, Version 13.313L; (*b*) J. D. C. Maia, G. A. Urquiza Carvalho, C. P. Mangueira, S. R. Santana, L. A. F. Cabral and G. B. Rocha, *J. Chem. Theory Comput.*, 2012, **8**, 3072.
22 (*a*) A. D. Becke, *J. Chem. Phys.*, 1993, **98**, 5648–5652; (*b*) C. Lee, W. Yang and R. G. Parr, *Phys. Rev. B: Condens. Matter Mater. Phys.*, 1988, **37**, 785–789; (*c*) B. Miehlich, A. Savin, H. Stoll and H. Preuss, *Chem. Phys. Lett.*, 1989, **157**, 200.
23 M. J. Frisch, G. W. Trucks, H. B. Schlegel, G. E. Scuseria, M. A. Robb, J. R. Cheeseman, G. Scalmani, V. Barone, B. Mennucci, G. A. Petersson, H. Nakatsuji, M. Caricato, X. Li, H. P. Hratchian, A. F. Izmaylov, J. Bloino, G. Zheng, J. L. Sonnenberg, M. Hada, M. Ehara, K. Toyota, R. Fukuda, J. Hasegawa, M. Ishida, T. Nakajima, Y. Honda, O. Kitao, H. Nakai, T. Vreven, J. A. Montgomery, Jr, J. E. Peralta, F. Ogliaro, M. Bearpark, J. J. Heyd, E. Brothers, K. N. Kudin, V. N. Staroverov, R. Kobayashi, J. Normand, K. Raghavachari, A. Rendell, J. C. Burant, S. S. Iyengar, J. Tomasi, M. Cossi, N. Rega, J. M. Millam, M. Klene, J. E. Knox, J. B. Cross, V. Bakken, C. Adamo, J. Jaramillo, R. Gomperts, R. E. Stratmann, O. Yazyev, A. J. Austin, R. Cammi, C. Pomelli, J. W. Ochterski, R. L. Martin, K. Morokuma, V. G. Zakrzewski, G. A. Voth, P. Salvador, J. J. Dannenberg, S. Dapprich, A. D. Daniels, Ö. Farkas, J. B. Foresman, J. V. Ortiz, J. Cioslowski and D. J. Fox, *Gaussian 09, Revision D.01*, Gaussian, Inc., Wallingford CT, 2009.
24 T. W. Kelley, P. F. Baude, C. Gerlach, D. E. Ender, D. Muyres, M. A. Haase, D. E. Vogel and S. D. Theiss, *Chem. Mater.*, 2004, **16**, 4413.

Faraday Discussions

PAPER

Supramolecular control of organic p/n-heterojunctions by complementary hydrogen bonding†

Hayden T. Black,[a] Huaping Lin,[a] Francine Bélanger-Gariépy[b] and Dmitrii F. Perepichka*[a]

Received 16th June 2014, Accepted 24th June 2014
DOI: 10.1039/c4fd00133h

The supramolecular structure of organic semiconductors (OSCs) is the key parameter controlling their performance in organic electronic devices, and thus methods for controlling their self-assembly in the solid state are of the upmost importance. Recently, we have demonstrated the co-assembly of p- and n-type organic semiconductors through a three-point hydrogen-bonding interaction, utilizing an electron-rich dipyrrolopyridine (P2P) heterocycle which is complementary to naphthalenediimides (NDIs) both in its electronic structure and H bonding motif. The hydrogen-bonding-mediated co-assembly between P2P donor and NDI acceptor leads to ambipolar co-crystals and provides unique structural control over their solid-state packing characteristics. In this paper we expand our discussion on the crystal engineering aspects of H bonded donor–acceptor assemblies, reporting three new single co-crystal X-ray diffraction structures and analyzing the different packing characteristics that arise from the molecular structures employed. Particular attention is given toward understanding the formation of the two general motifs observed, segregated and mixed stacks. Co-assembly of the donor and acceptor components into a single, crystalline material, allows the creation of ambipolar semiconductors where the mutual arrangement of p- and n-conductive channels is engineered by supramolecular design based on complementary H bonding.

Introduction

Organic semiconductors (OSCs) possesses tunable molecular properties, accessible *via* the versatile tools of synthetic organic chemistry, and are therefore attractive electronic components for a variety of device applications. Structure–property relationships in OSCs enable, for example, band-gap engineering, the

[a]*Department of Chemistry and Centre for Self-Assembled Chemical Structures, McGill University, Montreal, QC, Canada, H3A 0B8. E-mail: dmitrii.perepichka@mcgill.ca*

[b]*Département de Chimie, Université de Montréal, Montréal, QC, Canada, H3T 1J4*

† CCDC 990667, 990641–990643 and 1008322–1008324. For crystallographic data in CIF or other electronic format see DOI: 10.1039/c4fd00133h

Table 1 Reported donor–acceptor pairs used in co-assembly of bicomponent ambipolar semiconductors

Donor	Acceptor	Packing motif	$\mu_h{}^a$/cm^2 V^{-1} s^{-1}	$\mu_e{}^b$/cm^2 V^{-1} s^{-1}	Ref.
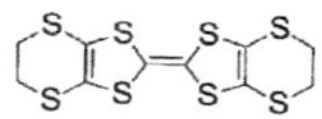	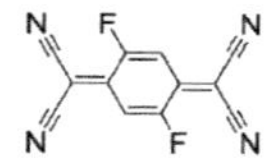	Mixed stack	$\sim 10^{-2}$	$\sim 10^{-2}$	3 and 4
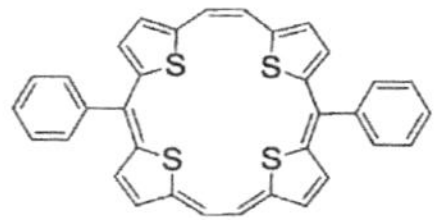	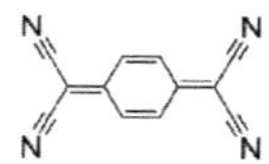	Mixed stack	4×10^{-2}	3×10^{-2}	5
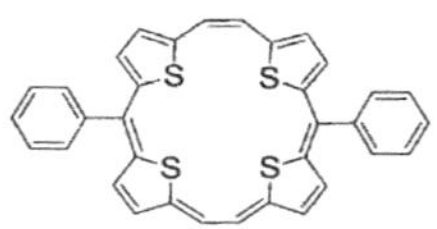		Segregated stack	3×10^{-1}	1×10^{-2}	6
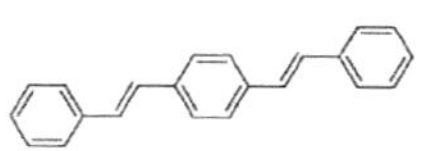	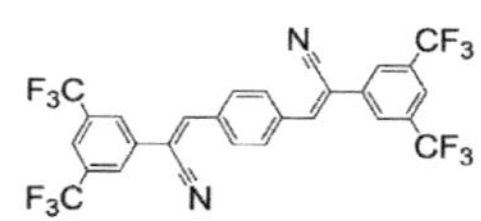	Mixed stack	7×10^{-3}	7×10^{-2}	7
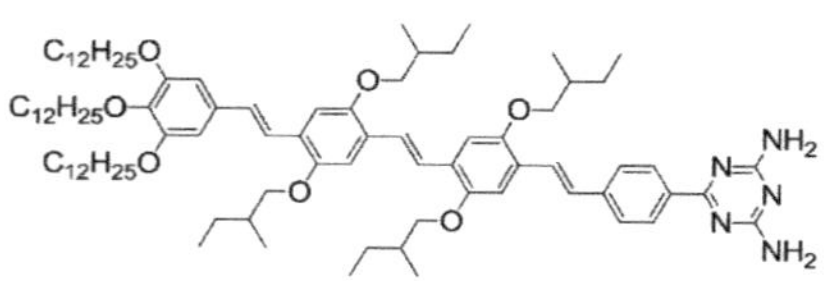	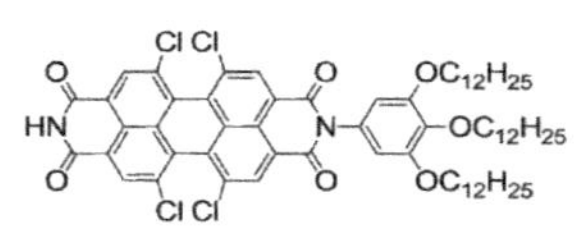	Unknown	2×10^{-7}	2×10^{-7}	14*b*

[a] μ_h = hole mobility. [b] μ_e = electron mobility.

control of redox properties, and tailoring of solubility. Thus, rational structural design and appropriate chemical substitution of OSCs has led to significant advancements in the performance of organic electronic devices. In organic photovoltaics (OPVs), band-gap engineering has led to improved solar absorption, providing increased short-circuit currents and enhanced power conversion efficiencies.[1] In organic field-effect transistors (OFETs), control of the OSC frontier molecular orbital energies has enabled tuning of the majority charge carrier, affording p-type, n-type, or ambipolar devices.[2]

In contrast to these molecular properties, the electrical properties of OSCs are intimately tied to the solid-state structure, which defines the efficiency of charge transport in specific directions.

Control of the solid-state structure requires a larger viewpoint than the conventional structure–property picture, which can only give insights into the isolated molecular properties (*i.e.* optical absorption, redox potentials, *etc.*). In order to truly control the electrical properties of OSCs, predictable intermolecular interactions are required in order to enable the rational construction of desirable morphologies which are conducive to charge transport.

Highly ordered bicomponent OSCs prepared through co-crystallization of donor–acceptor components have recently emerged as an interesting class of optoelectronic materials (Table 1). In the realm of p/n organic heterojunctions, donor–acceptor co-crystals represent the ultimate case for long-range molecular ordering. Early strategies for co-crystallization relied on strong donor–acceptor components in order to facilitate π-orbital interactions, leading to a mixed-stack solid-state arrangement of donor–acceptor components.[3–5] More recently, shape complementarity has been employed to help drive the co-crystallization of weaker donor–acceptor pairs. This approach has led to novel features in bicomponent semiconductors, including segregated stack structures with balanced charge mobilities of $\sim$0.1 $cm^2\ V^{-1}\ s^{-1}$,[6] and luminescent ambipolar co-crystals.[7] While these materials have shown promising optoelectronic properties, the ability to fine tune the packing characteristics has been absent, since the co-assembly of donor–acceptor through electronic- or shape-complementary interactions lacks directionality. New donor–acceptor co-crystal strategies which harness the predictive power of directed intermolecular interactions would greatly improve the structural control of these interesting electronic materials.

In recent years, hydrogen bonding (H bonding) has begun to emerge as a useful tool for manipulating the packing motifs of π-conjugated materials, and several studies have demonstrated efficient charge transport in OSCs possessing H bonding functionalities.[8] Notably, quinacridone-based semiconductors with two-point self-complementary H bonding have shown air-stable p-type OFETs with hole mobility values as high as 1.5 $cm^2\ V^{-1}\ s^{-1}$, representing the best performance to date in organic transistors employing H bonding interactions.[8h] In other reports, self-complementary H bonding has been associated with a tighter packing in the crystal structure, resulting from the synergistic effects of H bonding and π–π interactions.[8b,d] Still, the incorporation of H bonding functionalities into OSCs has, for the most part, been unsystematic, mainly approached *via* empirical investigations of isolated material systems. The field of organic electronics would benefit from an improved understanding of the relationship between molecular structures with H bonding functionality, and their supramolecular organization in the solid state.

Compared to self-complementary interactions, H bonding between two complementary partners is attractive for electronic applications where more than one semiconductor component is required (*i.e.* for p/n heterojunctions). In the field of supramolecular chemistry, cross-complementary H bonding stands as one of the most reliable interactions, and has been used for decades to construct a broad range of complex structures.[9] The self-assembly of DNA strands into the ubiquitous double-helix is perhaps most illustrative of the complexity and homogeneity obtainable through complementary H bonding, whereas the field of DNA nanotechnology[10] exemplifies the utility of complementary H bonding as a supramolecular synthon in materials science. In the field of organic metals, complementary H bonding has been explored for tuning both molecular packing and conductivity.[11,12] More recently, supramolecular ferroelectrics involving the H bonding of electron donor and acceptor components, have been demonstrated.[13]

With the exception of only a few papers,[14] complementary H bonding between donor–acceptor OSCs has remained largely unexplored. This is unfortunate, since the versatile intermolecular interaction could provide an effective handle for structural control of charge-transport pathways in p/n heterojunctions. Early reports of complementary H bonding in donor–acceptor OSCs have been limited to co-assemblies which lacked structural characterization, and therefore did not provide a clear picture as to what the role of the H bonding was in dictating the solid-state organization. Nonetheless, complementary H bonding between an oligophenylene-vinylene/perylenediimide donor–acceptor pair has afforded ambipolar thin-films, albeit with low charge mobilities of the order of 10^{-7} cm^2 V^{-1} s^{-1}.[14b]

Recently, we reported the co-crystallization of an electron-donating diphenyl-dipyrrolopyridine (DP-P2P) with electron-accepting naphthalenediimides (NDIs), mediated through a three-point complementary H bonding interaction (Scheme 1).[15] This approach led to crystalline materials with a well-defined donor–acceptor interface controlled by the H bonding interaction, and solid-state packing characteristics that could be tuned *via* peripheral structural changes to the individual components. In this paper, we expand our investigation of the

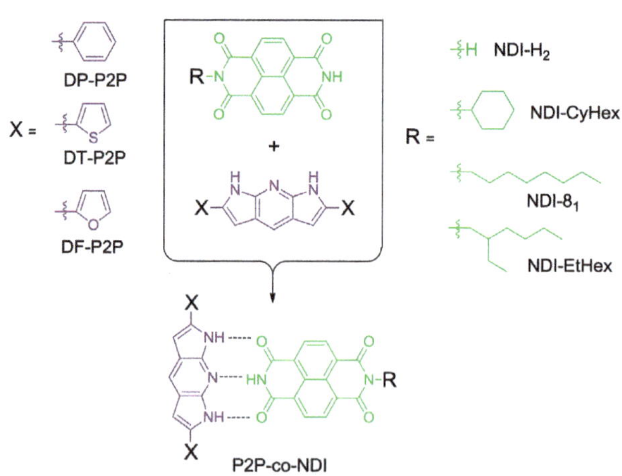

Scheme 1 P2P/NDI structures and their co-assembly.

H bonding co-crystallization strategy, exploring new structural variants of both the P2P donors and NDI acceptors. Through analysis of the X-ray diffraction structures of six different donor–acceptor H bonded co-crystals, we offer insights into the relationship between the individual molecular structures employed and the resulting co-crystal packing characteristics. In addition, the charge-transport properties of the individual components, as well as a selection of co-crystals, is examined through analysis of thin-film and single-crystal field-effect transistor devices.

Experimental section

Materials and methods

The syntheses of DP-P2P,[16] NDI-8_1, and NDI-CyHex were carried out following previously published procedures.[15] The syntheses of DT-P2P, DF-P2P and NDI-EtHex will be published elsewhere. Co-crystals were grown *via* a two-step process, in which a stoichiometric mixture of P2P donor and NDI acceptor were first co-sublimated in a horizontal glass tube. The tube was evacuated at room temperature (~0.1 mbar) and then placed in a furnace with a hot zone at 340 °C and a crystallization zone at ~100 °C. The dark co-assembled powder was then removed from the tube and used as source material for the second-stage growth, where physical vapor transport under a flow of either He or Ar (~50 mL min^{-1}) was performed with a hot zone at ~250 °C and a crystallization zone near the exit of the furnace.

FET device fabrication

Single-crystal devices were fabricated as previously described.[15] Thin-film transistors were fabricated using a heavily n-doped Si wafer as gate electrode (Sb, resistivity = 0.005–0.025 ohm·cm) with 200 nm SiO_x as dielectric (capacitance = 17.2 nF cm^{-2}) and an octyl(trichloro)silane (OTS) monolayer. The hydrophobic OTS monolayer was first formed on the SiO_x surface by submersion of the device substrate in a 0.1 M OTS solution in toluene at 80 °C for 30 min. Thin-films of DP-P2P or NDI-8_1 were then deposited *via* vacuum sublimation at a rate of 0.1–0.2 Å s^{-1}, monitored using a quartz microbalance. Au source/drain electrodes (50 nm) were deposited through a shadow-mask *via* thermal evaporation at pressures $<5 \times 10^{-6}$ Torr. The *I*–*V* measurements on the OFET devices were acquired under vacuum using a Keithley 4200-SCS semiconductor parameter analyzer.

DFT calculations

All calculations were performed using the B3LYP hybrid functional and 6-31G(d) basis set as implemented in Gaussian 09W.[17]

Results and discussion

Molecular design of complementary H bonding donor–acceptors

The design of electrically conducting supramolecular structures requires a careful selection of organic materials which meet the extensive requirements for semiconducting device applications. Specifically, we were focusing on identifying individual p/n components that combine the general materials requirements of

OSCs, *i.e.* appropriate π-conjugation, suitable frontier molecular orbital energies, chemical and thermal stability, with high-fidelity supramolecular motifs, such as complementary H bonding. With this in mind, naphthalenediimide was chosen as the model n-type semiconductor component. It is one of the simplest and most chemically stable n-type organic materials which displays sufficiently high electron mobility, often exceeding 0.1 $cm^2 V^{-1} s^{-1}$.[18] It also offers a diverse structural space with properties that can be easily tuned through N-[19,20] and/or π-substitution,[21,22] and/or π-extension (as in perylene[23] and terrylene[24] homologues). More importantly, the unsubstituted imide functionality can act as a hydrogen acceptor–donor–acceptor (A–D–A) motif, engaging in high-fidelity H bonding interactions with molecules containing the complementary proton D–A–D functionality. 2,6-Diaminopyridine and its derivatives are among the most studied H bonding receptors.[14b,25,26] We reasoned that, utilizing this structural element in a fused dipyrrolo[2,3-*b*:3′,2′-*e*]pyridine (P2P) heterocycle would lead to an ideal partner for co-assembly with NDIs because (1) it possesses the requisite complementary D–A–D motif and (2) the fused pyrrole rings impart an overall electron-rich character and rigid extended π-conjugated system, as required for an efficient p-type semiconducting material.

Proceeding from our initial studies on the co-assembly of diphenyl-P2P (DP-P2P) with NDIs,[15] we explored structural variations with regard to the P2P peripheral aromatic substituents as well as the aliphatic *N*-substituents of NDIs (Scheme 1). By replacing the phenyl substituents of DP-P2P with thienyl and furanyl groups, a stepwise increase in the electron-donating character of the peripheral rings is achieved. This causes a progressive shift of the DFT calculated highest occupied molecular orbital (HOMO) energy to higher values, moving from −5.10 eV for DP-P2P to −4.91 eV for DF-P2P (Fig. 1). The substitution also leads to crossover of the HOMO and HOMO−1 orbitals; the HOMO−1 of DP-P2P, which

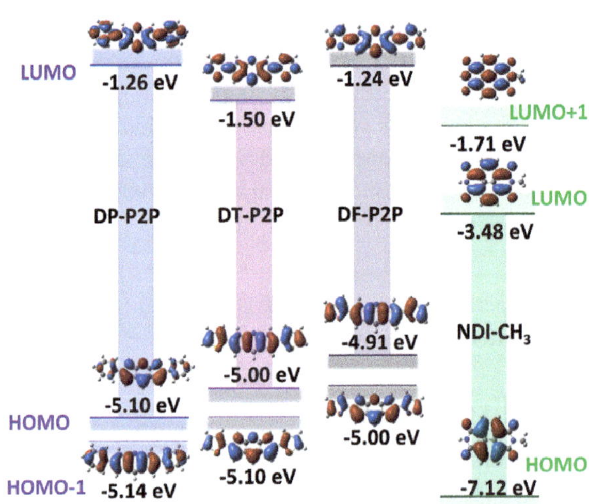

Fig. 1 DFT calculated (B3LYP/6-31G(d)) frontier molecular orbital topologies and corresponding orbital energies of monoalkylated NDI-CH_3 (green) and aryl-substituted P2P derivatives (purple).

Table 2 Single-crystal X-ray diffraction data

Crystal	Packing motif	a/Å(α/°)	b/Å(β/°)	c/Å(γ/°)	$d_{\text{H-bonding}}$ O–H/Å	$d_{\text{H-bonding}}$ N–H/Å	$d_{\text{P2P/NDI}\ \pi-\pi}$ C–C[a]/Å	$d_{\text{P2P/P2P}}$ C–C[b]/Å	$d_{\text{NDI/NDI}}$ C–C[c]/Å
DP-P2P	Herringbone	11.20(90.0°)	7.22(91.0°)	18.25(90.0°)	N/A	2.22, 2.16	N/A	3.43	N/A
DP-P2P-*co*-NDI-H_2	Segregated stack	18.83(90.0°)	5.11(91.9°)	21.29(90.0°)	2.13, 1.96	1.82	3.05	3.59	3.30
DP-P2P-*co*-NDI-8_1	Segregated stack	14.50(90.0°)	5.11(104.2°)	23.00(90.0°)	2.26, 2.08	1.91	3.13	3.43	3.41
DP-P2P-*co*-NDI-CyHex	Mixed stack	6.91(96.4°)	13.00(94.9°)	19.43(97.2°)	2.24, 2.14	1.95	3.17	4.32	3.48
DT-P2P-*co*-NDI-EtHex	Mixed stack	6.93(88.6°)	13.34(84.2°)	18.51(76.7°)	2.16, 2.13	1.95	3.22	4.33	3.33
DF-P2P-*co*-NDI-8_1	Mixed stack	22.00(90.0°)	9.38(106.0°)	31.70(90.0°)	2.24, 2.15(2.30, 2.05)	1.96(1.94)	3.17	7.47	3.25
DF-P2P-*co*-NDI-CyHex	Mixed stack	17.94(90.0°)	9.35(112.6°)	19.47(90.0°)	2.24, 2.09	1.94	3.26	7.05	3.25

[a] Shortest intermolecular P2P/NDI C–C distance. [b] Shortest intermolecular P2P/P2P C–C distance. [c] Shortest intermolecular NDI/NDI C–C distance.

has an appearance typical of linearly conjugated polyaromatics, increases in energy and becomes the HOMO in DT-P2P and DF-P2P.

Donor–acceptor interactions between NDI and P2P derivatives are manifested by the appearance of the new long-wavelength charge-transfer (CT) absorption band at ~500–700 nm, both in solution and in the solid state.[15] Also, while the studied NDI and P2P derivatives are highly fluorescent, both in solution and in the solid state, their H bonded complexes show no detectable emission, suggesting effective charge separation (photoinduced electron transfer) in these systems.

Single co-crystal X-ray diffraction analysis

The growth of P2P/NDI single co-crystals was achieved *via* a two-step process as previously described for the DP-P2P-*co*-NDI derivatives.[15] First, the two components were mixed in powder form and co-sublimed under vacuum to form the complex under fast growth conditions. This step is necessary to ensure a stoichiometric vapor of the two components and to obtain a high yield of the co-assembly, but leads to small crystals that are not suitable for X-ray diffraction. The co-crystal powder can then be used to grow larger single crystals of higher quality *via* a slow growth in a horizontal physical vapor transport process, using a flow of either He or Ar at ambient pressure. The black color of the co-crystals brought about by the charge-transfer interactions makes them distinctly different from the colorless individual donor–acceptor components, which facilitates optimization of the co-sublimation procedure.

The resulting co-crystals were analyzed using X-ray diffraction and, in some cases, single-crystal transistor devices. The high fidelity of the complementary H bonding provides a robust co-crystallization approach, tolerant to significant structural variations of each component. The single-crystal X-ray diffraction data is collected in Table 2 for the six co-crystals discussed here, as well as for pure DP-P2P. In all cases the co-crystallization afforded a predictable donor–acceptor interface defined by the three-point complementary H bonding. Fig. 2 shows the packing configurations for compounds DP-P2P-*co*-NDI-H_2 and DP-P2P-*co*-NDI-8_1,

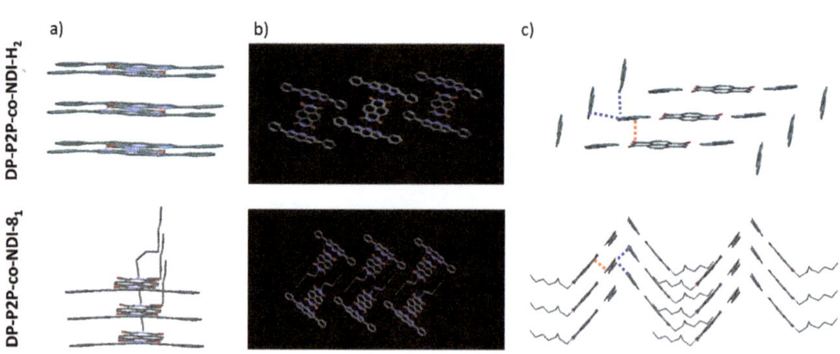

Fig. 2 Single co-crystal X-ray diffraction structures for segregated stacks DP-P2P-*co*-NDI-H_2 and DP-P2P-*co*-NDI-8_1, with slices depicting (a) $\pi-\pi$ stacking, (b) columnar arrangement of π stacks, and (c) lateral herringbone organization. Dotted blue and red lines show the shortest $\pi\cdots\pi$ contacts between the donor molecules and between the donor and acceptor, respectively.

both of which form segregated stacks with isolated columns of donor and acceptor. This type of packing motif is rarely observed in donor–acceptor co-crystals,[6] which usually prefer to organize into mixed stacks in order to maximize the π–π interactions between donors and acceptors.[3,5,7] The segregated stack motif is reminiscent of the ideal morphology for OPVs, where the alternation of donor–acceptor domains maximizes exciton dissociation and provides percolating channels for hole/electron transport. The stoichiometry of the p- and n-channel materials is precisely controlled by the molecular structure. Thus, removing the alkyl chain from the NDI component liberates the second H bonding functionality and predictably leads to 2 : 1 stoichiometry in the DP-P2P-*co*-NDI-H_2 co-crystals.

It should be noted that despite the segregation of donor and acceptors into separated p- and n-domains, donor–acceptor interactions are still observed (Fig. 2c, red dotted lines). Although the π–π overlap is small, the donor···acceptor contacts in these segregated stack structures are even shorter than those in the mixed-stack structures discussed below. This is likely due to weaker steric repulsion between the donor–acceptor molecules, which are significantly slipped from a co-facial arrangement. At the same time, this π–π interaction between DP-P2P and NDI polarizes DP-P2P, and the introduced Coulombic repulsion increases the separation of the DP-P2P molecules within the herringbone columns ($d_{P2P\ \pi\text{-CH}}$, Table 2).

Upon transition from the linear octyl substituent in DP-P2P-*co*-NDI-8_1 to the cyclohexyl substituent in DP-P2P-*co*-NDI-CyHex, a shift from segregated stack to mixed stack is observed. This can be explained by the steric effect of the cyclohexyl ring, which protrudes out of the π-orbital plane and prohibits co-facial approach of two NDI molecules (Fig. 3c). In this alternating donor–acceptor arrangement the NDI-CyHex settles slightly off-center from the core of DP-P2P, with several short contacts between the electron-deficient naphthalene and the electron-rich dipyrrolopyridine. A 40.5° twist of one peripheral phenyl group limits the π-conjugation within the DP-P2P molecules, but at the same time it allows secondary π-interactions between the phenyl rings in adjacent columns. Thus, the π-stacks are separated by regions of interacting phenyl rings and are arranged into parallel, co-planar columns.

An almost identical packing is observed in the DT-P2P-*co*-NDI-EtHex co-crystal, with alternating donor–acceptor π–π interactions giving rise to mixed-stack columns, separated by regions of interacting thiophene rings. Steric repulsion of the branched *N*-ethylhexyl substituent could again be responsible for alternating stacking in DT-P2P-*co*-NDI-EtHex, reiterating the importance of the *N*-substituent on NDI. Both thiophene rings are twisted out of plane by ~30° in order to accommodate the above-mentioned secondary π-interactions.

When strongly electron-donating furan rings are employed in the co-crystal DF-P2P-*co*-NDI-8_1, a mixed-stack arrangement is obtained, even with the sterically uninhibiting octyl substituent on the NDI component. The location of the donor–acceptor π–π overlap is shifted away from the P2P core and closer to the more electron-rich furan-containing periphery (Fig. 3a). In contrast to DP-P2P and DT-P2P derivatives, the peripheral furan rings are almost perfectly planar with respect to the P2P core (3–11°). The increased planarity of oligofurans (compared with oligothiophenes) has previously been noted and explained by the reduced steric repulsions (due to the smaller oxygen atom) and lower aromaticity of furan, which

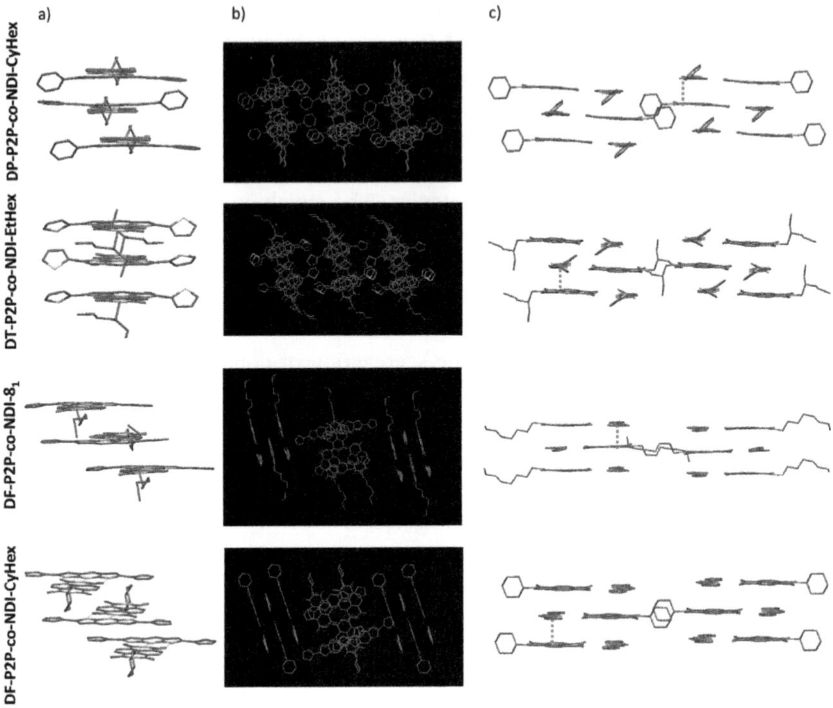

Fig. 3 Single co-crystal X-ray diffraction structures for mixed stacks DP-P2P-*co*-NDI-CyHex, DT-P2P-*co*-NDI-EtHex, DF-P2P-*co*-NDI-8_1 and DF-P2P-*co*-NDI-CyHex, with slices depicting (a) π–π stacking, (b) the columnar arrangement of π stacks, and (c) the lateral organization of mixed-stack columns.

increases inter-ring electron delocalization, enhancing the quinoidal character.[27] As a result of the planar DF-P2P structures in these co-crystals, the above-mentioned secondary π-interactions of terminal rings, observed for DP-P2P and DT-P2P co-crystals are not favored. In order to maximize the interaction between the π-columns, they pack in a 'herringbone-like' pattern, where the aromatic planes of the molecules from adjacent columns are not co-parallel (Fig. 3b).

Together, these six co-crystals enable the formulation of design rules for the rational construction of desirable donor–acceptor supramolecular structures. While a mixed-stack organization does not necessarily preclude efficient charge transport (see the discussion below), the segregated stacks are potentially of greater interest for semiconducting device applications, since clear hole/electron-transporting channels can be identified in these structures, and because segregated stacks are more likely to maintain the electronic properties of the individual semiconductor components. It is first noted that only co-crystals using DP-P2P as donor showed segregated stack organization, while those incorporating DT-P2P and DF-P2P all displayed a mixed stack. This can in part be explained by the electronic structure of the three donors: DP-P2P is the weakest donor with its HOMO level at −5.1 eV, and therefore possesses weaker donor–acceptor interactions (which favor co-facial mixed stacking). Thus, while DP-P2P-*co*-NDI-8_1 shows a segregated stack, the stronger donor characteristic of DF-P2P affords a mixed-stack arrangement in DF-P2P-*co*-NDI-8_1. A second important consideration arises from the steric constraints, as evidenced by the shift to a mixed-stack arrangement for DP-P2P-*co*-NDI-CyHex. In order to promote segregated stacks one must also ensure the facile approach of donor-to-donor and acceptor-to-acceptor.

Charge-transport properties of individual components and their co-assemblies

The hole- and electron-transport properties of DP-P2P and NDI derivatives were investigated using OFET devices. Device studies were focused only on DP-P2P as the p-type component, for which co-crystals containing both segregated and mixed stacks were obtained. Thin-film transistors (TFTs) were fabricated for DP-P2P and NDI-8_1 using Si(n++)/SiO_x as the gate electrode and dielectric, respectively, employing a hydrophobic self-assembled monolayer by reaction of the silanol surface groups with octyltrichlorosilane. The TFT device characteristics are summarized in Table 3 and the *I*–*V* curves are shown in Fig. 4. The DP-P2P transistors displayed typical p-type (only) characteristics with good saturation behavior. The devices afforded an average hole mobility (calculated from saturation regime) of $\mu_h = 0.045$ cm^2 V^{-1} s^{-1}, with a relatively low threshold voltage $V_T = -10$ V and on/off current ratios reaching $I_{on/off} = 10^5$. The NDI-8_1 TFT devices showed only an n-type response, with an electron mobility of $\mu_e = 0.047$ cm^2 V^{-1} s^{-1}, a higher threshold voltage of $V_T = 40$ V, and $I_{on/off} = 10^5$. Such balanced hole- and electron-transport characteristics of DP-P2P and NDI-8_1 makes them the ideal pair to explore the co-assembly of p/n heterojunctions. It is also important to note that the presence of acidic (imide, pK_a < 10) and basic (pyridine, pK_a > 5) functional groups in these materials does not appear detrimental for efficient hole/electron-transport, which is an important pre-requisite for using H bonding in the design of semiconducting materials. Indeed, dialkylated NDI-8_2 which possesses very similar electronic properties and similar herringbone packing in the solid state, but lacks the H bonding functionality of NDI-8_1, displays a similar electron mobility (0.16 cm^2 V^{-1} s^{-1}) in thin-film OFET devices.[28] Although some examples of

Table 3 Thin-film and single-crystal field-effect transistor properties of individual components and co-crystals

Semiconductor	μ_h [a] (cm^2 V^{-1} s^{-1})	μ_e [a] (cm^2 V^{-1} s^{-1})	V_T [b] (V) (hole/electron)	$I_{on/off}$ [c] (hole/electron)
TFT devices				
DP-P2P	0.045	—	−10/—	10^5
NDI-8_1	—	0.047	−/40	10^5
Single-crystal devices				
DP-P2P	0.027	—	−15/—	10^4/—
NDI-8_1	—	0.081	−/35	—/10^5
NDI-CyHex	—	0.17	−/20	—/10^5
DP-P2P-*co*-NDI-H_2	—	0.00034	−/7	—/10^2
DP-P2P-*co*-NDI-8_1	0.043	0.089	−94/7	10^1/10^3
DP-P2P-*co*-NDI-CyHex	0.00041	0.22	−71/4	10^3/10^6

[a] μ_h and μ_h are the field-effect hole and electron mobilities, respectively (measured in saturated regime). [b] V_T is the threshold voltage. [c] $I_{on/off}$ is the ratio of the maximum to minimum I_{SD}, obtained from the transfer characteristics.

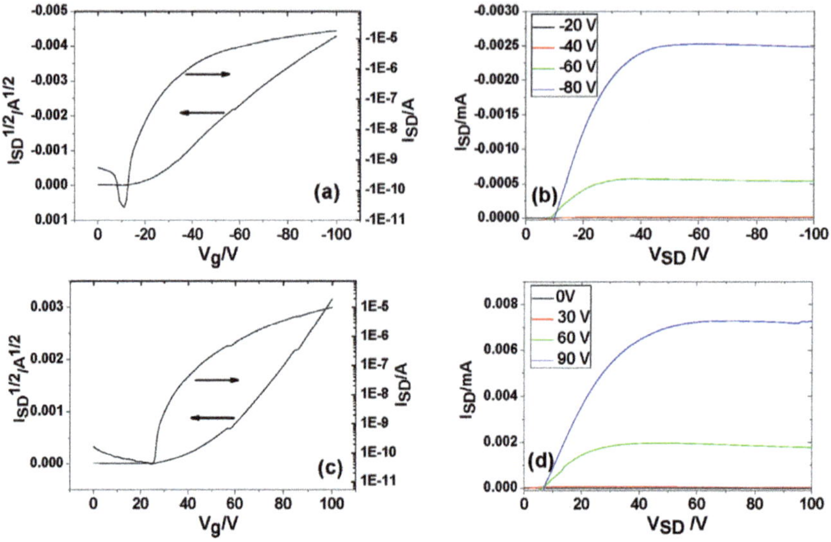

Fig. 4 Thin-film transistor transfer (a and c) and output (b and d) characteristics for DP-P2P (a and b) and NDI-8_1 (c and d).

efficient charge-transport in the presence of weakly acidic functionalities (NH) have been reported before,[8b,c,e–h] these are typically avoided in the design of an OSC due to concerns of charge trapping *via* the protonation/deprotonation of charge carriers.[2]

Note, that compared to the thin-film devices, single-crystal transistors showed a decrease in hole mobility for DP-P2P by a factor of ~2, whereas NDI-8_1 single-crystal devices showed an improvement in electron mobility by a factor of ~2. The inconsistent trend in going from thin-film to single-crystal device architecture reflects the balance of two effects: the increased molecular order/eliminated grain boundaries in the single-crystal device improves the charge transport within the channel, while the relatively high semiconductor thickness (which is difficult to control in a single-crystal device) increases the contact resistance in these top-contact devices, suppressing charge injection. Nevertheless, single-crystal OFETs have been chosen to explore the co-assembled semiconducting materials, as they eliminate the uncertainty of the composition and structure of the bottom-most layer of the semiconducting channel.

All of the co-assemblies employing DP-P2P exhibited a field-effect response in single-crystal devices, as previously reported.[15] However, the DP-P2P-*co*-NDI-H_2 co-crystal only showed n-type behavior with a low electron mobility of μ_e = 0.00034 $cm^2\ V^{-1}\ s^{-1}$, and both DP-P2P-*co*-NDI-8_1 and DP-P2P-*co*-NDI-CyHex demonstrated ambipolar transport. The DP-P2P-*co*-NDI-8_1 featuring segregated stacks in its crystal structure afforded balanced hole and electron mobilities (μ_h = 0.043 $cm^2\ V^{-1}\ s^{-1}$ and μ_e = 0.089 $cm^2\ V^{-1}\ s^{-1}$), which are very similar to the values obtained for the individual components. Efficient hole transport in DP-P2P-*co*-NDI-8_1 can be related to the tight herringbone packing of P2P molecules, with the shortest P2P/P2P C–C contact (3.43 Å) identical to that in pure DP-P2P. On the other hand, DP-P2P-*co*-NDI-CyHex afforded the highest electron mobility of μ_e =

0.22 $cm^2 V^{-1} s^{-1}$, but gave a lower hole mobility of $\mu_h = 0.00041$ $cm^2 V^{-1} s^{-1}$. Although such a large difference in electron- and hole-transporting properties is not desirable for most device applications, it is remarkable that a rather high electron mobility was achieved for a mixed-stack architecture, with no continuous π-interactions between the NDI molecules. This could potentially be explained by a superexchange mechanism (involving π-orbitals of the donor DP-P2P component).[29] In fact, moderately high charge mobility in mixed-stack co-crystals has been reported before (Table 1).

The unique character of each studied single-crystal (*i.e.* thickness, surface roughness) and manual device assembly process limits the reproducibility of single-crystal device studies, and the data does not provide a statistical measure of each co-assembly's electrical properties. On the other hand, important information is gained from these experiments, such as the confirmation of ambipolar transport in both segregated and mixed-stack co-crystal structures. Furthermore, the complementary H bonding interaction does not compromise the electrical performance of the individual components, as observed in DP-P2P-*co*-NDI-8_1 which supported hole and electron mobilities as high as the best performing devices from the pure materials. These results demonstrate the ability to construct bicomponent, crystalline donor–acceptor structures through predictable H bonding interactions, leading to electrically functional ambipolar p/n molecular heterojunctions.

Conclusions and outlook

By taking advantage of the high-fidelity complementary A–D–A···D–A–D H bonding motif, we have demonstrated rational molecular-level design of donor–acceptor co-crystals. Crystallographic analysis of six different H bonded donor–acceptor co-crystals has revealed the relationships which link the individual donor–acceptor structures with their supramolecular organization. Segregated stack structures were observed in two co-crystals, arising from the co-assembly of relatively weak donor–acceptor pairs that possess molecular structures which are compatible with self-aggregation after H bonding complexation. On the other hand, mixed-stack structures are obtained when the donor–acceptor electronic interaction is stronger, or when the steric hindrance imposed by the substituents prevents packing *via* self-aggregation of the donor and acceptor.

Field-effect transistor studies, performed in thin-film and single-crystal devices, show that the presence of acidic (NH) and basic (pyridine) functionalities in the H bonding components, generally thought of as charge-trapping centers, is not detrimental to either electron- or hole-mobility. Co-crystallization of the donor P2P and acceptor NDI components creates ambipolar materials with the electron and hole mobilities resembling those in the individual components.

One of the goals for our laboratory in moving forward with the H bonding-mediated co-assembly of donor–acceptors is to fabricate supramolecular structures with photovoltaic properties. For OPV applications, a large p/n interfacial area is required for efficient exciton dissociation, a feature that is omnipresent in P2P/NDI co-crystals. Indeed, complete quenching of the fluorescence of individual components in the co-assembled materials indicate an efficient exciton dissociation by an electron transfer from the donor to the acceptor. On the other hand, efficient OPV heterojunctions require spatial separation of hole/electron

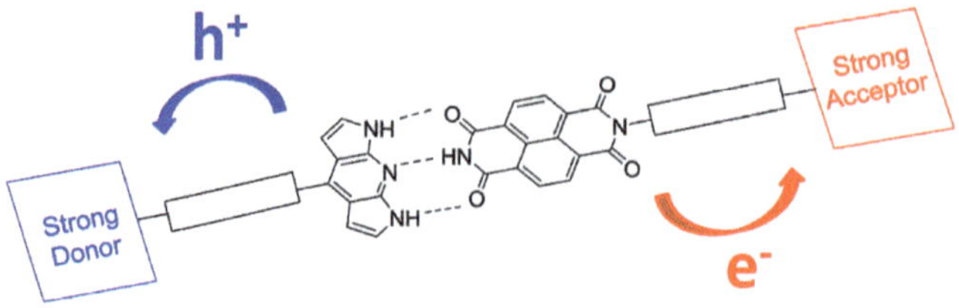

Fig. 5 Proposed P2P-NDI H bonded co-assembly capable of effective charge separation.

pairs in order to enhance charge collection and reduce charge recombination. For this reason, most bulk heterojunction OPVs feature p-/n-domain sizes of the order of 10–50 nm. This presents a problem for molecular heterojunction co-crystals, where the domain size cannot easily be tuned. One method for circumventing the problem of hole/electron proximity in donor–acceptor co-crystals is to construct a molecularly defined p/n interface possessing a redox gradient, which would shuttle holes and electrons away from the p/n interface after exciton dissociation (Fig. 5). In order to accomplish this using the P2P-NDI H bonded interface, a stronger electron donor (a lower oxidation potential than P2P) would be covalently linked to P2P, while a strong electron acceptor (a lower reduction potential than NDI) would be linked to NDI. This type of system would mimic certain aspects of the highly efficient supramolecular complexes involved in photosynthesis. One can also imagine the implementation of this concept (as well as P2P and NDI building blocks) in the design of donor and acceptor semiconducting polymers/long-chain oligomers, where the phase separation will be controlled by complementary H bonding between the donor and acceptor strands.

In conclusion, the prospects of using complementary H bonding to control the supramolecular organization of electronic materials are immense, and H bonding is perhaps the most versatile and reliable route towards solid-state structural control in organic electronics.

Acknowledgements

This work was supported by NSERC of Canada and NanoQuebec. H.L. is thankful to FRQS for the post-doctoral fellowship.

References

1 H. Zhou, L. Yang and W. You, *Macromolecules*, 2012, **45**, 607.
2 (*a*) A. R. Murphy and J. M. J. Fréchet, *Chem. Rev.*, 2007, **107**, 1066; (*b*) S. Z. Bisri, C. Piliego, J. Gao and M. A. Loi, *Adv. Mater.*, 2014, **26**, 1176.
3 T. Hasegawa, S. Kagoshima, T. Mochida, S. Sugiura and Y. Iwasa, *Solid State Commun.*, 1997, **103**, 489.
4 M. Sakai, H. Sakuma, Y. Ito, A. Saito, M. Nakamura and K. Kudo, *Phys. Rev. B: Condens. Matter Mater. Phys.*, 2007, **76**, 045111.
5 J. Zhang, H. Geng, T. Singh Virk, Y. Zhao, J. Tan, C. Di, W. Xu, K. Singh, W. Hu, Z. Shuai, Y. Liu and D. Zhu, *Adv. Mater.*, 2012, **24**, 2603.

6 J. Zhang, J. Tan, Z. Ma, W. Xu, G. Zhao, H. Geng, C. Di, W. Hu, Z. Shuai, K. Singh and D. Zhu, *J. Am. Chem. Soc.*, 2013, **135**, 558.
7 S. K. Park, S. Varghese, J. H. Kim, S.-J. Yoon, O. K. Kwon, B.-K. An, J. Gierschner and S. Y. Park, *J. Am. Chem. Soc.*, 2013, **135**, 4757.
8 (*a*) H.-L. Yip, H. Ma, A. K.-Y. Jen, J. Dong and B. A. Parviz, *J. Am. Chem. Soc.*, 2006, **128**, 5672; (*b*) M. Gsänger, J. H. Oh, M. Könemann, H. W. Höffken, A.-M. Krause, Z. Bao and F. Würthner, *Angew. Chem., Int. Ed.*, 2010, **49**, 740; (*c*) J. A. Oh, W.-Y. Lee, T. Noe, W.-C. Chen, M. Könemann and Z. Bao, *J. Am. Chem. Soc.*, 2011, **133**, 4204; (*d*) Z. He, D. Liu, R. Mao, Q. Tang and Q. Miao, *Org. Lett.*, 2012, **14**, 1050; (*e*) Y. Suna, J. Nishida, Y. Fujisaki and Y. Yamashita, *Org. Lett.*, 2012, **14**, 3356; (*f*) N. B. Kolhe, R. N. Devi, S. P. Senanayak, B. Jancy, K. S. Narayan and S. K. Asha, *J. Mater. Chem.*, 2012, **22**, 15235; (*g*) E. D. Głowacki, L. Leonat, M. Irimia-Vladu, R. Schwödiauer, M. Ullah, H. Sitter, S. Bauer and N. S. Sariciftci, *Appl. Phys. Lett.*, 2012, **101**, 023305; (*h*) E. D. Głowacki, M. Irimia-Vladu, M. Kaltenbrunner, J. Gsiorowski, M. S. White, U. Monkowius, G. Romanazzi, G. P. Suranna, P. Mastrorilli, T. Sekitani, S. Bauer, T. Someya, L. Torsi and N. S. Sariciftci, *Adv. Mater.*, 2013, **25**, 1563.
9 L. J. Prins, D. N. Reinhoudt and P. Timmerman, *Angew. Chem., Int. Ed.*, 2001, **40**, 2382.
10 F. A. Aldaye, A. L. Palmer and H. F. Sleiman, *Science*, 2008, **321**, 1795.
11 M. Fourmigué and P. Batail, *Chem. Rev.*, 2004, **104**, 5379.
12 T. Murata, Y. Morita, Y. Yakiyama, K. Fukui, H. Yamochi, G. Saito and K. Nakasuji, *J. Am. Chem. Soc.*, 2007, **129**, 10837.
13 A. S. Tayi, A. K. Shveyd, A. C.-H. Sue, J. M. Szarko, B. S. Rolczynksi, D. Cao, T. J. Kennedy, A. A. Sarjeant, C. L. Stern, W. F. Paxton, W. Wu, S. K. Dey, A. C. Fahrenbach, J. R. Guest, H. Mohseni, L. X. Chen, K. L. Wang, J. F. Stoddart and S. L. Stupp, *Nature*, 2012, **488**, 485.
14 (*a*) C.-H. Huang, N. D. McClenaghan, A. Kuhn, J. W. Hofstraat and D. M. Bassani, *Org. Lett.*, 2005, 7, 3409; (*b*) P. Jonkheijm, N. Stutzmann, Z. Chen, D. M. de Leeuw, E. W. Meijer, A. P. H. J. Schenning and F. Würthner, *J. Am. Chem. Soc.*, 2006, **128**, 9535.
15 H. T. Black and D. F. Perepichka, *Angew. Chem., Int. Ed.*, 2014, **53**, 2138.
16 C. Koradin, W. Dohle, A. L. Rodriguez, B. Schmid and P. Knochel, *Tetrahedron*, 2003, **59**, 1571.
17 M. J. Frisch, G. W. Trucks, H. B. Schlegel, G. E. Scuseria, M. A. Robb, J. R. Cheeseman, G. Scalmani, V. Barone, B. Mennucci, G. A. Petersson, H. Nakatsuji, M. Caricato, X. Li, H. P. Hratchian, A. F. Izmaylov, J. Bloino, G. Zheng, J. L. Sonnenberg, M. Hada, M. Ehara, K. Toyota, R. Fukuda, J. Hasegawa, M. Ishida, T. Nakajima, Y. Honda, O. Kitao, H. Nakai, T. Vreven, J. A. Montgomery, J. E. Peralta, F. Ogliaro, M. Bearpark, J. J. Heyd, E. Brothers, K. N. Kudin, V. N. Staroverov, R. Kobayashi, J. Normand, K. Raghavachari, A. Rendell, J. C. Burant, S. S. Iyengar, J. Tomasi, M. Cossi, N. Rega, J. M. Millam, M. Klene, J. E. Knox, J. B. Cross, V. Bakken, C. Adamo, J. Jaramillo, R. Gomperts, R. E. Stratmann, O. Yazyev, A. J. Austin, R. Cammi, C. Pomelli, J. W. Ochterski, R. L. Martin, K. Morokuma, V. G. Zakrzewski, G. A. Voth, P. Salvador, J. J. Dannenberg, S. Dapprich, A. D. Daniels, O. Farkas, J. B. Foresman, J. V. Ortiz,

J. Cioslowski and D. J. Fox, *Gaussian 09, Revision B.01*, Gaussian, Inc., Wallingford CT, 2009.

18 (*a*) B. A. Jones, A. Facchetti, T. J. Marks and M. R. Wasielewski, *Chem. Mater.*, 2007, **19**, 2703; (*b*) D. Shukla, S. F. Nelson, D. C. Freeman, M. Rajeswaran, W. G. Ahearn, D. M. Meyer and J. T. Carey, *Chem. Mater.*, 2008, **20**, 7486; (*c*) K. C. See, C. Landis, A. Sarjeant and H. E. Katz, *Chem. Mater.*, 2008, **20**, 3609.

19 H. E. Katz, J. Johnson, A. J. Lovinger and W. Li, *J. Am. Chem. Soc.*, 2000, **122**, 7787.

20 R. Schmidt, J. H. Oh, Y.-S. Sun, M. Deppisch, A.-M. Krause, K. Radacki, H. Braunschweig, M. Könemann, P. Erk, Z. Bao and F. Würthner, *J. Am. Chem. Soc.*, 2009, **131**, 6215.

21 Y. Zhao, C. Di, X. Gao, Y. Hu, Y. Guo, L. Zhang, Y. Liu, J. Wang, W. Hu and D. Zhu, *Adv. Mater.*, 2011, **23**, 2448.

22 B. A. Jones, A. Facchetti, M. R. Wasielewski and T. J. Marks, *J. Am. Chem. Soc.*, 2007, **129**, 15259.

23 C. Li and H. Wonneberger, *Adv. Mater.*, 2012, **24**, 613–636.

24 J. H. Oh, W.-Y. Lee, T. Noe, W.-C. Chen, M. Könemann and Z. Bao, *J. Am. Chem. Soc.*, 2011, **133**, 4204.

25 S.-K. Chang, D. V. Engen, E. Fan and A. D. Hamilton, *J. Am. Chem. Soc.*, 1991, **113**, 7640.

26 M. Gray, A. O. Cuello, G. Cooke and V. M. Rotello, *J. Am. Chem. Soc.*, 2003, **125**, 7882.

27 O. Gidron and M. Bendikov, *Angew. Chem., Int. Ed.*, 2014, **53**, 2546.

28 H. E. Katz, J. Johnson, A. J. Lovinger and W. Li, *J. Am. Chem. Soc.*, 2000, **122**, 7787.

29 L. Zhu, Y. Yi, Y. Li, E.-G. Kim, V. Coropoceanu and J.-L. Brédas, *J. Am. Chem. Soc.*, 2012, **134**, 2340.

Faraday Discussions

PAPER

Design of donor–acceptor star-shaped oligomers for efficient solution-processible organic photovoltaics†

S. A. Ponomarenko,*[ab] Y. N. Luponosov,[a] J. Min,[c] A. N. Solodukhin,[a] N. M. Surin,[a] M. A. Shcherbina,[a] S. N. Chvalun,[a] T. Ameri[c] and C. Brabec[c]

Received 1st July 2014, Accepted 21st July 2014
DOI: 10.1039/c4fd00142g

This contribution describes recent progress in the design, synthesis and properties of solution-processible star-shaped oligomers and their application in organic photovoltaics. Even though alternative chemistry has been used to design such oligomers, the most successful approach is based on a triphenylamine donor branching center, (oligo)thiophene conjugated spacers and dicyanovinyl acceptor groups. These are mainly amorphous low band-gap organic semiconductors, though crystalline or liquid crystalline ordering can sometimes be realized. It was shown that the solubility, thermal behavior and structure of such molecules in the bulk strongly depend on the presence and position of alkyl groups, as well as on their length. The photovoltaic properties of solution-processed molecules of this type are now approaching 5% which exceeds those of vacuum-sublimed devices. The design rules and future perspectives of this class of organic photovoltaic molecules are discussed.

Introduction

Organic photovoltaics is a fast growing area of science and technology due to possibility to create light weight large area flexible solar cells, which can be produced by low cost printing or roll-to-roll techniques.[1] The technology is based on organic semiconductors, either π-conjugated polymers[2] or small molecules,[3] capable of efficient light absorption in the visible spectral range with the creation of excitons. However, opposite to inorganic solar cells, charge separation in organic solar cells (OSC) leading to a photovoltaic effect takes place on the interface between the electron donor and the electron acceptor organic

[a]*Enikolopov Institute of Synthetic Polymeric Materials of the Russian Academy of Sciences, Profsoyuznaya st. 70, Moscow 117393, Russia. E-mail: ponomarenko@ispm.ru*

[b]*Chemistry Department, Moscow State University, Leninskie Gory 1-3, Moscow 119991, Russia*

[c]*Institute of Materials for Electronics and Energy Technology (I-MEET), Friedrich-Alexander-University Erlangen-Nuremberg, Martensstraße 7, 91058 Erlangen, Germany*

† Electronic supplementary information (ESI) available: ^{1}H NMR spectra and TGA curves. See DOI: 10.1039/c4fd00142g

semiconductors.[4] In order to maximize this interface area, "bulk heterojunction" OSCs, consisting of a phase-separated mixture of donor and acceptor organic semiconductors were suggested.[5] In the case of the solution-processing of organic semiconductors, which is considered to be a cost-effective alternative to expensive high vacuum processing techniques, the bulk heterojunction (BHJ) is formed by the spontaneous phase separation of organic semiconductors during solvent evaporation.[6] For the creation of the most efficient BHJ OSC, typical dimensions of the donor and the acceptor phases should be on the order of 10–20 nm, that is comparable to the exciton diffusion length. Moreover, these phases have to be highly interconnected to allow electrons and holes separated on their interphase to reach the electrodes and give a photocurrent. That is why the morphology of bulk heterojunctions plays a crucial role on the performance of BHJ OSCs.[7] The ability to control the morphology depends first of all on the chemical structure of a particular organic semiconductor, its solubility and the compatibility of the two types of organic semiconductors used in the BHJ.[8] Therefore, the design of novel solution-processible organic semiconductors is a important for the creation of highly efficient BHJ OSCs. Further improvements to the morphology and hence the efficiency of BHJ OSCs can be made by including various additives during solution processing, like 1,8-octanedithiol or 1,8-diiodooctane,[9] or by different post-treatment techniques, like thermal or solvent vapor annealing.[10] Introduction of special interfacial layers into the device structure of the OSC, which help optimize the electronic and electrical properties between the interfaces of the light-harvesting active layer and charge-collecting electrodes, can further improve the efficiency.[11]

Nowadays it is accepted that theoretically the efficiency of BHJ OSCs can reach as high as 10% in single junction structures and 15% in tandem devices.[12] Recent achievements in organic photovoltaic material design, morphology optimization and device architecture showed fast progress towards these values: power conversion efficiency (PCE) of 7–9% was reported for solution-processed single junction BHJ OSCs[13] and above 10% for tandem structures.[14] It is noteworthy that in all these high efficiency organic photovoltaic devices, a soluble fullerene derivative, $PC_{71}BM$, was used as the acceptor and different low band gap small molecules or polymers served as the donor organic semiconductors. The design of the latter is based on the concept of incorporation of different donor and acceptor units into the structure of the oligomeric or polymeric molecule, which allow tuning of their HOMO and LUMO levels, leading to a narrow band gap. This means that the less than 1.9 eV reported for the "standard" organic semiconductor poly(3-hexylthiophene) P3HT, is widely used in organic photovoltaics.[15] Although the record efficiencies of BHJ OSCs based on small molecules and polymers are comparable now, a clear advantage of the small molecules is their exact molecular structure, absence of any polydispersity and different chain ends, which lead to less batch to batch variations during their large scale synthesis and easier purification process. Among the small molecules used as donor materials in organic photovoltaics one can distinguish two different classes: linear and star-shaped molecules. The first class is widely investigated, which already led to impressive PCE values of the BHJ OSC devices based on them, mentioned above. The star-shaped molecules seems to be less researched and it is worth considering them in detail in this paper based on recent achievements made in our group in order to outline the design rules leading to the most efficient star-shaped molecules for organic photovoltaic devices and the future perspective of this class of materials.

Results and discussion

Overview of the literature data on different star-shaped molecules used in organic photovoltaic devices

The first star-shaped molecule, properties of which were checked in OSCs, was planarized star-shaped oligothiophene **1** reported by Roncali *et al.* (Fig. 1).[16] Under light illumination at 77 mW cm^{-2} irradiation intensity, a bilayer OSC, prepared by vacuum evaporation of **1** as the donor and *N*,*N*′-bis(tridecyl)-perylenedicarboxyimide (DP13) as the acceptor, showed an open circuit voltage (V_{oc}) of 0.86 V, short circuit current (J_{sc}) of 1.35 mA cm^{-2} and fill factor (FF) of 0.51, which gives an overall PCE of 0.77%. This value itself was not very high, albeit it was significantly better compared to a similar device made from its lineal analogue **2**, containing a central core and just one arm of the star, which under the same conditions showed only a PCE of 0.04%. This difference in behaviour was explained by the different orientation of the linear and the star-shaped compounds in thin films, supported by X-ray and UV-Vis optical spectroscopy investigations. The linear oligomer **2** has a preferential vertical orientation of the molecules onto the surface of the substrates, while the star-shaped compound **1** has a preferential horizontal orientation, which is more favourable for absorption of the incident light.

Similar conclusions were made after investigation of a series of quaterthiophene-based multipods **3–5** with linear and branched terminal alkyl groups (Fig. 1).[17] It was found that multipods with branched 2-ethylhexyl groups **4** and **5** showed better photovoltaic characteristics in solution-processable BHJ OSCs, as compared to multipod **3** with linear hexyl groups: PCE = 0.9% and 0.27% *vs.*

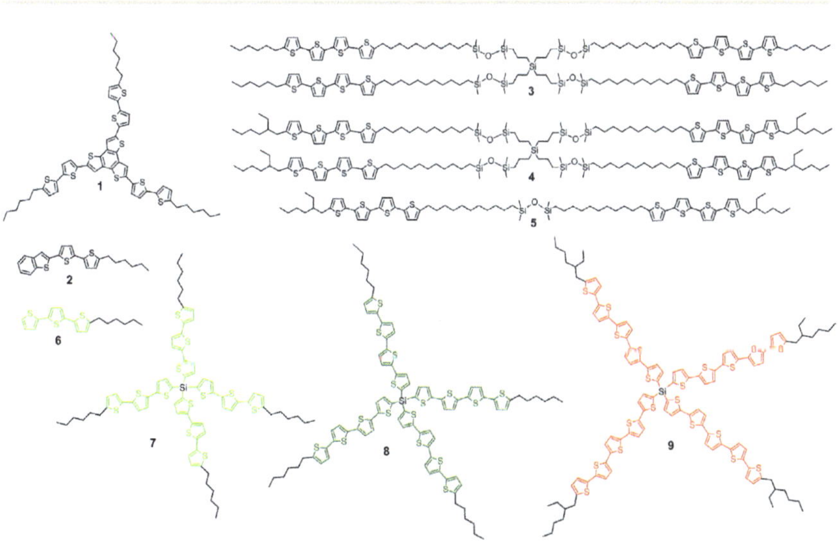

Fig. 1 Chemical structures of star-shaped oligothiophenes and their linear analogues used in organic photovoltaic devices.

Table 1 Photovoltaic characteristics of OSCs prepared from different star-shaped molecules and their linear analogues

Donor	λ_{max}(abs) in solution/film	Optical band gap, eV	Acceptor	Device type, D : A ratio	I,[b] mW cm^{-2}	V_{oc}, V	I_{sc}, mA cm^{-2}	FF, %	PCE, %	Ref.
1	405/420	2.9 (2.55[d])	DP13	Bilayer	77	0.86	1.35	51	0.77	16
2[a]	379/319	3.0	DP13	Bilayer	77	0.74	0.094	46	0.04	16
3	403/—[c]	2.6	$PC_{71}BM$	BHJ, 1 : 3	100	0.36	0.45	29	0.05	17*c*
4	403/380	2.6 (2.3[e])	$PC_{71}BM$	BHJ, 1 : 3	100	0.60	1.3	35	0.27	17*c*
5	402/399	2.6 (2.3[e])	$PC_{71}BM$	BHJ, 1 : 3	100	0.68	3.2	40	0.90	17*c*
6[a]	367/—	—	DP13	Bilayer	80	0.76	0.14	32	0.04	18
7	384/406	2.65[d]	DP13	Bilayer	80	0.99	0.36	42	0.20	18
7a	390/390	2.65[d]	$PC_{61}BM$	BHJ, 1 : 3	80	0.85	1.13	24	0.29	18
7	385/—	2.85	$PC_{71}BM$	BHJ, 1 : 2	100	0.83	2.9	25	0.6	19
8	415/—	2.6	$PC_{71}BM$	BHJ, 1 : 2	100	0.75	5.0	26	1.0	19
9	435/—	2.5	$PC_{71}BM$	BHJ, 1 : 2	100	0.80	6.0	28	1.4	19

[a] Linear analogue. [b] AM 1.5 illumination intensity. [c] No data reported. [d] From thin film measurements. [e] Electrochemical band gap.

0.05% were reported (Table 1).[17b] In all of these devices one of the multipods served as a donor and soluble fullerene derivative $PC_{71}BM$ as an acceptor organic semiconductor. This behaviour was attributed not only to improved solubility and morphology of multipods **4** and **5**, but also to a different orientation of their molecules relative to the surface of the substrate: star-shaped multipod **3** with linear end groups tended to have a vertical orientation, while its analogue **4** with branched end groups tended to have a more horizontal orientation. It was subsequently confirmed by X-ray analysis, which showed 22 and 55 degrees inclination of the quaterthiophene groups in the multipods **3** and **4**, respectively, in their unit cells.[17c]

Since the orientation of the chromophores in the space seems to be very important for the efficiency of photovoltaic devices, the other type of star-shaped molecules – tetrahedral oligothiophenesilanes **7–9** – seem to be very promising. Under any orientation of such molecule in the space, some the oligothiophene groups orient more or less horizontally to the surface of the cell. This leads to efficient light harvesting in the region of the spectral match of the oligothiophene and the solar light (Table 1). Moreover, the non-planar structure of these molecules makes them highly soluble in normal organic solvents like chloroform, toluene, THF, *etc.* The first oligothiophenesilane **7** was reported by Roncali *et al.*[18] Bilayer heterojunction solar cells prepared from **7** as a donor material and DP13 as an acceptor material showed a high V_{oc} of 0.99 V, reasonably good FF of 51%, but a low overall PCE of 0.2% mainly due to a low J_{sc} of 0.36 mA cm^{-2}. However, a clear advantage of the star-shaped structure over its linear analogue, α-hexylterthiophene **6**, was demonstrated: a similar device prepared with **6** as a donor showed reduced photovoltaic characteristics with a PCE of only 0.04% . In the same work terthiophenesilane **7a**, which has an *n*-hexylthiole instead of *n*-hexyl end-capping group, leading to a somewhat higher solubility, was reported.

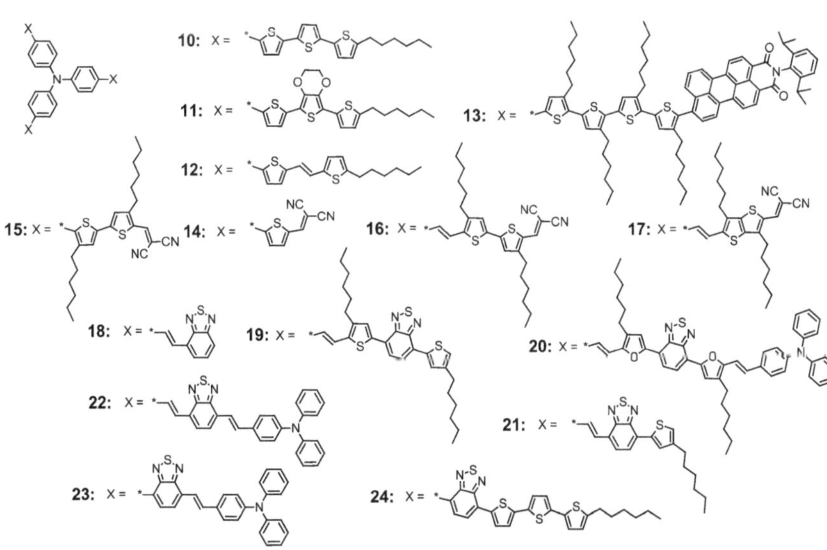

Fig. 2 Chemical structures of triphenylamine-based star-shaped molecules used in organic photovoltaic devices.

Table 2 Photovoltaic characteristics of OSCs prepared from triphenylamine-based star-shaped molecules

Donor	λ_{max} (abs) in solution/film	Band gap, eV	HOMO/LUMO, eV	Acceptor	Device type, D : A ratio	V_{oc}, V	I_{sc}, mA cm^{-2}	FF, %	PCE,[a] %	Ref.
10	429/440	2.55 (2.4[c])	−5.24[e]/—[b]	C_{60}	Bilayer	0.67	1.7	30	0.32	21
11	439/450	2.50 (2.4[c])	−5.18[e]/—	C_{60}	Bilayer	0.32	1.5	30	0.14	21
12	424/435	2.38	−5.33[e]/—	$PC_{61}BM$	BHJ, 1 : 3	0.60	2.43	28	0.41	22*a*
13	429, 495, 520/446, 490, 515	2.01 (1.48[d])	−5.04[e]/−3.36[e]	$PC_{61}BM$	BHJ, 1 : 4	0.60	1.4	29	0.25	23
14	368, 509/382, 538	1.78	−5.48[e]/—	C_{60}	Bilayer	0.96	3.65	29	1.02	22*a* and *d*
				C_{60}	Bilayer	1.15	4.59	28	1.85[f]	22*c*
15	296, 342, 504/312, 402, 563	1.83[c] (1.88[d])	−5.22/−3.34	$PC_{71}BM$	BHJ, 1 : 2	0.84	5.21	30.8	1.4	24*a*
16	298, 357, 518/320, 421, 585	1.65[c] (1.61[d])	−5.03/−3.42	$PC_{71}BM$	BHJ, 1 : 2	0.88	7.76	43.9	3.0	24*a*
17	429, 542/432, 580	1.77[c]	−5.13/−3.38	$PC_{61}BM$	BHJ, 1 : 2	0.9	4.66	38.8	1.63	24*b*
				$PC_{71}BM$	BHJ, 1 : 2	0.96	6.80	43.5	2.87	24*b*
18	320, 341, 460/324, 351, 486	2.14 (2.04)[d]	−5.41/−3.37	$PC_{61}BM$	BHJ, 1 : 3	0.89	1.66	0.41	0.61	25*a*
19	402, 544/414, 552	1.88	−5.30/−3.42	$PC_{71}BM$	BHJ, 1 : 3	0.75	6.41	39.0	1.90	25*b*
20	410, 560/416, 576	1.76	−5.14/−3.38	$PC_{71}BM$	BHJ, 1 : 3	0.71	5.41	32.0	1.23	25*b*
21	316, 361, 509/320, 370, 529	1.96 (2.11)[d]	−5.19/−3.08	$PC_{71}BM$	BHJ, 1 : 3	0.85	8.58	32.7	2.39	25*c*
22	375, 531/541	1.86	−5.3/−3.27	$PC_{61}BM$	BHJ, 1 : 3	0.81	4.18	39.0	1.33	25*d*
				$PC_{71}BM$	BHJ, 1 : 3	0.84	9.10	44.1	3.37	25*e*
23	328, 501/327, 509	2.05	−5.15/−2.99	$PC_{71}BM$	BHJ, 1 : 3	0.92	7.77	44.0	3.14	25*e* and *f*
24	328, 380, 512/338, 390, 538	1.9	−5.28/−3.11	$PC_{61}BM$	BHJ, 1 : 3	0.86	5.90	46.2	2.34	25*g*
				$PC_{71}BM$	BHJ, 1 : 2	0.87	9.51	52	4.3	25*g*

[a] Measured at AM 1.5G under an illumination intensity of 100 mW cm^{-2}, unless otherwise stated. [b] No data reported. [c] Estimated from the edge of the thin film absorption spectra. [d] Measured electrochemically. [e] Estimated in this paper from the CVA data given in the reference. [f] Measured at AM 1.5G under an illumination intensity of 80 mW cm^{-2}.

In combination with a $PC_{61}BM$ acceptor it was used as a donor in a solution-processed BHJ OSC device, which showed an improved J_{sc} of 1.13 mA cm^{-2}, but an only slightly higher PCE of 0.29% mainly due to a reduced fill factor to 24%.

Our group was able to synthesize soluble oligothiophenesilanes **7–9** with different conjugation lengths ranging from three to five conjugated thiophene rings.[19] Since solubility in this series of star-shaped molecules decreased with increasing conjugation length, it was necessary to use branched 2-ethylhexyl end-capping groups for quinquethiophenesilane **5** to make it soluble. The efficiency of solution-processible BHJ OSCs prepared from compounds **7–9** as donors and $PC_{71}BM$ as the acceptor improved with increasing conjugation length: from 0.6% for **7** to 1.4% for **9** (Table 1). A major improvement comes from the short circuit currents, which lay in the range 2.9–6.0 mA cm^{-2} with a clear tendency to increase for higher conjugation lengths of the donor, implying the improved efficiency of the light absorption for more conjugated molecules. Interestingly, the external quantum efficiency (EQE) reaches values above 20% in a broad wavelength range from 380 to 580 nm for quaterthiophenesilane **8** and above 30% in the region between 400 and 500 nm for quinquethiophenesilane **9**, *i.e.* in the region of efficient light absorbance of these two donor compounds, with a clear impart of the acceptor absorbance. Further improvements of the PCE are possible with donor molecules possessing a lower band gap than those of the oligothiophenes considered above.

A fruitful approach to the preparation of low band gap star-shaped molecules is based on the usage of triphenylamine (TPA) donor branching core and different accepting groups located in the arms of these molecules.[20] The first such molecules were reported by Roncali *et al.* (**10** and **11** in Fig. 2).[21] They contained α-hexylterthienyl or 3′,4′-ethylenedioxy-5″-hexyl-5-terthienyl groups, respectively. A bilayer OSC obtained by spin-coating of **10** or **11** donor material followed by vacuum evaporation of a C_{60} acceptor showed, however, low PCE values, 0.32 and 0.14% respectively (Table 2). BHJ OSCs prepared from spin-coating of a mixture of the TPA-based donor molecule **12** having three (2-thienylethen-2-yl)-2-thienyl groups and $PC_{61}BM$ acceptor showed similar photovoltaic behaviour with PCE = 0.41%.[22a] The similar TPA-based molecule **13** with pyrenylimide-quaterthienyl arms reported by Cremer and Bauerle,[23] albeit has a significantly lower optical band gap of 2.01 eV, in BHJ OSCs with a $PC_{61}BM$ acceptor showed a PCE of 0.25%. In fact, this molecule has several optical transitions, the shortest of which (at 429 nm) can be attributed to the conjugated quaterthiophene-phenylene system, while the longest (at 520 nm) to intramolecular charge transfer (ICT) from the TPA-donor group to a pyrenylimide acceptor group. A similar situation was identified for the TPA-molecule **14** with dicyanovinyl (DCN) acceptor groups by Roncali *et al.*[22] Initial efforts showed a PCE of 1.02% for a bilayer device with **14** as a donor and C_{60} as an acceptor.[22a,b] Further optimization of the device structure led to an increase in its efficiency to PCE = 1.85%.[22c] More elongated TPA–DCN molecules with bithienyl (**15**) or bithienylethenyl (**16**) conjugated spacers were reported by Jing Zhang *et al.*[24a] BHJ OSCs with **15** or **16** as donors and $PC_{71}BM$ as an acceptor showed a PCE of 1.4 and 3.0%, respectively. To make them soluble, 3-hexylthienyl instead of thienyl groups was used. The star-shaped oligomer **16** had a remarkable shift in the absorption edge of 78 nm compared to oligomer **15**, leading to a narrow band gap of 1.65 eV *vs.* 1.83 eV (Table 2). A combination of a good solubility (18 g L^{-1}) with a narrow band gap of **16** is the most probable factor, which

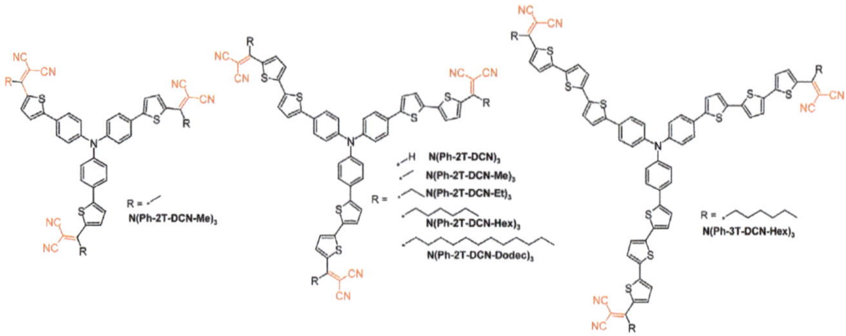

Fig. 3 Chemical structures of TPA-based star-shaped molecules with different conjugation length and both without alkyl groups **N(TPA-2T-DCN)$_3$** and with solubilizing alkyl chain ends **N(TPA-*n*T-DCN-Alk)$_3$**.

leads to a record PCE in this series of the star-shaped oligomers. Comparable results were obtained by the same group for the TPA star-shaped molecule **17** with thieno[3,2-*b*]thiophene–dicyanovinyl arms.[24b] BHJ OSC with **17** as the donor and $PC_{61}BM$ as the acceptor showed a PCE of 1.63%. Replacing $PC_{61}BM$ by $PC_{71}BM$ led to an increase in the PCE to 2.87% due to a wider absorption spectra of $PC_{71}BM$. The good solubility of **17** was governed by the presence of two hexyl groups at each thieno[3,2-*b*]thiophene unit.

One of the other strong donor groups is benzothiadiazole, which was also used in the design of TPA-based donor–acceptor molecules. Yongfang Li's group reported the synthesis of star-shaped molecule **18** with benzo[1,2,5]thiadiazol-vinylene arms.[25a] Its 1 : 3 mixture with $PC_{61}BM$ led to a BHJ OSC with a PCE of 0.61%. Elongation of these arms with two 3-hexylthienyl units (**19**) and further with ethynyltriphenylene (**20**) units led to lowering of the band gap from 2.04 to 1.88 and 1.76 eV, respectively (Table 2).[25b] BHJ OSCs prepared from **19** or **20** with a $PC_{71}BM$ acceptor led to an improved PCE of 1.90 and 1.23%, respectively. Attaching just a 3-hexylthyenyl unit to **18** led to molecule **21**, which in a mixture with a $PC_{71}BM$ acceptor showed a photovoltaic efficiency of 2.39%.[25c] Alternative elongation of benzo[1,2,5]thiadiazolvinylene arms with an ethylene-triphenylamine group (**22**) led to a BHJ OSC with a PCE of 1.33% using a $PC_{61}BM$ acceptor and with a PCE of 3.37% using a $PC_{71}BM$ acceptor.[25d,e] Direct linking of the benzothiadiazole-ethylene-triphenylamine arm to a TPA core in star-shaped molecule **23** allowed a PCE of 3.14%, using a $PC_{71}BM$ acceptor.[25f] Interestingly that simple incorporation of a benzothiadiazole unit between the TPA core and α-hexylterthiophene arm in **24** allowed further improvement of the photovoltaic properties in the BHJ OSC: to PCE = 2.34% using a $PC_{61}BM$ acceptor and to PCE = 4.3% using a $PC_{71}BM$ acceptor.[25g]

Thus, nowadays there is quite a lot of different TPA-based star-shaped molecules used in organic photovoltaic applications, the most efficient of which includes strong dicyanovinyl or benzothiadiazole units. However, a systematic investigation of the influence of their molecular structure on their photovoltaic performance has not yet been reported. Moreover, the photovoltaic properties of different molecules were measured in different device architectures, with different acceptors or donor : acceptor ratios, but this does not allow for their

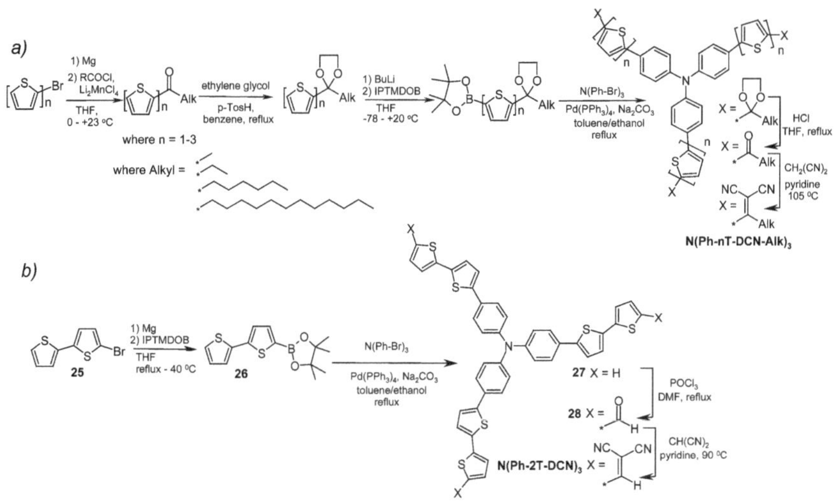

Fig. 4 Scheme for the synthesis of TPA-based star-shaped molecules with (a) and without (b) alkyl groups.

direct comparison. Below we compare a series of TPA-based star-shaped molecules with different conjugation lengths and solubilizing alkyl chain ends, **N(TPA-*n*T-DCN-Alk)$_3$** and the molecule without alkyl groups, **N(TPA-2T-DCN)$_3$** (Fig. 3). Their optical, electrochemical, photovoltaic and thermal properties, as well as bulk structure, measured under identical conditions, should allow for the elucidation of rules for the design of the most efficient structures of star-shaped molecules for organic photovoltaic applications.

Synthesis of star-shaped molecules with triphenylamine donor core and dicyanovinyl acceptor arms

The chemical structures of all TPA-based star-shaped molecules under consideration are presented in Fig. 3. A general scheme of the synthesis of the molecules with terminal alkyl chains, **N(TPA-*n*T-DCN-Alk)$_3$**, is outlined in Fig 4a. Their preparation consists of six steps.[26] The first step includes preparation of oligothienyl ketones with different numbers of conjugated thiophene rings by acylation of the magnesium derivatives of oligothienyl bromides with the corresponding alkanoyl chlorides. The second step requires preparation of the protected ketones in order to use them on the third stage of the reaction, which includes formation of their lithium derivatives followed by the reaction with 2-isopropoxy-4,4,5,5-tetramethyl-1,3,2-dioxaborolane (IPTMDOB) for the preparation of pinacol boronic esters. In the fourth step, a Suzuki cross-coupling reaction between the organoboron compounds prepared and tris(4-bromophenyl)amine is carried out to obtain the star-shaped precursors with protected ketone groups. Then the simple treatment of their solutions in THF with hydrochloric acid under reflux led to precipitation of the poorly soluble star-shaped ketones in high yields (95–99%). The successful substitution of three carbonyl groups with dicyanovinyl functions was required to elaborate and to optimize the conditions of the Knövenagel condensation. As a result, all TPA-based molecules, **N(TPA-*n*T-DCN-Alk)$_3$**, were synthesized in high reaction yield (90–95%) *via* the Knövenagel condensation between trifunctional ketones and an excess of malononitrile under microwave heating, using pyridine both as a base and as a solvent. Microwave heating for the last reaction step was found to decrease both the reaction time and the amount of by-products compared to conventional heating.

Synthesis of the star-shaped molecule without alkyl groups **N(TPA-2T-DCN)$_3$** was carried out in four steps using the synthetic approach developed earlier by Roncali *et al.* for similar molecules[22] (Fig. 4b). It should be noted, however, that synthesis of the same molecule *via* a scheme of protection/deprotection of carbonyl groups was done recently in 5 steps with 37% overall yield starting from 2,2′-bithiophene-5-carbaldehyde.[22b] In this work it was obtained in 52% overall yield by an alternative technique. In the first step 2-(2,2′-bithien-5-yl)-4,4,5,5-tetramethyl-1,3,2-dioxaborolane (**26**) was synthesized by the reaction between IPTMDOB and 2′-bithien-5-yl magnesium bromide, prepared *in situ* from 5-bromo-2,2′-bithiophene (**25**) and magnesium, in a 97% isolated yield. A Suzuki cross-coupling between **26** and tris(4-bromophenyl)amine gave star-shaped precursor **27** in a 90% yield. A Vilsmeier–Haack formylation of compound **27** led to tris[4-5-formyl-2,2′-bithiyenyl]amine **28** in a 90% yield. Finally, **N(Ph-2T-DCN)$_3$** was obtained in a 60% isolated yield by the Knövenagel condensation reaction of compound **28** with an excess of malononitrile in pyridine under microwave heating. The purity and molecular structure of all precursors and the final molecule was proved by ^{1}H-, and IR, spectroscopy, elemental analysis, and MALDI-TOF (see Experimental section and ESI, Fig. S1–S4†).

Optical and electrochemical properties of star-shaped molecules with triphenylamine donor core and dicyanovinyl acceptor arms

The optical properties of the seven TPA–DCN molecules synthesized were investigated both in dilute solutions and in thin films. Fig. 5 shows the UV-Vis absorption spectra of dilute solutions of these star-shaped small molecules in *o*-dichlorobenzene (ODCB) and thin films doctor bladed on quartz substrates. The corresponding optical data are summarized in Table 3. It should be noted that all of these compounds are soluble in polar organic solvents, like THF, chloroform, ODCB, *etc.* An estimation of their solubility was made by dissolving the

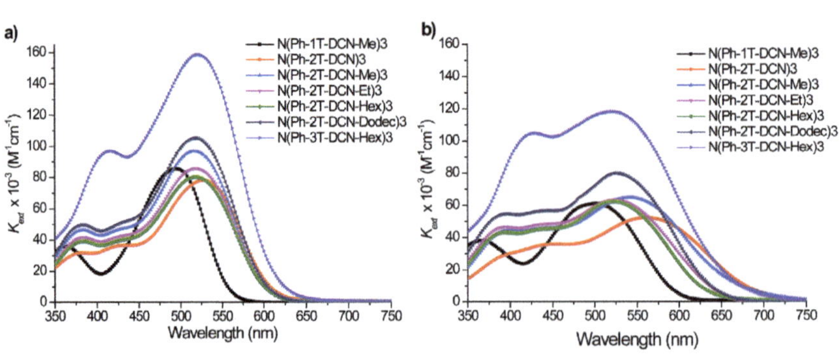

Fig. 5 Absorption spectra of TPA–DCN star-shaped molecules in ODCB solution (a) and in the film (b) shown as a distribution of their molar extinction coefficients, K_{ext}.

Table 3 Optical properties of the TPA–DCN star-shaped molecules in dilute ODCB solutions and thin films

Compound	Solution[a]					Film[b]			
	Solubility, $g\ L^{-1}$	λ_{max}, nm	K_{ext}, $M^{-1}\ cm^{-1}$	λ_{onset},[c] nm	E_g^{opt},[d] eV	λ_{max}, nm	K_{ext}, $M^{-1}\ cm^{-1}$	λ_{onset},[c] nm	E_g^{opt},[d] eV
N(Ph-1T-DCN-Me)$_3$	6	362 495	35 800 85 500	559	2.22	368 502	38 200 61 100	596	2.08
N(Ph-2T-DCN)$_3$	2	382 528	31 700 77 800	611	2.03	392 560	28 800 52 100	698	1.78
N(Ph-2T-DCN-Me)$_3$	3	384 517	46 300 96 800	602	2.06	400 541	44 800 64 700	690	1.80
N(Ph-2T-DCN-Et)$_3$	8	384 517	41 700 85 500	604	2.05	394 525	46 600 63 400	646	1.92
N(Ph-2T-DCN-Hex)$_3$	20	384 517	39 000 80 100	604	2.05	394 525	43 000 62 200	636	1.95
N(Ph-2T-DCN-Dodec)$_3$	52	382 517	49 600 105 000	603	2.06	396 525	54 700 79 700	641	1.93
N(Ph-3T-DCN-Hex)$_3$	12	416 522	96 800 158 000	615	2.02	428 522	105 000 118 000	668	1.85

[a] Measured in ODCB solution. [b] Cast from ODCB solution. [c] Estimated as the wavelength at which the absorption decreases to 10% from its maximum value. [d] Band gap estimated from the onset wavelength (λ_{onset}) of the optical absorption: $E_g^{opt} = 1240/\lambda_{onset}$.

compounds in ODCB (1 mL) until relevant molecules remained undissolved at room temperature and filtering the complete solution obtained through a 0.25 μm PTFE syringe filter. Immediately after filtration the solution was quickly and quantitatively transferred (500 μL) into another glass bottle of known mass, the solvent from the filtrate was evaporated and the residue was weighed. After the solubility of the relevant material was calculated. It was found that the solubility decreases with elongation of the arm conjugation length, when going from 1T to 2T and 3T-containing molecules, but it significantly improves with increasing the alkyl chain length or even adding methyl groups to the DCN groups. As a result, a reasonable compromise between the conjugation length and the alkyl chain end can be found to realize solution-processable star-shaped molecules.

Onsets of the absorption spectra of these molecules in solution (Fig. 5a) are clearly grouping in three locations, corresponding to the different conjugation lengths of the arms of the molecules: (1) at 560 nm for the 1T-containing molecule **N(Ph-1T-DCN-Me)$_3$**, (2) at *ca.* 603 nm for the 2T-containing molecules **N(Ph-2T-DCN-Alk)$_3$** and (3) at 615 nm for the 3T-containing molecule **N(Ph-3T-DCN-Hex)$_3$**. The absorption edge of the 2T-containing molecule without alkyl ends **N(Ph-2T-DCN)$_3$** lies at 611 nm, that is in between the second and the third cases, indicating a small influence of the alkyl substituents at dicyanovinyl groups on the optical properties. The absorption spectra of the five 2T-containing molecules in solution show similar absorption peaks and regions, exhibiting strong visible absorption peaks at *ca.* 382 nm (with molar extinction coefficients K_{ext} = 31 700–49 600 $M^{-1}\ cm^{-1}$) and *ca.* 517 nm (with 2–2.5 times higher K_{ext} = 77 800–105 000 $M^{-1}\ cm^{-1}$), as can be seen in Fig. 5a. The absorption peak at 382 nm in the short wavelength region corresponds to the π–π* transition. And the visible absorption peak at 517 nm can be attributed to the intramolecular charge-transfer (ICT) transition between the triphenylamine (TPA)-2T donor unit and the dicyanovinyl (DCN) acceptor unit. Compared to the five 2T-containing molecules, the absorption peaks of the 1T-containing **N(Ph-1T-DCN-Me)$_3$** are significantly blue-shifted to 363 nm and 495 nm. Their molar extinction coefficients are 35 800 M^{-1} cm^{-1} and 85 500 M^{-1} cm^{-1}, respectively. Probably, the shorter oligothiophene π-bridges linkage between TPA and DCN increases torsional interactions and reduces the strength of the ICT transition, which determines the band gap of the molecules under consideration. The optical band gaps of the 2T-containing molecules **N(Ph-2T-DCN-Alk)$_3$** in solution are 2.05–2.06 eV, it is somewhat smaller (2.03 eV) for **N(Ph-2T-DCN)$_3$**, while the corresponding band gap of **N(Ph-1T-DCN-Me)$_3$** is significantly larger, 2.22 eV. Considering these reasons, the shorter wavelength region peaks red shift from 363 nm/382 nm to 416 nm for **N(Ph-3T-DCN-Hex)$_3$**, which is attributed to the stronger localized π–π* transition arising from the longer oligothiophene π-bridge lengths. The peak at 416 nm has a remarkably high $K_{ext} = 96\ 800\ M^{-1}\ cm^{-1}$, while the ICT transition of this molecule at 522 nm has the highest $K_{ext} = 158\ 000\ M^{-1}\ cm^{-1}$ in the whole series of molecules under consideration. In addition, the similar absorption peaks for **N(Ph-3T-DCN-Hex)$_3$** in solution comparing with the 2T-contining molecules suggests that the longer oligothiophene π-bridge lengths of star-shaped molecules have no obvious influence on the ICT transition.

Comparative analysis of the solution and thin films' photophysical properties revealed that absorption spectra of these molecules become broader and their absorption peaks are significantly red-shifted, as shown in Fig. 5b and Table 3. Broadening the absorption peaks is accompanied by decreasing their absorption coefficients, especially of the long wavelength ICT transitions (at 502–560 nm) – they decrease by 1.3–1.5 times. However, changes in the short wavelength π–π* transitions are much smaller: their molar extinction coefficients K_{ext} changes just by 10%, and these changes can have either direction (increasing or decreasing in K_{ext}). Such changes in the K_{ext} make the absorption of these molecules in a thin film more smoothly distributed throughout the whole absorption spectra compared to their solution. The most significant changes can be seen for the onsets of the absorption spectra of these molecules in films (Fig. 5b). As in the case of solutions, they can also be grouped in three cases, albeit somewhat different than those in the solutions: (1) at *ca.* 600 nm for the 1T-containing molecule **N(Ph-1T-DCN-Me)$_3$**, (2) at *ca.* 630–650 nm for the 2T-containing molecules **N(Ph-2T-DCN-Alk)$_3$** with ethyl, hexyl and decyl groups and (3) at *ca.* 670–700 nm for the 2T-containing molecules **N(Ph-2T-DCN)$_3$** and **N(Ph-2T-DCN-Me)$_3$**, as well as for the 3T-containing molecule **N(Ph-3T-DCN-Hex)$_3$**. The absorption spectra of **N(Ph-2T-DCN)$_3$** is broader than those of **N(Ph-2T-DCN-Me)$_3$**, and even more than those of the other three 2T-containing molecules with the longer alkyl terminal chains. These results indicate that there are strong intermolecular interactions leading to aggregation in the **N(Ph-2T-DCN)$_3$** and **N(Ph-2T-DCN-Me)$_3$** films. It is likely that the long alkyl terminal chains could increase steric hindrance and torsional interactions between the DCN groups and 2T-bridges and shorten the conjugated length of each branch in the molecule. Interestingly, the absorption peak of **N(Ph-3T-DCN-Hex)$_3$** is negligibly shifted between solution and film. This may be attributed to the weak aggregation and intermolecular interactions of the molecules in the solid state. In addition, due to the decreased branch conjugation length and poor electronic communication between the TPA and DCN units, 1T-containing **N(Ph-1T-DCN-Me)$_3$** exhibits a high solid state optical band gap of 2.08 eV, while it is significantly lower (1.80 eV) for its 2T-homologue **N(Ph-2T-DCN-Me)$_3$** and 1.78 eV for 2T-containing star without alkyl groups **N(Ph-2T-DCN-Me)$_3$**. The other 2T-containing molecules with longer alkyl groups have somewhat larger optical band gaps of 1.92–1.95 eV in the solid state due to weaker interactions between the molecules. Finally, compared to the **N(Ph-2T-DCN-Hex)$_3$**, the absorption intensity of **N(Ph-3T-DCN-Hex)$_3$** in the shorter wavelength region is higher, illustrating that a longer oligothiophene π-bridge effectively releases the distortion induced by the hexyl terminal chains. Higher conjugation of the arms in **N(Ph-2T-DCN-Hex)$_3$** leads to a rather low optical band gap in the solid state (1.85 eV). It is noteworthy that optical band gaps in the solid state are *ca.* 0.10–0.26 eV smaller than those in the solution, indicating different aggregation of the molecules in the solid state.

The electrochemical properties of seven star-shaped molecules under consideration and their highest occupied molecular orbital (HOMO) and lowest unoccupied molecular orbital (LUMO) energy levels were examined by cyclic voltammetry (CVA). From the values of the standard formal reduction potentials (φ_{red}) and the standard formal oxidation potentials (φ_{ox}), the HOMO and LUMO energy levels of these compounds were calculated as presented in Table 3, according to the equations of LUMO $= -e(\varphi_{red} + 4.40)$ (eV) and HOMO $= -e(\varphi_{ox} + 4.40)$ (eV).[27] The results revealed that the measured LUMO energy levels of these seven molecules are very close, suggesting that the incorporation of various oligothiophene π-bridges and alkyl terminal chains has little effect on the LUMO level differences. However, it is interesting to note that the HOMO energy levels are significantly different due to the effect of the π-conjugated bridges with various electron donating characters.[28] Owing to the strong electron-donating terthiophene units as π-bridges, **N(Ph-3T-DCN-Hex)$_3$** exhibits a high HOMO energy level of −5.26 eV. And it is 0.34 eV higher than that of **N(Ph-1T-DCN-Hex)$_3$** (−5.60 eV) because of its relatively weak electron-donating thiophene bridges. In addition, the HOMO energy levels of the other four 2T-bridge based molecules

Table 4 Electrochemical properties of the TPA–DCN star-shaped molecules[a]

	Cyclic voltammetry data			Energy levels	
Compound	φ_{ox}, V	φ_{red}, V	E_g^{EC}, eV	HOMO, eV	LUMO, eV
N(Ph-1T-DCN-Me)$_3$	1.20	−0.94	2.14	−5.60	−3.46
N(Ph-2T-DCN)$_3$	0.77	—[b]	—	−5.17	−3.39[c]
N(Ph-2T-DCN-Me)$_3$	0.92	−0.99	1.91	−5.32	−3.41
N(Ph-2T-DCN-Et)$_3$	0.92	−1.00	1.92	−5.32	−3.40
N(Ph-2T-DCN-Hex)$_3$	0.94	−0.99	1.93	−5.34	−3.41
N(Ph-2T-DCN-Dodec)$_3$	0.94	−1.00	1.94	−5.34	−3.40
N(Ph-3T-DCN-Hex)$_3$	0.86	−1.00	1.86	−5.26	−3.40

[a] φ_{ox} – standard formal oxidation potential, φ_{red} – standard formal reduction potential, E_g^{EC} – electrical energy gap. [b] No reversible reduction can be observed. [c] Estimated as a difference between HOMO energy level and optical band gap in thin film.

Table 5 Thermal properties of the TPA–DCN star-shaped molecules[a]

Compound	T_g, °C	ΔC_p, J g^{-1} K^{-1}	T_m, °C	ΔH, J g^{-1}	Phase1	T1, °C	Phase2	T2, °C	Phase 3	T_d (air), °C	T_d (N_2), °C
N(Ph-1T-DCN-Me)$_3$	128	0.27	276	81.8	Crystal	128	Crystal	290	Iso[b]	401	402
N(Ph-2T-DCN)$_3$	146	0.29	276	27.8	Crystal	140	Crystal	276	Iso	405	405
N(Ph-2T-DCN-Me)$_3$	159	0.21	247	60.7	Amorphous crystal	247	Iso			416	416
N(Ph-2T-DCN-Et)$_3$	114	0.24	—	—	Amorphous					392	402
N(Ph-2T-DCN-Hex)$_3$	64	0.25	—	—	Amorphous					371	403
N(Ph-2T-DCN-Dodec)$_3$	31	0.26	—	—	Col[c]	100	Iso			370	402
N(Ph-3T-DCN-Hex)$_3$	72	0.29	—	—	Col	160	Iso			388	388

[a] T_g – glass transition, ΔC_p – heat capacity change at the glass transition, T_m – melting temperature, ΔH – melting enthalpy. [b] Isotropization temperature. [c] Col – columnar mesophase; T_d – decomposition temperature (at 5% weight-loss).

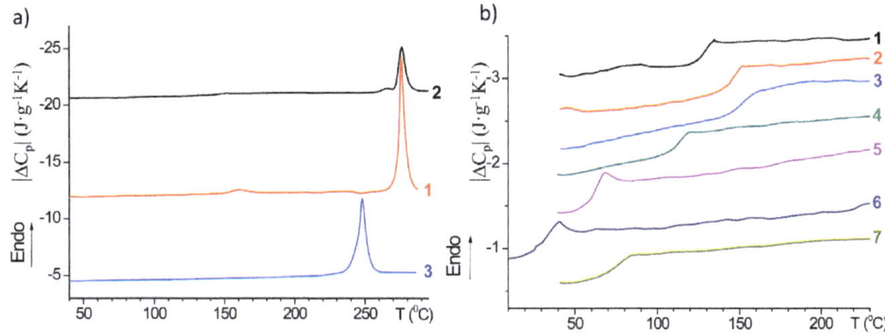

Fig. 6 DSC scans of the first (a) and the second (b) heating of (1) N(Ph-1T-DCN-Me)$_3$, (2) N(Ph-2T-DCN)$_3$, (3) crystalline fraction of N(Ph-2T-DCN-Me)$_3$, (4) N(Ph-2T-DCN-Et)$_3$, (5) N(Ph-2T-DCN-Hex)$_3$, (6) N(Ph-2T-DCN-Dodec)$_3$, and (7) N(Ph-3T-DCN-Hex)$_3$. For the sake of simplicity, scans are shifted along the heat capacity axis.

with various alkyl terminal chains are in between the HOMO energy levels of **N(Ph-1T-DCN-Me)$_3$** and **N(Ph-3T-DCN-Hex)$_3$** and similar, as shown in Table 4.

It should be noted that only the HOMO energy level of **N(Ph-2T-DCN)$_3$** can be estimated from the CVA measurements, since this compound shows a reversible formal oxidation potential at +0.77 V, but irreversible reduction. This gives the value of HOMO of this compound as −5.17 eV – the highest in the series of compounds under consideration. It is 0.15 eV higher as those for its methyl analogue **N(Ph-2T-DCN-Me)$_3$**, indicating less oxidation stability of the former, which can be attributed to the easy oxidation of the liable proton at the dicyanovinyl groups. A similar trend can be observed, when comparing HOMO energy levels of **N(Ph-1T-DCN)$_3$** (compound **14** in Table 2) and **N(Ph-1T-DCN-Me)$_3$**, which were estimated to be −5.48 and 5.60 eV, respectively, giving a difference of 0.12 eV. Moreover, for all alkyl-substituted stars **N(Ph-*n*T-DCN-Alk)$_3$** both oxidation and reduction potentials are reversible, while the reduction of their proton-containing analogs, **N(Ph-1T-DCN)$_3$** and **N(Ph-2T-DCN)$_3$**, was found to be irreversible not only in this work, but also in the literature.[22d] Thus, the presence of alkyl substituents at dicyanovonyl groups makes **N(Ph-*n*T-DCN-Alk)$_3$** stars not only more oxidation stable, but also makes their reduction reversible due to

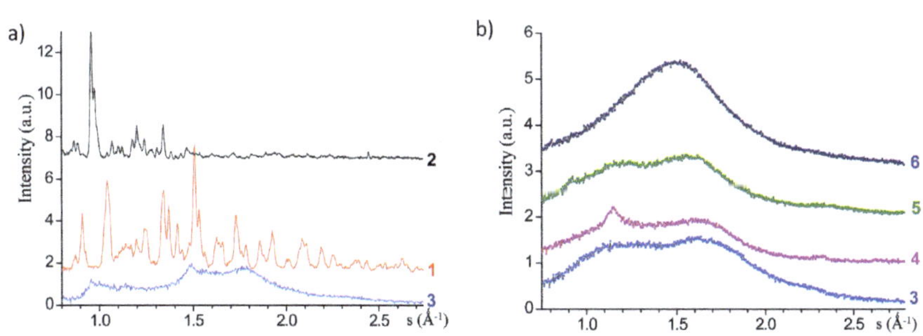

Fig. 7 WAXS patterns of (a) crystalline samples of (1) N(Ph-2T-DCN)$_3$, (2) N(Ph-1T-DCN-Me)$_3$ and (3) crystalline fraction of N(Ph-2T-DCN-Me)$_3$ and (b) amorphous samples of (3) N(Ph-2T-DCN-Me)$_3$, (4) N(Ph-2T-DCN-Et)$_3$, (5) N(Ph-2T-DCN-Hex)$_3$, and (6) N(Ph-2T-DCN-Dodec)$_3$. For the sake of simplicity, scans are shifted along the intensity axis.

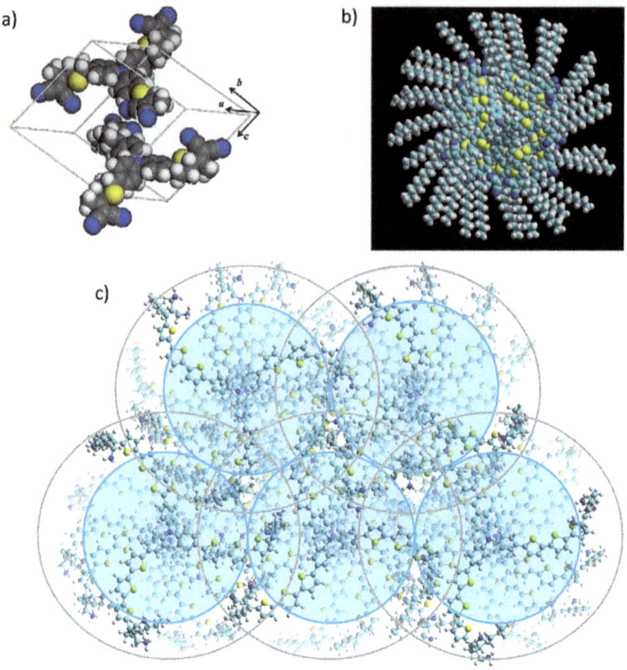

Fig. 8 (a) Packing of the molecules of compound **N(Ph-1T-DCN-Me)$_3$** in the triclinic crystal lattice with parameters $a = 11.47$ Å, $b = 18.17$ Å, $c = 18.30$ Å, and $\alpha = 98.3°$, $\beta = 52.2°$, $\gamma = 130.7°$. (b) Molecular modelling of the column formed by **N(Ph-2T-DCN-Dodec)$_3$** at low temperatures. (c) Molecular modelling of the packing of **N(Ph-3T-DCN-Hex)$_3$** molecules in a 2D hexagonal columnar mesophase. The blue solid circles correspond to 2D hexagonal columnar packing with a cylinder diameter $D = 23.7$ Å. The grey circles are the envelopes of molecular discs ($D_m = 30.6$ Å).

stabilization of anion-radicals by alkyl groups, that can positively influence the stability of organic photovoltaic devices made from these compounds.

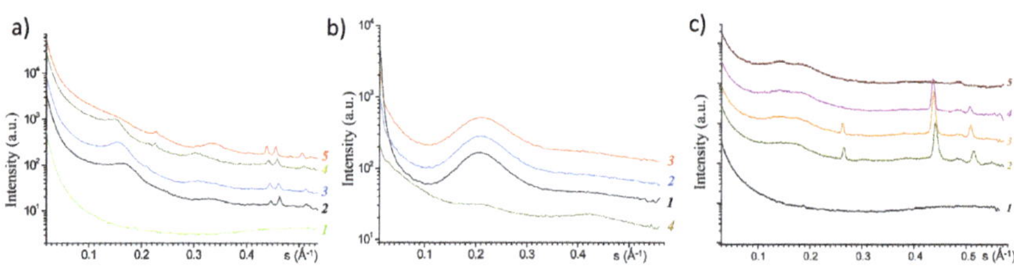

Fig. 9 SAXS patterns of (a) amorphous **N(Ph-2T-DCN-Me)$_3$** (1) and partially crystalline **N(Ph-2T-DCN)$_3$** (2–5) at room temperature (2), 100 °C (3), 180 °C (4), 250 °C (5); (b) forming columnar mesophase **N(Ph-2T-DCN-Dodec)$_3$** at room temperature (1, as-received), 40 °C (2), 60 °C (3), and at room temperature after heating to isotropic phase (4); (c) amorphous **N(Ph-2T-DCN-Me)$_3$** (1) and forming hexagonal columnar mesophase **N(Ph-3T-DCN-Hex)$_3$** (2–5) at room temperature (2), 120 °C (3), 140 °C (4), 160 °C (5). For the sake of simplicity, scans are shifted along the intensity axis.

Thermal properties and structure of star-shaped molecules with triphenylamine donor core and dicyanovinyl acceptor arms

Thermal stability, phase behaviour and structure of the series of TPA–DCN star-shaped molecules were investigated by TGA, DSC, and X-ray diffraction in small and wide angles (WAXS and SAXS). The results obtained are summarized in Table 5. TGA data (see Table 5 and Fig. S5†) indicate that all of these compounds are thermally stable and demonstrate a temperature of 5% decomposition (T_d) above 370 °C both under an inert atmosphere (under nitrogen) and under normal conditions (in the presence of oxygen and water vapours). The temperature of decomposition, T_d, is diminished with increasing alkyl chain length, which could be explained by the decomposition of the aliphatic tails.

DSC analysis revealed that both the methyl-containing molecules **N(Ph-1T-DCN-Me)$_3$** and **N(Ph-2T-DCN-Me)$_3$** as well as those without alkyl groups **N(Ph-2T-DCN)$_3$** can be crystalline at the first heating, but they become amorphous after melting, as can be seen on the cooling and the second heating scans (Fig. 6). Such behaviour was used for simplification of the dissolution of these molecules, since amorphous phases dissolve much faster and easier than the crystalline ones. Interestingly, the melting temperatures of **N(Ph-1T-DCN-Me)$_3$** and **N(Ph-2T-DCN)$_3$** are the same, 276 °C, but their melting enthalpies differ by more than three times: 81.1 *vs.* 27.8 J g^{-1}. However, it should be noted that the latter compound is only partly crystalline and shows both a glass transition and melting peak on the same first heating DSC curve. This can be explained by its low solubility, leading to fast solidification during solvent evaporation. Both the melting temperature and enthalpy of **N(Ph-2T-DCN-Me)$_3$** are lower compared to its 1T-containing analogue **N(Ph-1T-DCN-Me)$_3$**. Glass transition temperatures for the whole series of compounds investigated tend to increase when increasing the conjugation length of the molecule, but decrease when increasing the alkyl chain length, even though the appearance of a methyl group increases it compared to the non-alkyl analogue. This tendency is the reverse of the solubility trends discussed above. Heat capacity changes at the glass transition are quite close for all the compounds investigated and lies in the range of 0.21–0.29 J g^{-1} K^{-1}.

X-ray diffraction measurements confirmed the conclusions made from the DSC data. WAXS patterns of all three crystalline compounds – **N(Ph-1T-DCN-Me)$_3$**, **N(Ph-2T-DCN)$_3$**, and **N(Ph-2T-DCN-Me)$_3$** – reveal a number of narrow reflections, unequivocally confirming the presence of highly crystalline domains (Fig. 7a), while non-crystalline compounds showed only a diffuse halo (Fig. 7b). Crystal packing of the molecules of compound **N(Ph-1T-DCN-Me)$_3$** were recently resolved.[26c] It forms a triclinic crystal lattice with parameters $a = 11.47$ Å, $b = 18.17$ Å, $c = 18.30$ Å, and $\alpha = 98.3°$, $\beta = 52.2°$, $\gamma = 130.7°$ (Fig. 8a). Similar crystal packing can be expected for the other two crystalline star-shaped molecules, however their crystal lattice is not yet resolved. The temperature behaviour of **N(Ph-2T-DCN)$_3$** is presented in Fig. 9a. Glass transition manifests itself in the disappearance of the first of two diffuse halos and in the growth of a new reflection at $s = 0.227$ nm^{-1}.

The star-shaped compound with the longest alkyl group, **N(Ph-2T-DCN-Dodec)$_3$**, showed a changed shape of the amorphous halo compared to the other disordered compounds in Fig. 7b. Moreover, its SAXS patterns are characterized by a wide small-angle reflection corresponding to *d*-spacing of 30.3 Å (Fig. 9b, curves 1–3). With increasing temperature, the mesophase reflection of this compound shifts to wider angles, the d-spacing was decreased to 29.5 Å at 60 °C. This change corresponds to the thermal expansion coefficient $\beta = 6.6 \times 10^{-4}\ K^{-1}$ which is rather common for liquid crystalline materials and is due to the shrinking of alkyl end chains. With increasing temperature a transition to the isotropic phase is observed near 100 °C. Mesophase order was not restored on cooling. Unfortunately structural data is not enough for the unambiguous identification of the mesophase formed in the as-received samples of compound **N(Ph-2T-DCN-Dodec)$_3$**. However some remarks could be made. According to the results of molecular modeling, the radius of the molecular disc of **N(Ph-2T-DCN-Dodec)$_3$** is 26.5 Å provided the alkyl chains are in extended conformation, which coincides, with great precise, with the diameter of columns calculated from the observed small-angle reflection d-spacing. This supramolecular aggregate formed by twelve molecules is shown in Fig. 8b. Their arms are shifted relative to each other by several degrees providing π–π stacking between neighboring arms of different molecules forming a helical aggregate. However, as wide reflections as well as higher orders of a 10 columnar reflection are not observed, this interpretation is open for further studies.

In contrast to the crystalline **N(Ph-2T-DCN)$_3$**, and to the amorphous **N(Ph-2T-DCN-Hex)$_3$**, which does not reveal any reflections in the SAXS patterns (Fig. 9c, curve 1), **N(Ph-3T-DCN-Hex)$_3$** at room temperature is characterized by a set of three reflections with the ratio of corresponding d-spacings $d_1^2 : d_2^2 : d_3^2 = 1 : 3 : 4$ (Fig. 9c, curve 2). They can be indexed by a 2D hexagonal columnar phase with a cylinder diameter $D = 23.7$ Å. Moreover, a wide halo is observed at $s = 0.16$ Å^{-1}. Its origin relates to the Fourier transform of a single cylinder of columnar mesophase. With the temperature increasing, mesophase reflections of compound **N(Ph-3T-DCN-Hex)$_3$** shift to small angles, the diameter of the columns increased to 23.9 Å at 120 °C (Fig. 9c, curve 3). Such a change corresponds to the thermal expansion coefficient $\beta = 1.3 \times 10^{-4}\ K^{-1}$. As the diameter of the molecular envelope is 30.6 Å, substantial correlation between the relative positions of the molecules in neighbouring molecules should be suggested (Fig. 8c). There is no order in the packing of the molecules along the column axis (because of the absence of diffraction maxima on the wide-angle X-ray scattering patterns), as

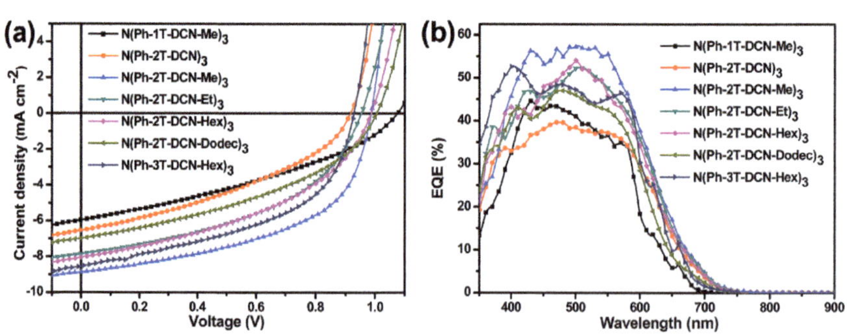

Fig. 10 Current density–voltage (J–V) curves (a) and EQE spectra (b) of solar cells based on **N(Ph-1T-DCN-Me)$_3$**:PC$_{70}$BM, **N(Ph-2T-DCN)$_3$**:PC$_{70}$BM, **N(Ph-2T-DCN-Me)$_3$**:PC$_{70}$BM, **N(Ph-2T-DCN-Et)$_3$**:PC$_{70}$BM, **N(Ph-2T-DCN-Hex)$_3$**:PC$_{70}$BM, **N(Ph-2T-DCN-Dodec)$_3$**:PC$_{70}$BM, and **N(Ph-3T-DCN-Hex)$_3$**:PC$_{70}$BM (1 : 2, wt%), respectively, under an illumination of AM 1.5, 100 mW cm^{-2}.

Table 6 Photovoltaic properties of star-shaped molecules: $PC_{71}BM$ (1 : 2, wt%) devices, under an illumination of AM 1.5, 100 mW cm^{-2}

Donor: $PC_{71}BM$	V_{oc}, V	J_{sc}, mA cm^{-2}	FF, %	PCE, %
N(Ph-1T-DCN-Me)$_3$	1.08	5.97	35.6	2.30
N(Ph-2T-DCN)$_3$	0.92	6.54	38.4	2.31
N(Ph-2T-DCN-Me)$_3$	0.98	8.87	54.3	4.72
N(Ph-2T-DCN-Et)$_3$	0.96	7.86	46.0	3.47
N(Ph-2T-DCN-Hex)$_3$	0.98	8.07	43.1	3.41
N(Ph-2T-DCN-Dodec)$_3$	1.00	7.00	41.0	2.87
N(Ph-3T-DCN-Hex)$_3$	0.93	8.51	50.0	3.96

well as no ordering between the neighbouring columns, summarizing that the pattern of molecular packing is irregular in each transverse section due to the steric necessity of mutual packing of long π-conjugated bridges, while column centres form 2D hexagonal lattice. Appearing steric limitations lead to the azimuthal irregularities in stacking of the molecules along the column, as every molecule should adjust its position not only with its top and bottom neighbours inside the cylinder, but with its side neighbours as well. The columnar mesophase formed in the samples of **N(Ph-3T-DCN-Hex)$_3$** looks to be very promising for directed charge transport along the columns as the oligothiophene fragments of neighbouring molecules are arranged to allow the best π–π stacking. However, an alignment of the columns is needed for the good semiconductor properties to manifest.

Thus, one can observe the tendency of the star-shaped compounds with short arms and short aliphatic tails to form crystalline structures. Moreover, it was shown before[29] that for the compound to form a liquid crystalline mesophase, optimal combination of the rigid core and soft side chains should exist, which is the case for **N(Ph-2T-DCN-Dodec)$_3$** and **N(Ph-3T-DCN-Hex)$_3$**, but not the case for **N(Ph-2T-DCN-Me)$_3$**, **N(Ph-3T-DCN-Et)$_3$** or **N(Ph-2T-DCN-Hex)$_3$**.

Photovoltaic properties of star-shaped molecules with a triphenylamine donor core and dicyanovinyl acceptor arms

To study the effects of the oligothiophene π-bridge length as well as alkyl terminal chain length on the photovoltaic properties of the star-shaped molecules, bulk heterojucntion organic solar cells (OSCs) with a configuration of ITO/PEDOT:PSS/TPA-based molecules: $PC_{70}BM$ (1 : 2, wt%)/Ca(15 nm)/Ag(80 nm) were fabricated. The best current density–voltage (*J*–*V*) curves are shown in Fig. 10a and the corresponding photovoltaic performance is summarized in Table 6. As expected, OSCs based on these small molecules show relatively high V_{oc} (>0.9 V), due to their low-lying HOMO energy levels. Encouragingly, OSCs based on **N(Ph-1T-DCN-Me)$_3$** with thiophene bridges exhibited a high V_{oc} up to 1.08 V. In contrast, OSCs based on **N(Ph-3T-DCN-Hex)$_3$** with terthiophene bridges showed a lower V_{oc} of 0.93 V, which are consistent with their HOMO level variations. In addition, the other four 2T-bridge based molecules with various alkyl terminal chains exhibited the expected V_{oc} of 0.98 V, as shown in Table 6. However, the device based on **N(Ph-2T-DCN)$_3$** without alkyl terminal chains showed a low V_{oc} of 0.92 V, which may be attributed to poor film quality. Following this reasoning, a significantly low J_{sc} of 6.54 mA cm^{-2} and a fill factor (FF) of 38.4% are observed for the **N(Ph-2T-DCN)$_3$**:PC$_{70}$BM device. In these seven molecule systems, the **N(Ph-2T-DCN-Me)$_3$**:PC$_{70}$BM device exhibited the highest power conversion efficiency (PCE) of 4.72% with a V_{oc} of 0.98 V, a J_{sc} of 8.87 mA cm^{-2}, and a FF of 54.3% under 100 mW cm^{-2} AM 1.5G illumination. When increasing the terminal alkyl chain length, the photovoltaic performance of the relevant 2T-bridge based systems gradually decreased, as shown in Table 6. These results suggest that the introduction of methyl terminal chains instead of the relatively long alkyl chains can enhance the intermolecular interactions, charge separation and transportation, charge carrier lifetime and consequently improving the photovoltaic performance. However, despite this the device based on **N(Ph-1T-DCN-Me)$_3$** with thiophene bridges exhibited a high V_{oc} of 1.08 V, the J_{sc} (5.97 mA cm^{-2}) and FF (35.6%) showing a significant drop due to the poor film quality and charge transport. In addition, compared to the **N(Ph-2T-DCN-Hex)$_3$**:PC$_{70}$BM device, the **N(Ph-3T-DCN-Hex)$_3$**:PC$_{70}$BM device showed a higher PCE of 3.96% with a slightly lower V_{oc} of 0.93 V. This suggests that subtly extending the π-bridge length in star-shaped molecules can represent an interesting method for the effective modulation of the device performance at a moderate cost. The external quantum efficiency (EQE) of the corresponding devices is shown in Fig. 10b. And the integrated EQE agrees well with the J_{sc} of the various devices mentioned above.

Experimental

General

^{1}H NMR spectra were recorded using a "Bruker WP-250 SY" spectrometer, working at a frequency of 250.13 MHz and utilising CDCl$_3$ signal (7.25 ppm) as the internal standard. In the case of ^{1}H NMR spectroscopy, the compounds to be analysed were taken in the form of 5% solutions in CDCl$_3$. The spectra were then processed on the computer using ACD Labs software.

Mass-spectra (MALDI) were registered on the Autoflex II Bruker (resolution FWHM 18000), equipped with a nitrogen laser (work wavelength 337 nm and time-of-flight mass-detector working in reflections mode). The accelerating voltage was 20 kV. Samples were applied to a polished stainless steel substrate. Spectra were recorded in the positive ion mode. The resulting spectrum was the sum of 300 spectra obtained at different points of the sample. 2,5-Dihydroxybenzoic acid (DHB) (Acros, 99%) and α-cyano-4-hydroxycinnamic acid (HCCA) (Acros, 99%) were used as matrices.

Elemental analysis of C, H, N elements was carried out using a CHN automatic analyzer CE 1106 (Italy). A settling titration using BaCl2 was applied to analyze sulphur. The spectrophotometry technique was used for Si analysis. The experimental error for the elemental analysis is 0.30–0.50%. The Knövenagel condensation was carried out in a microwave "Discovery", (CEM corporation, USA), using a standard method with the open vessel option, 50 watts.

Thermogravimetric analysis was carried out in dynamic mode in 30 ÷ 900 °C interval using the Mettler Toledo TG50 system equipped with a M3 microbalance allowing the measurement of the weight of samples in a 0–150 mg range with 1 μg precision. Heating/cooling rate was chosen to be 10 °C min^{-1}. Every compound was studied twice: in air and under a nitrogen flow of 200 mL min^{-1}. DSC scans were obtained with a Mettler Toledo DSC30 system with a 10 °C min^{-1} heating/cooling rate in the temperature range of +20 to +290 °C. A nitrogen flow of 50 mL min^{-1} was used.

Absorption profiles were recorded with a Perkin Elmer Lambda-35 absorption spectrometer from 350 to 1100 nm. The electrochemical properties were studied by cyclic voltammetry (CVA). The measurements were carried out in the 1,2-dichlorobenzene–acetonitrile (4 : 1) mixture of solvents using 0.1 M Bu_4NPF_6 as the supporting electrolyte. The glassy carbon electrode was used as the working electrode. Potentials were measured relative to a saturated calomel electrode.

X-ray diffraction. X-ray diffraction patterns in small and wide angle scattering regions were recorded by a S3-Micropix system (Hecus company), CuK_α-radiation, $\lambda = 1.5406$ Å with Xenocs Genix source (voltage and current – 50 kV and 1 mA respectively). A Pilatus 100K detector was employed, as well as a linear PSD 50 M gas detector (Ar–Me mixture at 8×10^5 Pa). A collimation system Fox 3D with Kratky collimation slits of 0.1 mm and 0.2 mm width was used, allowing stable measurements in a wave vector range from $s = 0.003$ Å^{-1} to $s = 1.9$ Å^{-1} where $s = 4\pi \sin \theta/\lambda$, 2θ is the scattering angle. To get rid of the scattering of X-Rays on air molecules, the Goebbel mirrors and scattering path are vacuumed at pressures 2.6 ÷ 5.0 Pa. Exposure times were from 600 to 5000 s providing the high reliability of scattering data. The temperature behavior of samples was studied using a Joule attachment in the temperature range of 23°C to 300°C. Transmission X-ray diffraction patterns at wide angles were also recorded using a Bruker D8 Advance powder diffractometer (CuK_α-radiation, $\lambda = 1.5406$ Å, Vantec 2D detector). The Accelrys Materials Studio® program set was employed for molecular modeling of the compounds studied. We used two sets of potentials, which allow taking into account non-covalent interactions: COMPASS (Condensed-phase Optimized Molecular Potentials for Atomistic Simulation Studies) and UFF (Universal Force Field). The COMPASS set is suitable for modeling isolated molecules and condensed phases of mainly organic, polymeric and of some inorganic compounds,[30] it also allows parameterization of the partial charges and valency *ab initio* with subsequent system optimization.

Fabrication and characterization of the OSCs. All of the devices were fabricated in the normal architecture. Photovoltaic devices were fabricated by doctor-blading on indium tin oxide (ITO)-covered glass substrates (from Osram). These substrates were cleaned in toluene, water, acetone, and isopropyl alcohol. After drying, the substrates were bladed with 50 nm PEDOT:PSS (HC Starck, PEDOT PH-4083). Photovoltaic layers, consisting of three different small molecules and $PC_{71}BM$ in 1 : 2 wt% ratios were dissolved at different concentrations in ODCB and bladed on top of the PEDOT:PSS layer. Finally, a calcium/silver top electrode of 15/80 nm thickness was evaporated. The typical active area of the investigated devices was 10.4 mm^2. The current–voltage characteristics of the solar cells were measured under AM1.5G irradiation on an OrielSol 1A Solar simulator (100 mW cm^{-2}). The EQE was detected with a Cary 500 Scan UV-Vis-NIR Spectrophotometer under monochromatic illumination, which was calibrated with a mono-crystalline silicon diode. Absorption profiles were recorded with a Perkin Elmer Lambda-35 absorption spectrometer from 350 to 1100 nm.

Materials

Tetrakis(triphenylphosphine)palladium(0)$Pd(PPh_3)_4$, magnesium, malononitrile, phosphorus(v) oxychloride, isopropoxy-4,4,5,5-tetramethyl-1,3,2-dioxaborolane (IPTMDOB), tris(4-bromophenyl)amine were obtained from Sigma-Aldrich Co. and used without further purification. 1,2-Dichloroethane, THF, DMF, pyridine were dried and purified according to the known techniques and then used as solvents.

Synthesis

N(Ph-1T-DCN-Me)$_3$, **N(Ph-2T-DCN-Me)$_3$**, **N(Ph-2T-DCN-Et)$_3$**, **N(Ph-2T-DCN-Hex)$_3$**, **N(Ph-2T-DCN-Dec)$_3$** and **N(Ph-2T-DCN-Hex)$_3$** were synthesized as described elsewhere.[26]

2-(2,2′-Bithien-5-yl)-4,4,5,5-tetramethyl-1,3,2-dioxaborolane (26). A solution of 5-bromo-2,2′-bithiophene (2.2 g, 9 mmol) in 22 mL of THF was added dropwise to a suspension of magnesium (0.27 g, 9.4 mmol) in 2 mL of THF. The Grignard reagent was refluxed for 2 h, then cooled to room temperature and 2-isopropoxy-4,4,5,5-tetramethyl-1,3,2-dioxaborolane (1.67 g, 9 mmol) was added in one portion. The reaction mixture was stirred for 1 h at room temperature, followed by stirring for 1 h at 40 °C. After completion of the reaction 75 mL of freshly distilled diethyl ether and 100 mL of distilled water and 9 mL of 1 M HCl were added to the reaction mixture. The organic phase was separated, washed with water, and dried over sodium sulfate and filtered. The solvent was evaporated to give 2.47 g (95%) of the product (purity was also 99% according to GPC and ^{1}H NMR) as a blue oil. The product was used in the subsequent synthesis without further purification. ^{1}H NMR (250 MHz, $CDCl_3$): 1.34 (s, 12H), 6.98–7.04 (m, 1H, $M = 4, J = 9.1$ Hz), 7.21 (m, 3H, $M = 4, J = 6.1$ Hz), 7.51 (d, 1H, $J = 3.7$ Hz). Calcd (%) for $C_{14}H_{17}BO_2S_2$: C, 57.54; H, 5.86; S, 21.94. Found: C, 56.91; H, 5.72; S, 21.83. MALDI-MS: found *m*/*z* 292.59; calculated for $[M]^+$ 292.07.

Tris[4-(2,2′-bithien-5-yl)phenyl]amine (27). In an inert atmosphere, degassed solutions of tris(4-bromophenyl)amine (1.08 g, 2.24 mmol) and 2-(2,2′-bithien-5-yl)-4,4,5,5-tetramethyl-1,3,2-dioxaborolane (**1**) (2.36 g, 8.07 mmol) in toluene–ethanol mixture (40/4 mL) and 2 M solution of aq. Na_2CO_3 (12.1 mL) were added to $Pd(PPh_3)_4$ (280 mg, 0.24 mmol). The reaction mixture was stirred under reflux for 9 h, and then it was cooled to room temperature and the organic phase was separated and the solvent was evaporated in vacuum and the residue was dried at 1 Torr. The product was passed through silica gel column (eluent hot toluene, 70 °C) and recrystallized from toluene to give pure compound **2** (1.48 g, 90%) as a yellow solid. M.p.: 209–210 °C. ^{1}H NMR (250 MHz, $CDCl_3$): δ [ppm] 7.01 (dd, 3H, $J_1 = 3.7$ Hz, $J_2 = 3.7$ Hz), 7.12 (d, 6H, $J = 4.3$ Hz), 7.14–7.23 (overlapping peaks, 12H), 7.49 (d, 3H, $J = 8.7$ Hz). Calcd (%) for $C_{42}H_{27}NS_6$: C, 68.35; H, 3.69; N, 1.90; S, 26.07. Found: C, 67.88; H, 3.77; N, 1.82; S, 26.03. MALDI-MS: found *m*/*z* 735.45; calculated for $[M]^+$ 737.04.

Tris[4-(5-formyl-2,2′-bithiophen-5-yl)phenyl]amine (28). Tris[4-(2,2′-bithien-5-yl)phenyl]amine (**2**) (760 mg, 1.03 mmol) is dissolved in 50 mL of 1,2-dichloroethane. DMF (753 mg, 10.3 mmol) and $POCl_3$ (1579 mg, 10.3 mmol) are added dropwise and the mixture is refluxed for 13 h under nitrogen. After cooling to room temperature and addition of 50 mL of dichloromethane and 150 mL of a saturated aqueous solution of sodium acetate, the mixture is stirred for 2 h at room temperature. The organic phase is then washed with water and the organic phase was separated and the solvent was evaporated in vacuum and the residue was dried at 1 Torr. The product was recrystallized from toluene to give 760 mg (90%) of an orange solid. M.p.: 145–146 °C. ^{1}H NMR (250 MHz, $CDCl_3$): δ [ppm] 7.14 (d, 6H, $J = 8.7$ Hz), 7.21 (d, 3H, $J = 3.7$ Hz), 7.32 (d, 3H, $J = 4.3$ Hz), 7.51 (d, 3H, $J = 8.7$ Hz), 7.66 (d, 6H, $J = 4.3$ Hz), 9.85 (s, 3H). IR (KBr) ν (cm^{-1}): 1660 (C=O). Calcd (%) for $C_{45}H_{27}NS_6$: C, 65.75; H, 3.31; N, 1.70; S, 23.40. Found: C, 65.88; H, 3.27; N, 1.66; S, 23.30. MALDI-MS: found *m*/*z* 820.25; calculated for $[M]^+$ 821.03.

2,2′,2″-[Nitrilotris(4,1-phenylene-2,2′-bithiene-5′,5-diylmethylylidene)]trimalononitrile (N(Ph-2T-DCN)$_3$). Compound **3** (690 mg, 0.8 mmol), malononitrile (273.2 mg, 4.1 mmol) and dry pyridine (15 mL) were placed in a reaction vessel and stirred under an argon atmosphere for 7 hours at 90°C using microwave heating. After completeness of the reaction the pyridine was evaporated in vacuum and the residue was dried at 1 Torr. This crude product was purified by preparative GPC chromatography. Further purification included precipitation of the product from its THF solution with toluene and hexane to give the pure product as a black solid (480 mg, 60%). ^{1}H NMR (250 MHz, DMSO-d6): δ [ppm] 7.06 (d, 6H, $J = 3.7$ Hz), 7.52 (d, 3H, $J = 3.7$ Hz), 7.58 (d, 3H, $J = 3.7$ Hz), 7.65 (m, 9H, $M = 3$, $J = 7.9$ Hz), 7.85 (d, 3H, $J = 3.7$ Hz), 8.59 (s, 3H). IR (KBr) ν (cm^{-1}): 2220 (C≡N). Calcd (%) for $C_{54}H_{27}N_7S_6$: C, 67.13; H, 2.82; N, 10.15; S, 19.91. Found: C, 66.98; H, 3.09; N, 9.78; S, 19.78. MALDI-MS: found *m*/*z* 963.41; calculated for $[M]^+$ 965.06.

Conclusions and outlook

Star-shaped oligomers present an emerging class of small molecules for efficient solution-processable organic photovoltaics. The most prospective approach for the design of such molecules is based on the "weak donor – strong acceptor" concept, similar to those used in the synthesis of low band gap copolymers.[15] In the case of star-shaped molecules the most successful weak donor is triphenylamine, which serves also as a convenient branching centre. Successful strong acceptors include benzo[1,2,5]thiadiazol and dicyanovinyl groups, which can be attached to the triphenylamine core either directly or *via* conjugated spacers from 2,5-thiphene or 1,2-vinylene units or their combination. Dicyanovinyl (DCN) groups seem to be more synthetically viable, however, typically they contain a liable imide proton –HC=C(DCN), originating from the aldehyde precursors, which can be easily oxidized. To avoid this drawback, we have developed a convenient synthetic approach, which allows using a ketone precursor in the Knövenagel condensation, to form stable alkyl derivatives of the dicyanovinyl groups.

Thus, alkyl groups, used also for increasing the solubility of the small molecules, can influence their stability, morphology and structure in the bulk. It was found that adding methyl groups to dicyanovinyl acceptor units of the TPA-based star-shaped molecules decreases the melting points, but increases the melting enthalpy. These findings indicate that methyl end-capped molecules have stronger interactions in the bulk structures, leading to improved charge carrier mobilities and better photovoltaic performance. Comparison of **N(Ph-2T-DCN-Me)$_3$** and **N(Ph-2T-DCN)$_3$** shows that the presence of methyl groups in the molecule can lead to more than a 2 times increase in PCE:

from 2.31 to 4.76%, when the BHJ OSCs were made under identical conditions. Such a strong effect is a result of improving all the parameters influencing the PCE: open circuit voltage increases from 0.92 to 0.98 V, short circuit current – from 6.54 to 8.87 mA cm^{-2}, and fill factor – from 38.4 to 54.3%. Enlarging the open circuit voltage can be related to lowering the HOMO energy level of the donor star-shaped molecule by 0.15 eV – a side effect of the improved oxidation stability. An improved short circuit current can be explained by higher crystallinity and hence improved charge carrier mobility. Improving the fill factor can be related to a better morphology of the bulk heterojunction in the case of methyl end-capped molecules.

Increasing the alkyl chain length leads to suppressing the crystallinity and lowering the glass transition temperatures of the molecules. This fact positively influences the solubility, but decreases the short circuit current and fill factors in BHJ OSCs, thus lowering the overall PCE to 3.47–2.87%. Nevertheless, the ethyl group in **N(Ph-2T-DCN-Et)$_3$** seems to be a good compromise between the crystallinity and the solubility. This star-shaped compound is soluble not only in the chlorinated solvents, like most conjugated oligomers and polymers, but also in more environmentally friendly solvents, like benzaldehyde, that allow the preparation of BHJ OSCs from green solvents with PCE = 3.6%.[31a] Moreover, this compound shows good PCE values up to 3.9% in the inverted structures of BHJ OSCs, which are more stable under normal conditions compared to the the standard ones.[31b] Finally, interface engineering of the devices based on this donor molecule and $PC_{71}BM$ acceptor improves the PCE up to 5.4%.[31c]

Further increasing the alkyl chain length led to a very interesting effect – **N(Ph-2T-DCN-Dodec)$_3$** forms a columnar ordering of the molecules. If it were not long dielectric dodecyl groups, it would lead to a record high PCE. Even stronger ordering was observed for **N(Ph-3T-DCN-Hex)$_3$** molecules – they form a 2D hexagonal columnar mesophase. This factor definitely influences the performance of the BHJ OSC based on this molecule, for which the PCE approaches 4%.

Further improvements in the efficiency of the BHJ OSCs based on star-shaped molecules are foreseen due to the design of more flattened TPA–DCN molecules capable of forming an ordered 2D hexagonal columnar mesophase, and ordering of these phases in the devices. In addition, improved interfaces in BHJ OSCs of **N(Ph-2T-DCN-Et)$_3$** (ref. 31c) could lead to even higher performance for its methyl analogue **N(Ph-2T-DCN-Me)$_3$**.

Acknowledgements

The authors would like to thank Dr P. V. Dmitryakov (NBICS center of Kurchatov Institute, Moscow, Russia) for DSC and TGA measurements and S. M. Peregudova (Institute of Organoelement Compounds of RAS, Moscow, Russia) for CVA measurements. This part of this work including the synthesis and characterization of the oligomers was carried out under financial support from the Russian Science Foundation (grant 14-13-01380). The preparation and characterization of the photovoltaic devices was funded by the Sonderforschungsbereich 953 "Synthetic Carbon Allotropes", the "Solar Technologies go Hybrid (SolTech)", and the China Scholarship Council (CSC).

Notes and references

1 R. Søndergaard, M. Hösel, D. Angmo, T. T. Larsen-Olsen and F. C. Krebs, *Mater. Today*, 2012, **15**, 36–49.
2 Y. G. Liang and L. P. Yu, *Polym. Rev.*, 2010, **50**, 454–473.
3 (*a*) A. Mishra and P. Bauerle, *Angew. Chem., Int. Ed.*, 2012, **51**, 2020–2067; (*b*) J. Roncali, P. Leriche and P. Blanchard, *Adv. Mater.*, 2014, **26**, 3821–3838.
4 C. J. Brabec, N. S. Sariciftci and J. C. Hummelen, *Adv. Funct. Mater.*, 2001, **11**, 15–26.
5 A. J. Heeger, *Adv. Mater.*, 2014, **26**, 10–28.
6 J. Peet, A. J. Heeger and G. C. Bazan, *Acc. Chem. Res.*, 2009, **42**, 1700–1708.
7 Y. Huang, E. J. Kramer, A. J. Heeger and G. C. Bazan, *Chem. Rev.*, 2014, **114**, 7006–7043.
8 P. A. Troshin, D. K. Susarova, E. A. Khakina, A. A. Goryachev, O. V. Borshchev, S. A. Ponomarenko, V. F. Razumov and N. S. Sariciftci, *J. Mater. Chem.*, 2012, **22**, 18433–18441.
9 (*a*) J. Peet, J. Y. Kim, N. E. Coates, W. L. Ma, D. Moses, A. J. Heeger and G. C. Bazan, *Nat. Mater.*, 2007, **6**, 497–500; (*b*) T.-Y. Chu, J. P. Lu, S. Beaupre, Y. G. Zhang, J.-R. Pouliot, S. Wakim, J. Y. Zhou, M. Leclerc, Z. Li, J. F. Ding and Y. Tao, *J. Am. Chem. Soc.*, 2011, **133**, 4250–4253; (*c*) J. A. Love, I. Nagao, Y. Huang, M. Kuik, V. Gupta, C. J. Takacs, J. E. Coughlin, L. Qi, T. S. van der Poll, E. J. Kramer, A. J. Heeger, T.-Q. Nguyen and G. C. Bazan, *J. Am. Chem. Soc.*, 2014, **136**, 3597–3606.
10 H.-C. Liao, C.-C. Ho, C.-Y. Chang, M.-H. Jao, S. B. Darling and W.-F. Su, *Mater. Today*, 2013, **16**, 326–336.
11 H.-L. Yip and A. K.-Y. Jen, *Energy Environ. Sci.*, 2012, **5**, 5994–6011.
12 (*a*) M. C. Scharber, D. Muehlbacher, M. Koppe, P. Denk, C. Waldauf, A. J. Heeger and C. J. Brabec, *Adv. Mater.*, 2006, **18**, 789–794; (*b*) G. Dennler, M. C. Scharber, T. Ameri, P. Denk, K. Forberich, C. Waldauf and C. J. Brabec, *Adv. Mater.*, 2008, **20**, 579–583.
13 (*a*) Y. G. Liang, Z. Xu, J. B. Xia, S.-T. Tsai, Y. Wu, G. Li, C. Ray and L. P. Yu, *Adv. Mater.*, 2010, **22**, 135–138; (*b*) H. J. Son, W. Wang, T. Xu, Y. G. Liang, Y. Wu, G. Li and L. P. Yu, *J. Am. Chem. Soc.*, 2011, **133**, 1885–1894; (*c*) J. Zhou, Y. Zuo, X. Wan, G. Long, Q. Zhang, W. Ni, Y. Liu, Z. Li, G. He, C. Li, B. Kan, M. Li and Y. Chen, *J. Am. Chem. Soc.*, 2013, **134**, 8484–8487; (*d*) Z. A. Tan, S. S. Li, F. Z. Wang, D. P. Qian, J. Lin, J. H. Hou and Y. F. Li, *Sci. Rep.*, 2014, **4**, 4691, DOI: 10.1038/srep04691.
14 (*a*) J. B. You, L. T. Dou, K. Yoshimura, T. Kato, K. Ohya, T. Moriarty, K. Emery, C.-C. Chen, J. Gao, G. Li and Y. Yang, *Nat. Commun.*, 2013, **4**, 1446, DOI: 10.1038/ncomms2411; (*b*) Y. S. Liu, C.-C. Chen, Z. Hong, J. Gao, Y. M. Yang, H. P. Zhou, L. Dou, G. Li and Y. Yang, *Sci. Rep.*, 2013, **3**, 3356, DOI: 10.1038/srep03356; (*c*) J. B. You, C.-C. Chen, Z. Hong, K. Yoshimura, K. Ohya, R. Xu, S. L. Ye, J. Gao, G. Li and Y. Yang, *Adv. Mater.*, 2013, **25**, 3973–3978.
15 H. X. Zhou, L. Q. Yang and W. You, *Macromolecules*, 2012, **45**, 607–632.
16 R. de Bettignies, Y. Nicolas, P. Blanchard, E. Levillain, J. M. Nunzi and J. Roncali, *Adv. Mater.*, 2003, **15**, 1939–1943.
17 (*a*) S. A. Ponomarenko, E. A. Tatarinova, A. M. Muzafarov, S. Kirchmeyer, L. Brassat, A. Mourran, M. Moeller, S. Setayesh and D. de Leeuw, *Chem.*

Mater., 2006, **18**, 4101–4108; (*b*) P. A. Troshin, S. A. Ponomarenko, Y. N. Luponosov, E. A. Khakina, M. Egginger, T. Meyer-Friedrichsen, A. Elschner, S. M. Peregudova, M. I. Buzin, V. F. Razumov, N. S. Sariciftci and A. M. Muzafarov, *Sol. Energy Mater. Sol. Cells*, 2010, **94**, 2064–2072; (*c*) D. A. Ivanov, D. V. Anokhin, M. Defaux, A. Mourran, M. Möller, Y. N. Luponosov, O. V. Borshchev, A. V. Bakirov, M. A. Shcherbina, S. N. Chvalun, T. Meyer-Friedrichsen, A. Elschner, S. Kirchmeyer and S. A. Ponomarenko, *J. Phys. Chem. C*, 2012, **116**, 22727–22736.

18 S. Roquet, R. Bettignies, P. Leriche and J. Roncali, *J. Mater. Chem.*, 2006, **16**, 3040–3045.

19 E. A. Kleimyuk, P. A. Troshin, E. A. Khakina, Y. N. Luponosov, Y. L. Moskvin, S. M. Peregudova, S. D. Babenko, T. Meyer-Friedrichsen and S. A. Ponomarenko, *Energy Environ. Sci.*, 2010, **3**, 1941–1948.

20 Y. N. Luponosov, A. N. Solodukhin and S. A. Ponomarenko, *Polym. Sci., Ser. C*, 2014, **56**, 105–135.

21 A. Cravino, S. Roquet, O. Aleveque, P. Leriche, P. Frere and J. Roncali, *Chem. Mater.*, 2006, **18**, 2584–2590.

22 (*a*) S. Roquet, A. Cravino, P. Leriche, O. Aleveque, P. Frere and J. Roncali, *J. Am. Chem. Soc.*, 2006, **128**, 3459–3466; (*b*) A. Cravino, S. Roquet, P. Leriche, O. Aleveque, P. Frere and J. Roncali, *Chem. Commun.*, 2006, 1416–1418; (*c*) A. Cravino, P. Leriche, O. Aleveque, S. Roquet and J. Roncali, *Adv. Mater.*, 2006, **18**, 3033–3037; (*d*) E. Ripaud, Y. Olivier, P. Leriche, J. Cornil and J. Roncali, *J. Phys. Chem. B*, 2011, **115**, 9379–9386.

23 J. Cremer and P. Bauerle, *J. Mater. Chem.*, 2006, **16**, 874–884.

24 (*a*) J. Zhang, D. Deng, C. He, Y. He, M. Zhang, Z. G. Zhang, Z. Zhang and Y. Li, *Chem. Mater.*, 2011, **23**, 817–822; (*b*) D. Deng, S. L. Shen, J. Zhang, C. He, Z. J. Zhang and Y. F. Li, *Org. Electron.*, 2012, **13**, 2546–2552.

25 (*a*) G. L. Wu, G. J. Zhao, C. He, J. Zhang, Q. G. He, X. M. Chen and Y. F. Li, *Sol. Energy Mater. Sol. Cells*, 2009, **93**, 108–113; (*b*) J. Zhang, J. T. Yu, C. He, D. Deng, Z.-G. Zhang, M. J. Zhang, Z. B. Li and Y. F. Li, *Org. Electron.*, 2012, **13**, 166–172; (*c*) J. Zhang, Y. Yang, C. He, Y. J. He, G. J. Zhao and Y. F. Li, *Macromolecules*, 2009, **42**, 7619–7622; (*d*) C. He, Q. G. He, Y. P. Yi, G. L. Wu, F. L. Bai, Z. G. Shuai and Y. F. Li, *J. Mater. Chem.*, 2008, **18**, 4085–4090; (*e*) J. Zhang, Y. Yang, C. He, D. Deng, Z. B. Li and Y. F. Li, *J. Phys. D: Appl. Phys.*, 2011, **44**, 475101; (*f*) Y. Yang, Y. Zhou, Q. G. He, C. He, C. H. Yang, F. L. Bai and Y. F. Li, *J. Phys. Chem. B*, 2009, **113**, 7745–7752; (*g*) H. X. Shang, H. J. Fan, Y. Liu, W. P. Hu, Y. F. Li and X. W. Zhan, *Adv. Mater.*, 2011, **23**, 1554–1557.

26 (*a*) J. Min, Y. N. Luponosov, T. Ameri, A. Elschner, S. M. Peregudova, D. Baran, T. Heumuller, N. Li, F. Machui, S. Ponomarenko and C. J. Brabec, *Org. Electron.*, 2013, **14**, 219–229; (*b*) J. Min, Y. N. Luponosov, A. Gerl, M. S. Polinskaya, S. M. Peregudova, P. V. Dmitryakov, A. V. Bakirov, M. A. Shcherbina, S. N. Chvalun, S. Grigorian, N. Kaush-Busies, S. A. Ponomarenko, T. Ameri and C. J. Brabec, *Adv. Energy Mater.*, 2014, **4**, 1301234; (*c*) J. Min, Y. N. Luponosov, D. Baran, S. N. Chvalun, M. A. Shcherbina, A. V. Bakirov, P. V. Dmitryakov, S. M. Peregudova, N. Kausch-Busies, S. A. Ponomarenko, T. Ameri and C. J. Brabec, *J. Mater. Chem. A*, 2014, **2**, 16135–16147.

27 (*a*) S. A. Ponomarenko, S. Kirchmeyer, A. Elschner, N. M. Alpatova, M. Halik, H. Klauk, U. Zschieschang and G. Schmid, *Chem. Mater.*, 2006, **18**, 579–586;

(*b*) S. A. Ponomarenko, N. N. Rasulova, Y. N. Luponosov, N. M. Surin, M. I. Buzin, I. Leshchiner, S. M. Peregudova and A. M. Muzfarov, *Macromolecules*, 2012, **45**, 2014–2024.

28 J. Roncali, *Macromol. Rapid Commun.*, 2007, **28**, 1761–1775.

29 (*a*) C. Destrade, N. H. Tinh, H. Gasparoux, J. Malthete and A. M. Levelut, *Mol. Cryst. Liq. Cryst.*, 1981, **71**, 111–135; (*b*) C. Destrade, P. Foucher, H. Gasparoux, N. H. Tinh, A. M. Levelut and J. Malthete, *Mol. Cryst. Liq. Cryst.*, 1984, **106**, 121–146.

30 (*a*) H. Sun, *J. Phys. Chem. B*, 1998, **102**, 7338–7364; (*b*) H. Sun, *Comput. Theor. Polym. Sci.*, 1998, **8**, 229–246; (*c*) D. Rigby, H. Sun and B. E. Eichinger, *Polym. Int.*, 1998, **44**, 311–330.

31 (*a*) I. Burgués-Ceballos, F. Machui, J. Min, T. Ameri, M. M. Voigt, Y. N. Luponosov, S. A. Ponomarenko, P. D. Lacharmoise, M. Campoy-Quiles and C. J. Brabec, *Adv. Funct. Mater.*, 2014, **24**, 1449–1457; (*b*) J. Min, H. Zhang, T. Stubhan, Y. N. Luponosov, M. Kraft, S. A. Ponomarenko, T. Ameri, U. Scherf and C. J. Brabec, *J. Mater. Chem. A*, 2013, **1**, 11306–11311; (*c*) J. Min, Y. N. Luponosov, Z. G. Zhang, S. A. Ponomarenko, T. Ameri, Y. F. Li and C. J. Brabec, *Adv. Energy Mater.*, 2014, DOI: 10.1002/aenm.201400816.

Faraday Discussions

DISCUSSIONS

Organic photovoltaics and energy: general discussion

Justin Hodgkiss, Gary Conboy, Graham Hutchings, Simon Higgins, Ian Galbraith, Neil Findlay, Rupert Taylor, Marc Salle, Shuyu Zhang, Gitti Frey, Ioan Bâldea, Ifor Samuel, Bernd Ebenhoch, Fred Wudl, Ayse Turak, Bingqian Xu, Enrico Angioni, Ravindra Venkatramani, Uli Lemmer, Dmitrii Perepichka, Alexander Kanibolotsky, Peter Skabara, Jack Wildman, Eli Zysman-Colman, Donal Bradley, Iain McCulloch, Joseph McDouall, Sergey Ponomarenko, Jenny Nelson and Ji-Seon Kim

DOI: 10.1039/c4fd90050B

Justin Hodgkiss opened the discussion of the paper by Jenny Nelson: A viable explanation for the initial long-range charge separation step in organic photovoltaic cells is *via* extremely high non-equilibrium charge mobilities of charge pairs initially formed high in the density of states (DOS). In this model, charge separation would be enhanced by some degree of disorder, including possibly the disorder induced by the packing with the fullerene at the heterojunction.

Can you comment on the implications of your work on this model of charge separation? Do you see sufficient disorder higher in the DOS to expect high initial hole mobilities?

Jenny Nelson answered: The effect of enhanced disorder near the interfaces in driving charge away from the interfaces towards regions where the materials are more ordered and the electronic states lie deeper into the band gap seems a very plausible one to me. The model would be compatible with this because the deeper lying states result from more extended regions of conjugation that would be more likely to occur away from the interfaces, where the second material in a heterojunction is less able to disrupt packing. In the work done so far, we concentrated mainly on the polymer domain and so were not able to study the interfaces in detail. The model could however be applied to binary systems, and this effect of a destabilisation of the electronic states due to packing disorder close to the interface could be studied. The limiting factor will be the time taken to carry out molecular dynamics (MD) of many representative binary systems.

Donal Bradley queried: Did you assume that the P3HT polymer is 100% regioregular and, if so, what will happen if you consider a non-regioregular polymer structure?

Jenny Nelson responded: Yes, we modelled a 100% regioregular material. The case of the regiorandom material would be very interesting to study in comparison, but we have not done that yet. This is a good idea for further study.

Uli Lemmer asked: How much do you learn about the delocalisation of the wave functions using this approach?

Jenny Nelson replied: The approach delivers information on the delocalisation of the wavefunction, through the form of the eigenfunctions obtained from the tight binding Hamiltonian, and on the relationship between localisation and energy levels. This aspect of the approach is being studied in more detail in a follow-on project.

Bernd Ebenhoch said: There are the two concepts for the density of states: a Gaussian distribution and an exponential tail. Could you please explain the differences and implications for charge transport and recombination?

Jenny Nelson responded: The density of states could have a wide variety of different forms. Our results could be approximated by an exponential or a Gaussian depending on the range of energies considered. Both functional forms would lead to phenomena such as charge density-dependent mobility, but with different dependences. For example, an exponential tail leads to a power law in the density dependence of mobility, and a Gaussian leads to a more complex form. Qualitatively the effects would be similar. The best approach would be to use an approximation to the DOS that gives the best fit over the range of Fermi levels that is of interest experimentally.

Gitti Frey remarked: You showed that the tail in the DOS is due to torsion, which has a small activation energy. What is the effect of temperature (mainly working device conditions, 50–60 °C) on this DOS?

Jenny Nelson answered: The effect of temperature on the tail of the DOS is shown in Fig. 3 in the manuscript.[1] The temperature dependence is very weak and no significant change is seen between 300 K and 400 K. This surprising result was attributed to the constraining effect of the close chain packing on the polymer conformation. However, there may be other factors than torsion affecting the DOS in real life, and those factors may be temperature-dependent.

1 J. M. Frost, J. Kirkpatrick, T. Kirchartz and J. Nelson, *Faraday Discuss.*, 2014, DOI: c4fd00153b

Ifor Samuel asked: In your conclusion you said that intramolecular interactions in P3HT dominate over intermolecular interactions.[1] How should that be reconciled with the considerable work by Spano and others explaining the properties of P3HT in terms of aggregate formation? Is it that you are studying different properties from those explained by aggregation?

1 J. M. Frost, J. Kirkpatrick, T. Kirchartz and J. Nelson, *Faraday Discuss.*, 2014, DOI: c4fd00153b

Jenny Nelson answered: We decided to focus our attention on the more amorphous region of the P3HT film, as we expect these regions to dominate charge transport, and most experimental estimates of the DOS are obtained from methods that probe transport in some manner, and because we expect the amorphous regions to be rate limiting for charge transport. The result for intermolecular interactions may be different in regions with closer and more ordered packing. Also, the effect of packing on the energies of polaron states may be different from that on neutral excited states.

Ian Galbraith asked: Can you provide more details of the calculations which underpin the tight-binding parameters, and also whether changing the relative magnitude of these parameters would/could lead to a qualitative change in the calculated DOS?

Jenny Nelson responded: The electronic coupling between monomers and different chains is obtained by a projective method calculation where the HOMO of the first monomer is projected onto the HOMO of the second, see for example ref. 46 of the paper.[1]

The coupling between neighbouring monomers on the same chain is approximated by the formula $J_0 \cos T$, where T is the torsion angle between the monomers and J_0 is obtained by calculating the splitting of the HOMO for planar oligomers of increasing length and fitting the results to the form expected for a linear tight binding model. A change in the value of J_0—that might result from using different quantum chemical methods or basis sets to calculate the energy levels of the oligomers—would result in a proportionate change in the energy scale of the density of states, for systems with no intrinsic energetic disorder.

1 J. Kirkpatrick, *Int. J. Quantum Chem.*, 2008, **108**, 51–56.

Ravindra Venkatramani commented: The final result that the DOS profile for P3HT polymers is mostly defined by intra-chain couplings and not inter-chain couplings is very surprising and interesting. Since you are using a minimal basis set (INDO/S-based calculations), it is likely that the nearest neighbour couplings might be underestimated. Have you checked for the basis set size dependence of your results?

Jenny Nelson responded: INDO/S should not seriously underestimate the coupling for intermolecular transfer integrals. The molecular orbital overlap (MOO) implementation of INDO/S that is used to find the intermolecular couplings in this paper was previously tested in comparison with the projective method using a medium-sized basis set (6-31G(d)) in ref. 46 of the paper.[1] That study found that the two methods agreed to within less than 10–20%.

1 J. Kirkpatrick, *Int. J. Quantum Chem.*, 2008, **108**, 51–56.

Jack Wildman said: How valid is the assumption that the transfer integral and, thus, the hopping term, of intrachain coupling between monomer units depends solely on the dihedral angle between these units? Would you, for instance, expect any effect to arise due to sharp bends between monomers?

Jenny Nelson replied: In principle, yes, we would expect other factors such as the dihedral to affect the transfer integral. In the case of P3HT where the optimum configuration of neighbouring monomers is the same as the one with the highest J, then the effect of a bend and a high torsion on J would be similar (they would both reduce J), and there may not be much effect from neglecting bends, as a bend would typically coincide with a non-zero torsion. For other polymers with more complex structures, we would need to look at which other degrees of freedom are important.

Peter Skabara remarked: Can you introduce solvent molecules to emulate the effect of solvent annealing on conformation and self-assembly?

Jenny Nelson replied: In principle, yes, but I doubt if we could reliably model the different effects of different solvents at the present stage. It would nevertheless be a good idea to compare the structures generated with and without solvent molecules.

Rupert Taylor communicated: The use of hexadecamers was stated to be a compromise between reality and computational time. Were shorter chains simulated, and if so how consistent were the results? Was the length of 16 reached by a trial and error approach or selected due to known validity from previous studies?

Jenny Nelson communicated in reply: It was not possble to test the effect of the oligomer length in the limit of longer oligomers because of computer time constraints. We tested different lengths mainly to find an oligomer length at which the MD would equilibrate. We did not systematically test shorter oligomers, and we agree that this would be a very good idea for future work.

Gitti Frey opened the discussion of the paper by Ji-Seon Kim: What is the mechanism of the *in situ* conversion which requires the spacial accumulation of many molecules, their interaction, particle nucleation and growth? Does the required temperature, 160 °C, reflect the precursor decomposition or molecule assembly through the polymer film?

Ji-Seon Kim replied: The mechanism of decomposition and formation of the morphology has been discussed more fully previously using TGA and SEM.[1] The decomposition occurs *via* a Chugaev rearrangement over the temperature range 50–150 °C. The first stage (loss of ethene) occurs below 100 °C and results in large CdS domains (~100nm), at higher temperature the other parts of the ligands are lost and the length scale of phase separation decreases with temperature. It is suggested that the evolution of gases during this later stage of decomposition ruptures the aggregates, leading to smaller domains.

1 S. Dowland, T. Lutz, A. Ward, S. P. King, A. Sudlow, M. S. Hill, K. C. Molloy and S. A. Haque, *Adv. Mater.*, 2011, **23**, 2739–2744.

Gitti Frey commented: We have recently shown that inorganic nanoparticles can be formed inside P3HT pre-formed films without disturbing the ordered domains.[1] This is done by ALD using ZnO precursors in the gas phase. The precursors, diethyl zinc and water, diffuse into a pre-formed P3HT film and *in situ* convert to ZnO. The diffusion is through the disordered domains of P3HT and hence the ordered domains are not disrupted.

1 S. Obuchovsky, I. Deckman, M. Moshonov, T. Segal-Peretz, G. Ankonina, T. J. Savenije and G. L. Frey, *J. Mater. Chem. C*, 2014, **2**, 8903–8910.

Fred Wudl noted: As shown by various techniques, PCBM will diffuse through a planar film device, as published recently by my colleague Michael Chabinyc.

Ji-Seon Kim responded: It is entirely possible that nanoparticles of CdS may be dispersed throughout the blend in addition to forming aggregated domains. This might explain the disruption of the polymer morphology, and is also expected to be detrimental to electron transport.

Simon Higgins asked: When you anneal at 160 °C to convert the Cd(xanthate)$_2$ species to CdS, what happens to the organic residue? I assume there are gases produced? Does the outgassing also affect the film morphology?

Ji-Seon Kim responded: The products are C_2H_4, COS, and H_2S, which are all volatile and not expected to remain in the film. This outgassing is probably a factor in the development of the film morphology; it correlates with a decrease in the size of the CdS aggregates and also the disruption of the P3HT molecular order described here.[1,2]

1. S. Dowland, T. Lutz, A. Ward, S. P. King, A. Sudlow, M. S. Hill, K. C. Molloy and S. A. Haque, *Adv. Mater.*, 2011, **23**, 2739–2744.
2 H. C. Leventis, S. P. King, A. Sudlow, M. S. Hill, K. C. Molloy and S. A. Haque, *Nano Lett.*, 2010, **10**, 1253–1258.

Iain McCulloch commented: Can you pre-aggregate your solutions using a theta or non-solvent, rather than aggregate them during film formation and annealing?

Ji-Seon Kim responded: It is certainly worth considering this; generally the use of different solvents to control blend thin film morphology is under-exploited in the field. Our study begins to address this using a "poor solvent" to produce the polymer nanowires in solution before deposition, but we can also produce a similar effect using a non-solvent like cyclohexanone. I assume that we would see similar effects. The problem we have identified is that the thermal annealing step required to produce the inorganic component is the dominant step in determining the final morphology of the blend film, so we would have to combine solvent control with a different method of producing the CdS nanoparticles.

Graham Hutchings remarked: I am going to ask a naive question as this is not my field. However, I am very interested in the synthesis and use of nanoparticles for catalysis. In our case we use organic capping agents and we use reducing agents to form the nanoparticles at temperatures of *ca.* 70–80 °C.[1] The nanoparticles can be thermally treated as well. In your case you are using temperature alone to form the nanoparticles and I am interested to know if you could trigger the formation of nanoparticles at a temperature lower than the 160 °C you use, as you indicate that high temperatures can have adverse effects on the polymer.

1 J. A. Lopez-Sanchez, N. Dimitratos, S. White, G. Brett, L. Kesavan, P. Miedziak, R. Tiruvalam, R. L. Jenkins, A. F. Carley, D. Knight, C. J. Kiely and G. J. Hutchings, *Nat. Chem.*, 2011, **3**, 551–556.

Ji-Seon Kim replied: The idea of triggering nanoparticle formation at lower temperatures is attractive for our study, since we have found that high temperatures are problematic for the polymer morphology. If this approach can be applied to our system (or a similar one), then it would be interesting to study. Our work clearly shows the need to develop novel approaches, so this suggestion presents a good option.

Graham Hutchings commented: In response to the point that you can trigger nanoparticle formation using photolysis, is this possible in your case as then the system could be irradiated at low temperature, and you could then probe the effect of temperature as a separate variable?

Ji-Seon Kim responded: Using light to trigger nanoparticle formation is also an attractive approach for decoupling the desired nanoparticle formation process from the undesirable high temperature. However, if it is the outgassing of volatile products that causes the disruption, this effect would not be overcome. We might also expect photodegradation of the polymer to become a significant effect: the precursor used here has a large energy gap so high energy photons would be needed, which pose a risk to the polymer.

Shuyu Zhang remarked: In the field of hybrid photovoltaics, several approaches have demonstrated impressive device efficiencies, such as the use of perovskite materials. So compared to other approaches, what is the advantage of this new P3HT:CdS blend-based photovoltaic material? Where will this work go in the future?

Ji-Seon Kim responded: Clearly perovskite (and other) approaches have yielded more impressive device efficiencies, but none of these satisfactorily meet the requirements of high efficiency, solution processability and good stability. The limitation for perovskites is stability, whilst the limitation for our approach is efficiency. Our work identifies the interaction between inorganic and organic materials as the key challenge to be addressed, particularly the morphology of the interface. If we are able to develop techniques to control this interface then we will unlock the potential of polymer/nanoparticle photovoltaics.

Iain McCulloch asked: What are the bottlenecks preventing hybrid devices from achieving high performance? Electron transport or voltage?

Ji-Seon Kim answered: It is hard to identify a single cause without a more detailed study of the device performance. Our devices have a reasonable V_{oc} (around 0.8V) but poor J_{sc} and FF, which indicates that the problem is with charge generation and transport. The high degree of polymer disorder in our optimised devices would suggest that hole transport is going to be a big problem, so I would suggest that this is one of the primary factors to address.

Uli Lemmer commented: Would it be possible to build devices just using the cadmium xanthate precursor solution, *i.e.* without using any conjugated polymer?

Ji-Seon Kim replied: We know that the CdS absorption contributes to the photocurrent in the blend film and there is some evidence of excited state delocalisation in the CdS nanoparticles, so it might be worth a try.[3,4] However, the lack of visible absorption would be limiting for solar cell applications.

3. S. A. Dowland, L. X. Reynolds, A. MacLachlan, U. B. Cappel and S. A. Haque, *J. Mater. Chem. A*, 2013, **1**, 13896.
4. N. Bansal, L. X. Reynolds, A. MacLachlan, T. Lutz, R. S. Ashraf, W. Zhang, C. B. Nielsen, I. McCulloch, D. G. Rebois, T. Kirchartz, M. S. Hill, K. C. Molloy, J. Nelson and S. A. Haque, *Sci. Rep.*, 2013, **3**, 1531.

Gary Conboy opened the discussion of the paper by Joseph McDouall: These calculations work well with simplified, well established single crystal molecules. Do you envisage this technique being applicable to larger oligomeric molecules and, eventually, polymers?

Joseph McDouall responded: The treatment of polymers on a model surface is quite challenging but we are looking at ways that we might approach it, although we have nothing to report on it at the present time. For cluster-type calculations of mobilities in polymers (without a surface), a way to proceed is to perform a periodic calculation on the polymer and then remove "oligomeric" units for calculating the electronic couplings and reorganisation energies. This must involve some experimentation to ensure that in the cation calculation (necessary for evaluating the reorganisation energy), the size of the oligomeric unit is sufficiently large that it fully encapsulates the distribution of the cation charge.

Donal Bradley remarked: I would be grateful if you would confirm the role of the surface in your calculations. Is it primarily to provide a suitable dielectric environment? Or is there another reason? How is it included in the calculations?

Joseph McDouall replied: The role of the surface was to provide a dielectric platform for the clusters to sit on. Importantly, it has the property that as more units of pentacene are added to the surface, the energetically preferred mode of attachment is with the long axis of the molecule perpendicular to the surface. The first unit attaches itself parallel to the surface. The second unit still very slightly favours the face down arrangement, but for three units (or more) the preferred mode of attachment is to have the pentacene molecules perpendicular to the surface. This has been observed experimentally in thin films of pentacene and it is an important feature for the model surface to be able to reproduce it.

Ravindra Venkatramani commented: The reorganization energy calculation, as outlined in eqn (6a) and (6b),[1] accounts for only inner sphere contributions. It has been shown that for solvent-accessible electronic sites, the outer sphere contribution is the dominant contribution to the overall reorganization energy.[2] Neglecting the outer sphere contributions (from both neighboring molecules and the solvent) would significantly underestimate the reorganization energy cost and therefore over-estimate the calculated hole mobilities in Table 4.[1] From the point of view of developing a computational framework to reliably estimate hole mobilities in a broad range of molecular clusters/aggregates, including outer-sphere contributions to reorganization energy calculations is very important.

1 R. A. Krawczuk, S. Tierney, W. Mitchell and J. J. W. McDouall, *Faraday Discuss.*, 2014, DOI: c4fd00082j.
2 V. Tipmanee, H. Oberhofer, M. Park, K. S. Kim and J. Blumberger, *J. Am. Chem. Soc.*, 2010, **132**, 17032–17040.

Joseph McDouall replied: The outer sphere reorganisation energy will certainly play a role. We have not included it in our calculations. The paper by Blumberger *et al.*[1] refers to solvent rearrangement around a protein, and it may be expected that the solvent reorganisation will be rather greater in such a process than for the systems we have studied. However, it remains an important consideration and will need to be addressed to refine the model.

1 V. Tipmanee, H. Oberhofer, M. Park, K. S. Kim and J. Blumberger, *J. Am. Chem. Soc.*, 2010, **132**, 17032–17040.

Ravindra Venkatramani remarked: The off-diagonal terms of the Fock matrix in eqn (10) in the manuscript[1] are not directly the non-adiabatic electronic couplings. To extract the electronic couplings from eqn (10) a transformation has to be applied to make the diagonal energy terms equal. Can you verify this? The sentence below eqn (10) in the manuscript[1] suggests that such a transformation was likely made ($\tilde{\mathbf{F}}$ was "symmetrically orthogonalized").

1 R. A. Krawczuk, S. Tierney, W. Mitchell and J. J. W. McDouall, *Faraday Discuss.*, 2014, DOI: c4fd00082j.

Joseph McDouall replied: The reference for the procedure used is Valeev *et al*,[1] the pertinent equations in that paper being eqn (8) and (9). The symmetric orthogonalisation is applied to the back transformed Fock (or Kohn–Sham) matrix and then the electronic coupling is obtained. This step is essential otherwise a very large overestimate of the coupling will be obtained.

1 E. F. Valeev, V. Coropceanu, D. A. da Silva Filho, S. Salman and J.-L. Brédas, *J. Am. Chem. Soc.*, 2006, **128**, 9882–9886.

Bingqian Xu said: The computation was obtained on a model surface. If you introduce defects on the surface, what will happen to the hole mobilities?

Joseph McDouall answered: We have not investigated that so far. It would certainly be possible to introduce geometric defects into the model surface, but it would probably require a much larger surface to be used in the calculations so that local defects did not preclude the formation of clusters away from the defect.

Dmitrii Perepichka commented: It is not very clear from Fig. 14 in your paper,[1] and I would like to confirm whether the pentacene molecules in your calculated clusters are oriented with their long molecular axis perpendicular to the film plane.

1 R. A. Krawczuk, S. Tierney, W. Mitchell and J. J. W. McDouall, *Faraday Discuss.*, 2014, DOI: c4fd00082j.

Joseph McDouall responded: The figure is slightly tilted to show some perspective on the pentacene units.[1] In the global minimum the pentacene units are aligned perpendicular to the surface. There is however also a local minimum in which the monomers tilt a few degrees away from perpendicular.

1 R. A. Krawczuk, S. Tierney, W. Mitchell and J. J. W. McDouall, *Faraday Discuss.*, 2014, DOI: c4fd00082j.

Dmitrii Perepichka commented: Do you use reorganization energies for individual molecules in the gas phase? Do you know how significantly would it change if calculated for a cluster?

Joseph McDouall replied: The reorganisation energies we use are for the individual molecules. We are investigating the reorganisation energy for the cluster but have nothing to report as yet.

Ayse Turak said: In your paper,[1] you state that the monolayer structure that you used is representative of the bulk structure of the organic material. However, this is generally not the case, especially for small oligomers such as pentacene. See, for example, Hayakawa *et al.*,[2] Casu *et al.*,[3] Witte *et al.*,[4] Durr *et al.*,[5] Hiroshiba *et al.*,[6] Krauss *et al.*,[7] Koch *et al.*,[8] Kury *et al.*,[9] Amassian *et al.*,[10] Brauer *et al.*,[11] amongst many others. Given that the first monolayer is generally not representative of the overall film structure, is it possible to extend your calculations to the 2nd or 3rd layer, incorporating a change in the morphology above the first monolayer to give a more realistic model of the interfacial structure, and therefore a better basis to understand the charge transport?

1 R. A. Krawczuk, S. Tierney, W. Mitchell and J. J. W. McDouall, *Faraday Discuss.*, 2014, DOI: c4fd00082j.
2 R. Hayakawa, A. Turak, X. Zhang, N. Hiroshiba, H. Dosch, T. Chikyow and Y. Wakayama, *J. Chem. Phys.*, 2010, **133**, 034706.
3 M. B. Casu, I. Biswas, B.-E. Schuster, M. Nagel, P. Nagel, S. Schuppler and T. Chassé, *Appl. Phys. Lett.*, 2008, **93**, 024103.
4 G. Witte and C. Wöll, *J. Mater. Res.*, 2004, **19**, 1889–1916.
5 A. C. Dürr, B. Nickel, V. Sharma, U. Täffner and H. Dosch, *Thin Solid Films*, 2006, **503**, 127–132.
6 N. Hiroshiba, J. P. Hill, R. Hayakawa, K. Ariga, K. Matsuishi and Y. Wakayama, *Thin Solid Films*, 2014, **554**, 74–77.
7 T. N. Krauss, E. Barrena, D. G. de Oteyza, X. N. Zhang, J. Major, V. Dehm, F. Würthner and H. Dosch, *J. Phys. Chem. C*, 2009, **113**, 4502–4506.
8 N. Koch, I. Salzmann, R. L. Johnson, J. Pflaum, R. Friedlein, J. P. Rabe, *Org. Electron.*, 2006, **7**, 537–545.
9 P. Kury, K. R. Roos, D. Thien, S. Möllenbeck, D. Wall, M. Horn-von Hoegen and F.-J. Meyer zu Heringdorf, *Org. Electron.*, 2008, **9**, 461–465.
10 A. Amassian, V. A. Pozdin, T. V. Desai, S. Hong, A. R. Woll, J. D. Ferguson, J. D. Brock, G. G. Malliaras and J. R. Engstrom, *J. Mater. Chem.*, 2009, **19**, 5580–5592.
11 B. Bräuer, A. Virkar, S. C. B. Mannsfeld, D. P. Bernstein, R. Kukreja, K. W. Chou, T. Tyliszczak, Z. Bao and Y. Acremann, *Chem. Mater.*, 2010, **22**, 3693–3697.

Joseph McDouall responded: It is certainly true that the first monolayer is not representative of the overall film structure. However, within the computational schemes that we have employed, we are limited to how many molecules we can treat and hence the number of layers we can treat. Our model surface is able to reproduce the well-known property of pentacene thin films that as the number of molecules adsorbed increases, the energetic balance favours the adsorption of the pentacenes end-on to the surface rather than face-down or edge-on. Much larger simulations of the thin film (monolayer) structure,[1] obtained with classical force fields, also confirm this structural motif. The hole mobility, which we calculate is dependent on all nearest neighbour interactions, and small variations in specific monomer–monomer interactions will not significantly change the overall value. Additionally, in preparing organic field effect transistor (OFET) devices, engineers are often able to alter the characteristics of surfaces so that thin films form in particular motifs. Our calculations, while idealised, are nonetheless able to distinguish between synthetic targets, and which can be expected to yield good hole mobilities, and this is our principal aim in this work.

1 L. Viani, C. Risko, M. F. Toney, D. W. Breiby and J.-L. Brédas, *ACS Nano*, 2014, **8**, 690–700.

Ioan Baldea asked: In your approach, you use the intramolecular reorganisation energy λ computed according to the four-point scheme sketched in Fig. 7 in your manuscript:[1]

$$\lambda = \lambda_1 + \lambda_2,$$

$$\lambda_1 = E_0(\mathbf{Q}_+) - E_0(\mathbf{Q}_0),$$
$$\lambda_2 = E_+(\mathbf{Q}_0) - E_+(\mathbf{Q}_+).$$

Considering differences between the total energies of the neutral (E_0) and cationic (E_+) species taken at the corresponding optimised geometries ($\mathbf{Q} = \mathbf{Q}_0$ and $\mathbf{Q} = \mathbf{Q}_+$, respectively), you (implicitly) assume that all molecular vibrational modes $\mathbf{Q}$ can be treated classically, or, rephrasing, that all vibrational modes reorganize classically. A fact on which extensive work drew attention to recently[2–6] is that, in reality, only low frequency modes, which are thermally activated, can be considered as classically reorganizable. Higher frequency modes reorganize quantum mechanically. So, like in other cases,[2–5] I would expect that, by using the above equations, you substantially overestimated the intramolecular reorganization energy λ. To properly estimate λ, you need an analysis of the individual vibrational modes.

Further, computing the rate of charge transfer W *via* your Marcus-type eqn (1),[1] you (implicitly) assume a parabolic $\mathbf{Q}$-dependence of $E_{0,+}$. Based on previous experience,[2–6] for the low-frequency intramolecular modes that reorganize classically I would expect substantial anharmonic contributions, resulting in a modified formula for W.

1 R. A. Krawczuk, S. Tierney, W. Mitchell and J. J. W. McDouall, *Faraday Discuss.*, 2014, DOI: c4fd00082j.
2 I Bâldea, 2012, arXiv: 1204.4023v1.
3 I Bâldea, *Europhys. Lett.*, 2012, **99**, 47002.
4 I Bâldea, H. Köppel and W. Wenzel, *Phys. Chem. Chem. Phys.*, 2013, **15**, 1918–1928.
5 I Bâldea, *J. Phys. Chem. C*, 2014, **118**, 8676–8684.

6 I Bâldea, *RSC Adv.*, 2014, **4**, 33257–33261.

Joseph McDouall responded: We have not applied the vibrational terms to our calculations at all, but I do accept that their treatment would alter the reorganisation energies. Similarly the Marcus treatment, which we employ, is based on parabolic surfaces which would again be modified by anharmonic terms. However, for our purposes, the net effect on the calculated mobilities is expected to be small. The difference in the vibrational frequencies is likely to be significant for only one or two modes, and so the net effect on the overall energy differences will be small.

Enrico Angioni remarked: In your work you have reported the values of some binding energies. Which method did you use to calculate these energies?

Joseph McDouall replied: The semiempirical PM7 method was used. All geometries of the clusters and surface were also obtained at this level. PM7 is parameterised to reproduce non-covalent interactions, as encountered in the S22 dataset, with some success and hence it is an appropriate choice.

Fred Wudl opened the discussion of the paper by Dmitrii Perepichka: Very nice work. I have three remarks. 1. In the case of your donor molecule, it should be more akin to an acceptor since pyridine is an electronegative moiety. 2. What was the conductivity of the solid? 3. The angle of the hydrogen bonds of the pyrrole components of the donor appear to be too wide, would a piperidine equivalent not be better?

Dmitrii Perepichka answered: 1. Indeed, pyridine nitrogen stabilizes both frontier orbitals of the compounds, making it a weaker electron donor and a stronger electron acceptor. However, owed to the presence of two electron-rich pyrrole rings, it still acts as an electron donor towards NDI. 2. The cocrystal solids are neutral materials and do not have intrinsic charge carriers to expect high conductivity. The conductivity value can be deduced from the off current of the OFET measurements. The two-probe conductivity of the OFET channel at zero gate voltage, as expected, shows non-Ohmic behavior and at 100V bias across a 50 μm channel is on the order of 10^{-5} S cm^{-1} for DP-P2P:NDI-81. 3. If I correctly understand your idea to replace pyrrole rings with piperidine, it would lead to a more favorable direction of NH bonds but you would lose π-conjugation, which is necessary for the transport properties. In any case, stronger binding H-bond components with a suitable electronic structure could be designed (and we indeed are working on creating DDD···AAA pairs), however, I should comment that the strength of H-bonding is not at all a limiting factor for the applications of these materials. The complementary H-bonding in the current system is already sufficiently robust to ensure the interaction between both components in the solid state.

Gitti Frey remarked: Are there design rules to ensure that the H-bonding does not disturb the π-π interactions required to support charge transport?

Dmitrii Perepichka replied: Our approach was based on making the H-bonding functionality a part of the π-conjugated framework. In this case H-bonding interactions occur in the direction perpendicular to the π-stacking and thus are not expected to disturb it.

Gitti Frey asked: Self assembly of small molecules directs nanostructures that are useful for FETs, but it is difficult to form the continuous films necessary for PV devices. Are there design rules to direct the formation of thin films from self-assembled small molecule structures?

Dmitrii Perepichka responded: Yes, there are standard design rules used to improve the solubility/processability of π-conjugated materials. We introduced long branched alkyl chains on both the donor and the acceptor components, and are currently in the process of investigating the device properties of the H-bonded donor/acceptor films prepared by spin-coating. Inducing liquid-crystalline behavior in these materials, needed for finer control of molecular organization, is in our future plans.

Eli Zysman-Colman said: Is it your long term plan to make bifunctional materials with donor/acceptor units on both sides of the molecule?

Dmitrii Perepichka answered: We do not have immediate plans to prepare covalently linked donor–acceptor molecules with H-bonding functionality, if that's what you mean. Covalent D–A dyads have been studied as potential photovoltaics materials since the late 1990s, with very limited success. What we do work on is the synthesis of larger donor and acceptor oligomers with complementary H-bonding. This should allow us to improve light absorption in the NIR region and increase the separation between the hole and the electron, to slow down charge recombination, which is essential for photovoltaic applications.

Eli Zysman-Colman responded: Assuming you are able to do this, how easily will the materials be processable? How will you address this point?

Dmitrii Perepichka responded: You are quite right, processability is an important limitation while working with organic semiconductors. We are addressing the issue through the most standard approach: putting branched long chain substituents on both the donor and acceptor components. Thus, our latest donor–acceptor co-assemblies, based on π-extended perylenediimide, are fully soluble for spin-coating film processing.

Neil Findlay asked: Have you looked at OFET device measurements for the thiophene and furan analogues of DP-P2P?

Dmitrii Perepichka responded: We have indeed attempted to measure field mobility for some difuryl- and dithienyl-P2P cocrystals. We have seen low electron mobility in a few devices, but never observed ambipolar transport as for DP-P2P. Since our X-ray crystallographic studies showed mixed stacking for these cocrystals, good transport properties were not expected, and we have not pursued it further.

Alexander Kanibolotsky asked: You have got proton donor and acceptor groups in the molecule of the semiconductor, which can facilitate the formation of charge carrier traps. Did you observe any hysteresis in the transfer characteristics of your OFETs? In other words, did you notice any difference in the characteristics when sweeping the voltage forward and backwards? For an OFET with hysteresis, the calculated values of charge carrier mobility and threshold voltage can depend heavily on whether one uses forward or reverse scanning.

Dmitrii Perepichka replied: We almost always observe some hysteresis, even in OSCs without labile protons. We have not paid close attention to the extent of hysteresis in this specific case (the mobility value was calculated for the forward IV scan), and I agree that it can provide additional information about possible charge trapping in our H-bonded systems. However, because the measured electron mobility for disubstituted NDI-R2 (without labile protons) was similar to that of NDI-R1 and NDI···P2P, we concluded that charge trapping due to proton transfer is not significant.

Iain McCulloch remarked: Can you comment on whether you have the optimal geometry to facilitate the hydrogen bonding required to form the dimer molecule. The molecule has a fused structure with a fixed distance between bonding sites, with no rotational/conformational freedom. In nature, this is not the case, for example in proteins. Is this the ideal geometry or could hydrogen bonding be increased?

Dmitrii Perepichka replied: Finding a molecular structure with an optimal geometry for three-point H-bonding to NDI was an essential part of this work. I believe that the dipyrrolopyridine structural motif is not too far from that optimum geometry. From X-ray structural analysis, the N···H hydrogen bond with the pyridine nitrogen is *ca.* 1.8 Å which is quite short (Table 2 in the manuscript[1]). Indeed, as was suggested by Prof. Wudl, the geometry of the five-membered pyrrole rings places the NH hydrogen bond donors a little too far from the C=O acceptors (O···H distance 2.0–2.2 Å). However, the overall DAD···ADA hydrogen bonding in the P2P/NDI pairs is sufficiently strong to provide reliable binding. In fact, the calculated gas-phase binding energy of P2P/NDI (19.6 kcal mol^{-1}) exceeds that of the A···T nucleobase pairing (16.2 kcal mol^{-1}). It is also stronger than the binding of NDI with diaminopyridine (17.2 kcal mol^{-1}), which does have a perfect geometric match with NDI (but less acidic NH protons).

Regarding the flexibility of naturally-occurring H-bonding systems, this is certainly true at the macromolecular level, such as in DNA or proteins. However, the elementary multicenter H-bonding motifs, such as nucleobases or individual amides are in fact as rigid as our structures. When we start incorporating our H-bonding semiconducting motif in polymers and other soft materials, I agree that some conformational flexibility between the building blocks will need to be designed for the self-assembly to take place.

1 H. T. Black, H. Lin, F. Bélanger-Gariépy and D. F. Perepichka, *Faraday Discuss.*, 2014, DOI: c4fd00133h.

Eli Zysman-Colman asked: Why did you choose these donor/acceptor units in particular? There are many examples of donor/acceptor motifs in the literature, some with more favourable binding than your system. Can you describe more precisely the criteria for your choice of donor and acceptor molecules?

Dmitrii Perepichka responded: Our choice was dictated by (i) simplicity and (ii) the established semiconducting properties of the selected components, because we wanted to ensure the crystallinity of the co-assembled solids (for X-ray structural analysis) and to be able to correlate their structure with the charge transport properties. Naphthalenediimides (NDI, and related rylene dyes) are some of the most established and synthetically most accessible n-type organic semiconductors, which have an intrinsic (but hidden) three-point hydrogen bonding motif. They are expected (and known) to form complementary H-bonding with diaminopyridine. Thus, we rationalized that synthetically accessible dipyrrolopyridine should also form complementary H-bonding with NDI and would have a sufficiently extended and rigid π-electron structure to support effective hole transport.

Certainly, there are stronger H-bonding motifs that could be used in D···A coassembly, but I do not think this is required. The binding should be strong enough to ensure the D···A contacts in the solid state (and it already is in our systems), but not too strong, so that the dynamics of association/dissociation are fast and the self-assembly pathway with defect-corrections can be followed.

Marc Salle remarked: Firstly, a key issue with the system presented is that it relies on H-bond interactions. Are there any data available to quantify the magnitude of this interaction in solution, for instance from an Isothermal Microcalorimetry (ITC) study? Secondly, a two-step experimental procedure is used to grow single crystals of the CT complex. Why not use a direct diffusion method of both components in solution?

Dmitrii Perepichka replied: We have studied the binding in solution by fitting NMR and UV-Vis data at different concentrations. The data analysis is somewhat complicated by the coexistence of H-bonding and π-stacking association, however, we can tell that the binding constant is on the order of 100 M^{-1} in chloroform, as expected for such DAD···ADA complementary H-bonding. Regarding the second question, we can easily grow crystals by the simple mixing of hot solutions of the individual components, followed by cooling to room temperature. However, the size and quality of the resulting crystals is not adequate for either X-ray analysis or OFET fabrication. Generally speaking, vapor transport almost always produces higher quality crystals.

Dmitrii Perepichka opened the discussion of the paper by Sergei Ponomarenko: Your molecular design replaces an unstable vinylene hydrogen with an alkyl group R. However, the high acidity of the α-methylene in this group R, brought about by resonance stabilization of the corresponding anion by two CN groups, potentially creates another site of chemical vulnerability. Could you please comment on that?

Sergey Ponomarenko replied: I would compare the oxidation stability of these two units with their oxygen precursors: aldehyde and acetyl groups. It is well known that the former can be easily oxidized to the carboxyl group, but the latter is much more stable to oxidation. It probably comes from the fact that the oxidation proceeds through the formation of radical rather then anionic species. Perhaps under basic conditions, the alkyl-substituted molecules are indeed less stable to oxidation, but we have not observed it under normal conditions. Moreover, all alkyl-substituted molecules have higher oxidation potentials, indicating their better stability to oxidation as compared to their analogs without alkyl groups containing vinylene hydrogens.

Iain McCulloch remarked: From the DFT model it appears that the LUMO is on the periphery of the molecule and the HOMO in the centre. Is this the preferred way/optimal design for a small donor molecule?

Sergey Ponomarenko replied: If we are talking about amorphous materials, like most of the molecules considered in this paper, it does not actually matter whether they have the LUMO or HOMO on the periphery, since there is efficient contact between the acceptor and donor phases with different orientations of the star-shaped molecules. However, if we are going to make a columnar phase from such molecules, having the HOMO in the center could be helpful for efficient hole transfer through such a column to the anode. However, there might be a problem with the charge separation between the columns of the donor and acceptor phases, if the donor molecules form columns with the HOMO in the centre. In this sense it would also be interesting to check this molecular design for small acceptor molecules.

Rupert Taylor communicated: Can the central silicon atom in the tetrahedral oligothiophene silanes be replaced by other group 4 atoms (C, Ge, Sn...), and how would this alter the properties of the molecules?

Sergey Ponomarenko communicated in reply: Star-shaped oligothiophene silanes with central C atoms are known up to quaterthiophenes in one arm on the molecule, albeit their synthesis is more complicated then the corresponding oligothiophene silanes. However, neither their semiconducting or photovoltaic properties have been reported so far. Similar molecules with a central Ge atom have not been reported yet, but I expect that they can be easily synthesized, similar to the oligothiophene silanes. In a hypothetical series of star-shaped oligothiophenes with C, Si and Ge central atoms, I would expect an increased red shift when going from C to Ge for the same arm conjugation length. Therefore, such molecules with Ge could show improved semiconducting and photovoltaic properties, as some other Ge-containing organic semiconductors do.

Faraday Discussions

PAPER

Novel electrochemiluminescent materials for sensor applications†

Lynn Dennany,* Zahera Mohsan, Alexander L. Kanibolotsky and Peter J. Skabara

Received 2nd May 2014, Accepted 19th June 2014
DOI: 10.1039/c4fd00090k

Electrochemiluminescence (ECL) uses redox reactions to generate light at an electrode surface, and is gaining increasing attention for biosensor development due to its high sensitivity and excellent signal-to-noise ratio. ECL studies of monodisperse oligofluorene–truxenes (T4 series) have been reported previously, showing the production of stable radical cations and radical anions, generating blue ECL. The compound in this study differs from the original structures, in that there are 2,1,3-benzothiadazole (BT) units inserted between the first and second fluorene units of the quarterfluorenyl arms. It was therefore anticipated that the incorporation of these highly luminescent and ECL-active compounds into sensor development would lead to significant decreases in detection limits. In this contribution, we report on the impact of incorporating these novel complexes into sensor devices on the ECL efficiency, as well as the ability of these to improve the detection sensitivity and decrease the limit of detection using the reagent-free detection of model analytes. The real world impact of these compounds is elucidated through the comparison with more standard ECL materials such as ruthenium-based compounds. The potential for multiple applications is to be examined within this contribution.

Introduction

Electrochemiluminescence (ECL) represents a powerful analytical approach that combines simple equipment with inherent sensitivity, and a wide dynamic range for many analytes.[1,2,3,4,5] ECL has been the subject of extensive study for the past three decades.[6–8] The production of light from intermediates generated during electrolysis occurs when the energy liberated by the reaction between the electrogenerated precursors is sufficient to generate a product in an electronically excited form.[9] Studies of inorganic ECL have been dominated by transition metal complexes,[10–12] particularly ruthenium poly(pyridyl) species, *e.g.*, those of the general formula $Ru(L)_3{}^{2+}$, where L = 2,2′-bipyridine,[13–15] 4,7-diphenyl-1,10-

WestCHEM, Department of Pure and Applied Chemistry, University of Strathclyde, 295 Cathedral Street, Glasgow, G1 1XL, UK. E-mail: lynn.dennany@strath.ac.uk; Fax: +141 548 2535; Tel: +141 548 4322

† Electronic supplementary information (ESI) available. See DOI: 10.1039/c4fd00090k

phenanthroline[16] or 2,2′-bipyrazine.[17] This is due to the attractive photophysical and electrochemical properties that these compounds typically exhibit. ECL is a remarkably versatile, highly sensitive[18] and selective technique that has emerged in various research fields.[19] In recent years, particularly fascinating electronic properties have been discovered for various star-shaped conjugated oligomers with different core arms.[20]

As with all sensors, there is a constant drive to improve their sensitivity and selectivity. The future health industry in conjunction with point of care health monitoring will demand the detection of life-threatening diseases before critical stages have been reached. This will require a new breed of biomedical sensors capable of measuring disease biomarkers down to the picomolar range. In order to design such sensors, it is necessary to consider the limitations of the current biomedical sensors.

Depending on the nature of a specific ECL reaction, the sensitivity of ECL-based systems may be limited by certain factors. These include slow rates of charge transfer (D). A slow rate of diffusion will lead to slow regeneration of the excited state, typically Ru^{3+}, which is required to react with an analyte to produce ECL. A fast diffusion rate to the electrode will lead to an increase in the amount of the excited state available to react with a specific analyte. This should lead to increased ECL production and improved sensitivity. One appealing possibility is to wire the luminophores using a conjugated polymer backbone which can have a positive effect on ECL efficiencies.[21] An alternative to this is to create an entirely new compound with high luminescence quantum yields to allow for the determination of lower concentrations of the target analyte.

ECL can be produced by two dominant pathways, *i.e.* annihilation and co-reactant pathways.[8] The co-reactant pathway is predominantly utilised for sensor applications, and the common oxidation mechanism for inorganic compounds corresponding to the production of ECL through the interaction of an ECL-active material with a co-reactant is well established, and is as follows:[2,4,5,7,8]

$$\mathrm{ML}^{n+} \rightarrow \mathrm{ML}^{n+1} + \mathrm{e}^{-} \quad (1)$$

$$\mathrm{ML}^{n+1} + \mathrm{analyte} \rightarrow \mathrm{ML}^{n+} + \mathrm{analyte}_{\mathrm{Ox}} \quad (2)$$

$$\mathrm{Analyte}_{\mathrm{Ox}} \rightarrow \mathrm{analyte}_{\mathrm{Med.Ox}} + \mathrm{analyte}_{\mathrm{radical}} \quad (3)$$

$$\mathrm{Analyte}_{\mathrm{radical}} + \mathrm{ML}^{n+} \rightarrow \mathrm{analyte}_{\mathrm{Med.Ox}} + \mathrm{ML}^{n+}* \quad (4)$$

$$\mathrm{ML}^{n+}* \rightarrow \mathrm{ML}^{n+} + h\nu \quad (5)$$

where ML is the ECL-active material and the analyte can refer to either the target compound being analysed or the co-reactant undergoing mediated oxidation, thereby producing an ECL response. The ECL produced is often governed by the concentration of the excited state form of the ECL-active material, usually determined in the solution phase by the diffusional coefficient observed.

This work will examine the applicability of star-shaped conjugated oligomers, specifically truxene–oligofluorene with 2,1,3-benzothiadazole units (T4BT–B), to facilitate future ECL sensor development. This system will be compared to a common ECL system, namely utilising $[Ru(bpy)_3]^{2+}$ for the determination of tri-propylamine (TPA) concentrations. TPA is often used as a co-reactant for ECL production and by comparing the ECL response of T4BT–B with that of $[Ru(bpp)_3]^{2+}$ to TPA, we will evaluate the ECL production and efficiency of T4BT–B. Many clinically relevant biomarkers are detected through the co-reactant pathway and many are based upon the interaction of amine structures. Therefore, TPA is being utilised as a model for how other proteins would interact with T4BT–B. In addition, there are a considerable number of illicit substances which also have amine-based structures, and T4BT–B could also be utilised for their detection following the same reaction pathways as shown for TPA. Electrochemical sensors are not entirely focused on amine detection, but ECL sensors are particularly prevalent in biosensor applications and can also be utilised as co-reactants for label-based ECL systems.[2,4,5,7,8] Since TPA represents a particularly well-established system, both for biosystems detection and label-based ECL systems, it was utilised as a model system to be investigated in this study.

Experimental section

Materials and reagents

The star-shaped truxene compound with additional 2,1,3-benzothiadiazole units (T4BT–B, shown in Fig. 1) has been synthesised as previously described.[22–25] $[Ru(bpy)_3]^{2+}$ was purchased from Sigma Aldrich and utilised without purification.

All solvents used were of spectroscopic grade. The working electrodes were prepared by polishing with alumina (1.0–0.3 μm) on a felt pad. All solutions were deoxygenated using nitrogen or argon prior to measurement.

Apparatus

Electrochemical experiments were performed in a standard electrochemical cell using a CH instruments (Memphis, TN) model 760D potentiostat. Cyclic voltammetry experiments were carried out using a 3 mm diameter glassy carbon working electrode in a conventional three electrode assembly using a platinum flag as the counter electrode and an Ag wire as the reference electrode. Potentials were controlled *versus* a standard Ag wire reference electrode. Measurements

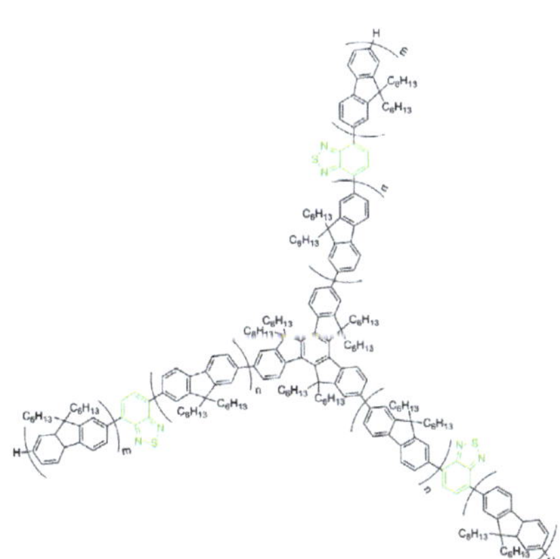

Fig. 1 Chemical structure of truxene–quarterfluorenes with BT (green) units, $n = 1$, $m = 3$ for T4BT–B.

involving simultaneous detection of light and current utilised a CH instrument model 760D connected to a Hamamatsu (H6780-20) photomultiplier tube (PMT). The PMT was biased at −800 or −400 V (indicated where utilised) by a high voltage power supply. During the experiments, the cell was kept in a light-tight box in a specially designed holder where the working electrode was positioned directly opposite to the fibre optic bundle, the other end of which was coupled to the PMT.

All measurements were made at room temperature (20 °C). For ECL experiments, concentrations of up to 2 mM tripropylamine (pH 6) were used as the co-reactant. For a comparison of ECL systems, 1×10^{-4} M solutions of T4BT–B in anhydrous dichloromethane (CH_2Cl_2) and separate solutions of 1×10^{-4} M $[Ru(bpy)_3]^{2+}$ in 0.1 M tetrabutylammonium hexafluorophosphate ($(TBA)PF_6$) in anhydrous dichloromethane (CH_2Cl_2) electrolyte with varying concentrations of TPA were used. All other reagents used were of analytical grade.

Results and discussion

Electrochemical properties of T4BT–B

The electrochemical behaviour of the T4BT–B compound and $[Ru(bpy)_3]^{2+}$ were examined using cyclic voltammetry. Fig. 2 shows the typical voltammetric behaviour of both compounds. By analogy with the voltammetric behaviour of $Ru^{2+/3+}$ complexes,[2,4,7,14] the peaks centred at $\sim E_{1/2} = 1.2$ V are attributed to the $Ru^{2+/3+}$ couple while those at −1.44 and −1.64 V are bpy-based reductions to the +1 and 0-charged species respectively. In addition, the peak-to-peak separation between the anodic and cathodic waves is close to zero at low scan rates, the full width at half maximum (FWHM) is close to the theoretically value of 59 mV, and the peak current varies linearly with the square root of the scan rate.

Fig. 2 also shows the voltammetric behaviour of T4BT–B, consisting of three overlapping anodic waves. The peak-to-peak separation of this first anodic wave is close to zero, as anticipated. This first anodic wave is fully reversible, followed by two quasi-reversible waves. The reduction of T4BT–B shows a stable single

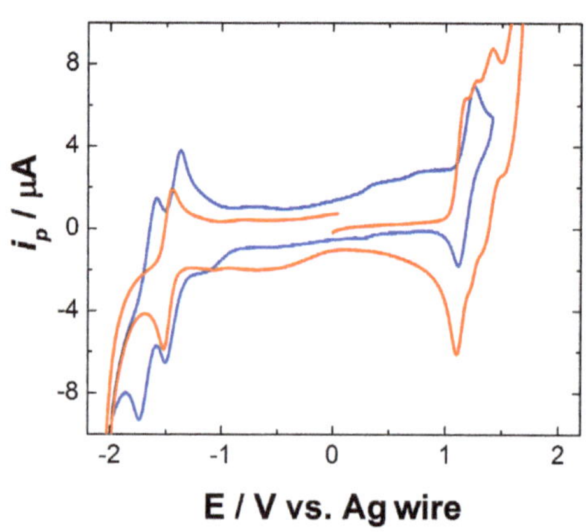

Fig. 2 Solution phase cyclic voltammetric behaviour of 1×10^{-4} M T4BT–B (red line) and 1×10^{-4} M $[Ru(bpy)_3]^{2+}$ (blue line). The electrolyte was 0.1 M $(TBA)PF_6$ in CH_2Cl_2, and a scan rate of 100 mV s^{-1} was used.

reversible cathodic wave at ~−1.49 V. For each of these compounds, the peak current increases linearly with the square root of the scan rate for both the oxidation and reduction processes. However, given the overlapping anodic waves, the FWHM cannot be determined. This behavior is consistent with semi-infinite linear diffusion limiting the redox process, and under these conditions the response can be described by the Randles–Sevçik equation:

$$i_p = 0.4463nFAC\left(\frac{nF\nu D}{RT}\right)^{\frac{1}{2}} \quad (6)$$

where n is the number of electrons transferred, A is the area of the working electrode, ν is the scan rate, D is the diffusion coefficient, F is the Faraday constant, R is the gas constant, T is the temperature in Kelvin, and C is the concentration of the redox centers. Thus, eqn (6) allows D to be estimated for both the oxidation and reduction processes of each compound, provided that n is known (or assuming a value for n; for T4BT–B in DCM an n value of 1 is assumed). Table 1 shows the D values for the oxidation process for T4BT–B and $[Ru(bpy)_3]^{2+}$.

ECL production can be highly dependent upon the rate of generation of the excited state, which is often based on the diffusion coefficient (D) that is calculated for the ECL-active species. The faster the diffusion coefficient, the more ML^{n+1} is available for interaction with the co-reactant or analyte of interest, thereby producing an ECL response. A slower diffusional rate and insufficient amounts of ML^{n+1} will result in decreased ECL emission and hence decreased sensitivity. A faster diffusion rate would thereby significantly increase the production of ECL, leading to enhanced sensitivity. This increase coupled with an increased luminescence yield should lead to favourable characteristics necessary for enhanced ECL production. The diffusion coefficients for $[Ru(bpy)_3]^{2+}$ and T4BT–B are given in Table 1. Different solvents are utilised for this comparison, with the optimal conditions for each material being utilised and therefore facilitating comparison of the most efficient systems for ECL production.

This slightly faster regeneration diffusion coefficient of T4BT–B may ultimately increase the efficiency of an ECL system based on this star-shaped compound. This could impact greatly on sensor design, as the faster regeneration of the excited state species should drastically improve the sensitivity of systems involving these star-shaped compounds; although other factors may also have an influence, the diffusion rates are a factor solely influenced by the properties of the material itself. These values were calculated using cyclic voltammetry and will be confirmed in future work by investigation using chronoamperometry. Both the influence of the rate of reaction and diffusion rates of the electrogenerated

Table 1 D for 1×10^{-4} M T4BT–B in DCM and 1×10^{-4} M $[Ru(bpy)_3]^{2+}$ in 0.1 M $(TBA)PF_6$ in CH_2Cl_2, based on the results from Fig. 2[a]

Compound	$D \pm 0.08$ ($cm^2\ s^{-1}$)
$[Ru(bpy)_3]^{2+}$	2.3×10^{-6}
T4BT–B	1.63×10^{-5}

[a] All values are based on averaged results from 3 independent measurements.

product will impact on the efficiency of ECL production. To contrast the efficiency of the materials themselves, the diffusion coefficient can be an important parameter to investigate when examining the potential of ECL-generating materials.

Electrochemiluminescence properties

One of the key objectives of this work was to investigate whether T4BT–B exhibits ECL in comparative dynamic linear ranges compared to ruthenium-based systems and to subsequently compare the efficiency of ECL generation of these two systems for the detection of TPA. Fig. 7 shows the potential dependence of the ECL response for both T4BT–B and $[Ru(bpy)_3]^{2+}$ using TPA as the co-reactant. Both systems produced ECL upon electrochemical production of the cation radicals $T4BT–B^{+}$ and $[Ru(bpy)_3]^{3+}$ respectively, in the presence of TPA. However, these intensities were considerably higher for the T4BT–B system in comparison to the $[Ru(bpy)_3]^{2+}$-based system under the same experimental conditions.

For the detection of amino acids, specifically TPA in this study, the electrogenerated $[Ru(bpy)_3]^{3+}$ has been shown to undergo a mediated oxidation reaction with the amino acids. Significantly, the chemiluminescence intensity is maximised at pH values near the pK_a of the N-terminal amine site.[2] The T4BT–B compound undergoes a similar process with the reaction between the electrogenerated $T4BT–B^{+\cdot}$ and the TPA being sufficiently energetic to produce an ECL response. Direct oxidation of TPA occurs at ~2.1 V. As with the reaction mechanism for ruthenium, T4BT–B and TPA are simultaneously oxidised during a single potential step to produce $(T4BT–B)^{+\cdot}$ and TPA^{+} respectively. The TPA radical then loses a proton to form the strongly reducing radical species $TPA^{\cdot}$, which reacts with the $(T4BT–B)^{+\cdot}$, producing the excited state that emits light, according to eqn (7)–(11), which is similar to the mediated oxidation of oxalate by TPA.[26] This is a proposed scheme for the production of ECL from this new material, T4BT–B, and current investigations are ongoing to establish an exact mechanism.

$$\text{T4BT–B} - e^{-} \rightarrow (\text{T4BT–B})^{+\cdot} \tag{7}$$

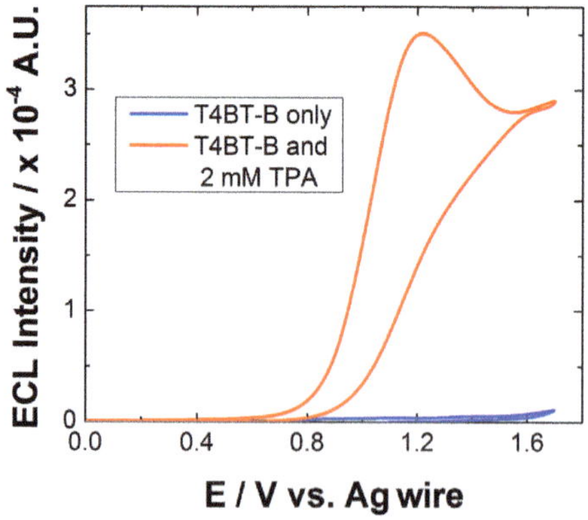

Fig. 3 Potential dependence of the ECL emission intensity of 1×10^{-4} M T4BT–B in the presence (red line) and absence (blue line) of 2 mM TPA, monitored at a scan rate of 100 mV s^{-1}. The electrolyte was 0.1 M $(TBA)PF_6$ in anhydrous DCM.

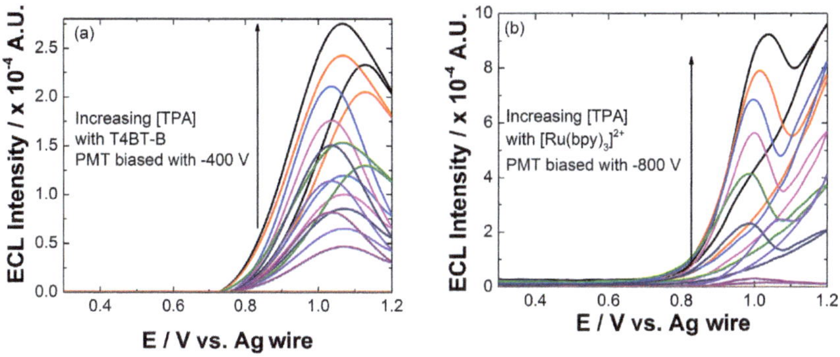

Fig. 4 ECL response of 1×10^{-4} M (a) T4BT–B and (b) $[Ru(bpy)_3]^{2+}$ on the concentration of [TPA] at a scan rate of 100 mV s^{-1} over the potential range $0.2 V \leq E \leq 1.2 V$ *vs.* Ag wire. The PMT was biased at −400 V for (a) and −800 V for (b).

$$\text{TPA} - e^- \rightarrow \text{TPA}^{+} \tag{8}$$

$$\text{TPA}^{+} - \text{H}^+ \rightarrow \text{TPA}^{\cdot} \tag{9}$$

$$(\text{T4BT–B})^{+\cdot} + \text{TPA}^{\cdot} \rightarrow (\text{T4BT–B})^{*} + \text{Pr}_2\text{N}^+ = \text{CHEt} \tag{10}$$

$$(\text{T4BT–B})^{*} \rightarrow \text{T4BT–B} + h\nu \tag{11}$$

Fig. 3 shows the potential dependence of the ECL emission intensity of T4BT–B. In the absence of TPA, there is no ECL response. However, in the presence of 2 mM TPA, the onset of light emission coincides closely with the onset of the oxidative current where the potential of T4BT–B^{+} is generated, as can be seen in the inset of Fig. 7. This behaviour is consistent with TPA reducing the electrogenerated T4BT–B^{+} and creating the excited state species which subsequently results in the generation of ECL emission. The ECL emission intensity is centred

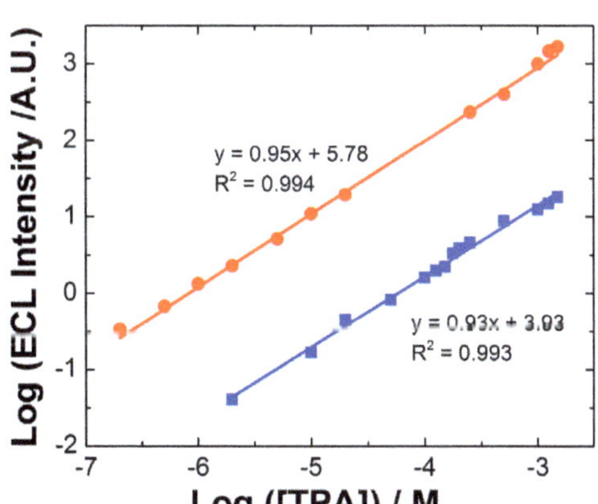

Fig. 5 Typical calibration plot of the dependence of the logarithm of the maximum ECL intensity on log[TPA] response for 1×10^{-4} M T4BT–B (red line) and $[Ru(bpy)_3]^{2+}$ (blue line). The responses for T4BT–B have been adjusted for changes in the PMT bias.

at 564 nm, which is slightly shifted compared to the luminescence intensity which is centred at 551 nm (see ESI, Table S1 and Fig. S1†). In comparison to other T4 compounds,[20] where there was no fluorescence or ECL observed above 450 nm, for T4BT–B there was luminescence observed even above 550 nm, which can be related to the addition of BT units, attributing to longer wavelength emission. However, the ECL monitored is total light produced rather than monitoring a specific wavelength. Therefore, the advantages of a different wavelength of ECL emission are only assessed for systems of coloured matrices, and therefore it can be useful to have a variety of ECL emitters capable of producing ECL at a variety of different wavelengths.

The onset of the ECL emission is at approximately 0.8 V, which is significantly less positive than the standard redox potential of TPA of ~2.1 V.[2,27] This behaviour arises because a significant overpotential must be applied before TPA can be oxidised at a bare electrode. The large increase in emission intensity at circa 1.1 V clearly demonstrates that TPA is oxidised at potentials when the electrogenerated T4BT–B^{+} sites exist, allowing interaction and subsequent ECL production to occur. This response can also be seen in the current response recorded simultaneously. Fig. 4 shows that the ECL response increases with increasing concentrations of TPA over the concentration range 2×10^{-8} M ≤ [TPA] ≤ 2.0×10^{-3} M for T4BT–B and over the range 1 10^{-4} M ≤ [TPA] ≤ 2.0×10^{-3} M for $[Ru(bpy)_3]^{2+}$.

Fig. 4 highlights the fact that although both compounds produced an ECL response in the presence of TPA, the magnitude of the response was dramatically higher for T4BT–B compared with $[Ru(bpy)_3]^{2+}$. In fact, to record responses, the settings of the PMT needed to be doubled for $[Ru(bpy)_3]^{2+}$ compared to T4BT–B to obtain similar responses. This result is consistent with the higher diffusion rates of T4BT–B, as well as the enhanced luminescence properties. Fig. 5 illustrates a typical calibration plot for [TPA] ranging from 0.2 μM to 2 mM based on the data obtained from Fig. 4; this cannot be compared directly to Fig. 6 since different PMT parameters were utilised. However, it does highlight the enhanced response

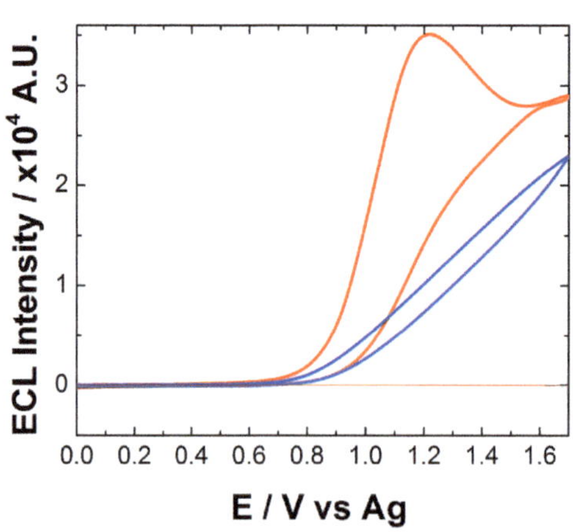

Fig. 6 ECL response of 1×10^{-4} M T4BT–B only (red line) and with 2×10^{-8} M TPA (blue line) at a scan rate of 100 mV s^{-1} over the potential range 0 V ≤ ν ≤ 1.2 V *vs.* Ag^{+}. The PMT was biased at −750 V.

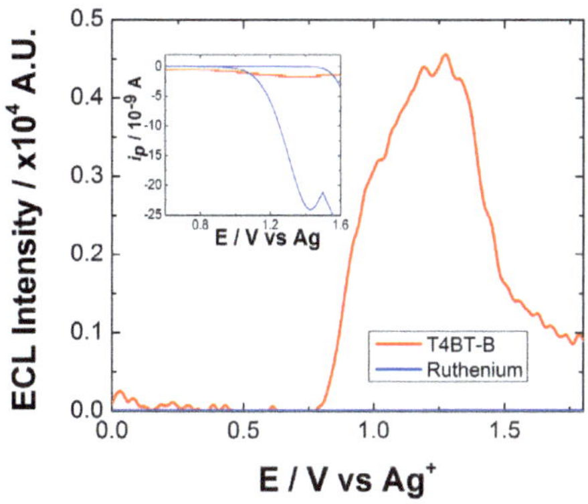

Fig. 7 Potential dependence of the ECL emission intensity of 1×10^{-4} M T4BT–B (red line) and 1×10^{-4} M $[Ru(bpy)_3]^{2+}$ (blue line) in the presence of 2 mM TPA, monitored at a scan rate of 100 mV s^{-1}. The electrolyte was 0.1 M $(TBA)PF_6$ in anhydrous DCM. The PMT was biased at −750 V. The inset shows the amperometric response recorded simultaneously.

for TPA when using the T4BT–B system. For each concentration, a well-defined response is observed with a high signal-to-noise ratio even at 20 pM, as can be seen in Fig. 7. This calibration plot demonstrates that log(I_{ECL}) depends linearly on log[TPA] for both compounds. The logarithm of ECL intensity is almost directly proportional to log[TPA] (slope of 0.95), suggesting that although the TPA concentration does influence the light emission, other factors such as the diffusion of both species to the electrode do not play a major role in this system. The wide dynamic range and high S/N ratio makes T4BT–B attractive for reagent-free, ultrasensitive detection of chemical and biological analytes.

In terms of developing sensors, a key issue is the relative quantum efficiency of the materials. To evaluate this parameter, it is usual to compare the quantum efficiency of the new compound to a standard; for ECL production $[Ru(bpy)_3]^{2+}$ is most commonly used. The overall ECL efficiency (ϕ_{ECL}) is defined as the number of photons emitted per Faradaic electron passed during the chemiluminescence reaction. It is a product of the efficiency of populating the excited state (ϕ_{EX}) and the quantum yield of emission from the excited state (ϕ_P), with $[Ru(bpy)_3]^{2+}$ normally used as the relative standard.[27] This investigation will be undertaken in future research, however, as an indication, the intensity of ECL produced under the same experimental conditions was examined to provide a quick insight into the efficiency of T4BT–B in comparison to $[Ru(bpy)_3]^{2+}$ utilising TPA as the co-reactant.

Although not as accurate as determining the ϕ_{ECL} for each system, Fig. 7 does highlight that the overall efficiency for T4BT–B would be higher than that obtained for $[Ru(bpy)_3]^{2+}$, which is usually taken as 5%.[2,4,28] Fig. 7 shows the response of both materials under identical conditions and can therefore be utilised for direct comparison. For T4BT–B a substantially lower current response is observed compared to that of the ruthenium system. Despite this, T4BT–B shows a dramatically higher ECL response. This also means that although T4BT–B has a slightly higher diffusion coefficient, it is more likely that the higher luminescence quantum yield (see Table S1†) is responsible for the greater ECL response. Fig. 6 shows that the ECL intensity obtained for T4BT–B is approximately twice that observed for $[Ru(bpy)_3]^{2+}$. This enhanced efficiency is most likely due to a combination of the faster diffusion rates as well as the higher luminescence quantum yields (see ESI, Table S1 and Fig. S1†) of T4BT–B compared to $[Ru(bpy)_3]^{2+}$, and for these systems the higher luminescence quantum yields are most likely to contribute significantly more to the enhanced ECL. This is in conjunction with the limit of detection, LOD, defined as $I_{ECL.BLANK} + 3\sigma_{BLANK}$, which is 3 fM for the T4BT–B system, where $I_{ECL.BLANK}$ refers to the response when no co-reactant is added to the solution or noise inherent to the system and σ refers to the standard deviation of this noise. This confirms that T4BT–B can enhance the ECL emission intensity. The wide dynamic range and low limit of detection makes T4BT–B attractive for biosensor and chemical sensor development. The additional benefits of lower power requirements for PMT settings due to the higher ECL intensity produced from equivalent concentrations of co-reactants also means that T4BT–B would be more viable for miniaturisation, which would be highly beneficial for biomedical, forensic and food monitoring applications.

Conclusions

The electrochemical and ECL properties of the trigonal quarterfluorene–truxene oligomer with fused 2,1,3-benzothiadazole units, T4BT–B, have been presented here. T4BT–B shows good electrochemistry in CH_2Cl_2, with excellent stability properties of both the anion and cation radicals. T4BT–B also shows a yellow emission when generated by reaction with TPA, consistent with its photophysical properties. The production of ECL from the reaction of T4BT–B with TPA was shown to be linearly dependant over a wide dynamic linear range, with low limits of detection and good reproducibility, significantly producing almost double the ECL intensity compared to $[Ru(bpy)_3]^{2+}$ under the same experimental conditions. This work illustrates the proof of concept of the promising ECL sensing applications of this new family of compounds in a wide variety of applications ranging from biomedical diagnostics to forensic science. These possibilities are open due to the highly luminescent and fast charge transfer properties of these compounds which can be utilised to detect low levels of co-reactant without increasing the background noise levels, thereby allowing for the development of an ultrasensitive and selective ECL sensor.

Acknowledgements

The authors acknowledge funding support from the Marie Curie Reintegration Grant (PIRG-2010-268236). PJS thanks the Royal Society for a Wolfson Research Merit Award.

References

1 J. Li, S. Guo and E. Yang, *RSC Adv.*, 2012, **2**, 3579–3586.
2 L. Dennany, E. J. O'Reilly, T. E. Keyes and R. J. Forster, *Electrochem. Commun.*, 2006, **8**, 1588–1594.
3 H. Wei and E. Wang, *Chem. Soc. Rev.*, 2011, **26**, 77–85.
4 A. Devadoss, L. Dennany, C. Dickinson, T. Keyes and R. J. Forster, *Electrochem. Commun.*, 2012, **19**, 43–45.

5 L. Dennany, M. Gerlach, S. O'Carroll, T. E. R. J. Forster and P. Bertoncello, *J. Mater. Chem.*, 2011, **21**, 13984–13990.

6 L. Hu and G. Xu, *Chem. Soc. Rev.*, 2010, **39**, 3275–3304.

7 A. W. Knight and G. M. Greenway, *Analyst*, 1994, **119**, 879–890.

8 P. Bertoncello and R. J. Forster, *Biosens. Bioelectron.*, 2009, **24**, 3191–3200.

9 J. K. Lee, D. S. Yoo, E. S. Handy and M. F. Rubner, *Appl. Phys. Lett.*, 1996, **69**, 1686–1688.

10 W. K. Ward, *ASAIO J.*, 1999, **45**, 555–561.

11 A. Weisner, *Curr. Pharm. Biotechnol.*, 2004, **5**, 45–67.

12 M. S. Wilson, *Anal. Chem.*, 2005, **77**, 1496–1502.

13 A. E. Cass, G. Davis, G. D. Francis, H. A. O. Hill, W. J. Aston, J. Higgins, E. V. Plotkin, L. D. L. Scott and P. F. Tuner, *Anal. Chem.*, 1984, **56**, 667–671.

14 L. Dennany, E. J. O'Reilly, P. C. Innis, G. G. Wallace and R. J. Forster, *J. Phys. Chem. B*, 2008, **112**, 12907–12912.

15 O. Haas, M. Kierns and J. G. Vos, *J. Am. Chem. Soc.*, 1981, **103**, 1318–1319.

16 P. Qu, D. W. Thompson and G. J. Meyer, *Langmuir*, 2000, **16**, 4662–4671.

17 R. P. Swager, *J. Am. Chem. Soc.*, 1999, **121**, 1100–1104.

18 E. Rampazzo, S. Bonacchi, D. Genovese, J. Riccardo, M. Marcaccio, M. Montalti, F. Paolucci, M. Sgarzi, G. Valenti, N. Zaccheroni and L. Prodi, *Coord. Chem. Rev.*, 2012, **256**, 1664–1681.

19 W. Miao, *Chem. Rev.*, 2008, **108**, 2506–2553.

20 A. L. Kanibolotsky, F. Vilela, J. C. Forgie, S. E. T. Elmasly, P. J. Skabara, K. Zhang, B. Tieke, J. McGurk, C. R. Belton, P. N. Stavrinou and C. D. D. Bradley, *Adv. Mater.*, 2011, **23**, 2093–2097.

21 C. G. Cameron and P. G. Pickup, *J. Am. Chem. Soc.*, 1999, **121**, 11772–11779.

22 C. R. Belton, A. L. Kanibolotsky, J. Kirkpatrick, C. Orofini, S. E. T. Elmasly, P. N. Stavrinou, P. J. Skabara and D. D. C. Bradley, *Adv. Funct. Mater.*, 2013, **23**, 2792–2804.

23 E. J. O'Reilly, L. Dennany, D. Griffith, F. Moser, T. E. Keyes and R. J. Forster, *Phys. Chem. Chem. Phys.*, 2011, **13**, 7095–7101.

24 L. Dennany, T. E. Keyes and R. J. Forster, *Analyst*, 2008, **133**, 753–759.

25 L. Dennany, E. J. O'Reilly, P. C. Innis, G. G. Wallace and R. J. Forster, *Electrochim. Acta*, 2008, **53**, 4599–4605.

26 K. M. Omer, A. L. Kanibolotsky, P. J. Skabara, I. F. Perepichka and A. J. Bard, *J. Phys. Chem. B*, 2007, **111**, 6612–6619.

27 A. Venkatanarayanan, A.-M. Spehar-Délèze, L. Dennany, Y. Pellegrin, T. E. Keyes and R. J. Forster, *Langmuir*, 2008, **24**(19), 11233–11238.

28 E. J. O'Reilly, T. E. Keyes, R. J. Forster and L. Dennany, *Analyst*, 2013, **138**(2), 677–682.

Faraday Discussions

PAPER

Hybrid organic semiconductor lasers for bio-molecular sensing

Anne-Marie Haughey,[a] Caroline Foucher,[a] Benoit Guilhabert,[a] Alexander L. Kanibolotsky,[b] Peter J. Skabara,[b] Glenn Burley,[b] Martin D. Dawson[a] and Nicolas Laurand*[a]

Received 2nd May 2014, Accepted 16th May 2014
DOI: 10.1039/c4fd00091a

Bio-functionalised luminescent organic semiconductors are attractive for biophotonics because they can act as efficient laser materials while simultaneously interacting with molecules. In this paper, we present and discuss a laser biosensor platform that utilises a gain layer made of such an organic semiconductor material. The simple structure of the sensor and its operation principle are described. Nanolayer detection is shown experimentally and analysed theoretically in order to assess the potential and the limits of the biosensor. The advantage conferred by the organic semiconductor is explained, and comparisons to laser sensors using alternative dye-doped materials are made. Specific biomolecular sensing is demonstrated, and routes to functionalisation with nucleic acid probes, and future developments opened up by this achievement, are highlighted. Finally, attractive formats for sensing applications are mentioned, as well as colloidal quantum dots, which in the future could be used in conjunction with organic semiconductors.

A Introduction

The development of personalised medicine requires that specific diagnostic tools be made available for use at the point-of-care. In particular, sensitive and compact platforms that can flexibly and accurately identify disease biomarkers, *e.g.* from a patient's blood sample, and do so in the shortest time, are critically needed.[1,2] These diagnostic platforms should be simple to implement and operate in order to maximise practicality and minimise the overall cost. Label-free optical biosensors enabled by advanced functional materials, including bio-functionalised organics, have the potential to address the challenge. Here, we describe an organic semiconductor (OS) laser as a technology platform for such label-free biosensing.

Label-free biosensors in general use evanescent optical waves in order to measure minute changes in the dielectric permittivity caused by the presence of

[a]*University of Strathclyde, Institute of Photonics, Glasgow, UK. E-mail: nicolas.laurand@strath.ac.uk*
[b]*University of Strathclyde, WestCHEM, Department of Pure and Applied Chemistry, Glasgow, UK*

analytes at an interface between the device and the biological environment.[3–11] Such sensors are attractive because of their non-electrical detection, sensitivity and specificity. Several label-free optical sensor technologies currently exist.[3] Each has its advantages and downsides but not all of them are easily scalable to the level of miniaturisation required for future point-of-care use. It is often necessary to sacrifice the detection resolution and/or sensitivity for ease of implementation, and *vice versa*. This is because the transduction mechanisms for reading the dielectric permittivity changes, *i.e.* for sensing, are based on resonating structures.[3–11] The sensing function is realised by monitoring the change of an optical resonance, for example by the direct reading of the optical spectrum. The overall sensitivity of such sensor systems is then a combination of the magnitude of a resonance shift (caused by the presence of analytes) and of the intrinsic resonance bandwidth, which determines the minimum detectable shift. Optimising the sensitivity of an evanescent wave biosensor therefore means maximising the resonance shift (which can be done by maximising the optical wave intensity in the region where the analytes are present), while minimising the resonance bandwidth. Doing both simultaneously is not straightforward, as there exists a trade-off between the resonance bandwidth and the strength of the wave/analyte interaction.

A plastic laser sensor platform, which can take the form of a distributed feedback[12,13] or a photonic crystal active resonator,[14,15] can mitigate this trade-off while maintaining a simple architecture for ease of application. In such a device, the narrow resonance bandwidth is obtained through the coherence of laser action without sacrificing the strength of the field (in this case the laser mode) interaction with the sensing region. The great potential of this solution for the detection of bio-molecules has been shown. However, the first generation of plastic laser sensors utilised a dye-doped composite, *i.e.* dyes diluted in a polymeric matrix, as the optical gain region. The relatively low dielectric permittivity contrast between such a diluted medium ($n \approx 1.5$) and a biological environment ($n \approx 1.4$) limits the strength of the laser mode intensity at the sensing interface. Therefore, the approach necessitates the addition of a thin inorganic layer (TiO_2 $n \approx 1.8$ to 2.4) on top of the surface of the laser to boost the field at the interface and hence the sensitivity. This additional step adds complexity to the fabrication and means that it is not entirely done by solution processing, thereby cancelling some of the advantages of the technology.

We propose a novel embodiment of the plastic laser sensor, where the laser material is made of a neat OS medium. This approach leads to a simple device architecture where no additional inorganic layer is necessary to obtain high sensing performance[16,17] thanks to the relatively high dielectric permittivity of the dense OS medium. In turn, the fabrication of sensors can be entirely based on solution-processing techniques. Importantly for point-of-care, an OS laser configuration can be made suitable for compact optical-pumping arrangements using either InGaN laser diodes or LEDs,[18–20] thereby paving the way for miniaturisation of the biosensor system. Other key attributes of the OS laser biosensor include (i) wavelength flexibility across the visible region by molecular engineering of the OS material, an attractive feature for multiplexing; (ii) compatibility with a variety of material platforms, thereby facilitating integration in legacy sensing equipment as well as the development of new sensor formats; and (iii) the possibility to envision direct hybridisation of the OS material at the molecular level to combine the functionalities of light amplification with those of bio-molecular interactions.

In this paper, we start in part B by presenting the design, concept and characteristics of the OS laser biosensor. In part C, we show that this laser platform can detect a nanolayer that is adsorbed onto the surface of the OS and explain the sensitivity advantage of using a dense OS as the gain medium. In part D, we demonstrate specific bio-molecular sensing, show DNA functionalisation of the OS laser and discuss opportunities opened up by these demonstrations. We briefly discuss possible OS lasers in the array format which are attractive for implementing particular sensing concepts. We end by mentioning another promising solution-processed material for hybrid laser biosensors, colloidal quantum dots, which could be used in the future in conjunction with the OS.

B Organic laser sensor: design, materials and concept

The OS laser sensor considered here is based on a planar distributed-feedback structure, as shown in Fig. 1(a). It consists of a transparent, nanopatterned substrate overcoated with a thin film of OS that acts as the laser material and provides the interface with the sensing region. For sensing, the OS layer is put into contact with the biological environment (*e.g.* a drop of fluid to be tested can simply be deposited or flowed over the laser surface). The laser mode oscillates in the plane of the OS layer and, as can be seen in Fig. 1(b), interacts with the biological environment ('superstrate' and 'biolayer') through its evanescent tails. The red curve in Fig 1(b) represents the laser mode intensity profile overlapped with the different regions of the planar laser. For ease-of-implementation, the nano-pattern period is chosen to equal one wavelength of the desired laser emission. This laser emission has a vertically outcoupled component that can be collected for detection through the substrate of the device, *i.e.* on the same side as the pump excitation (Fig 1(a)). For biosensing, the sensor needs to be able to detect analytes specifically. Therefore, the surface of the laser is functionalised with molecular probes whose role is to capture and immobilise the molecules to be detected that might be present in the biological medium. Once captured, these immobilised analytes form a (possibly non-homogeneous) layer on the laser surface. The dielectric permittivity of this 'biolayer' is slightly higher than the more diluted biological medium, referred to as the superstrate in Fig. 1(b). This in turn affects the laser wavelength, which can be monitored for biosensing.

The laser wavelength is given by the product of the pattern periodicity, Λ, with the value of the mode effective index, n_{eff}. The latter is obtained by averaging the refractive indices of the substrate, OS, biolayer and superstrate weighted by the respective overlap values of these regions with the laser mode. The formation of a biolayer on the surface of the laser modifies the effective index and hence changes the laser wavelength. The magnitude of this change is dependent on the mode overlap with the biolayer and the refractive index of the latter. Hence, it depends on the amount of analyte, and monitoring the wavelength enables the sensing function (although other parameters could be monitored as well). The mode overlap with the biolayer can be optimised for sensitivity by maximising the refractive index of the laser gain material and minimising its thickness. An OS

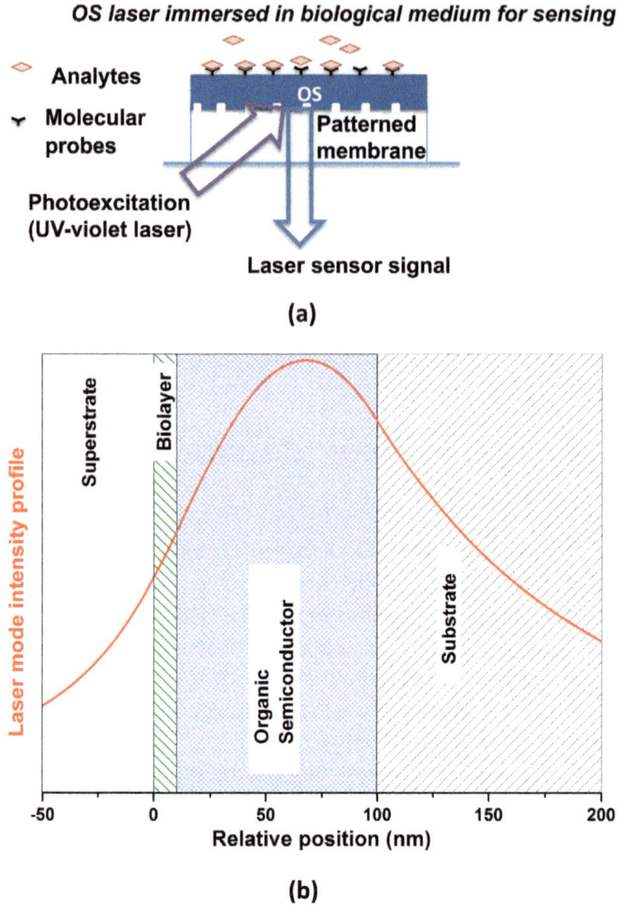

Fig. 1 (a) Schematic representing the implementation of the organic laser sensor; (b) laser mode intensity cross-section showing the overlap with the different regions of the laser structure.

laser material is therefore advantageous because it forms a 'dense' laser medium that has a high refractive index and can accommodate a low thickness without compromising the laser performance.

We use star-shaped oligofluorene truxene molecules to form the OS layer of our laser sensors (see Fig. 2(a)). Each of the three arms of the molecule, attached to a truxene core, consists of repeated fluorene units.[21] In our case the number of units per arm is three, and we therefore refer to this truxene-core material as T3. T3 macromolecules are efficiently excited by light in the 350–390 nm range and emit in the blue part of the spectrum. They can also form high-quality solid-state thin films with a high T3 density ideal for planar lasers.[22,23] The thickness of the T3 layer is set to be 70 nm ± 10 nm, thin enough to maximise sensitivity and thick enough to sustain a TE-polarised optical mode. This layer of T3 is deposited onto a nanopatterned imprinted epoxy substrate. The periodicity of the pattern is $\Lambda = 277$ nm. Such all-organic lasers can be easily incorporated into cuvettes or integrated into microtiter plates.

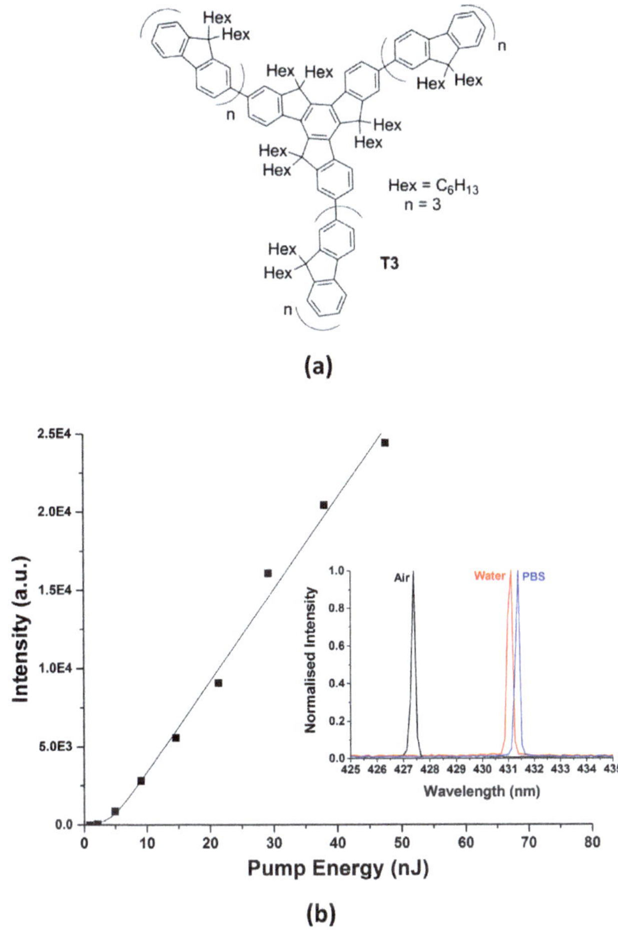

Fig. 2 (a) Representation of the T3 molecule; (b) laser transfer function (in air); inset: laser spectrum with the superstrate being, respectively, air, water and PBS.

The laser sensors are excited with 5 ns-long pulses of 355 nm pump light. A typical laser transfer function of such a laser (operated in air) is plotted in Fig. 2(b). Despite the low T3 thickness, the oscillation threshold is 5 nJ and this value is even lower (a couple of nJ) when the laser is operated in a liquid superstrate such as water or buffer. The inset of Fig. 2(b) displays the laser spectra when the superstate is, respectively, air, water and phosphate buffered saline (PBS). The laser linewidth is limited by the spectrometer resolution in the three cases. The laser wavelength is seen to red shift for the increasing refractive index of the superstrate: $\Delta\lambda$ (air-to-water) = 3.73 ± 0.06 nm and $\Delta\lambda$ (air-to-PBS) = 4.00 ± 0.06 nm. The respective refractive indices of air, water and PBS solution are approximately 1, 1.34 and 1.35. The measurements of the refractive index changes of the superstrate in such a way are repeatable and have been used to measure the concentration of glycerol–water solutions and have served to validate the sensing approach.[16]

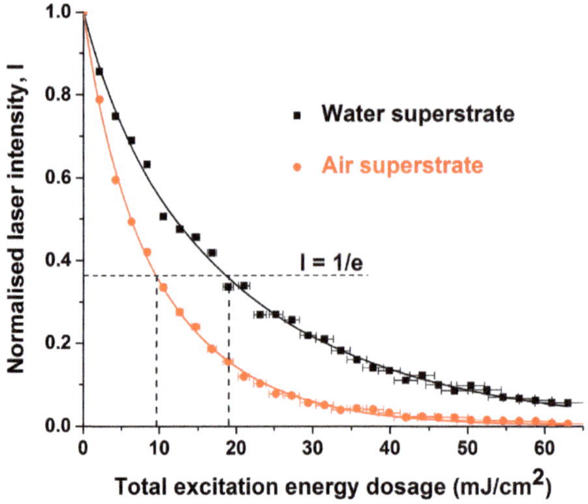

Fig. 3 Laser output intensity *versus* the total pump energy dosage when the device is operated in air and in water.

To operate as a biosensor, the photostability of the OS laser over the duration of an experiment is critical. T3 lasers operated in air have a $1/e$ degradation dosage of 10 J cm^{-2}, while it is close to 20 J cm^{-2} when operated in a water environment (see Fig. 3). In the experimental conditions of our sensor, this corresponds to more than 10 to 20 $\times$ 10^{5} pulses. Over such a number of pulses the oscillation wavelength does not vary. Because a single sensing measurement requires less than 10 to 20 pulses, there are no parasitic errors due to photodegradation effects. In fact, the T3 laser photostability is high enough that it should enable dynamic/continuous sensing over tens of minutes if exciting the laser sensor at a pulse repetition rate of 10 Hz.

C Nanolayer detection and comparison with dye-doped lasers

The biosensing principle of the laser platform is to detect molecules immobilised onto the surface of the OS material. This is different than sensing a change in the bulk refractive index of the superstrate, as was shown in the inset of Fig. 2. To observe the physical effect that a biolayer will have on the laser biosensor response, as well as to assess the capability and get a first idea of the sensing limit of the technology platform independently of the biochemical surface functionalisation, nm-thick layers of polyelectrolytes can be built up onto the OS surface. For this, we use poly(ethyleneimine) (PEI), poly(allylamine hydrochloride) (PAH) and poly(sodium 4-styrenesulfonate) (PSS). These are well studied and form monolayers of known thickness (5 nm) and refractive index (1.49, close to that of most proteins) after a few initial layers of varying thicknesses.[24,25] To adsorb each polyelectrolyte layer, the laser biosensor is immersed in solutions of the corresponding polyelectrolytes (all at 5 mg mL^{-1} in NaCl) for 10 minutes with intermediate washing steps. The first deposited layer is PEI and the successive layers

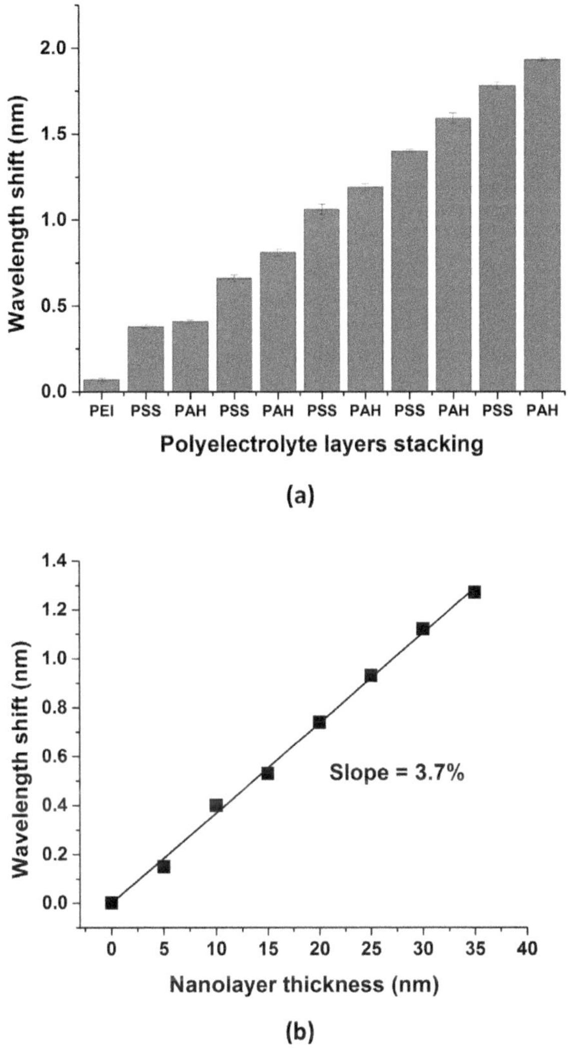

Fig. 4 (a) Laser wavelength shift upon polyelectrolyte layer stacking. The first layer is made of PEI, while each successively adsorbed polyelectrolyte layer is PSS and PAH respectively. (b) Laser wavelength shift *versus* adsorbed nanolayer thickness.

form a stack of PSS and PAH. After each layer deposition and surface washing, the laser wavelength is measured when immersed in a reference NaCl solution. The wavelength shift of the laser induced by this growth of the nanolayers on top of the OS laser is shown in Fig. 4(a). After the first 3 layers, the wavelength red shifts linearly up to the 11th layer. The slope of the wavelength shift per nm of adsorbed material is 3.7%, as indicated in Fig. 4(b). This means that the laser sensor can detect thicknesses of the adsorbed material from the nanometre scale up to several tens of nanometres. In principle, the smallest thickness that could be detected is limited by the resolution of the current system ($\pm$0.06 nm), corresponding to <2 nm. This equates to proteins in the range of 5 kDa,[26] *i.e.* the OS

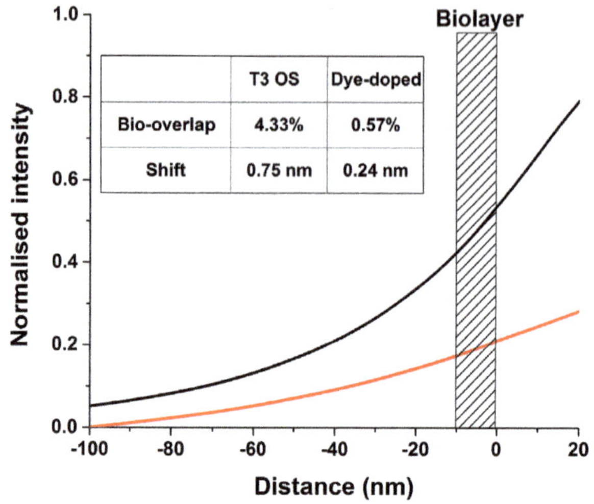

Fig. 5 Mode intensity profile in the vicinity of the laser/biolayer interface (position at 0 nm on the graph). Red curve: case of the dye-doped laser; black curve: case of the T3 OS laser. A 10 nm thick biolayer is considered (patterned area between −10 nm and 0 nm). The mode overlap, and hence the wavelength shift, is higher for the T3 OS laser.

laser sensor has the potential to detect most of the relevant biomarkers. This assumes that proteins cover the whole of the laser surface, which might be the case only in the limit of high concentration. Nevertheless, it demonstrates the intrinsic capability of the sensor while ignoring any amplification stage (which could be used to further boost the sensitivity).

Adsorption of a nanolayer is also a useful tool to compare different surface sensing platforms independent of considerations for the chemistry of surface molecular probes. Here, we model our OS laser sensor and a similar laser structure that instead uses a dye-doped polymer for the gain region. Our model is based on a modified planar waveguide approach where we considered the dispersion of the different layers of the structure. This model has been shown to accurately predict and replicate the experimental results for our OS laser.[16,17] We have verified that it replicates published results of dye-doped laser sensors as well. We consider our T3 OS laser (details given previously) and a dye-doped laser emitting at 590 nm with a refractive index for the gain region of 1.51 and a pattern periodicity Λ = 400 nm. The model determines in both cases the effective refractive index for the TE_0 mode, the mode profile, the overlap with the biolayer and the expected wavelength shift. The mode intensity profiles are represented in Fig. 5 along with the overlap values and the wavelength shifts for a 10 nm biolayer with a refractive index of 1.45. Both the overlap and wavelength shifts are higher (4.37% *vs.* 0.57% and 0.75 nm *vs.* 0.24 nm) for the OS laser. The higher sensitivity of the OS laser is due to the higher index contrast between the laser and the biolayer. The dye-doped laser cannot match this sensitivity without an additional TiO_2 layer. Overall this section demonstrates that our T3 laser (and by extension other lasers made from neat OS) represents a simple yet potentially sensitive biosensing platform.

D Specific bio-molecular detection and DNA functionalisation

The specific detection of molecules requires surface functionalisation of the laser with molecular probes (Fig. 1(a)). We show, as a proof-of-principle, specific detection of avidin by functionalising the T3 laser surface with biotin as the molecular probe. The approach for functionalisation utilises the adsorption of a monolayer of a polyelectrolyte, polyphenylalanine lysine, or PPL (see the inset of Fig. 6 for a schematic representation of PPL). For this, a similar approach to the layer adsorption described in Section C was taken. Specifically, a solution of PPL in PBS (10 mM PBS, pH of 7.4) was prepared at a concentration of 1 mg mL^{-1}. The laser was immersed in this solution for 10 minutes and then washed and rinsed with PBS. The nanolayer of PPL provides accessible primary amine functional groups that can be used to anchor molecular probes on the laser surface. Here, the coated laser was immersed in a *N*-hydroxysuccinimide (NHS)–biotin/PBS solution for 20 minutes in order to attach biotin molecules onto the laser. The laser surface was then washed again in PBS. For sensing experiments, the biotin-functionalised laser was immersed in solutions of avidin/PBS of different concentrations for 20 minutes before a washing step to remove unbound avidin molecules.

The results of the wavelength shift due to avidin attachment to the biotin laser as a function of the avidin concentration are plotted in Fig. 6. Between 100 ng mL^{-1} and 1000 ng mL^{-1} the wavelength is seen to red shift with increasing avidin concentration. The behaviour saturates at a concentration above 25 mg mL^{-1}. This saturating behaviour follows a Langmuir equation that relates to the coverage of molecules on a surface, and is caused in this case by avidin molecules occupying all available biotin sites on the laser. The limit of detection corresponds to a

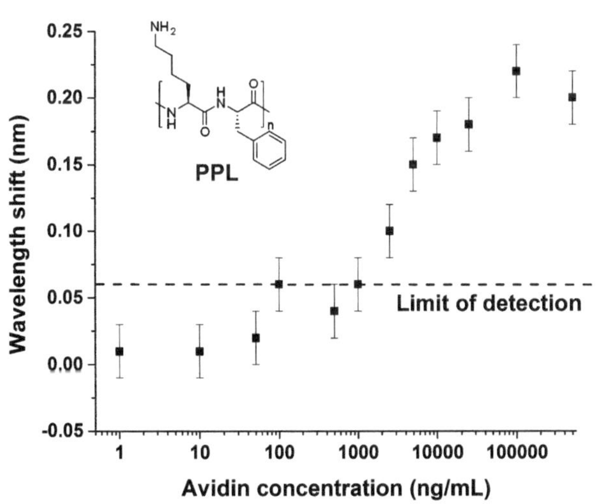

Fig. 6 Laser wavelength shift caused by avidin–biotin binding onto the laser surface as a function of avidin concentration. Inset: schematic of the PPL monomer. PPL is used as an intermediate functionalisation layer on the T3 laser surface. It presents amine groups which are then reacted with NHS-functionalised molecules (biotin in this case).

concentration of 1000 ng mL^{-1} (shift of 0.06 nm). Currently, the limiting factor is the resolution of the system (convolution of the laser linewidth and of the spectrometer intrinsic resolution). Further refinement of the laser structure and the use of a spectrometer with a higher resolution will improve this limit of detection. We have also verified that this sensing capability is unaffected by the presence of albumin at a concentration equivalent to that found in blood.[17] This is significant as albumin molecules tend to non-specifically stick to surfaces, and this is known to cause problems in some biosensors.

This biomolecular sensing demonstration proves the potential of OS lasers for such applications. The next step is to devise functionalisation pathways for the specific sensing of relevant biomarkers and to assess the limit of the technology in this case. The most common commercial biosensors (*e.g.* ELISA assays) use antibodies, but we think nucleic acid (NA) probes are well suited for functionalising OS lasers. NA probes have the advantage of being more environmentally stable than antigens/antibodies, and they can also be synthetically programmed. With this approach, the NA probes could immobilise the mRNA of protein biomarkers on the laser surface. Alternatively, aptamers could be used to directly immobilise proteins. As a first step on this route, we demonstrate NA attachment onto our T3 OS laser. We react an amide-to-maleimide crosslinker (sulfosuccinimidyl-4-[*N*-maleimidomethyl]cyclohexane-1-carboxylate or SMCC) with the amine functional groups available onto the PPL-coated OS laser. This makes maleimide moieties available on the laser surface for further functionalisation. Fig. 7 represents the wavelength of such an SMCC-functionalised laser before and after treatment with a solution of thiol–DNA. The shift (0.17 ± 0.06 nm) is attributed to the attachment of DNA onto the laser through the thiol–maleimide reaction. This represents a possible route to sensing with NA probes.

An alternative and exciting avenue, albeit possibly synthetically challenging, will be to directly hybridise the NA probes with the OS molecules. Such a step, if viable, would enable an OS laser which is intrinsically bio-functionalised and

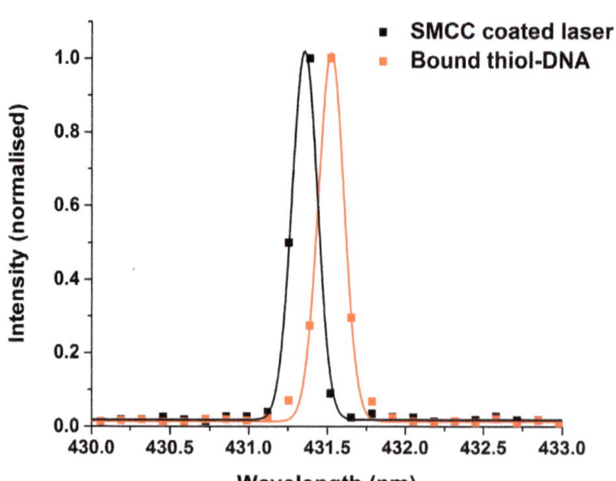

Fig. 7 Laser wavelength shift caused by the binding of thiol–DNA onto the surface of an SMCC-coated laser.

'sensing ready' after fabrication. This will create interesting opportunities for photonics and biochemistry and could lead to improved sensitivities.

Increasing the throughput of a biosensor can be obtained by (i) spatial multiplexing and by (ii) spectral multiplexing, thereby enabling the parallel detection of different biomarkers. Both approaches could be combined. A possible way to realise spatial multiplexing is to have an array of OS lasers with each laser element being functionalised with a different type of molecular probes. The simplicity of the all-solution processing fabrication of the OS lasers lends itself to the assembly of such an array of lasers. OS lasers in such a geometry, which are compatible for integration with arrays of InGaN light-emitting diode excitation sources, have been demonstrated.[27] Another method of spatial multiplexing is to functionalise a single OS laser, featuring a wide sensing area, with different molecular probes at different positions. Again, OS lasers are well-suited for this as their active sensing area can be scaled from below 1 mm^2 up to several cm^2.[13] Spectral multiplexing would enable simultaneous reading of closely assembled lasers, each emitting at a different wavelength. This would necessitate the use of lasers with different nanopattern periodicity as well as the use of different OS materials for further wavelength coverage.

We will end this paper by mentioning an alternative solution-processing material which could also be utilised for such laser sensing in the future: colloidal quantum dots (CQDs). CQDs are inorganic semiconductor nanocrystals with surfaces that are coated with organic ligands. CQDs can cover a wide wavelength range (and are therefore attractive for spectral multiplexing) by tuning both the size of the CQDs and their alloy composition. Optically-pumped CQD lasers have great potential for sensing and their performance is improving.[28,29] Furthermore, CQD laser sensing will benefit from the advancing functionalities which are being developed for their use as bio-imaging labels. However, the threshold performance of CQD lasers is still two orders of magnitude higher than OS lasers, and further improvement is needed. Nevertheless, it is a technology worth watching as in the future it could complement OS in such laser sensing applications.

E Conclusions

Light-emitting organic semiconductors are already being investigated for gas sensing using their photoluminescent and/or optical gain properties.[30,31] Here, we see that they also have great potential for biosensing in a laser format. OS can act as the laser material as well as the interface with the biological environment. The resulting OS lasers could lead to simple and compact label-free biosensing platforms. The DNA functionalization of the OS laser shown here opens up opportunities using the power of NA technology, both for device research and bio-applications. The challenge is now to assess the capability and limits of this approach in true applications. The all-solution-processing fabrication capability of OS lasers is one of their advantages. Other solution-based laser materials (*e.g.* CQDs) may also have a role to play.

Acknowledgements

The authors acknowledge the support of EPSRC for this work through the grant 'Hybrid Colloidal Quantum Dot Lasers for Conformable Photonics' (EP/J021962/1) and the platform grant 'Gallium nitride enabled hybrid and flexible photonics' (EP/I029141/1).

Notes and references

1 C. A. Holland and F. L. Kiechle, *Curr. Opin. Microbiol.*, 2005, **8**, 504–509.
2 A. J. Tudos, G. A. J. Besselink and R. B. M. Schasfoort, *Lab Chip*, 2001, **1**, 83–95.
3 X. Fan, I. M. White, S. I. Shopova, H. Zhu, J. D. Suter and Y. Sun, *Anal. Chim. Acta*, 2008, **620**, 8–26.
4 M. Kristensen, A. Krüger, N. Groothoff, J. García-Rupérez, V. Toccafondo, J. García-Castelló, M. J. Banuls, S. Peransi-Llopis and A. Maquieira, *Optical Sensors*, Optical Society of America, 2011, SWB1, http://www.opticsinfobase.org/abstract.cfm?URI=Sensors-2011-SWB1.
5 S. M. Shamah and B. T. Cunningham, *Analyst*, 2011, **136**, 1090–1102.
6 V. Koubová, E. Brynda, L. Karasová, J. Škvor, J. Homola, J. Dostálek, P. Tobiška and J. Rošický, *Sens. Actuators, B*, 2001, **74**, 100–105.
7 M. Baaske and F. Vollmer, *ChemPhysChem*, 2012, **13**(2), 427–436.
8 J. Homola, S. S. Yee and G. Gauglitz, *Sens. Actuators, B*, 1999, **54**, 3–15.
9 B. Cunningham, J. Qiu, P. Li and B. Lin, *Sens. Actuators, B*, 2002, **87**(2), 365–370.
10 F. Volmer and S. Arnold, *Nat. Methods*, 2008, **5**(7), 591–596.
11 M. Armani, R. P. Kulkarni, S. E. Fraser, R. C. Flagan and K. J. Vahala, *Science*, 2007, **317**, 783–787.
12 M. Lu, S. S. Choi, C. J. Wagner, J. G. Eden and B. T. Cunningham, *Appl. Phys. Lett.*, 2008, **92**(26), 261502.
13 Y. Tan, C. Ge, A. Chu, M. Lu, W. Goldshlag, C. S. Huang, A. Pokhriyal, S. George and B. Cunningham, *IEEE Sens. J.*, 2012, **12**(5), 1174–1180.
14 C. L. C. Smith, J. U. Lind, C. H. Nielsen, M. B. Christiansen, T. Buss, N. B. Larsen and A. Kristensen, *Opt. Lett.*, 2011, **36**(8), 1392–1394.
15 C. Vannahme, C. L. C. Smith, M. B. Christiansen and A. Kristensen, *Appl. Phys. Lett.*, 2012, **101**, 151123.
16 A.-M. Haughey, B. Guilhabert, A. L. Kanibolotsky, P. J. Skabara, G. A. Burley, M. D. Dawson and N. Laurand, *Sens. Actuators, B*, 2013, **185**, 132–139.
17 A.-M. Haughey, B. Guilhabert, A. L. Kanibolotsky, P. J. Skabara, G. A. Burley, M. D. Dawson and N. Laurand, *Biosens. Bioelectron.*, 2014, **54**, 679–686.
18 T. Riedl, T. Rabe, H.-H. Johannes, W. Kowalsky, J. Wang, T. Weimann, P. Hinze, B. Nehls, T. Farrell and U. Scherf, *Appl. Phys. Lett.*, 2006, **88**, 241116.
19 A. E. Vasdekis, G. Tsiminis, J.-C. Ribierre, L. O'Faolain, T. F. Krauss, G. A. Turnbull and I. D. W. Samuel, *Opt. Express*, 2006, **14**(20), 9211–9216.
20 Y. Yang, G. A. Turnbull and I. D. W. Samuel, *Appl. Phys. Lett.*, 2008, **92**(16), 163306.
21 A. L. Kanibolotsky, R. Berridge, P. J. Skabara, I. F. Perepichka, D. D. C. Bradley and M. Koeberg, *J. Am. Chem. Soc.*, 2004, **126**(42), 13695–13702.
22 G. Tsiminis, Y. Wang, P. E. Shaw, A. L. Kanibolotsky, I. F. Perepichka, M. D. Dawson, P. J. Skabara, G. A. Turnbull and I. D. W. Samuel, *Appl. Phys. Lett.*, 2009, **94**(24), 243304.
23 J. Herrnsdorf, B. Guilhabert, Y. Chen, A. Kanibolotsky, A. Mackintosh, R. Pethrick, P. Skabara, E. Gu, N. Laurand and M. D. Dawson, *Opt. Express*, 2010, **18**(25), 25535–25545.

24 J. Schmitt, T. Gruenewald, G. Decher, P. S. Pershan, K. Kjaer and M. Loesche, *Macromolecules*, 1993, **26**(25), 7058–7063.
25 M. Lösche, J. Schmitt, G. Decher, W. G. Bouwman and K. Kjaer, *Macromolecules*, 1998, **31**(25), 8893–8906.
26 H. P. Erickson, *Biol. Proced. Online*, 2009, **11**(1), 32–51.
27 B. Guilhabert, N. Laurand, J. Herrnsdorf, Y. Chen, A. L. Kanibolotsky, C. Orofino, P. J. Skabara and M. D. Dawson, *IEEE Photonics Journal*, 2012, **4**(3), 684–690.
28 B. Guilhabert, C. Foucher, A.-M. Haughey, E. Mutlugun, Y. Gao, J. Herrnsdorf, H. D. Sun, H. V. Demir, M. D. Dawson and N. Laurand, *Opt. Express*, 2014, **22**, 7308–7319.
29 C. Foucher, B. Guilhabert, N. Laurand and M. D. Dawson, *Appl. Phys. Lett.*, 2014, **104**(14), 141108.
30 A. Rose, Z. Zhu, C. Madigan, T. Sager and V. Bulovic, *Nature*, 2005, **434**(7035), 876–879.
31 Y. Yang, G. A. Turnbull and I. D. W. Samuel, *Adv. Mater.*, 2010, **20**(13), 2093–2097.

Faraday Discussions

PAPER

Bio-sorbable, liquid electrolyte gated thin-film transistor based on a solution-processed zinc oxide layer

Mandeep Singh,[a] Gerardo Palazzo,[b] Giuseppe Romanazzi,[c] Gian Paolo Suranna,[d] Nicoletta Ditaranto,[e] Cinzia Di Franco,[f] Maria Vittoria Santacroce,[g] Mohammad Yusuf Mulla,[h] Maria Magliulo,[i] Kyriaki Manoli[j] and Luisa Torsi*[k]

Received 25th April 2014, Accepted 9th May 2014
DOI: 10.1039/c4fd00081a

Among the metal oxide semiconductors, ZnO has been widely investigated as a channel material in thin-film transistors (TFTs) due to its excellent electrical properties, optical transparency and simple fabrication *via* solution-processed techniques. Herein, we report a solution-processable ZnO-based thin-film transistor gated through a liquid electrolyte with an ionic strength comparable to that of a physiological fluid. The surface morphology and chemical composition of the ZnO films upon exposure to water and phosphate-buffered saline (PBS) are discussed in terms of the operation stability and electrical performance of the ZnO TFT devices. The improved device characteristics upon exposure to PBS are associated with the enhancement of the oxygen vacancies in the ZnO lattice due to Na^+ doping. Moreover, the dissolution kinetics of the ZnO thin film in a liquid electrolyte opens the possible applicability of these devices as an active element in "transient" implantable systems.

[a]*Dipartimento di Chimica Università degli Studi di Bari Aldo Moro, Via Orabona 4, 70126 Bari, Italy*
[b]*Dipartimento di Chimica Università degli Studi di Bari Aldo Moro, Via Orabona 4, 70126 Bari, Italy*
[c]*DICATECh: Dipartimento di Ingegneria Civile, Ambientale, del Territorio, Edile e di Chimica, Polytechnic of Bari, Campus Universitario, via Orabona 4, 70125 Bari, Italy*
[d]*DICATECh: Dipartimento di Ingegneria Civile, Ambientale, del Territorio, Edile e di Chimica, Polytechnic of Bari, Campus Universitario, via Orabona 4, 70125 Bari, Italy*
[e]*Dipartimento di Chimica Università degli Studi di Bari Aldo Moro, Via Orabona 4, 70126 Bari, Italy*
[f]*CNR-IFN and Dipartimento Interateneo di Fisica, Università degli Studi di Bari "A. Moro", Via Orabona, 4, 70126, Italy*
[g]*CNR-IFN and Dipartimento Interateneo di Fisica, Università degli Studi di Bari "A. Moro", Via Orabona, 4, 70126, Italy*
[h]*Dipartimento di Chimica Università degli Studi di Bari Aldo Moro, Via Orabona 4, 70126 Bari, Italy*
[i]*Dipartimento di Chimica Università degli Studi di Bari Aldo Moro, Via Orabona 4, 70126 Bari, Italy*
[j]*Dipartimento di Chimica Università degli Studi di Bari Aldo Moro, Via Orabona 4, 70126 Bari, Italy*
[k]*Dipartimento di Chimica Università degli Studi di Bari Aldo Moro, Via Orabona 4, 70126 Bari, Italy. E-mail: luisa.torsi@uniba.it; Fax: +39 0805442092; Tel: +39 0805442092*

1. Introduction

Over the years, semiconductor oxide-based thin-film transistors have attracted a great deal of attention due to their high charge-carrier mobility, transparency and excellent chemical and mechanical stability.[1–3] Indeed, transparent electronics have gained attention during the last few years and are today established as one of the most promising technologies for the next generation of flat panel displays due to the excellent electronic performance of many wide band-gap semiconductor oxides. In particular, zinc oxide (ZnO) is an interesting and promising material as it holds a favorable combination of properties, including excellent transparency in the visible range, good electronic properties, even when processed from solution, and strong piezoelectric properties.[4,5] Last but not least, ZnO is biocompatible[6,7] and biodegradable, while zinc is among the most abundant elements in the earth's crust. All these make ZnO a key functional material with versatile properties for important applications in sensing, catalysis, optical emission, piezoelectric transduction and actuation.[8] It would be also suitable as an active layer in electronic devices that can be integrated in a living organism, even in the human body.

High-performance metal oxide TFTs are typically manufactured by vacuum-processing methods, such as radio frequency magnetron sputtering and pulsed laser deposition, which pose some limitations in mass production for the realization of low-cost optoelectronic applications. Solution-processable oxide semiconductors indeed represent a fast-emerging technology that holds promise in the area of TFT applications. Recently, significant research effort has been devoted to developing novel materials and methodologies to improve electronic transport figures of merit, as well as to actuate processing at low temperature to ensure compatibility with plastic or even biodegradable paper or silk substrates.[3,9,10] For this reason, a large research effort has been dedicated to the development of solution-processing methods for the fabrication of oxide TFTs, including ZnO TFTs.[11–14] Studies in this direction involve ultraviolet photochemical activation of sol–gel precursors,[15,16] sol–gel on chip,[17] along with a number of colloidal nanoparticle-based approaches.[18,19] The use of preformed colloidal metal oxide nanocrystals and nanoparticles (NPs) offers advantages, but the high melting temperatures, along with the presence of highly insulating carbon-based surfactants, lead to transistors with modest performance levels even at relatively high sintering temperatures.[18] Recent advances in sol–gel-based methods have, on the other hand, enabled the development of oxide-based TFTs with electron mobilities as high as 11 $cm^2 V^{-1} s^{-1}$ at 250 °C.[17] A very interesting approach has been also proposed that sees high-performance ZnO transistors processed *via* an aqueous carbon-free metal oxide precursor route at temperatures between 80 and 180 °C.[20]

Lately, the concept of gating TFTs with an ionic conducting electrolyte,[21,22] rather than with an insulating dielectric, has been put forward as it allows for low voltage operation directly through a droplet of pure water,[21] but also by means of a high ionic strength fluid in contact with a TFT channel, this making such devices very promising also for bio-organic electronic applications.[23] In electrolyte-gated TFTs (EG-TFTs), the electrolyte serves as a dielectric being positioned, as customary, between the gate electrode and the semiconductor active channel material. In a typical n-type EG-TFT device, when a positive gate bias is applied, the anions migrate towards the gate–electrolyte interface and the cations migrate towards the semiconductor–electrolyte interface, leaving a charge-neutral electrolyte in between them. This ion migration leads to a net charge accumulation at the gate–electrolyte (negative charges) and semiconductor–electrolyte (positive charges) interfaces, generating, at the steady state, two electrical double layers (EDLs), one at each interface, with all the applied voltage that drops across the EDLs. The total capacitance of the electrolyte gating layer, determined by the capacitance of the two EDLs connected in series, is dominated by the smaller of the two single EDLs. The value of capacitance is typically on the order of 10 μF cm^{-2}, thus making it possible to induce a very large charge carrier concentration (~10^{15} cm^{-2}) in the transistor channel at relatively low applied gate voltages (<3 V).[21] In addition, as the static capacitance of the electrolyte is virtually independent from the thickness of the electrolyte layer, even a thick layer will attain low-voltage operation, but also the device figures of merit are independent from thickness inhomogeneities.

This approach has also been used to gate ZnO-based TFTs, mostly through solid ionic polymers and gels, as well as ionic liquids. Interestingly, solid electrolyte-gated ZnO TFTs have been used to fabricate an inverter circuit by printing on a plastic substrate.[24] In this case, high-mobility, low-voltage ZnO TFTs on a kapton substrate were fabricated by aerosol jet printing, this being a versatile method for patterning a wide range of electronically functional liquid inks.[25] To achieve low-voltage operation a printable high-capacitance (3.80 μF cm^{-2}) ion-gel solid electrolyte was used, as previously proposed.[26] Such gating materials allow operation of the device below 2 V in a nitrogen environment, reaching a field effect mobility and threshold voltage of 1.67 cm^2 V^{-1} s^{-1} and 0.97 V, respectively. A similar contribution has been published regarding an amorphous ZnO channel gated through an ionic liquid tape.[27] Other contributions concerning ZnO-based TFTs gated through solid electrolytes such as ionic liquids[28,29] or composite solid polymers can be found. In the latter case a ZnO single nanowire is used as the active element of the channel, and the field-effect mobility and the threshold voltage of a typical ZnO single-nanowire transistor are 62 cm^2 V^{-1} s^{-1} and 0.93 V, respectively.[30] Interestingly, the electrochemical gating approach is shown not to limit the switching speed of the device, as a frequency in the range of 100 kHz is already achievable with an in-plane transistor geometry. Moreover, unlike the ionic liquids, the composite solid polymer electrolyte studied in this work is found to be stable under ambient conditions and the transistors have shown no noticeable degradation after an exposure of 3–4 weeks in air. More recently, an innovative contribution proposed by Grell's group reported a ZnO electron transporting thin-film transistor directly gated through a drop of pure water.[31] To minimize the impact of water on the stability of the thermally converted precursor route ZnO, the surface was modified by an hexamethyldisilazene hydrophobic layer. Assuming a capacitance per unit area (C_i) of 3 μF cm^{-2} measured on a water-gated polythiophene TFT,[21] an electron mobility as high as 8.8 cm^2 V^{-1} s^{-1} was found, similar to the 5.25 cm^2 V^{-1} s^{-1} reported for a "dry-gated" precursor-route ZnO TFT.[32] On the other hand, much inferior figures of merit are found by using a small gate area. This can be rationalized considering that a gate capacitance, comparatively higher than the ratio between the channel and gate contact area, should be in place to assure the same level of performance.[33]

As already outlined, one of the relevant aspects of ZnO structures is their biocompatibility. ZnO nanoparticles are indeed believed to be nontoxic, bio-safe and biocompatible, as proven by their use for many every-day life applications, such as drug carriers and cosmetics. For instance, a study on the biocompatibility of ZnO nanowires at the cellular level has shown that they are completely biocompatible, bio-safe and are ranked as reliable and trustworthy for applications in biomedicine and engineering.[8] Moreover, the biodegradability and biocompatibility of ZnO nanowires or nano-belts, which is crucial for the application of ZnO nanostructures in biosensing, has been also assessed.[34] A novel and extremely interesting approach foresees the use of water-sorbable devices and circuits[35–37] to be used as implantable "transient" opto-electronic systems. Not only has ZnO been shown to dissolve in water and in an electrolyte,[34] but ZnO devices have even been fabricated on silk or paper substrates that quickly vanish by simple dissolution. Each constituent material disappears due to hydrolysis at different rates, and the time frames for dissolution can be programmed not only by encapsulation and packaging methods, but also by choices of dimensions, thicknesses and configurations in the materials for the device structures. These "transient" devices and circuits could be used to regulate body functions, record biometrics or dispense medicines.

In this paper, we are proposing solution-processed ZnO-based TFTs gated through a phosphate-buffered saline (PBS) solution 10 mM (pH 7.4) put in direct contact with the ZnO surface. The ionic strength of the PBS electrolyte used is 162 mM, while that of human blood serum is 160 mM. As previously outlined, while a few reports are available on water-gated ZnO TFT,[31] to the best of our knowledge, this is the first time that a liquid electrolyte with an ionic strength comparable to that of a biological fluid is used to gate a ZnO TFT. A full morphological and chemical surface characterization is reported. In addition, a study of the optical and electronic properties of such devices is provided along with the dissolution kinetics, showing how this technology can be useful for "transient" implantable devices.

2. Experimental

2.1 Preparation of the ZnO precursor solution

The ZnO precursor solution (0.1 M) was prepared by dissolving zinc acetate dihydrate, $Zn(CH_3COO)_2 \cdot 2H_2O$, in ethanol. The solution was stirred for 7 h and at 80 °C, until a transparent, homogenous solution was obtained. The as-prepared precursor solution was deposited *via* spin-coating onto thermally oxidized silicon wafers (Si/SiO_2) for device fabrication.

2.2 Device fabrication

Prior to spin-coating, the Si/SiO_2 substrate was photolithographically patterned with source (S) and drain (D) electrodes (Au/Ti = 50 nm/5 nm). The channel length and width were, unless stated, 10 μm and 10 000 μm respectively. The substrates were cleaned with a piranha solution (3 : 1 mixture of sulfuric acid and hydrogen peroxide) and thoroughly rinsed with solvents of increased polarity (acetone, isopropyl alcohol and deionized water). The precursor solution was deposited on the S–D patterned substrate by spin coating with a Laurell Technologies WS-650MZ-23NPP/LITE instrument. The thin-film deposition was performed by dispensing a 70 μL drop of precursor solution on the SiO_2 surface and subsequent spinning at 2000 RPM for 50 seconds. The desired film thickness was obtained by repeating the deposition procedure 3 times. Before each further coating, the films were pre-heated at 150 °C to remove the organic residue. The resulting deposit was calcined at 450 °C for five hours on a hot plate. All the chemicals and solvents (high quality grade) were purchased by Sigma-Aldrich and used without any further treatment.

2.3 ZnO thin-film characterization

A thermogravimetric analyzer (Perkin Elmer Pyris 6 TGA) was used to study the thermal behavior of $Zn(CH_3COO)_2 \cdot 2H_2O$ in the temperature range from 30 to 500 °C with a heating rate of 10 °C min^{-1} in a 40 mL min^{-1} dinitrogen flow.

The topographical and cross-sectional view micrographs of the films were acquired by a scanning electron microscope (Carl Zeiss SEM system, mod. Sigma). The thickness of the film was evaluated from the cross-sectional SEM image.

The surface chemical composition of the ZnO films, before and after the exposure to water and PBS solutions, was realized with an X-ray photoelectron Thermo VG Theta Probe spectrometer equipped with a micro-spot monochromatized Al Kα source. Both survey scans and high resolution spectra were acquired in the fixed analyzer transmission mode with pass energies of 150 and 100 eV, respectively. The X-ray Photoelectron Spectroscopy (XPS) data were processed with the Avantage software.[38] The aliphatic C 1s component fixed at BE values of 284.8 ± 0.1 eV is referenced for calibration of the binding energy (BE) scale. The surface chemical speciation (in at%) of each sample was determined by evaluating the integrated peak areas of the principle photoelectron peaks and the respective sensitivity factor.[39]

2.4 TFT electrical characterization

The transistor current–voltage (*I*–*V*) characteristics were measured, through a pure deionized water or PBS solution gating, with Agilent 4155C and Keithley 4200-SCS Semiconductor Parameter Analyzers. It is worth mentioning that 10 mM PBS solution (pH 7.4) contains a great deal of Na^+ ions. In Fig. 1 a scheme of the device structure is reported.

The output characteristics (I_{DS} *vs.* V_{DS}) were measured, ranging the source–drain voltage (V_{DS}) between 0 and 0.5 V, while the gate bias (V_G) was kept constant for each curve and varied from one curve to the other in 0.1 V steps. The transfer characteristics (I_{DS} *vs.* V_G) were recorded, keeping the drain voltage (V_D) constant at 0.5 V. Each current trace was measured in the forward and reverse mode to evidence the occurrence of any hysteresis. The transistor electrical parameters, namely the field-effect mobility (μ), the threshold voltage (V_T) and the *on*/*off* current ratio (I_{on}/I_{off}), were extracted from the characteristic curves in the saturation regime following an assessed procedure.[40] All measurements were carried out in ambient air (relative humidity *ca.* 36%, $T = 25$ °C).

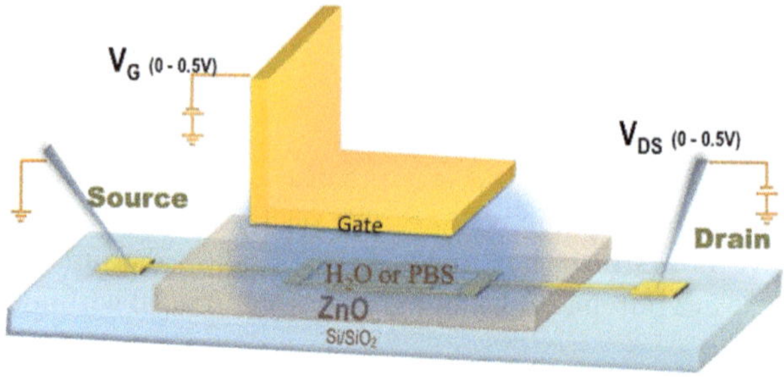

Fig. 1 Electrolyte-gated TFT device structure.

3. Results and discussion

3.1 Thermogravimetric analysis

Thermogravimetric analysis (TGA) was performed in order to study the thermal decomposition of zinc acetate dihydrate ($Zn(CH_3COO)_2 \cdot 2H_2O$) and the formation of a stable ZnO phase. The thermogram trace and its first derivative at temperatures ranging from 30 to 500 °C are shown in Fig. 2 and in the inset, respectively. The first thermal event is observed at 102.6 °C and is completed at 119 °C with a ~16.7% weight loss. This weight loss can be attributed to the thermal dehydration of zinc acetate dehydrate and more specifically to the loss of the two water molecules. By further increasing the temperature, the decomposition of anhydrous $Zn(CH_3COO)_2$ into ZnO occurs at ~316.9 °C, and is completed at ~353 °C. Beyond this temperature, and up to 500 °C, the formation of a stable ZnO crystal takes place, and no further weight loss is observed. Based on these results, we decided to carry out the calcination at 450 °C, since a stable ZnO phase is obtained. These observations are also in good agreement with previous reports.[32,41]

3.2 Optical properties

The optical transparency of the ZnO thin film was evaluated on film deposited on a quartz substrate, whose UV-Vis transmittance spectrum is shown in Fig. 3a. The

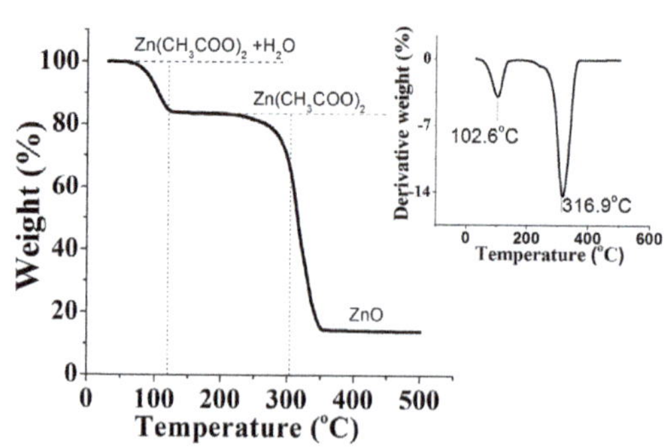

Fig. 2 TG thermogram (30–500 °C) of $Zn(CH_3COO)_2 \cdot 2H_2O$. Inset: first derivative of the TG thermogram.

optical absorption peak was observed at a wavelength of *ca.* 354 nm which falls in the ultraviolet region of the spectrum, proving the ZnO transparency in the visible region. Considering that ZnO holds a direct band gap transition, the energy gap (E_g) was obtained by using the Bardeen equation:[42]

$$\alpha h\nu = \mathrm{B}(h\nu - E_g)^{1/2} \quad (1)$$

where α is the absorption coefficient and B is a constant.[43] The extrapolation from the linear portion to $\alpha h\nu = 0$ (Fig. 3b) results in an $E_g \sim 3.29$ eV.

3.3 X-ray photoelectron spectroscopy surface characterization

X-ray photoelectron spectroscopy (XPS) characterization of the ZnO film was carried out to investigate how the water or PBS solutions influence the surface chemistry. Since the zinc binding energy is not affected by chemical speciation, the O 1s spectral region was studied for the ZnO film, before and after exposure to either water or PBS solution, and the results are shown in Fig. 4. Deconvolution of the O 1s signals showed the presence of four component peaks centered at 530.1 $\pm$ 0.2 eV, 531.1 $\pm$ 0.2 eV, 532.2 $\pm$ 0.2 eV, and 533.3 $\pm$ 0.2 eV, labeled as A, B, C and D in Fig. 4. Peak A was assigned to oxygen in the ZnO lattice, peak B was attributed to oxygen in the oxygen-deficient region in the ZnO matrix,[44,45] peak C was associated to oxygen adsorbed on the surface of the ZnO film, and peak D was assumed to come from the SiO_2 substrate. Interestingly, the four peak components are positioned at the same binding energies in all the samples, and the O 1s XP spectrum before and after water exposure is exactly the same. When the film is exposed to the PBS solution, a different O 1s shape is observed: in particular the intensity ratio *B*/*A* is increased, indicating an increase in the number of oxygen vacancies.[44,45]

3.4 Scanning electron microscopy

The surface morphology and the thickness of the ZnO thin film was evaluated by means of scanning electron microscopy (SEM) analysis. The SEM images, showing the morphological and cross-sectional views of a ZnO thin film deposited on the Si/SiO_2, are shown in Fig. 5a and b, respectively. The film clearly exhibits a granular morphology and has a densely packed structure.[46] The film thickness

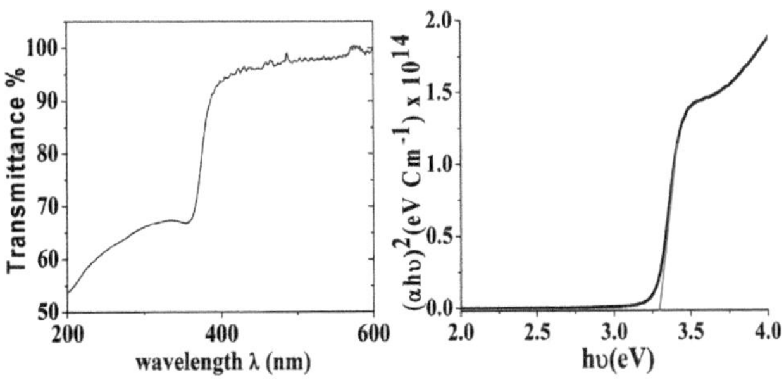

Fig. 3 (a) Optical transmittance spectra of the ZnO thin film deposited on a quartz substrate; (b) optical band gap of the ZnO thin film.

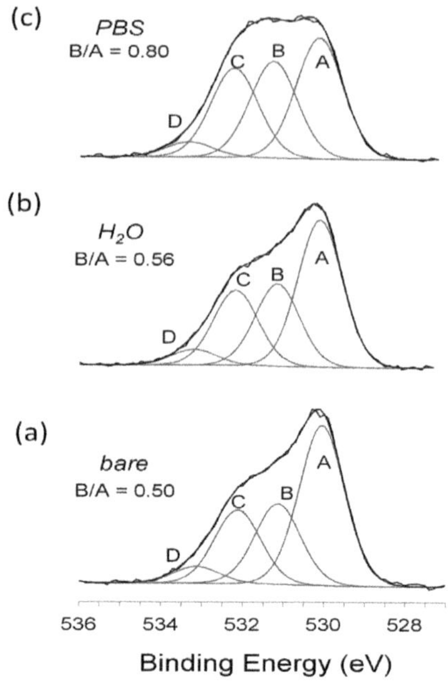

Fig. 4 O 1s XP spectra of the (a) ZnO film and the (b) ZnO sample kept in water, or (c) PBS, for 30 min.

was evaluated from the cross-sectional image as 46 ± 5 nm. In Fig. 5c and d, the SEM images of the ZnO films exposed to water and PBS solutions, respectively, for 14 h under stirring conditions, are reported. ImageJ Software[47] was used to evaluate the effect of the different electrolytes on the size of the grains. The average grain size, as calculated, was found to be 29 ± 9 nm, 28 ± 6 nm and 26 ± 7 nm for bare ZnO and for the films exposed to water and PBS, respectively. Analysis of the images shows that an overall similar granular morphology is retained after exposure to the two electrolytes, although a slight decrease in the average grain size can be seen for the PBS-exposed film.

3.5 Electrolyte-gated TFT characteristics

Fig. 6 shows the output and transfer characteristics of the pure deionized water and PBS (10 mM, pH = 7.4)-gated ZnO-based thin-film transistors. In Fig. 6a and c the I_{DS} current exhibits well-shaped linear and saturation regions as the V_{DS} bias is swept, and also shows good modulation with the gate bias V_G. The hysteresis shown by the water-gated device is visibly larger. The transfer characteristics in both cases (Fig. 6b and d) were measured at three different integration times, and also in this case the dependence of the integration time is larger for the water-gated device compared to the PBS device. Both these evidences concur to show that more stable electronic behavior is shown by the device operated through the PBS solution. This is an important aspect that will be addressed later on in the text. In the saturation regime the curves follow the well-known analytical expression:

$$I_{DS} = \frac{W}{2L}\mu C_i(V_G - V_T)^2 \tag{2}$$

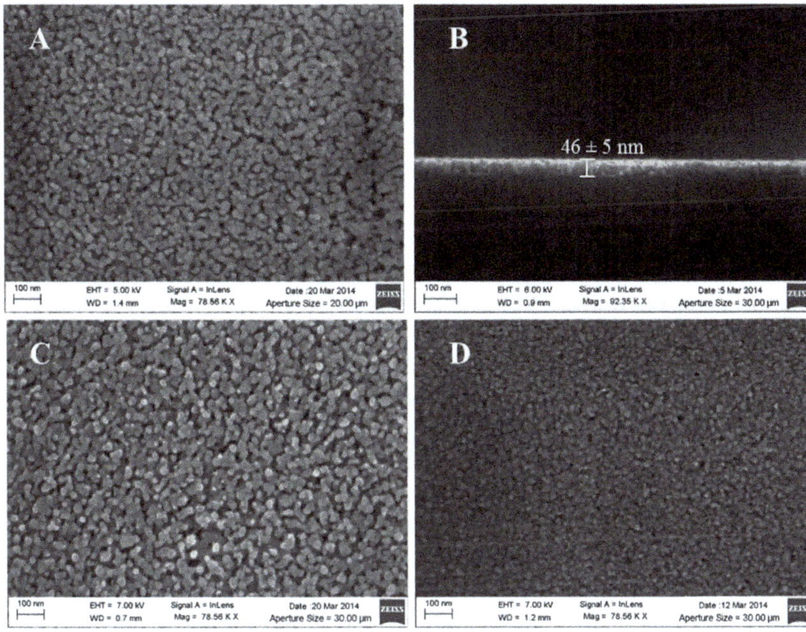

Fig. 5 Topographical scanning electron microscope view of (A) the ZnO thin film, (B) cross-sectional view of the ZnO thin film, (C) ZnO sample kept in water and (D) PBS, for 14 hours under stirring conditions.

The TFT field-effect mobility multiplied by the capacitance per unit area, μC_i, and the threshold voltage (V_T) were derived from the slope (B) and intercept (A) of the linear $\sqrt{I_{DS}}$ *vs.* V_G plot, according to eqn (3) and eqn (4) reported below.

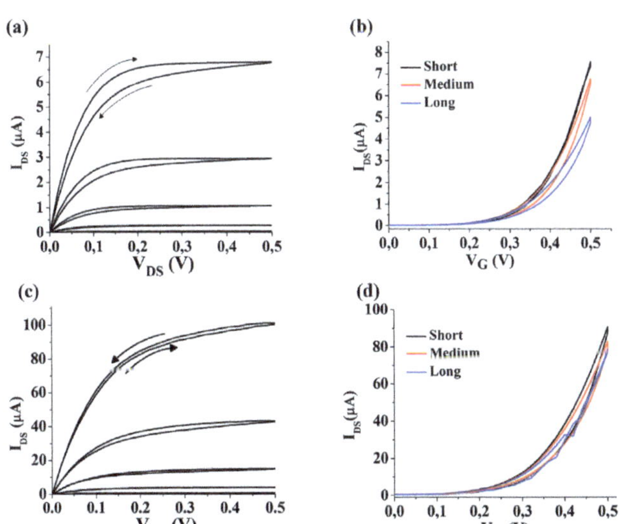

Fig. 6 Current–voltage (I–V) characteristics of the electrolyte-gated ZnO TFT: output (I_{DS} *vs.* V_{DS}) for (a) water, (c) PBS. Transfer (I_{DS} *vs.* V_G) for (b) water, (d) PBS at $V_{DS} = -0.5$ V.

$$\mu C_i = \frac{B^2 2L}{W} \tag{3}$$

$$V_T = -\frac{A}{B} \tag{4}$$

The device figures of merit averaged over 3–4 different devices with the relevant uncertainties (one standard deviation) are reported in Table 1. Table 1 also offers a comparison with the literature data concerning ZnO TFTs gated through different solid electrolytes. The comparison is also made with the sole report already published on water-gated ZnO TFTs. To allow for a fair comparison, rather than the mobility data, the μC_i figure of merit is proposed. This is necessary as fully reliable values for the capacitance of liquid or solid electrolytes are not always available.

The measured electrical characteristics obtained from the proposed ZnO devices, summarized in Table 1, present descending field effect behavior. The devices operate at voltages lower than 0.5 V, which is attributed to the electrolyte gating layer. The threshold voltage (V_{th}) is ~0.15 V and it is among the lowest reported. The I_{on}/I_{off} ratio is 10^2 for water and 10^3 for PBS. These values are comparable to that of liquid-gated ZnO TFTs.[31,29] Moreover, the on/off current ratio can be improved by patterning the active channel area. The value of μC_i with PBS as the electrolyte is 1.4×10^{-6} F V^{-1} s^{-1} at the saturation regime, which is similar to that of devices fabricated using a solid electrolyte as the dielectric layer.[24]

It is worth commenting that the *I–V* curves of the ZnO thin film exposed to PBS, which is rich in Na^+ ions, exhibit improved electrical characteristics in terms of hysteresis, the I_{on}/I_{off} current ratio and μC_i. More specifically, the hysteresis gap observed in the case of the water-gated ZnO TFT device, reveals the occurrence of charge trapping in the reverse voltage sweep in both the output and transfer characteristics (see Fig. 6a and b). In all cases, the reverse sweep current curve is lower than the forward sweep curve. Furthermore, by reducing the integration time of the gate voltage sweep, an increase in the hysteresis gap is evident, along with a decrease in the maximum on-current (Fig. 6b). It is probable that, for a longer duration of biasing, more charges have time to reach lower energy states and occupy trap sites. On the other hand, when the device is gated with the PBS electrolyte solution, almost hysteresis-free *I–V* curves are obtained (see Fig. 6c and d). In addition, a higher on current is achieved, meaning that a larger amount of free carriers accumulate in the semiconductor–electrolyte interface. This can be associated to the higher ionic strength of the PBS buffer solution compared to water.

It has been recently demonstrated that Na^+ ions can act as dopants in ZnO, tending to occupy interstitial sites rather than substitutional sites, therefore acting as donors and not as acceptors.[48] It has also been suggested that the presence of sodium ions can cause a lack of stoichiometry in the ZnO lattice and therefore induce oxygen vacancies.[45] This is confirmed by the XPS analysis reported in Fig. 4, where an increase in oxygen vacancies is observed in the case of ZnO films exposed to the PBS solution. Indeed, the exposure to Na^+ present in the PBS solution seems to alter the oxide semiconductor lattice on the surface, giving

Table 1 Comparison of the electrical figures of merit of solution-processed electrolyte-gated ZnO TFTs

ZnO	Gate dielectric	μ ($cm^2 V^{-1} s^{-1}$)	V_{DS} (v)	V_G (V)	C_i ($\mu F cm^{-2}$)	μC_i ($F V^{-1} s^{-1}$)	V_T (V)	Slope ($A V^{-2} cm^{-2}$)	I_{on}/I_{off}	Ref.
Inkjet printing	PS–PMMA–PS and [EMIM] [TFSI]	1.61	0–1.2	0–2	3.8	6.1×10^{-6}	0.97	0.0078	N/A	24
Spray-coated thin film	[HMIM] [FAP]	15.9	0–1.5	−3–1	3.3	5.2×10^{-5}	N/A	0.036	10^3	29
Spin-coated thin film	PS-*b*-PMMA-*b*-PS	13	0–0.7	0–1.8	5.08	6.6×10^{-5}	0.62	0.022	10^6	26
Spin-coated thin film	Water	8.8, 1.32	0–0.7	0–0.8	3.20	2.6×10^{-5}	0.36	0.051	10^3	31
Spin-coated thin film	Water	—	0–0.5	0–0.5	—	$(2.3 \pm 0.5) \times 10^{-7}$	0.15 ± 0.04	0.011	3×10^2	This paper
Spin-coated thin film	PBS	—	0–0.5	0–0.5	—	$(1.4 \pm 0.1) \times 10^{-6}$	0.14 ± 0.07	0.028	10^3	This paper

rise to oxygen vacancies. Although oxygen vacancies can produce potential wells that can act as trapping sites,[44] the excess of free electrons originating from interstitial doping due to Na^+ is such that they can fill the existing trap sites. Moreover, oxygen vacancies can induce more charges in the semiconductor's channel.[45]

3.6 ZnO film dissolution kinetics in water and PBS

The kinetics of ZnO dissolution in aqueous environments were determined by anodic stripping voltammetry (ASV). The Si/SiO_2 substrates, coated by a ZnO film according to the procedure used in the TFT fabrication, were placed in vials containing 3 mL of water or PBS and left under mild stirring on a roller mixer for different time lapses. Reference (blank) slides were left out of the liquids. The amount of residual ZnO at a given time lapse, after the immersion, is defined as the weight percentage of ZnO with respect to the blank. At a given time the slide was removed from the liquid and the coating was dissolved using a wet digestion procedure with nitric–perchloric acids. The solutions obtained after digestion were diluted with deionized water, adjusted to a pH in the 5.0–5.5 range with acetate buffer and dehydrated with a stream of nitrogen for 100 s. The amount of zinc cations dissolved in solution was measured by a differential pulse ASV electrochemical procedure using a three-electrode cell equipped with a hanging mercury drop electrode (HMDE) used as the working electrode, an Ag/AgCl with saturated 3 M KCl reference electrode, and a platinum wire counter electrode. Electrochemical deposition was performed at −0.8 V for 45 s under stirring. After this time the stirrer was switched off, and the oxidation curves were recorded in the potential range from −1.15 to −0.75 V (sweep rate of 0.06 V s^{-1}). Zinc was quantified by the addition of a Zn^{2+} standard solution to the cell.

In Fig. 7 the data are reported for ZnO dissolution in water over a period of 25 h. The zinc actually remaining on the sample after the exposure to water decreases to 46% after 16 h in water, while it decreases down to 25% after exposure to PBS. On the other hand, stable device operation will be shown over a half-hour time lapse later in this manuscript, however the limiting factor is the gating droplet evaporation, not the ZnO film degradation. The actual stable operation time of the device is in fact much longer when comparing the "transient time" of these devices with literature data.[37]

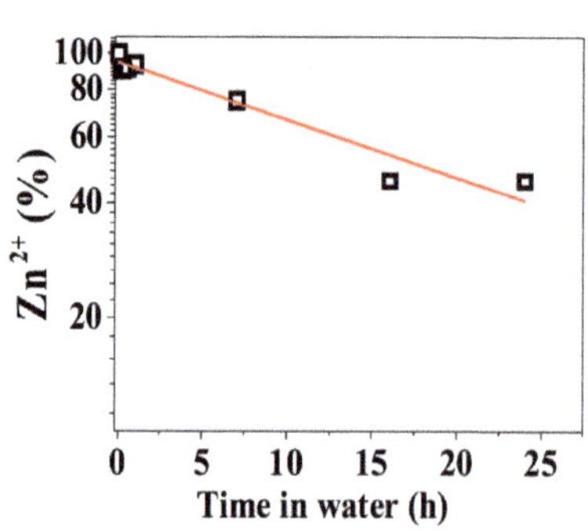

Fig. 7 Dissolution profile of the ZnO film in water.

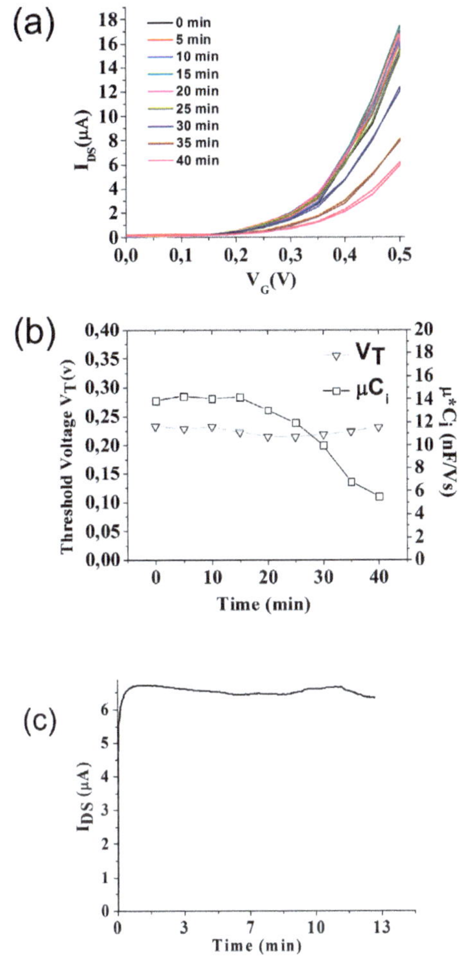

Fig. 8 (a) *I*–*V* transfer characteristics of the water-gated ZnO TFT measured repeatedly for 40 min, (b) the relevant V_T, μC_i values, and (c) pulsed transient current measurement of water gated ZnO TFT device.

ZnO appears to react differently under water and PBS solution. The different dissolution processes can be also accounted for by evaluating the degree of hydrophobicity of the ZnO surface after exposure to the two electrolytes. Improved wettability has been ascribed to enhance the dissolution of solid dispersions. Static contact angle evaluation was performed by using a CAM 200 goniometer (KSV Instruments Ltd, Finland) and placing a 2 μL droplet of water or 10 mM PBS on the ZnO film surface at ambient conditions ($T = 25$ °C). The measurement was repeated three times by placing the drop on different areas of the same sample, measuring average contact angles of 57 ± 2 for water and 39 ± 6 for PBS, this being more hydrophilic. Interestingly, such a difference cannot be ascribed to differences in the surface tension in the liquid phase because water and PBS have the same value (71.5 ± 0.2 mJ m^{-2}) measured using a Du Noüy ring tensiometer (Krüss Gmbh, Germany). The difference in the contact angle, therefore, must be ascribed to differences in the solid phase in contact with the liquid droplet, and this occurrence strongly suggests a doping of the ZnO by ions (presumably Na^+) present in PBS and not in pure water.

3.7 Operational stability of water-gated ZnO TFTs

Another important aspect to be addressed is the operational stability of the proposed ZnO TFT devices. Current instability in thin film transistors under gate voltage biasing is crucial for the practical application of such devices. For this purpose, double transfer characteristic curves were recorded repeatedly, every 5 min, from the moment the device was gated with a water drop until 40 min after. It is important to mention that the channel length and width of the device, used for these measurements, are 2 μm and 60 mm, respectively. As shown in Fig. 8a, the device exhibits stable performance for 25 min. A current decay is observed after half an hour, which can be attributed to possible evaporation of the electrolyte solution. This is further confirmed by the corresponding V_T and μC_i values as a function of time reported in Fig. 8b, where a clear decrease of the μC_i value is in fact observed after 20 min. Pulsed transient measurements[49] were carried out to examine the dynamic bias stress current instability as well. We used a rectangular wave as a dynamic pulse. The pulse-on time was 10 ms ($V_G = 0.5$ V), while the pulse-off time was 1 s ($V_G = 0$ V) at a constant $V_{DS} = 0.5$ V. As illustrated in Fig. 8c, a stable output current signal in time is observed for approximately more than 10 min. As previously mentioned, the limiting factors to record further the drain current as a function of time are the evaporation of the water droplet and not the degradation of the device.

4. Conclusions

In this study, we have operated an electrolyte-gated ZnO TFT, using a liquid electrolyte with an ionic strength similar to that of physiological fluids. The ZnO precursor solution was prepared by the sol–gel method, while calcination was realized at 450 °C. Low-voltage operation (0.5 V) was achieved, with the electrical figures of merit being in good agreement with those previous reported for electrolyte-gated ZnO TFTs. The carrier density induced in the case of the PBS electrolyte was proven to fill possible trap sites, leading to on-current values higher than the water-gated values. The *I–V* curves were also hysteresis-free, also exhibiting a slight shift of V_T to a lower voltage value. The enhanced performance of the PBS gating was attributed to interstitial doping of the ZnO matrix by Na^+ ions present in PBS and not in pure water. The proposed mechanism was further confirmed by XPS analysis of the surface chemical composition of ZnO films exposed to water and PBS. Finally, the dissolution profile of ZnO in water and PBS, along with the operational stability of the devices, was studied, showing that this technology can be useful for "transient" implantable devices.

Acknowledgements

Prof. Gaetano Scamarcio is acknowledged for useful discussions. Dr Francesca Intranuovo is acknowledged for the contact angle measurements. The PON project "Laboratorio per lo Sviluppo Integrato delle Scienze e delle Tecnologie dei Materiali Avanzati e per dispositivi innovativi-LABORATORIO SISTEMA" by the

Italian MIUR (Ministry of Education, Universities and Research) is acknowledged for partial support.

References

1 R. A. Street, *Adv. Mater.*, 2009, **21**, 2007.
2 G. Adamopoulos, S. Thomas, P. H. Wobkenberg, D. D. Bradley, M. A. McLachlan and T. D. Anthopoulos, *Adv. Mater.*, 2011, **23**, 1894.
3 E. Fortunato, *Adv. Mater.*, 2012, **24**, 2945.
4 D. C. Look, *Mater. Sci. Eng., B*, 2001, **80**, 383.
5 D. G. Thomas, *J. Phys. Chem. Solids*, 1960, **15**, 86.
6 C. Czekalla, J. Guinard, C. Hanisch, B. Q. Cao, E. M. Kaidashev, N. Boukos, A. Travlos, J. Renard, B. Gayral, D. L. S. Dang, M. Lorenz and M. Grundmann, *Nanotechnology*, 2008, **19**, 115202.
7 Z. Li, R. Yang, M. Yu, F. Bai, C. Li and Z. L. Wang, *J. Phys. Chem. C*, 2008, **112**, 20115.
8 C. Dagdeviren, S. W. Hwang, Y. Su, S. Kim, H. Cheng, O. Gur, R. Haney, F. G. Omenetto, Y. Huang and J. A. Rogers, *Small*, 2013, **9**, 3398.
9 D. H. Kim, Y. S. Kim, J. Amsden, B. Panilaitis, D. L. Kaplan, F. G. Omenetto, M. R. Zakin and J. A. Rogers, *Appl. Phys. Lett.*, 2009, **95**, 133701.
10 G. Amin, S. Zaman and A. Zainelabdin, *Phys. Status Solidi Rapid Res. Lett.*, 2011, **5**, 71.
11 J. Nishi, F. M. Hossain, A. Takagi, T. Aita, K. Saikusa, Y. Ohmaki, I. Ohkubo, S. Kishimoto, A. Ohtomo, T. Fukumura, F. Matsukura, Y. Ohno, H. Koinuma, H. Ohno and M. Kawasaki, *Jpn. J. Appl. Phys.*, 2003, **42**, L347.
12 P. Barquinha, L. Pereira, G. Goncalves, R. Martins and E. Fortunato, *J. Electrochem. Soc.*, 2009, **156**, H161.
13 L. Wang, M. H. Yoon, G. Lu, Y. Yang, A. Facchetti and T. J. Marks, *Nat. Mater.*, 2006, **5**, 893.
14 P. K. Nayak, J. Jang, C. Lee and Y. Hong, *Appl. Phys. Lett.*, 2009, **95**, 193503.
15 Y. H. Hwang, S. J. Seo, J. H. Jeon and B. S. Bae, *Electrochem. Solid-State Lett.*, 2012, **15**, H91.
16 Y. H. Kim, J. S. Heo, T. H. Kim, S. Park, M. H. Yoon, J. Kim, M. S. Oh, G. R. Yi, Y. Y. Noh and S. K. Park, *Nature*, 2012, **489**, 128.
17 K. K. Banger, Y. Yamashita, K. Mori, R. L. Peterson, T. Leedham, J. Rickard and H. Sirringhaus, *Nat. Mater.*, 2010, **10**, 50.
18 B. Sun and H. Sirringhaus, *Nano Lett.*, 2005, **5**, 2413.
19 H. Faber, M. Burkhardt, A. Jedaa, D. Kälblein, H. Klauk and M. Halik, *Adv. Mater.*, 2009, **21**, 3099.
20 Y. H. Lin, H. Faber, K. Zhao, Q. Wang, A. Amassian, M. McLachlan and T. D. Anthopoulos, *Adv. Mater.*, 2013, **25**, 4346.
21 E. Said, X. Crispin, L. Herlogsson, S. Elhag, N. D. Robinson and M. Berggren, *Appl. Phys. Lett.*, 2006, **89**, 143507.
22 T. Uemura, M. Yamagishi, S. Ono and J. Takeya, *Appl. Phys. Lett.*, 2009, **95**, 103301.
23 M. Magliulo, A. Mallardi, M. Y. Mulla, S. Cotrone, B. R. Pistillo, P. Favia, I. V. Lundin, G. Palazzo and L. Torsi, *Adv. Mater.*, 2013, **25**, 2094.
24 K. Hong, S. H. Kim, K. H. Lee and C. D. Frisbie, *Adv. Mater.*, 2013, **25**, 3418.

25 Y. Xia, W. Zhang, M. Ha, J. H. Cho, M. J. Renn, C. H. Kim and C. D. Frisbie, *Adv. Funct. Mater.*, 2010, **20**, 587.
26 H. Bong, W. H. Lee, D. Y. Lee, B. J. Kim, J. H. Cho and K. Cho, *Appl. Phys. Lett.*, 2010, **96**, 192115.
27 S. Bubel, S. Meyer, F. Kunze and M. L. Chabinyc, *Appl. Phys. Lett.*, 2013, **103**, 152102.
28 H. Yuan, H. Shimotani, A. Tsukazaki, A. Ohtomo, M. Kawasaki and Y. Iwasa, *Adv. Funct. Mater.*, 2009, **19**, 1046.
29 S. Thiemann, S. Sachnov, S. Porscha, P. Wasserscheid and J. Zaumseil, *J. Phys. Chem. C*, 2012, **116**, 13536.
30 B. Nasr, D. Wang, R. Kruk, H. Rösner, H. Hahn and S. Dasgupta, *Adv. Funct. Mater.*, 2013, **23**, 1758.
31 A. A. Naim and M. Grell, *Appl. Phys. Lett.*, 2012, **101**, 141603.
32 B. S. Ong, L. Chensha, L. Yuning, Y. Wu and R. Loutfy, *J. Am. Chem. Soc.*, 2007, **129**, 2750.
33 O. Yaghmazadeh, F. Cicoira, D. A. Bernards, S. Y. Yang, Y. Bonnassieux and G. G. Malliaras, *J. Polym. Sci., Part B: Polym. Phys.*, 2011, **49**, 34.
34 J. Zhou, N. Xu and Z. L. Wang, *Adv. Mater.*, 2006, **18**, 2432.
35 M. Irimia Vladu, P. A. Troshi, M. Reisinger, L. Shmygleva, Y. Kanbur, G. Schwabegger, M. Bodea, R. Schwödiauer, A. Mumyatov, J. W. Fergus, V. F. Razumov, H. Sitter, N. S. Sariciftci and S. Bauer, *Adv. Funct. Mater.*, 2010, **20**, 4069.
36 C. J. Bettinger and Z. Bao, *Adv. Mater.*, 2010, **22**, 651.
37 C. Dagdeviren, S. W. Hwang, Y. Su, S. Kim, H. Cheng, O. Gur, R. Haney, F. G. Omenetto, Y. Huang and J. A. Rogers, *Small*, 2013, **9**, 3398.
38 *Thermo Avantage Software v.4.75 copyright*, Thermo Fisher Scientific, 1999–2010.
39 R. Pilolli, N. Ditaranto, C. Di Franco, F. Palmisano and N. Cioffi, *Anal. Bioanal. Chem.*, 2012, **404**, 1703.
40 L. Torsi, M. Magliulo, K. Manoli and G. Palazzo, *Chem. Soc. Rev.*, 2013, **42**, 8612.
41 A. V. Ghule, B. Lo, S. H. Tzing, K. Ghule, H. Chang and Y. C. Ling, *Chem. Phys. Lett.*, 2003, **381**, 270.
42 J. Bardeen, F. J. Blatt and L. H. Hall, *Photoconductivity Conf.*, ed. R. Breckenridge, B. Russel and T. Hahn, John-Weiley, New York, 1956.
43 S. Ilican, Y. Caglar and M. Caglar, *J. Optoelectron. Adv. Mater.*, 2008, **10**, 2583.
44 H. Bong, W. H. Lee, D. Y. Lee, B. J. Kim, J. H. Cho and K. Cho, *Appl. Phys. Lett.*, 2010, **96**, 192115.
45 K. Kim, S. Y. Park, K.-H. Lim, C. H. Shin, J.-M. Myoung and Y. S. Kim, *J. Mater. Chem.*, 2012, **22**, 23120.
46 S. Singh and P. Chakrabarti, *Superlattices Microstruct.*, 2013, **64**, 283.
47 http://imagej.nih.gov/ij/index.html.
48 C. H. Park, S. B. Zhang and S. H. Wei, *Phys. Rev. B: Condens. Matter Mater. Phys.*, 2002, **66**, 073202.
49 K. Manoli, M. M. Patrikoussakis, M. Magliulo, L. M. Dumitru, M. Y. Mulla, L. Sabbatini and L. Torsi, *Org. Electron.*, 2014, **15**, 2372.

Faraday Discussions

PAPER

Characterization and simulation of electrolyte-gated organic field-effect transistors

Katharina Melzer,* Marcel Brändlein, Bogdan Popescu, Dan Popescu, Paolo Lugli and Giuseppe Scarpa

Received 5th May 2014, Accepted 16th May 2014
DOI: 10.1039/c4fd00095a

In this work we fabricate and characterize field-effect transistors based on the solution-processable semiconducting polymer poly(3-hexylthiophene) (P3HT). Applying two independent gate potentials to the electrolyte-gated organic field-effect transistor (EGOFET), by using a conventional SiO_2 layer as the back-gate dielectric and the electrolyte-gate as the top-gate, allows the measurement of the electrical double layer (EDL) capacitance at the semiconductor–electrolyte interface. We record the transfer curves of the transistor in salt solutions of different concentration by sweeping the bottom gate potential for various constant electrolyte-gate potentials. A change of the electrolyte-gate potential towards more negative voltages shifts the threshold voltage of the bottom-gate channel towards more positive back-gate potentials, which is directly proportional to the capacitive coupling factor. By operating the EGOFET in the dual-gate mode, we can prove the dependency of the EDL capacitance on the molarity of the electrolyte according to the Debye–Hückel theory, and additionally show the difference between a polarizable and non-polarizable electrolyte-gate electrode. With the experimentally obtained values for the EDL capacitance at the semiconductor–electrolyte interface we can model the electrolyte-gate transfer characteristics of the P3HT OTFT.

1 Introduction

Organic thin-film transistors (OTFTs) have been explored for various sensing applications in recent years due to their easy and low-cost fabrication (for recent reviews refer to[1–3] and references therein). In particular, the semiconducting polymer poly(3-hexylthiophene) (P3HT) is well-established as the active semiconducting component in OTFTs for biosensing applications[4–9] due to several key features offered by the material: its outstanding biocompatibility,[10] its solution

Institute for Nanoelectronics, TU München, Arcisstrasse 21, 80333 München, Germany. E-mail: katharina.melzer@nano.ei.tum.de

processability and subsequent applicability in large-scale, flexible and cost-effective electronics, and its relatively high mobility.[11,12]

In organic thin-film transistors, in general a thin film of an organic semiconducting material is utilized as the active layer. Two contacts, the source and drain electrodes, which are in direct contact with the semiconducting material, are used to apply the source–drain voltage and to measure the current flowing between them through the organic semiconducting channel. As the third electrode, the gate electrode is used to control the number of charge carriers, in p-type material holes, in the semiconducting channel at the interface between the semiconductor and the gate dielectric through an electric field applied across the insulating gate dielectric. Therefore one can modulate the source–drain current flowing between the source and drain contacts by applying a given potential to the gate electrode. For a p-type material the transistor is turned "ON" for negative gate voltages, whereas a positive gate bias switches the device "OFF".[13]

In recent years the usage of solid and liquid electrolytes as gate dielectrics in OTFTs has gained great interest. They display a huge capacitance (C_G) compared to the conventional silicon dioxide (SiO_2), which is typically in the range of 1–100 μF cm^{-2}, ~1000 times higher than the one offered by a nm-thick SiO_2-layer.[1,12,14,15] Especially important in EGOFETs is the formation of an electrical double layer (EDL) at the electrolyte–semiconductor interface as well as the gate electrode–electrolyte interface. At the gate electrode–electrolyte interface the electronic charges close to the electrode surface are compensated by oppositely charged ions in the electrolyte close to the interface. At the electrolyte–semiconductor interface the EDL consists of ions forming the inner Helmholtz plane in the electrolyte and charge carriers in the semiconducting channel. Additionally, at both interfaces, hydrated ions of both charges accumulate and form the outer Helmholtz plane. In the bulk of the electrolyte, the potential stays constant as the diffuse layer of the hydrated positive and negative ions ensures charge neutrality.[12,15,16] The charge distribution in an EDL can be described by the Gouy–Chapman–Stern model.[17–20] In the case of a flat electrode surface, the electrostatic capacitance of the EDL can be related to the Debye length λ_D, which defines the distance d between the interface and the outermost layer of ions which is necessary to compensate the gate surface charge fully. The Debye length is inversely proportional to the square root of the ionic strength, according to:

$$C_{DL} \sim \frac{1}{\lambda_D} \tag{1}$$

$$\lambda_D = \sqrt{\frac{\varepsilon_r \varepsilon_0 k_B T}{2 C_0 N_A e^2}} \tag{2}$$

where C_{DL} represents the capacitance of the electrical double layer, λ_D is the Debye length, the parameters ε_r and ε_0 are the relative and the vacuum permittivity, respectively, k_B is the Boltzmann constant, T is the temperature, C_0 is the ionic strength of the electrolyte, N_A is the Avogadro constant and e is the elementary charge.[19] Since the thickness of the electrical double layer is on the sub-nm length scale (a few Å to nm), the capacitance per unit area is typically in the range of tens of μF cm^{-2}. This huge gate capacitance allows a transistor to operate at very low voltages: already gate potentials of around 0.5 V are sufficient to accumulate enough charge carriers in the semiconducting channel to turn the device "ON". By exploiting the electrical double layer capacitance one can modulate the transistor current response extremely effectively. Another very important advantage of electrolyte-gated organic thin-film transistors is the convenient integration into biosensors operated in liquid environments. One is in fact independent from the choice of the substrate,[21,22] and the analyte is already in direct contact with the active semiconducting layer. Since the accumulation of charge carriers in the semiconducting channel and therefore also the source–drain current can be altered by the adsorption of charged analytes on the semiconductor surface or a change in the amount of solute ions in the electrolyte, electrical detection of certain species is possible.

Therefore it is of great interest for various sensing applications with EGOFETs to obtain information about how changes to the electrolyte–semiconductor interface, especially to the electrical double layer and its capacitance, influence the transistor characteristics.

In this work we concentrate on the influence of the electrical double-layer capacitance (C_{DL}) and try to gain information about the dependency of C_{DL} on the ionic strength of the electrolyte. We operate the OTFT based on the semiconducting polymer poly(3-hexylthiophene) (P3HT) in the so-called dual-gate mode[23] by using both a conventional back-gate as well as an electrolyte-gate as the top-gate. Exploiting the capacitive coupling factor[23,24] allows us to extract the double-layer capacitance from several back-gate sweeps with different constant electrolyte-gate potentials. Changing the ionic strength of the aqueous potassium chloride (KCl) solution used as the gate dielectric for the electrolyte-gate gives us the double-layer capacitance as a function of the salt concentration. We demonstrate that the capacitance of the electrical double layer at the semiconductor–electrolyte interface increases with an exponential increase of the ionic strength of the electrolyte. In addition to the electrical characterization of the organic thin-film transistors, we compare the results to theoretical data gained from simulations with a modified drift–diffusion model.

2 Experimental

2.1 Device fabrication

Organic thin-film transistors (OTFTs) were processed using highly p-doped silicon wafers (Si-Mat) as a substrate with a thermally grown 65 nm thick silicon dioxide layer (SiO_2) as the conventional gate dielectric. Source and drain electrodes (5 nm Cr as adhesion layer, 40 nm Au) were patterned on top of the oxide layer by a conventional negative optical photolithography process. An interdigitated finger structure (IDES) was used as the electrode structure, with a channel length of 50 μm and a width-to-length ratio of 900. A 1 wt% solution of regioregular poly(3-hexylthiophene) (P3HT, Rieke Metals) in anhydrous 1,2-dichlorobenzene (DCB, VWR) was prepared and stirred for 30 min to obtain a homogeneous solution. The solution was then filtered through a polytetrafluoroethylene (PTFE) syringe filter with a pore size of 200 nm (VWR) and spin-coated (1000 rpm, 90 s) onto the surface of the wafer in a glove box system (N_2). After spin-coating, the film was annealed on a hot plate under a nitrogen atmosphere at 150 °C for 20 min. The obtained polymer layer was approximately 50 to 70 nm thick. Atomic force microscopy (AFM, Jeol, JSPM-5200) measurements yielded an R_q value of 3.4 nm, which is a measure for the surface roughness of the polymeric film, and a peak-to-valley distance of about 22 nm.

2.2 Electrical characterization

Transfer characteristics (I_D *vs.* V_G) and output characteristics (I_D *vs.* V_{DS}) of the devices were recorded using a Keithley 4200-SCS semiconductor parameter analyser system. To operate the transistor in liquid, measurements were performed after mounting the devices in a PDMS flow-chamber, which has a liquid compartment volume of 100 μL. The transfer characteristics of the OTFT were obtained by sweeping either the back-gate potential ($V_{DS} = -10$ V) or the electrolyte-gate potential ($V_{DS} = -0.1$ V), while a constant drain voltage V_{DS} was applied. For the back-gate sweep no electrolyte was on top of the semiconducting layer, whereas for the electrolyte-gate sweep the back-gate electrode was left floating. The electrolyte-gate potential was chosen from the relatively small voltage regime between $V_{TgS} = +0.6$ V and $V_{TgS} = -0.6$ V to avoid any water hydrolysis or any other electrochemical reaction at the gate electrode or the semiconductor surface.[12]

For the measurement of the electrical double layer capacitance, the potential at the electrolyte-gate electrode, either a Pt wire or a homemade Ag/AgCl electrode, was kept constant while a scan of the back-gate voltage between $V_{BgS} = +20$ V and $V_{BgS} = -7.5$ V was performed. The back-gate sweep was recorded for six different electrolyte-gate potentials between $V_{TgS} = +0.1$ V and $V_{TgS} = -0.4$ V, whereas the drain voltage was kept constant for each back-gate sweep at $V_{DS} = -0.1$ V. All characterization measurements were performed in ambient conditions. The electrolyte was exchanged manually using a Gilson pipette. After exchanging the electrolyte the system was left to equilibrate for 5 minutes before the next measurement was started.

2.3 Preparation of electrolytes

The electrical double layer capacitance was characterized using either deionized water (DI-H_2O, classified with a resistivity of 18.2 MΩ cm). Potassium chloride solutions were prepared by dissolving 100 mM KCl (≥99.0%, Sigma Aldrich) in DI-H_2O and further diluting it with DI-H_2O to the desired KCl concentrations.

3 Results and discussion

3.1 Device characteristics

The transfer characteristics of the OTFT were obtained by sweeping either the back-gate potential or the electrolyte-gate potential, while a constant drain voltage V_{DS} was applied. For the back-gate sweep no electrolyte was on top of the semiconducting layer, whereas for the electrolyte-gate sweep the back-gate electrode was left floating. Fig. 1(a) shows the source–drain current I_D *versus* the back-gate potential; the drain voltage was kept constant at $V_{DS} = -10$ V. Fig. 1(b) shows I_D *versus* the electrolyte-gate potential V_{TgS} of the OTFT gated with a Pt electrode in a 10 mM solution of KCl in DI-water as the electrolyte; the drain voltage was kept constant at $V_{DS} = -0.1$ V. Both transfer curves exhibit typical p-type field-effect characteristics with ON/OFF ratios of around 100. When the OTFT is operated with the conventional silicon dioxide (SiO_2, 65 nm, $C_{Ox} = 53.2$ nF cm^{-2}) back-gate,

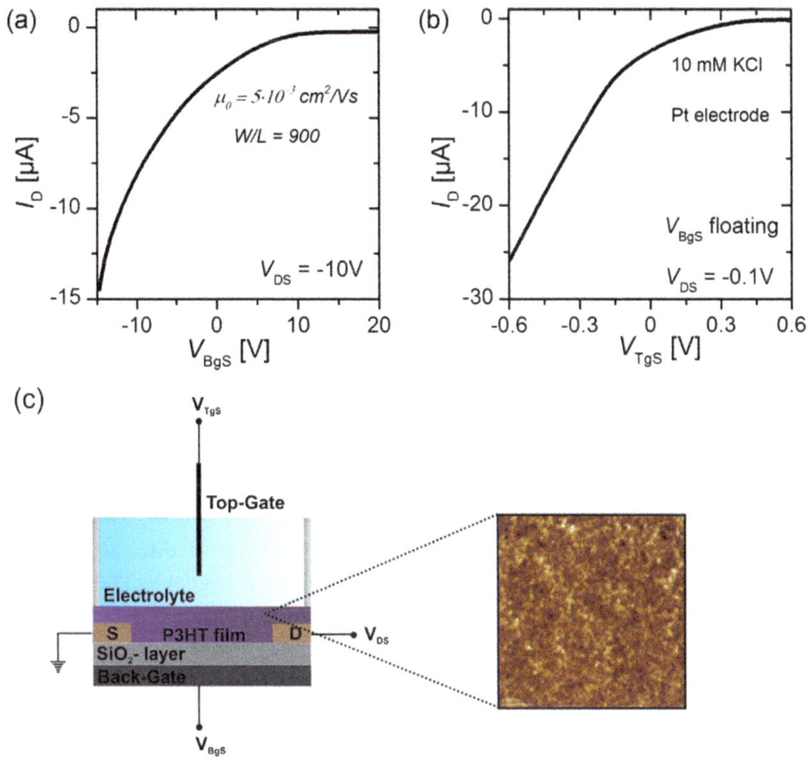

Fig. 1 (a) Transfer characteristics of a back-gated P3HT OTFT. (b) Transfer characteristics of an electrolyte-gated P3HT OTFT recorded in a 10 mM KCl solution. A Pt wire was used as the electrolyte-gate electrode. The back-gate electrode was left floating. (c) Schematic representation of a dual-gated P3HT OTFT. A 65 nm thick SiO_2-layer was used as the back-gate dielectric, whereas various electrolytes serve as the top-gate dielectric. The magnification shows an AFM scan (scan size 5 μm × 5 μm) of the spin-coated P3HT layer on a SiO_2 surface. The obtained polymer layer was approximately 50 to 70 nm thick. It yielded an R_q value of 3.4 nm and a peak-to-valley distance of about 22 nm.

the transfer curve reveals a transconductance of g_m = 2.4 μS. The transconductance of organic thin-film transistors is usually given by

$$g_m = \frac{W}{L}\mu_0 C_G V_{DS}, \tag{3}$$

where W and L are the channel width and length, respectively, μ_0 is the charge carrier mobility, C_G is the capacitance of the gate dielectric, and V_{DS} is the drain voltage.

The mobility calculated according to eqn (3) equals $\mu_0 = 5 \times 10^{-3}$ cm² V⁻¹ s⁻¹ and is comparable to the literature values obtained with OTFTs based on P3HT.[11,25] From Fig. 1(b) it is obvious that the electrolyte-gated OTFT delivers higher I_D currents at an electrolyte-gate voltage of $V_{TgS} = -0.6$ V and a drain bias of $V_{DS} = -0.1$ V, as compared to the SiO_2-gated device with V_{BgS} potentials in the range of 10 V, although the same width-to-length ratio (W/L = 900) was used. From a linear fit of the transfer curve we can extract a transconductance value of g_m = 46.0 μS. The higher the current, the higher the transconductance, and the lower operating voltages reflect the greatly enlarged gate capacitance of the electrical double layer relative to the 65 nm thick SiO_2-layer. In order to derive the mobility from the transfer curves of the electrolyte-gated OTFT, we first have to measure the capacitance of the electrical double layer.

3.2 Measurement of the electrical double layer capacitance

To measure the capacitance of the electrical double layer, we operate the OTFT in the so-called dual-gate mode[23] by using both the conventional silicon dioxide (SiO_2, 65 nm) back-gate as well as an additional electrolyte-gate. The transfer curves (I_D *vs.* V_{BgS}) of the back-gated transistor can be shifted by applying various fixed voltages to the electrolyte-gate electrode immersed in an aqueous solution (see Fig. 1(c)). Here the potential at the electrolyte-gate electrode was kept constant, while a sweep of the back-gate voltage was performed between $V_{BgS} =$ +20 V and $V_{BgS} = -7.5$ V. If the applied electrolyte-gate voltage is negative, a semiconducting channel at the electrolyte–polymer interface is formed and the charge carrier accumulation in the semiconducting layer can be controlled by sweeping the back-gate voltage. A back-gate sweep was recorded for six different electrolyte-gate potentials between $V_{TgS} = 0.1$ V and $V_{TgS} = -0.4$ V. The resulting transfer curves with DI-water as the electrolyte and a Pt wire as the gate electrode are shown in Fig. 2(a). The obtained transfer curves first exhibit a short charging phase and then follow a linear dependence for I_D *vs.* V_{BgS}. According to a theoretical model for dual gate transistors,[24] the source–drain current of a dual-gated OTFT in the linear regime can be expressed as follows:[23]

$$I_D = \frac{W}{L}\mu_0\left(C_{DL}V_{TgS} + C_{Ox}V_{BgS} + \sigma_i\right)V_{DS}, \quad (4)$$

where W is the width and L is the length of the channel, μ_0 is the mobility, C_{Ox} is the capacitance of the SiO_2 layer, C_{DL} is the electrical double layer capacitance, and V_{DS}, V_{TgS} and V_{BgS} are the drain voltage, the electrolyte-gate potential and the back-gate potential, respectively. According to eqn (4), the threshold voltage of the back-gate sweep $V_{BG,th}$, which is defined as the back-gate voltage where the

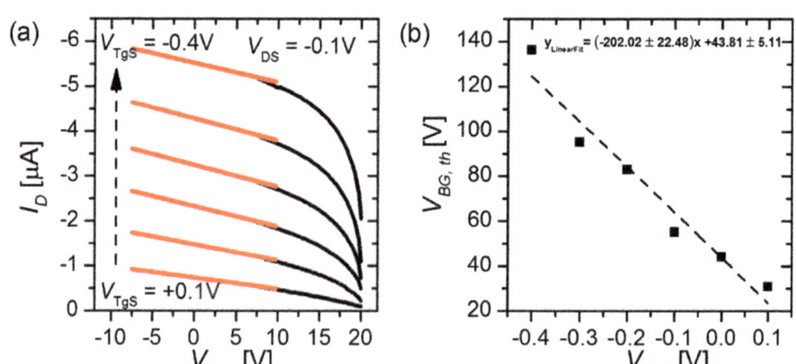

Fig. 2 (a) Back-gate transfer curves (black lines) at $V_{DS} = -0.1$ V recorded in DI-water with a Pt gate electrode and linear fits (red lines) at different electrolyte-gate voltages V_{TgS}. (b) Threshold voltages extracted from the linear fit curves of the back-gate sweeps as a function of the electrolyte-gate potential and the corresponding linear fit curve.

source–drain current equals zero, is directly proportional to the applied electrolyte-gate potential modulated by the capacitive coupling factor C_{DL}/C_{OX}:[23]

$$V_{BG,th} \propto -\frac{C_{DL}}{C_{OX}} V_{TgS}. \quad (5)$$

The threshold voltages were extrapolated from a linear fit (red lines, Fig. 2(a)) of the various transfer curves and plotted as a function of the applied top-gate voltage V_{TgS}, as depicted in Fig. 2(b).

By using the slope k obtained from the linear fit of the $V_{BG,th}$ *vs.* V_{TgS} dependence and taking into account the capacitance of the back-gate dielectric (thickness $d_{Ox} = 65$ nm, $C_{Ox} = 53.2$ nF cm^{-2}), the capacitance of the electrical double layer can be calculated according to eqn (5) (ref. 23).

We obtain a double layer capacitance of C_{DL} (DI-H_2O) = 10.8 ± 1.2 μF cm^{-2}, which exceeds the literature values for P3HT in contact with deionized water (3–6 μF cm^{-2}), as measured by impedance spectroscopy, by a factor of 2, but they are consistent with values obtained for metallic surfaces in contact with an electrolyte, ~20 μF cm^{-2} for gold in contact with DI-water.[12] Cramer *et al.*[23] determined the double layer capacitance with a pentacene thin-film transistor with deionized water as the electrolyte. They obtained values of C_{DL} (DI-H_2O) = 7.8 μF cm^{-2},[23] which is generally lower than the results obtained with the P3HT OTFT. One possible explanation could be that their semiconducting layer consists of less than six monolayers of pentacene and therefore has a much smoother surface than the P3HT film (see the AFM image in Fig. 1(c)) used in this work. In simplified terms, the huge surface roughness of the P3HT film results in an increased capacitance of the electrical double layer at the semiconductor–electrolyte interface since the geometrical size of the active area (4 mm × 4 mm) stays constant, whereas the effective contact area between the electrolyte and semiconductor increases with increasing surface roughness.

In general, the capacitance of the Debye–Helmholtz layer depends on the ionic strength of the electrolytic solution (see eqn (1) and (2)), therefore back-gate sweeps were not only recorded for different top-gate voltages but also in KCl solutions with varying ion concentrations. The obtained values for the double layer capacitance measured with a Pt wire as the gate electrode are shown in Fig. 3(a). The reduction of the capacitance to lower ionic strengths is in general agreement with theoretical models describing the Debye–Helmholtz layer.[17–19] Again, the value obtained by Cramer *et al.*[23] for a 100 mM solution of a monovalent salt (sodium chloride, NaCl) C_{DL} (100 mM NaCl) = 14.6 μF cm^{-2} is lower than the EDL capacitance measured with the P3HT OTFT.

Additionally, the same measurements were performed with an Ag/AgCl electrode. The main difference between these two gate materials is that platinum can be seen as an ideally polarizable electrode when operated in an electrolytic solution with these relatively low voltages ($|V_{TgS}| < 1$ V), which means that an electrical double layer forms at the interface between the gate-material and the electrolyte. However, due to the following reaction constantly taking place at the surface of the Ag/AgCl electrode, a steady-state current flows from the gate electrode to the electrolytic solution, and therefore no electrical double layer forms at the interface between the electrode and the electrolyte.[26]

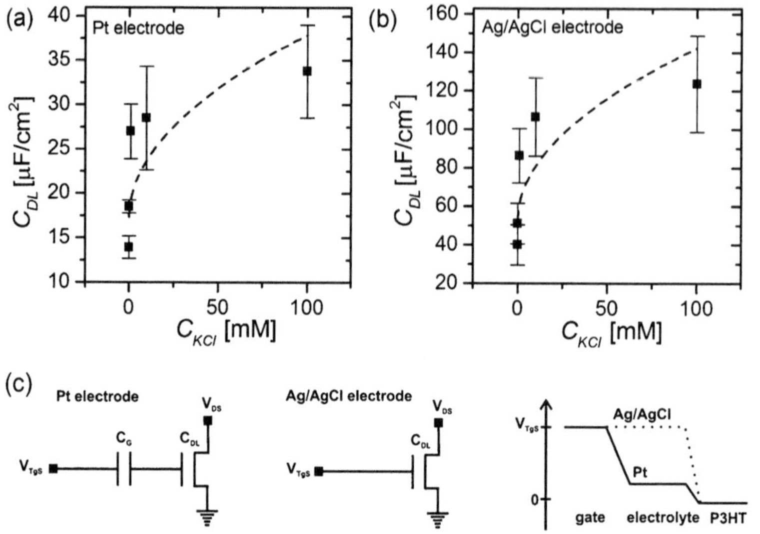

Fig. 3 Capacitance of the electrical double layer C_{DL} obtained (a) for a Pt-gated transistor and (b) with an Ag/AgCl electrode as the gate electrode as a function of the salt concentration. Panel (c) shows the equivalent circuits[28] of a Pt gate electrode and an Ag/AgCl electrode immersed in the electrolytic solution. The right graph depicts the potential distribution between the gate electrode and the semiconducting channel for both electrode types.[26]

$$\text{Ag} + \text{Cl}^- \leftrightarrow \text{AgCl} + \text{e}^- \quad (6)$$

This means that with a Pt wire immersed as the gate electrode in the electrolytic solution, the applied gate potential drops across two electrical double layers formed both at the electrode–electrolyte interface as well as at the electrolyte–semiconductor interface. On the contrary, for an Ag/AgCl electrode the whole potential has to drop across the electrical double layer formed at the electrolyte–semiconductor interface (see Fig. 3(c)).[19,20,27,28]

In a simple equivalent circuit, the Pt wire immersed in the electrolytic solution can be regarded as a capacitor C_G (see Fig. 3(c)), which is connected in series with the capacitor C_{DL} of the interface between the P3HT layer and the electrolytic solution, as shown in Fig. 3(a). Thus the effective gating capacitance of the OTFT operated with a Pt electrode is given by:[28]

$$C_{\text{eff,Pt}} = \frac{C_G C_{DL}}{C_G + C_{DL}}. \quad (7)$$

Therefore one would expect to measure a higher double layer capacitance C_{DL} by operating the device with an Ag/AgCl electrode. Fig. 3(b) shows the double layer capacitance measured with an Ag/AgCl electrode as a function of the salt concentration present in the electrolyte.

For a KCl concentration of 100 mM we obtain a double layer capacitance of C_{DL} (100 mM KCl, Ag/AgCl) $= 123.77 \pm 2.50$ μF cm^{-2}, which is about 4 times higher than with a Pt electrode, where C_{DL} (100 mM KCl, Pt) $= 33.78 \pm 5.25$ μF cm^{-2}.

3.3 Simulation of electrolyte-gated field-effect transistors

For the simulation of the back-gated and electrolyte-gated OTFT devices we use a modified drift–diffusion model implemented into the commercial TCAD software Sentaurus. For a given device geometry, this software simulates charge transport by solving a set of equations, including the drift–diffusion equations, the continuity equations and the Poisson equation

$$J_n = -e\mu_n \nabla\phi + eD_n\nabla n, \tag{8}$$

$$J_P = -e\mu_p \nabla\phi - eD_p\nabla p, \tag{9}$$

$$\nabla \cdot J_n = e(R - G), \tag{10}$$

$$\nabla \cdot J_p = -e(R - G), \tag{11}$$

$$\nabla \cdot (\varepsilon_0\varepsilon_r \nabla\phi) = e(n - p + N_D^+ - N_A^- + p_{At}), \tag{12}$$

where J_n (J_p) is the electron (hole) current density, n (p) is the electron (hole) density, μ_n (μ_p) is the electron (hole) mobility, D_n (D_p) is the electron (hole) diffusion coefficient, ϕ is the electrostatic potential, G and R are the net generation and recombination rates and p_{At} is the density of trapped holes. To account for the particularities of organic materials, three modifications are implemented.[29,30] Firstly, a low contact barrier height from the source to HOMO level of the semiconductor material ensures easy injection of holes, giving the device p-type conductivity. Secondly, a Poole–Frenkel-type field-dependent mobility needs to be considered:

$$\mu = \mu_0(T)\exp\left(\sqrt{E/E_0}\right), \tag{13}$$

where $\mu_0(T)$ is the low field mobility, E is the electric field and E_0 is an effective parameter of approximately 3×10^5 V cm^{-1}. Thirdly, an interface trap model at the oxide–P3HT interface and the electrolyte–semiconductor interface effectively models the hopping-assisted charge transport. Here, an exponential energy distribution of two different trap levels, a shallow trap at 0.1 eV and a deeper trap at 0.4 eV, from the HOMO level is considered. Such a broad energy distribution is justified by the use of standard spin-coating techniques for the deposition of the organic semiconductor, inevitably introducing defects into the P3HT active region.

In a first approximation, the electrolyte will be taken into account as an insulating layer between the electrolyte-gate contact and the organic semiconductor region, according to the simple Helmholtz model (see Fig. 4(a)).[31]

That way, the electrical double layer is modelled as a parallel plate capacitor (from here on referred to as the Helmholtz layer) accounting for a linear potential drop. The capacitance of this Helmholtz layer is taken from the experimentally extracted data to 28 μF cm^{-2} for an ion concentration of 10 mM, according to Section 3.2. As can be seen in Fig. 4(b), when operating with the top-gate contact the conductive channel is spatially shifted towards the electrolyte–semiconductor interface. This can be intuitively understood considering the electric field that is generated by the applied negative top-gate voltage and which attracts positive charge carriers towards this interface. The considerably low top-gate voltage is

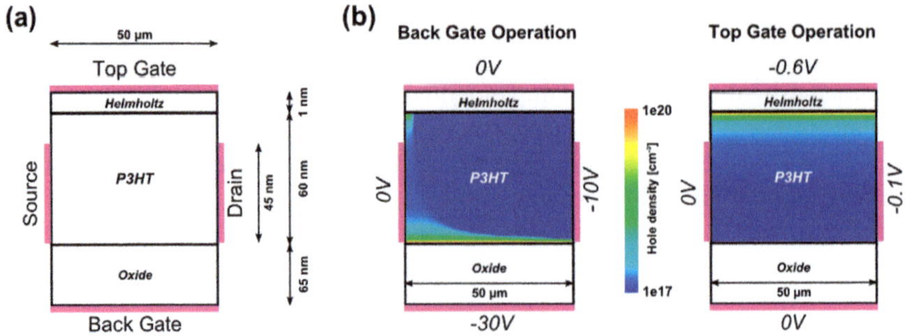

Fig. 4 (a) Drift–diffusion simulation layout for the Helmholtz model. The electrolyte is taken into account by an insulating Helmholtz layer. (b) Comparison of back-gate operation and top-gate operation. The values of the boundary conditions are given at the respective contacts and are typical for the different operation modes. The hole density gives information as to where the conducting channel is formed.

due to the large capacitive value of the Helmholtz layer compared to the oxide layer of the back-gate contact.

With this simulation setup, we simulate transfer and output curves in both back-gate and top-gate operation and compare them to experimental data to adjust the simulation parameters. The simulation of the back-gate operation has already been reported in the literature,[32] and serves as a reference for the simulation parameters of the organic semiconductor material, such as density of states, unintentional doping, trap level distribution, low field mobility *etc.* These parameters are then used for the simulation of top-gate operation to give a conclusive self-consistent picture. All relevant simulation parameters are given in Table 1. The results of such a fitting procedure are shown in Fig. 5 for one selected device. The simulation of transfer curves in back-gate operation (Fig. 5(a)) shows good agreement with the experimental data in the accessible voltage range, and also the output curve is in the right order of magnitude. The low field mobility in that case is $\mu_0 = 5 \times 10^{-3}$ cm^2 V^{-1} s^{-1}, which is in good agreement with literature values.[11,25] When simulating the top-gate operation (Fig. 5(b)), one can see that using this value for the low field mobility leads to a drastic underestimation of the drain current. There are two parameters in our model which affect the transconductance of the device, *i.e.* the slope of the linear transistor behaviour, the first being the capacitance of the Helmholtz layer and the second being the low field mobility. By changing the Helmholtz layer capacitance one would have to use unphysical values that are several orders of magnitude larger than what has been extracted experimentally, in order to fit the experimental data. Furthermore, this leads to large deviations from the experimental data in the sublinear regime as the curve becomes heavily kinked.

On the other hand, when changing the mobility by about one order of magnitude, the transconductance of the experimental data can be matched correctly and also the deviations in the sublinear regime are not as pronounced. Such an elevated low field mobility hints at a physical change in the transport properties at the electrolyte–semiconductor interface. Cramer *et al.*[33] stated that the reorientation of water dipoles at the electrolyte–semiconductor interface stabilizes positive charge carriers and leads to a polaronic trap state which

Table 1 Simulation parameters for back-gate and top-gate operation

		Device T#01
P3HT	LUMO	3 eV
	HOMO	5 eV
	ε_r	3
	DOS	5×10^{20} cm^{-3}
Doping concentration	N_D	1.2×10^{17} cm^{-3}
Poole–Frenkel mobility	μ_0 (back)	5×10^{-3} cm^2 V^{-1} s^{-1}
	μ_0 (top)	3×10^{-2} cm^2 V^{-1} s^{-1}
	E_0	3×10^5 V cm^{-1}
Shallow level trap	$c_{0,\text{shall}}$	5×10^{12} cm^{-2}
	E_{mid}	0.1 eV
	E_{sig}	0.2 eV
Deep level trap	$c_{0,\text{deep}}$	5×10^{12} cm^{-2}
	E_{mid}	0.4 eV
	E_{sig}	0.3 eV
Fixed charges	c_{fix} (back)	-2.1×10^{12} cm^{-2}
	c_{fix} (top)	-2.1×10^{12} cm^{-2}

remains highly mobile in the plane parallel to the interface. Their findings based on pentacene show a mobility elevated by about one order of magnitude.

They mention that the polaron binding energy depends only on water polarization and the spatial extension of the charge carriers in the organic semiconductor. Therefore, these findings can be generalized to other organic semiconductors and might give an explanation for the different mobilities for back-gate operation *versus* top-gate operation in our sensor devices. However, further investigations are needed in that direction.

In order to refine the simple approximation of the EDL as a parallel plate capacitor, one needs to account for a diffusive layer, the Gouy–Chapman layer, in

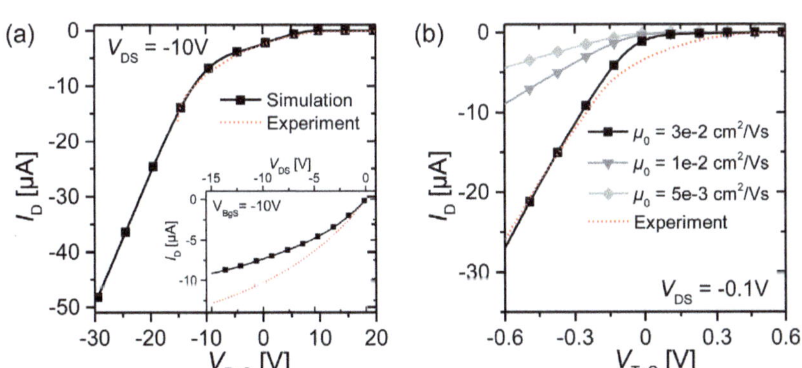

Fig. 5 (a) Best fit of the back-gate operation transfer curve for one selected device. The inset shows an output curve at $V_{BgS} = -10$ V. The mobility is $\mu_0 = 5 \times 10^{-3}$ cm^2 V^{-1} s^{-1}. (b) Fit of the top-gate operation transfer curve for device T#01 for different low field mobilities according to the Helmholtz model. The ion concentration in the electrolyte is 10 mM. All simulation parameters can be found in Table 1.

the form of the Poisson–Boltzmann equation (PBe) according to the Stern model.[34] The PBe reads

$$\nabla(\varepsilon_0\varepsilon_r\nabla\phi) = -\sum_{i=1}^{N} z_i e c_{i,0}\exp\left(-\frac{z_i e(\phi - V_{TgS})}{k_B T}\right), \tag{14}$$

where the sum goes over all ion species present in the electrolyte, z_i is the valency and $c_{i,0}$ the bulk concentration of ion species i. This is already under investigation and will be the topic of a future contribution.

4 Conclusion

In summary, we reported the measurement of the electrical double layer capacitance of a dual-gated P3HT OTFT by sweeping the back-gate potential with fixed electrolyte-gate potentials at the top-gate. We proved the dependency of the EDL capacitance on the ionic strength of the electrolyte and additionally we demonstrated the effect of a non-polarizable gate electrode in comparison to a polarizable electrode on the gating mechanism. With the experimentally obtained values for the capacitance of the Debye-Helmholtz layer at the semiconductor–electrolyte interface, we can model the electrolyte-gate transfer characteristics of the P3HT OTFT. Our study suggests that the mobility of the active material close to the semiconductor–electrolyte interface is larger than the bulk mobility.

Acknowledgements

This work was partially supported by the International Graduate School for Science and Engineering (IGSSE), the Graduate School (TUM-GS) at the Technische Universität München, and the European Integrated Training Networks "Olimpia" and "OrgBio".

References

1 L. Kergoat, B. Piro, M. Berggren, G. Horowitz and M.-C. Pham, *Anal. Bioanal. Chem.*, 2012, **402**, 1813–1826.
2 P. Lin and F. Yan, *Adv. Mater.*, 2012, **24**, 34–51.
3 J. T. Mabeck and G. G. Malliaras, *Anal. Bioanal. Chem.*, 2006, **384**, 343–353.
4 A. M. Münzer, K. Melzer, M. Heimgreiter and G. Scarpa, *Biochim. Biophys. Acta*, 2013, **1830**, 4353–4358.
5 C. Bartic, B. Palan, A. Campitelli and G. Borghs, *Sens. Actuators, B*, 2002, **83**, 115–122.
6 M. E. Roberts, S. C. B. Mannsfeld, N. Queraltó, C. Reese, J. Locklin, W. Knoll and Z. Bao, *Proc. Natl. Acad. Sci. U. S. A.*, 2008, **105**, 12134–12139.
7 S. Casalini, F. Leonardi, T. Cramer and F. Biscarini, *Org. Electron.*, 2013, **14**, 156–163.
8 M. D. Angione, S. Cotrone, M. Magliulo, A. Mallardi, D. Altamura, C. Giannini, N. Cioffi, L. Sabbatini, E. Fratini, P. Baglioni, G. Scamarcio, G. Palazzo and L. Torsi, *Proc. Natl. Acad. Sci. U. S. A.*, 2012, **109**, 6429–6434.
9 M. Magliulo, A. Mallardi, M. Y. Mulla, S. Cotrone, B. R. Pistillo, P. Favia, I. Vikholm-Lundin, G. Palazzo and L. Torsi, *Adv. Mater.*, 2013, **25**, 2090–2094.

10 G. Scarpa, A.-L. Idzko, S. Götz and S. Thalhammer, *Macromol. Biosci.*, 2010, **10**, 378–383.

11 S. M. Goetz, C. M. Erlen, H. Grothe, B. Wolf, P. Lugli and G. Scarpa, *Org. Electron.*, 2009, **10**, 573–580.

12 L. Kergoat, L. Herlogsson, D. Braga, B. Piro, M.-C. Pham, X. Crispin, M. Berggren and G. Horowitz, *Adv. Mater.*, 2010, **22**, 2565–2569.

13 H. Klauk, *Chem. Soc. Rev.*, 2010, **39**, 2643–2666.

14 S. H. Kim, K. Hong, W. Xie, K. H. Lee, S. Zhang, T. P. Lodge and C. D. Frisbie, *Adv. Mater.*, 2013, **25**, 1822–1846.

15 S. H. Kim, K. Hong, W. Xie, K. H. Lee, S. Zhang, T. P. Lodge and C. D. Frisbie, *Adv. Mater.*, 2013, **25**, 1822–1846.

16 T. Cramer, A. Campana, F. Leonardi, S. Casalini, A. Kyndiah, M. Murgia and F. Biscarini, *J. Mater. Chem. B*, 2013, **1**, 3728–3741.

17 P. J. Boddy, *J. Electroanal. Chem.*, 1965, 199–244.

18 D. C. Grahame, *Chem. Rev.*, 1947, **41**, 441–501.

19 A. J. Bard and L. R. Faulkner, *Electrochemical Methods-Fundamentals and Applications*, Wiley, New York, 2001.

20 M. Waleed Shinwari, M. Jamal Deen and D. Landheer, *Microelectron. Reliab.*, 2007, **47**, 2025–2057.

21 M. E. Roberts, S. C. B. Mannsfeld, R. M. Stoltenberg and Z. Bao, *Org. Electron.*, 2009, **10**, 377–383.

22 C. J. Bettinger and Z. Bao, *Adv. Mater.*, 2010, **22**, 651–655.

23 T. Cramer, A. Kyndiah, M. Murgia, F. Leonardi, S. Casalini and F. Biscarini, *Appl. Phys. Lett.*, 2012, **100**, 143302.

24 M.-J. Spijkman, K. Myny, E. C. P. Smits, P. Heremans, P. W. M. Blom and D. M. de Leeuw, *Adv. Mater.*, 2011, **23**, 3231–3242.

25 A. Salleo, T. W. Chen, A. R. Völkel and R. A. Street, *Phys. Rev. B: Condens. Matter Mater. Phys.*, 2004, **70**, 1–10.

26 G. Tarabella, C. Santato, S. Y. Yang, S. Iannotta, G. G. Malliaras and F. Cicoira, *Appl. Phys. Lett.*, 2010, **97**, 123304.

27 G. Tarabella, C. Santato, S. Y. Yang, S. Iannotta, G. G. Malliaras and F. Cicoira, *Appl. Phys. Lett.*, 2010, **97**, 123304.

28 P. Lin, F. Yan and H. L. W. Chan, *ACS Appl. Mater. Interfaces*, 2010, **2**, 1637–1641.

29 A. Bolognesi, A. Di Carlo and P. Lugli, *Appl. Phys. Lett.*, 2002, **81**, 4646.

30 A. Bolognesi, M. Berliocchi, M. Manenti, A. Di Carlo, P. Lugli, K. Lmimouni and C. Dufour, *IEEE Trans. Electron Devices*, 2004, **51**, 1997–2003.

31 H. Helmholtz, *Ann. Phys. Chem.*, 1853, **165**, 211–233.

32 A. Khaliq, F. L. Xue and K. Varahramyan, *Microelectron. Eng.*, 2009, **86**, 2312–2315.

33 T. Cramer, T. Steinbrecher, T. Koslowski, D. A. Case, F. Biscarini and F. Zerbetto, *Phys. Rev. B: Condens. Matter Mater. Phys.*, 2009, **79**, 155316.

34 O. Stern, *Z. Elektrochem. Angew. Phys. Chem.*, 1924, **30**, 508–516.

Faraday Discussions

DISCUSSIONS

Organic bioelectronics: general discussion

Simon Higgins, Justin Hodgkiss, Natalie Plank, Luisa Torsi, Paul Cowper, George Malliaras, Yasmine Braeken, Nicolas Laurand, Ifor Samuel, Agata Pomorska, Fred Wudl, Bingqian Xu, Uli Lemmer, Dmitrii Perepichka, Andrew Mount, Lynn Dennany, Donal Bradley and Katharina Melzer

DOI: 10.1039/c4fd90048k

George Malliaras opened the discussion of the paper by Lynn Dennany: Compared to labels using ruthenium complexes, how big are your labels?

Lynn Dennany replied: This label would actually fall within the mid range of current labels. It is smaller than those that only use ruthenium complexes but is considerable smaller than the microparticle labels that are also being utilised as antibody and DNA labels for immunoassays. Microparticle labels include silicon dots which are loaded with ruthenium labels, Au and Pt particles and also CNTs.

Fred Wudl asked: With regard to the electrochemiluminescence (ECL) mechanism, the usually accepted organic ECL mechanism involves oxidizing to the radical cation at the anode, reduction of the latter to a neutral molecule leaves it in an excited state which leads to light upon return to the ground state with emission of a photon, is the scheme in your paper for the inorganic system? What was the electrolyte of your actual device? What was your supporting electrolyte?

Lynn Dennany responded: The scheme given in the paper is that for inorganic systems. The scheme for an organic system would be as follows:

$\text{T4BT–B} - e^- \rightarrow (\text{T4BT–B})^{\bullet +}$ (1)

$\text{TPrA} - e^- \rightarrow \text{TPrA}^{\bullet +}$ (2)

$\text{TPrA}^{\bullet +} - H^+ \rightarrow Pr_2\text{N–CHEt}^{\bullet}$ (3)

$(\text{T4BT–B})^+ + \text{TPrA}^{\bullet} \rightarrow (\text{T4BT–B})^* + Pr_2N^+ = \text{CHEt}$ (4)

$(\text{T4BT–B})^* \rightarrow \text{T4BT–B} + h\nu$ (5)

The supporting electrolyte for this scheme, 0.1 M tetrabutylammonium hexafluorophosphate ($(TBA)PF_6$), in CH_2Cl_2, this was used for figure 2[1] as well as for the ECL analysis using T4BT–B.

1 L. Dennany, Z. Mohsan, A. L. Kanibolotsky and P. J. Skabara, *Faraday Discuss.*, 2014, DOI: 10.1039/c4fd00090k.

Nicolas Laurand asked: Could you comment on the use of your materials for bio-sensing applications? How would you ensure specificity for biomolecule detection?

Lynn Dennany responded: The truxenes are a very new set of compounds used for ECL analysis. As it currently stands we would need to address their solubility in aqueous solutions before they could be regularly used for bio-sensing. When this is achieved, these materials could be used as labels attached to antibodies which would give them their specificity.

In addition, there is a need for a family of compounds to be developed, each with different redox properties. This would be to fit into a multiplexed device in the future, but on the short term this would be how to tune the complex to a specific target allowing for specificity within biomolecule detection. The direct detection would most likely be within a film construct which would need more than one type of T4BT to address concerns relating to specificity.

Dmitrii Perepichka commented: Currently sensing in solution, in a real device you will likely want to place your sensing molecules on a metallic surface. Would surface quenching be an issue?

Lynn Dennany replied: Yes, in the long term we would hope to develop some sort of surface or film approach. To address the issue of surface quenching it is likely that the sensing molecules would be tethered to the surface rather than directly attached and that this would be sufficiently long enough to offset any surface quenching.

Dmitrii Perepichka asked: In determination of the diffusion coefficients, are you assuming a single electron process?

Lynn Dennany replied: Yes, we did assume a one electron transfer for this determination. This was based upon previous work on similar compounds by Bard *et al.*[1] There is still some research being carried out in relation to this but currently we are basing our interpretation on this assumption.

1 K. M. Omer, A. L. Kanibolotsky, P. J. Skabara, I. F. Perepichka and A. J. Bard, *J. Phys. Chem. B.*, 2007, **111**, 6612–6619.

Andrew Mount said: The Randles–Sevcik equation can only be applied to electrochemically reversible redox systems, and variation of peak current with the square root of scan rate is not a sufficient diagnosis of this (for example electrochemically irreversible reactions also show this dependence of peak current on sweep rate but their response is not determined by this equation). What further evidence is there which demonstrates that these reactions are electrochemically reversible?

Lynn Dennany responded: In addition to the linear dependence of the peak current on the square root of the scan rate, the full width at half maximum (FWHM) is close to the theoretical value of 59 mV. For the first oxidation we've got $E_{pa} = +0.83$ V, $E_{pc} = +0.75$ V, $\Delta E = 80$ mV, for the reduction, $E_{pa} = -1.80$,

$E_{pc} = 1.88$ V, $\Delta E = 80$ mV. The ΔE is higher than 59 mV probably due to some uncompensated resistance but this still indicates an electrochemically reversible reaction. As mentioned previously this also assumes a one electron transfer for this determination and although this is an assumption, we are currently undertaking further investigations to probe this assumption.

Andrew Mount asked: It is interesting that direct TPA oxidation at the glassy carbon electrode (GCE) is reported at ~ +2.1 V with respect to the silver pseudoreference electrode. Given that a thermodynamically favourable reaction between oxidised T4BT–B and TPA is postulated, and T4BT–B is oxidised at much lower potentials, this suggests a very large kinetic inhibition of direct TPA electrochemical oxidation at glassy carbon. It is interesting that, notwithstanding differences in the reference electrodes and solvents used, previous work has reported much lower direct oxidation potentials.[1,2] Have the authors tried measuring direct oxidation of TPA at other electrodes, and do they have an explanation for the differences in the observed direct oxidation of amines in these and their measurements?

1 Work reporting amine oxidation on glassy carbon electrodes: R. S. Deinhammer, M. Ho, J. W. Andereff and M. D. Porter, *Langmuir*, 1994, **10**, 1306–13.
2 Work reporting electrochemical oxidation of polyamines at diamond thin film electrodes: M. D. Koppang, M. Witek, J. Blau and G. M. Swain, *Anal. Chem.* 1999, **71**, 1188–1195.

Lynn Dennany responded: Yes, there can be quite a change in amine oxidation at different electrodes and also it can be shifted upon changes in pH as well. Although we have not tried measuring the direct oxidation of TPA at other electrodes it is something that we intend to look at in the future. We also anticipate that there may be changes with the oxidation of T4BT–B under different conditions as well. Within our current system both T4BT–B and ruthenium peaks appear at 1.2 V and if the scan is continued at 2.1 V the direct oxidation of TPA occurs. This dual response has also been seen in several other ECL studies, although I believe it is most documented with quantum dots,[1,2] although these do refer to similar trends for amines. This value for the direct oxidation of TPA has also been reported for several ruthenium based systems using different electrodes.[3]

1 G. Zu and H. Ju, *Anal. Chem.*, 2004, **76**, 6871–6876.
2 H. Jiang and H. Ju, *Anal. Chem.*, 2007, **79**, 6690–6696.
3 R. Huang, M.-Y. Wei and L.-H. Guo, *J. Electroanal. Chem.*, 2011, **656**, 136–139.

Andrew Mount asked: Fig 5 produces a comparison of the ECL response for the T4BT–B system and the $[Ru(bpy)_3]^{2+}$ system.[1] Firstly, how have the responses been "adjusted for changes in the PMT bias" so that comparison can be made, and secondly, isn't the statement in conclusions that "T4BT–B with TPA......significantly producing almost double the ECL intensity compared to $[Ru(bpy)_3]^{2+}$" system rather underselling the relative difference, as Fig 5 seems to show a difference in ECL offset of $10^{(5.78-3.93)}$ ~ 70 at the same TPA concentration?

1 L. Dennany, Z. Mohsan, A. L. Kanibolotsky and P. J. Skabara, *Faraday Discuss.*, 2014, DOI: 10.1039/c4fd00090k.

Lynn Dennany responded: The ECL intensity is recorded in arbitrary units which are essentially controlled by the PMT settings without the actual ECL reaction. If the voltage applied to the PMT is increased then the ECL response will also show an increase. Since the ECL responses for T4BT–B were quite substantial it was necessary to reduce the voltages applied to the PMT to prevent saturation of the PMT detector. Therefore to try and make a comparison, ECL responses were recorded for both ruthenium and T4BT–B with the same concentration of TPA at different PMT settings. These differences were then utilised to adjust the responses over the entire range accordingly. A more robust method is currently being investigated to allow for a more accurate comparison and assessment of the enhancement seen for the T4BT–B compounds. This was also the reason for our more conservative estimation of the ECL enhancement.

George Malliaras asked: Concerning the dependence of your signal on TPA concentration: Do you see a tail-off at low concentrations and a saturation at high concentrations?

Lynn Dennany replied: Yes, depending on the setting we do see a saturation at high concentrations which can be overcome by lowering the PMT. There is probably a point where this would no longer be possible but as yet, we haven't reached that concentration. Using ruthenium at low concentrations of TPA, the ruthenium alone becomes indistinguishable from the ruthenium with TPA. A similar trend is seen for the T4BT–B but at a much lower concentration.

Simon Higgins asked: I notice there are multiple closely-spaced anodic redox waves for the truxene, which I guess means that there are several redox sites that are effectively decoupled from each other. So is it known which part of the molecule is oxidised first? The core or the external oligofluorene 'arms'?

Lynn Dennany responded: The reduction of T4BT–B is a $3e^-$ transfer and the first oxidation is attributed to the BT acceptor unit. The three oxidations are not the same and come from the fluorene units (arms) but not the same fluorene unit hence why they are different. However, we do not know which arm oxidised first or why this may be the case.

Andrew Mount remarked: In the introduction there is mention of the ECL method potentially leading to a new breed of more sensitive biosensors, capable of measuring disease markers down to the picomolar range. Many biosensor systems already exist (*e.g.* using fluorescence or impedance transduction) with such low limits of detection. Why would ECL biosensor systems be expected to have enhanced sensitivity over *e.g.* fluorescence biosensors and/or are there other limitations with existing transduction approaches which can be potentially overcome by using ECL?

Lynn Dennany replied: ECL offers the ability to control the timing and position of the response since it is amperometrically controlled, which offers a benefit over fluorescence biosensors. In addition, since ECL does not require a excitation light source as the light produced is electrochemically generated, this offers another advantage over fluorescence biosensors. This means that there are no background light signals or light scattering effects that can be seen with fluorescence biosensors which should translate to greater sensitivities. The ability to control where the response is produced and also the elimination of a light source also means that ECL has benefits for miniturisation for the production of point of care devices, although this would be something for future work.

Yasmine Braeken asked: If you are sensing an analyte in a complex medium, is it possible to distinguish between different analytes if the peaks in the spectrum overlap?

Lynn Dennany answered: Although I don't know for every analyte, there are some subtle differences between some of the analytes we have checked. For some analytes we see more than one oxidation or an earlier onset of the amperometric response and therefore by combining both the ECL and simultaneously recorded amperometric response it is possible to distinguish between analytes.

Luisa Torsi opened the discussion of the paper by Nicolas Laurand: How does this approach compare to plasmon resonance detection? What are the main advantages and how does it compare in terms of sensitivity?

Nicolas Laurand replied: The distributed feedback (DFB) laser platform discussed here is a special type of evanescent wave sensor. The main advantages of the DFB laser approach lie in its high detection resolution combined with its ease of implementation and the possibility to set the position and size of the sensing area almost at will. The high resolution is a result of the laser oscillation, which enables narrow resonance without trade-off on the sensitivity. The sensitivity of plasmon resonance (SPR) sensors is typically higher than for other evanescent wave technologies, including DFB lasers. This is mainly because of the phase matching condition on plasmon excitation.[1] Nevertheless, SPR devices suffer from a low Q-factor (broad resonance linewidth), which is detrimental to the resolution and hence to the overall limit of detection. For comparison, typical Q-factors for SPR sensors and DFB laser sensors are 15 and 10 000, respectively. Losses in metals also limit the propagation of surface plasmons to around 20 microns, which in turn restricts the sensing area. Obviously, the organic DFB laser biosensor is still a very young technology compared to SPR and its real capacity and limits are still to be properly set. It might or might not match the detection levels of SPR systems but its key features make it an attractive solution for particular applications, *e.g.* point-of-care sensing.

1 M. Curie and C. Olson, *Opt. Express*, 2011, **19**, 10029–10930.

Andrew Mount asked: Can you explain fundamentally why the wavelength is linearly dependent on film thickness (Fig. 4)?[1] When will this dependency be lost (*e.g.* how thick a layer can be measured?) It is stated that other parameters could also be monitored other than wavelength; what are these? Finally, what are the strengths and weaknesses of this measurement approach compared to surface plasmon resonance?

1 A.-M. Haughey, C. Foucher, B. Guilhabert, A. L. Kanibolotsky, P. J. Skabara, G. Burley, M. D. Dawson and N. Laurand, *Faraday Discuss.*, 2014, DOI: 10.1039/c4fd00091a.

Nicolas Laurand responded: The laser oscillation wavelength is a linear function of the effective refractive index of the laser mode. Molecule binding to the laser surface locally increases the refractive index and in turn increases this effective refractive index, resulting in a red-shift of the wavelength. This red-shift is proportional to the overlap of the laser mode with the region where adsorbed or bound molecules are present multiplied by the local refractive index increase caused by molecule binding/adsorption. Therefore, for a homogeneous film the shift is proportional to the thickness, as long as the latter is small compared to the evanescent wave decay away from the laser surface. As the thickness of the film increases to several tens of nms the latter condition is not strictly valid and the wavelength shift will 'saturate' at some point. But the modal profile also changes and is pulled towards the surface for increasing layer thickness and this partly compensates the expected saturation effect. These are the reasons why the experimental data in Fig. 4 show a linear dependency over a wide thickness range.[1] We have shown that this behavior can be simply modeled.[2] When this linearity is lost depends on the refractive index profile of the whole laser structure and on the wavelength of operation. For the blue-emitting laser reported here, experimental data show that the behavior can be considered linear up to polyelectrolyte thicknesses of 60 nm but detection of materials above 100 nm is still possible.

1 A.-M. Haughey, C. Foucher, B. Guilhabert, A. L. Kanibolotsky, P. J. Skabara, G. Burley, M. D. Dawson and N. Laurand, *Faraday Discuss.*, 2014, DOI: 10.1039/c4fd00091a.
2 A.-M. Haughey, B. Guilhabert, A. L. Kanibolotsky, P. J. Skabara, G. A. Burley, M. D. Dawson and N. Laurand, *Sens. Actuators, B*, 2013, **185**, 132–139.

Andrew Mount asked: Fig 6 is stated to show a "saturating behaviour" which "follows a Langmuir equation".[1] In fact the change in the fraction of surface avidin-biotin surface complexes could take place over a wider range of avidin concentration (greater than 2 orders of magnitude) than expected by Langmuir binding thermodynamics. In our previous paper on immunosensor development,[2] a relatively large dynamic range for sensor response was found, consistent with increasing repulsive interactions between neighbouring complexes in the close-packed surface layer on binding. Given the inherently much larger cross-sectional area of avidin compared to biotin, wouldn't a similar increasing interaction between neighbouring avidin molecules bound to a monolayer of biotin be expected as a function of coverage? Can the much larger avidin molecules really "occupy all available biotin sites on the laser"? If so, is there evidence that the biotin functionalisation results in a relatively low density of surface biotin functionalisation?

1 A.-M. Haughey, C. Foucher, B. Guilhabert, A. L. Kanibolotsky, P. J. Skabara, G. Burley, M. D. Dawson and N. Laurand, *Faraday Discuss.*, 2014, DOI: 10.1039/c4fd00091a.
2 I. Ciani, H. Schulze, D. K. Corrigan, G. Henihan, G. Giraud, J. G. Terry, A. J. Walkton, R. Pethig, P. Ghazal, J. Crain, C. J. Cambell, T. T. Bachmann and A. R. Mount, *Biosens. Biolelectron.*, 2012, **31**, 413–418.

Nicolas Laurand replied: We believe that the biotin functionalization results in a relatively high surface coverage (because of the wavelength shift observed after biotin-treatment). It is possible that repulsive interactions between neigbouring avidin molecules bound to the biotin-surface increase as a function of coverage and these might have an effect on further surface immobilization of avidin. Furthermore, all biotin sites on the laser are not really 'occupied'. An avidin molecule bound to one biotin site will shield hundreds of other biotin sites that are then not available for further avidin immobilisation. This effect leads to an apparently lower dissociation constant as inferred from the sensor response.

Uli Lemmer said: For any sensing application, the linewidth of the DFB laser is very important. What do you consider as the lower limit for the linewidth of your DFB lasers?

Nicolas Laurand responded: It is true that the linewidth of the laser is an important parameter, as it ultimately sets the resolution of the sensor. The linewidth is dependent on a number of factors including the 'quality' of the DFB resonator and its lateral extension.[1,2] In principle, we see no reason why the linewidth could not reach values below 0.01 nm. In our current platform, the resolution is limited by the spectrometer used and not the linewidth – the latter is below 0.1 nm but further investigations need to be carried out.

1 A. Reichel, U. LEmmer, J. Feldmann, T. Benstem, W. Kowalsky, U. Scherf, A. Combert and V. Wittwer, *Appl. Phys. B*, 2000, **71**, 897–900.
2 B. Guilhabert, N. Laurand, J. Herrnsdorf, Y. Chen, A. L. Kanibolotsky, C. Orofino, P. J. Skabara and M. D. Dawson, *IEEE Photonics Journal*, 2012, **4**, 684–690.

Uli Lemmer said: As an alternative to DFB lasers for sensing other groups are using whispering gallery mode lasers,[1] which have the advantage of higher Q-factors and smaller linewidths. How do you see the competition between WGM lasers and DFB lasers?

1 U. Bog, T. Laue, T. Grossmann, T. Beck, T. Wienhold, B. Richter, M. Hirtz, H. Fuchs, H. Kalt and T. Mappes, *Lab Chip*, 2013, **13**, 2701–2707.

Nicolas Laurand replied: WGM sensors predate organic DFB laser biosensors. Both approaches are viable for biosensing and each has its advantages. While a sensing platform with high-Q WGM resonators (passive or active) can have very high resolving power, its implementation is more complex because of tighter tolerances on optical alignment for the interrogation and reading of the sensor. This is detrimental to the miniaturization potential, although it could be addressed with some trade-off.[1] A DFB laser on the other hand can be probed anywhere on its surface, which can be up to several cm^2.[2] It is also important to note that a higher Q-factor in a passive resonator comes at the expense of sensitivity and does not automatically lead to better detection levels. The bulk refractive index sensitivity of WGM lasers reported in the paper you mentioned[3] is comparable but slightly lower than for organic DFB lasers.[4,5]

1 A. Wellr, F. C. Liu, R. Dahint and M. Himmelhaus, *Appl. Phys. B*, 2008, **90**, 561–567.
2 Y. Tan, C. Ge, A. Chu, M. Lu, W. Goldshlag, C. S. Huang, A. Pokhriyal, S. George and B. Cunningham, *IEEE Sens. J.*, 2012, **12**, 1174–1180.
3 U. Bog, T. Laue, T. Grossmann, T. Beck, T. Wienhold, B. Richter, M. Hirtz, H. Fuchs, H. Kalt and T. Mappes, *Lab Chip*, 2013, **13**, 2701–2707.
4 M. Lu, S. S. Choi, C. J. Wagner, J. G. Eden and B. T. Cunningham, *Appl. Phys. Lett.*, 2008, **92**, 261502.
5 A.-M. Haughey, B. Guilhabert, A. L. Kanibolotsky, P. J. Skabara, G. A. Burley, M. D. Dawson and N. Laurand, *Sens. Actuators, B*, 2013, **185**, 132–139.

Ifor Samuel asked: You quoted a sensitivity for streptavidin detection of 100 ng ml^{-1}. Can you explain whether that number is good or bad *i.e.* how does it compare with other detection schemes? I ask because the streptavidin–biotin interaction is extremely strong.

Nicolas Laurand replied: This is a very good question but one that does not have a simple answer. First of all, this is a first proof-of-principle. We have shown that a DFB laser can detect biomolecules, specifically. The system is not totally optimized yet for sensitivity – both in terms of the optical platform and of the surface biochemistry. Secondly, avidin was detected with a biotin-functionalised laser and the difference in size of these molecules results in an apparent decrease of the avidin–biotin interaction. A better test to gauge the capability of the platform will be to detect biotin from a surface functionalised with avidin probes. Such a test will necessitate an optimization of the sensitivity first. But the best method for assessing the performance of the sensor will be to detect relevant biomarkers at different levels of detection.

Andrew Mount asked: From Fig 6,[1] the apparent dissociation constant for biotin–avidin appears to be around 300 ng mL^{-1} (where the fractional coverage is around 0.5). Given its tetrameric form, avidin is estimated to be around 60 kDa in size, so this corresponds to a dissociation around the order of nanoMolar. In solution, the dissociation constant for biotin–avidin is of the order of femtoMolar. Such an effect has been seen previously.[2–4] This work instead involves surface biotin functionalisation rather than surface streptavidin functionalisation, but as the relative change is similar are similar factors at play? Is there independent evidence for the value of the surface dissociation constant?

1 A.-M. Haughey, C. Foucher, B. Guilhabert, A. L. Kanibolotsky, P. J. Skabara, G. Burley, M. D. Dawson and N. Laurand, *Faraday Discuss.*, 2014, DOI: 10.1039/c4fd00091a.
2 C. E. Chivers, E. Crozat, C. Chu, V. T. Moy, D. J. Sherratt and M. Howarth, *Nat. Methods*, 2010, 7, 391–393.
3 J. L. Swift, R. Heuff, D. T. Cramb, *Biophys. J.*, 2006, **90**, 1396–1410.
4 T. Buranda, G. P. Lopez, J. Keij, R. Harris, L. A. Sklar, *Cytometry*, 1999, **37**, 21–31.

Nicolas Laurand responded: This is a possibility and could be another factor in the reduced, apparent dissociation constant.

Luisa Torsi asked: Having this dissociation constant, you should also go down to low detection because surface bound streptavidin–biotin systems have been shown to go down to picoMolar.[1] Your data present higher detection limits. Do you have an explanation for this occurrence?

1 M. D. Angione, S. Cotrone, M. Magliulo, A. Mallardi, D. Altamura, C. Gianni, N. Cioffi, L. Sabbatini, E. Fratini, P. Baglioni, G. Scamarcio, G. Palazzo and L. Torsi, *PNAS*, 2012, **109**, 6429–6434.

Nicolas Laurand responded: This is a very interesting comment. We note that in the above reference, surface-immobilized strepdavidin is used to detect biotin, whereas our proof-of-concept laser device does the contrary – it utilizes a biotin-functionalised surface to detect avidin. The significant difference in size between these two molecules means that in the latter case only a few biotin probes can effectively bind to avidin molecules. An immobilized avidin molecule 'shields' hundreds of biotin probes resulting in an 'apparent' reduction in the dissociation constant by the same amount. The homogeneity of surface functionalization might play a role as well but this warrants further investigation.

Andrew Mount said: Another interesting possibility is that the coverage of avidin molecules on the surface is limited by the formation of a close packed monolayer of bound avidin, where due to their mismatch in size not every biotin molecule has an avidin molecule bound to it. We have previously considered a similar case, the binding of a streptavidin-alkaline phophatase (AP) enzyme to a surface dsDNA layer; the AP amplifies the signal due to dsDNA formation (through subsequent enzyme reaction forming an insoluble surface product which is detected); the target strand of DNA is biotinylated in this case to enable surface AP binding.[1] As expected, this does not give a simple Langmuir binding response, but rather the start of a Langmuir response at low coverage (with the amount of bound AP and hence insoluble product being determined by the amount of dsDNA), truncated by the response due to close-packed AP monolayer formation at higher coverage. This coverage, although higher, is still well below Langmuir monolayer dsDNA formation, as once the amount of dsDNA bound is sufficient to form a monolayer of AP, further dsDNA binding does not lead to any more surface bound AP or any more insoluble product to detect. Although the binding and detection mechanism are different, might similar effects be at play in this binding study?

1 M. Kaatz, H. Schulze, I. Ciani, F. Lisdat, A. R. Mount and T. T. Bachmann, *Analyst*, 2012, 137, 59–63

Nicolas Laurand replied: Thank you for mentioning this possibility. We agree; the difference in size between avidin and biotin probably affects the detection sensitivity of our experiment where biotin attached to the laser surface is used to detect avidin. We expect this to lead to an apparent dissociation constant that is lower than that of avidin/biotin in solution as was discussed previously.

Ifor Samuel commented: How well does the laser work in water? Is the stability satisfactory? Does any water go into the organic film and swell it? Is there a shift of wavelength?

Nicolas Laurand answered: The laser operates very well in water. In fact, the threshold is lower and the photostability improves compared to operation in air. There is no evidence of swelling (probably thanks to the hydrophobicity of the active layer) and the wavelength shift observed upon immersion of the laser is as expected for the corresponding change in the refractive index of the medium. It was observed that the wavelength returns to its original value when the water is removed.

Justin Hodgkiss remarked: Do you see a different lasing threshold for bare semiconductor *versus* one saturated with biomolecules (or other test layers)? If so, would this be a possible alternative sensing method that exploits the inherent non-linearity of the system, *i.e.*, if you sit around the threshold such that analyte binding changes the balance to push the system above (ideally) or below the lasing threshold?

Nicolas Laurand responded: We have so far focused on sensing by monitoring the shift of the laser wavelength. It is true that other parameters, such as the threshold, could be used for sensing in principle but the practicability in doing so has not been demonstrated yet. A thin layer of transparent, small biomolecules should not significantly alter the laser threshold. Thicker layers might although it is difficult to predict *a priori* the trend, as there will be induced changes in the laser mode profile and in the associated modal losses and gain. The homogeneity of the layer could be a factor too. This is not something we have really looked at yet but it is certainly interesting. We have thought about non-linear, 'on/off'' detection methods, *e.g.* by having a gain layer thickness just at cut-off so that only in the event of molecule binding would laser oscillation be triggered. This could possibly be very sensitive but it would be difficult to obtain an exact reading of the amount of analytes due to the inherent non-linearity of such an approach.

Simon Higgins commented: I presume that you have checked that your sensors do not show a response for non-specific protein binding using, for instance a serum albumin?

Nicolas Laurand answered: Yes, detection of avidin was also performed in the presence of albumin at levels comparable to those found in blood. No significant non-specific binding was observed. It is assumed that the biotin-functionalisation helps as without biotin, albumin is found to severely adsorb onto the laser surface.

Yasmine Braeken said: Is it possible with this technique to make compact point-of-care devices, leading to results within a short time span?

Nicolas Laurand replied: Yes, in principle. Organic semiconductor lasers can be pumped with GaN laser diodes or even light-emitting diodes as reported in the literature,[1–3] *e.g.* refs [18–20]) and the detection system can be operated with a compact, palm-size spectrometer. Indeed, demonstrating a compact point-of-care device with this technology is one of our motivations.

1 T. Riedl, T. Rabe, H.-H. Johannes, W. Kowalsky, J. Wang, T. Weimann, P. Hinze, B. Nehls, T. Farrell and U. Scherf, *Appl. Phys. Lett.*, 2006, **88**, 241116

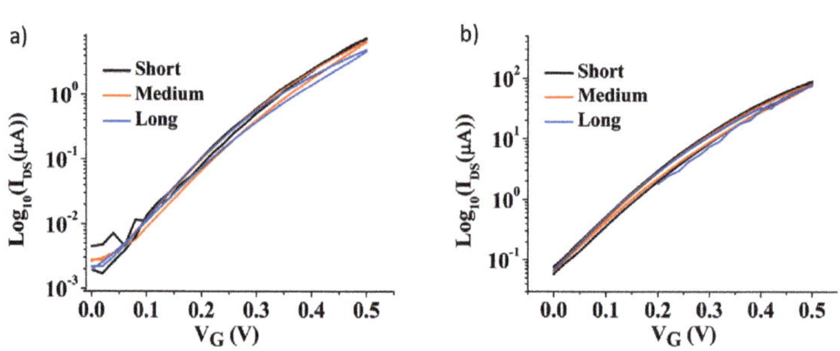

Fig. 1 Transfer characteristics of a) water and b) PBS gated ZnO TFT.

2 A. E. Vasdekis, G. Tsiminis, J.-C. Ribierre, L. O'Faolain, T. F. Krauss, G. A. Turnbull and I. D. W. Samuel, *Opt. Express*, 2006, **14**, 9211–9216.
3 Y. Yang, G. A. Turnbull and I. D. W. Samuel, *Appl. Phys. Lett.*, 2008, **92**, 163306.

Natalie Plank opened the discussion of the paper by Luisa Torsi: In the paper you show transfer characteristics for the ZnO nanowire field effect transitors (NW FETs) but do not quote any leakage currents in the system or the value of the off current. The phosphate-buffered saline (PBS) conducts, and your electrodes are higher than your ZnO films thickness, plus the thin films are porous. Are there passivation steps on electrodes to prevent leakage currents from the gate to the source/drain? Can you plot the transfer characteristics on a log scale to show the value of the off currents?

Luisa Torsi replied: No passivation step is implemented during the fabrication of the device, the only caution is to keep the driving potentials into the window water of electrochemical inertness. The off current is in the order of 10^{-7} and 10^{-9} A for PBS and water, respectively, resulting in an I_{on}/I_{off} ratio of ~103. In Fig. 1,[1] the graph of the transfer characteristic curve for both water and PBS is illustrated as log IDS *vs.* VG .

1 M. Singh, G. Palazzo, G. Romanazzi, G. P. Suranna, N. Ditaranto, C. Di Franco, M. V. Santacroce, M. Y. Mulla, M. Magliulo, K. Manoli and L. Torsi, *Faraday Discuss.*, 2014, DOI: 10.1039/c4fd00081a.

Natalie Plank commented: In the paper and discussion a flat gate electrode is placed in direct contact with the PBS solution gate. How is this gate electrode held in position and is the height well controlled?

1 M. Singh, G. Palazzo, G. Romanazzi, G. P. Suranna, N. Ditaranto, C. Di Franco, M. V. Santacroce, M. Y. Mulla, M. Magliulo, K. Manoli and L. Torsi, *Faraday Discuss.*, 2014, DOI: 10.1039/c4fd00081a.

Luisa Torsi replied: The metal gate is attached to a probe so that controlled movements can be made. In this way, it is possible to maintain the gate at the same height in all measurements. Nevertheless, we have seen that minor deviations from the initial position of the gate are causing changes in the performance of the device. This is because a double layer charge capacitance is the gating system.

Fred Wudl said: The figure of the standard device (Fig. 1) in your paper shows a water droplet.[1] Is the surface as hydrophobic as picture implies?

1 M. Singh, G. Palazzo, G. Romanazzi, G. P. Suranna, N. Ditaranto, C. Di Franco, M. V. Santacroce, M. Y. Mulla, M. Magliulo, K. Manoli and L. Torsi, *Faraday Discuss.*, 2014, DOI: 10.1039/c4fd00081a.

Luisa Torsi replied: The ZnO surface, particularly after exposure to PBS, can be slightly hydrophilic and the measured contact angles given in the manuscript provide a quantitative estimate for the actual hydrophilic/hydrophobic character.

Fred Wudl commented: In your work with liquid electrolytes (ionic liquids). What do droplets look like if ionic liquids are used? Do they spread out more?

Luisa Torsi responded: Ionic liquids do spread out more than water or PBS. But if you adjust the volume of the delivered electrolyte (2–4 μl) and accurately apply the gating to the device, then you will not face a spreading problem.

Agata Pomorska asked: Have you ever tried to synthesize a ZnO layer with another morphology in your system?, Do you expect any difference in the performance as a function of ZnO morphology?

Luisa Torsi responded: We have not yet tried to synthesize ZnO following a different procedure in order to obtain layers with a different morphology. We certainly believe that morphology affects the electrical characteristics of the device and it would be interesting to investigate this effect. But this was not the focus of the present study.

Bingqian Xu asked: The threshold voltage (gate voltage) is low, around 0.3 V. Does the position of the gate have any effect on the threshold voltage and the source–drain current? Is the real voltage drop only over the double layer?

Luisa Torsi responded: Before starting the measurements, we adjusted the position of the gate so that it could not affect the characteristics of the device. Given the relatively high currents measured we can infer that the voltage drop mostly occurs over the double layer.

Uli Lemmer asked: Is your transistor a real 2D transistor or do you expect that you somehow have 3D effects in the sense that the electrolyte is penetrating into the oxide semicondutor layer?

Luisa Torsi responded: The answer to this question is nicely given by the simulation performed in the work by Katharina Melzer,[1] showing that indeed an electrolyte gated device shows a better 2D confinement respect to a back-gated OFET involving the same organic semiconductor.

1 K. Melzer, M. Brändlein, B. Popescu, D. Popescu, P. Lugli and G. Scarpa, *Faraday Discuss.*, 2014, DOI: 10.1039/c4fd00095a.

Justin Hodgkiss asked: You described that either the gate or the surface of the reactive channel can be functionalised with bioreceptors to induce change of potential in channel, and thereby sensitivity towards the analyte. What are advantages and disadvantages of both approaches?

Luisa Torsi replied: In principle both approaches shall generate an impact on the electrical double layer capacitance. In the device with the bio-functionalized active channel, also an effect on the organic semiconductor field-effect mobility should be seen. The two effects, capacity change and mobility change, are however, not easy to decouple in this second configuration.

Andrew Mount remarked: With reference to the change in B/A intensity ratio on exposure to phosphate buffered saline in Fig. 4(c),[1] has the possibility of an O1s signal due to O in adsorbed/absorbed phosphate been excluded (*e.g.* by measuring the P XPS spectrum)? Also, given the attribution of the change to be an increase in the number of oxygen vacancies, is the response (*via* the number of vacancies) dependent on PBS solution purging/outgassing (the amount of oxygen in the solution)?

1 M. Singh, G. Palazzo, G. Romanazzi, G. P. Suranna, N. Ditaranto, C. Di Franco, M. V. Santacroce, M. Y. Mulla, M. Magliulo, K. Manoli and L. Torsi, *Faraday Discuss.*, 2014, DOI: 10.1039/c4fd00081a

Luisa Torsi replied: Based on the XPS spectrum, P is not present in the ZnO film after exposure to PBS. Moreover, when the sample is taken out of the buffer solution, it is rinsed with water so as to remove any remaining salts on the surface of the film. The XPS analysis also showed that the total relative percentage of oxygen does not change when we go from the bare ZnO film to the ZnO film after exposure to PBS solution. Furthermore, the observed increase in oxygen vacancies is not correlated to the presence of oxygen in the buffer.

Paul Cowper asked: On page 11, line 42 of your paper, it is stated that exposure to Na^+ ions in PBS solution leads to oxygen vacancies in the ZnO film.[1] What happens to the displaced oxygen atoms?

1 M. Singh, G. Palazzo, G. Romanazzi, G. P. Suranna, N. Ditaranto, C. Di Franco, M. V. Santacroce, M. Y. Mulla, M. Magliulo, K. Manoli and L. Torsi, *Faraday Discuss.*, 2014, DOI: 10.1039/c4fd00081a

Luisa Torsi replied: From quantification of the XPS analysis of all three cases, we found that there is no considerable difference in the total relative percentage of oxygen. This infers that the displaced oxygen remains within the matrix of ZnO. Furthermore, as mentioned in the literature,[1] the Na^+ occupy interstitial and not substitutional sites.

1 C.H Park, S. B.Zhang and S. H. Wei, *Phys. Rev. B: Condens. Matter Mater. Phys.*, 2002, **66**, 073202.

Luisa Torsi opened the discussion of the paper by Katharina Melzer: From your simulation it looks like the water gated device gives better 2D confinement than the other device. Why is this?

Katharina Melzer replied: For the back gate operation mode the conducting channel is confined to the oxide interface. A small pocket can also be seen at the electrolyte interface which is due to the fact that the parameters of the simulation are chosen such that even at zero top gate bias the device already starts to switch to the ON-state. However, this is only a small portion and the semiconducting channel at the electrolyte interface is not fully formed over the whole distance. Due to the high source drain potential a large concentration of holes is injected at the source contact, which leads to a stronger charge carrier gradient along the conducting channel. When operating the device with the electrolyte gate, it can be seen that the conducting channel is spatially shifted from the oxide interface towards the electrolyte interface. The hole density is practically symmetric with respect to an axis along the *z*-direction, which is due to the lower source drain voltage compared to the back gate operation. All of the holes that are injected at the source contact are directly moving towards the upper conducting channel and are confined to the first few nanometers from the interface.

George Malliaras remarked: Have you compared the capacitances you obtain experimentally with what would be expected for a double layer formed at the electrolyte/channel interface?

Katharina Melzer answered: For a P3HT film in contact with deionized water literature values reported for the double layer capacitance obtained by impedance spectroscopy are between 3 and 6 μF cm^{-2}.[1] We obtained a value of approximately 10 μF cm^{-2} in deionized water, which exceeds the literature values by a factor of 2. For a pentacene film in contact with deionized water Cramer *et al.* extracted from dual-gate measurements a little bit lower value of 7.8 μF cm^{-2}.[2]

1 L. Kergoat, L. Herlogsson, D. Braga, B. Piro, M.-C. Pham, X. Crispin, M. Berggren and G. Horowitz, *Adv. Mater*, 2010, **22**, 2565–2569.
2 T. Cramer, A. Kyndiah, M. Murgia, F. Leonardi, S. Casalini and F. Biscarini, *Appl. Phys. Lett.*, 2012, **100**, 143302.

Natalie Plank said: We work on FETs with thin films of carbon nanotubes (CNTs) as the active channel. These CNT films have interpenetrating networks with gaps between tubes. How appropriate would your method of measuring the electrical double layer be for calculating characteristics for such a system?

Katharina Melzer answered: Especially in field-effect transistors often CNT networks with a very low tube density are necessary to achieve a good transistor performance. Therefore most of the bottom-gate dielectric will not be covered with the semiconducting material and will therefore also be in direct contact with the electrolyte. This fact has to be taken into account. Moreover, surface groups of the gate dielectric can be ion- or pH-sensitive. Also the quantum capacitance of the carbon nanotubes should be considered.[1,2]

1 T. Haeberle, A. M. Münzer, F. Buth, J. A. Garrido, A. Abdellah, B. Fabel, P. Lugli and G. Scarpa, *Appl. Phys. Lett.*, 2012, **101**, 223101.
2 M. D. Angione, S. Cotrone, M. Magliulo, A. Mallardi, D. Altamura, C. Giannini, N. Cioffi, L. Sabbatini, E. Fratini, P. Baglioni, G. Scamarcio, G. Palazzo and L. Torsi, *PNAS*, 2012, **109**, 6429–6434.

Natalie Plank said: If you were characterizing a more porous film like the CNT network, do you have to be careful about source and drain potentials affecting capacitance/conductance? For example, in comparison the P3HT OFETs encapsulate the source and drain electrodes and the applied potential would not directly interact with the PBS unlike the case for CNT thin films with either exposed electrodes or thinner encapsulation.

Katharina Melzer replied: In the case of a very thin CNT network, which is most often used for field-effect transistors based on CNT networks almost the complete surface of the metallic source and drain electrodes would be directly exposed to the PBS buffer. Our group always encapsulated the source and drain electrodes with a μm-thick photo-resist layer in a positive photolithography step after spin-coating the CNT network on the source and drain electrodes.[1,2] This was necessary to avoid a huge gate leakage current flowing between the source and drain contacts and the gate electrode in electrolyte-gated CNT FETs. To minimize any influence of the metallic source and drain electrodes or of high gate leakage currents on the measurements and therefore on the extraction of the values of the electrical double layer capacitance it could actually be helpful to passivate the source drain contacts to avoid a direct contact of them with the electrolyte.

1 T. Haeberle, A. M. Münzer, F. Buth, J. A. Garrido, A. Abdellah, B. Fabel, P. Lugli and G. Scarpa, *Appl. Phys. Lett.*, 2012, **101**, 223101.
2 A. M. Münzer, K. Melzer, M. Heimgreiter, G. Scarpa, *BBA*, 2013, **1830**, 4353–4358.

Uli Lemmer asked: In Fig. 5,[1] the agreement of your experimental and simulated transfer curves is very good even your transport model looks rather simple as compared to other transport models discussed *e.g.* in Jenny Nelson's presentation.[2]

Your equations (8) to (12) for modelling transistor operation look like a conventional drift-diffusion modelling.[1] Table 1, however, indicates that you consider different trap states *etc.* How do you actually include the carrier dynamics related to the trap states in your simulations? You are also indicating that you are using a Poole–Frenkel type mobility in your model. Do you also consider a density dependent mobility?

1 K. Melzer, M. Brändlein, B. Popescu, D. Popescu, P. Lugli and G. Scarpa, *Faraday Discuss.*, 2014, DOI: 10.1039/c4fd00095a.
2 Jarvist M. Frost, James Kirkpatrick, Thomas Kirchartz and Jenny Nelson, *Faraday Discuss.*, 2014, DOI: 10.1039/c4fd000153b.

Katharina Melzer responded: We modelled the semiconductor region as divided into grains of a certain size and applied trap states at the grain boundaries. Since the semiconducting channel is confined to a few nm from the semiconductor/insulator interface, on trap states at this interface were considered with an effective trap concentration. A Gaussian distribution was assumed for the trap levels.[1] To better account for the experimental results two different levels for the charged trap states were assumed, one shallow level at 0.1 eV and one deep level at 0.4 eV. No, actually we only consider a Poole-Frenkel mobility to account for the experimentally observed dependency of the mobility in organic semiconductors on the electrical field.

1 A. Bolognesi, A. Di Carlo, P. Lugli, G. Conte, *Synth. Met.*, 2003, **138**, 95–100.

George Malliaras remarked: Concerning the determination of capacitance at the interface between the electrolyte and the channel: Am I right to say that the difference between a Ag/AgCl and a Pt gate is in the fraction of the applied gate potential that drops at that interface? Did you compare your capacitance values with values obtained by electrochemical impedance measurements?

Katharina Melzer responded: Yes, if a Pt-wire is immersed in the electrolyte as the gate electrode, the applied gate potential drops across two electrical double layers: one forming at the gate electrode/electrolyte interface and the other one at the electrolyte/semiconductor interface.[1] For a Ag/AgCl electrode the complete gate potential has to drop across the electrical double layer forming at the semiconductor/electrolyte interface, since no electrical double layer can form at the gate electrode/electrolyte interface due to the following reaction taking place at the electrode surface:[1]

$Ag + Cl^- \rightarrow AgCl + e^-$

We did not perform impedance measurements with the P3HT films fabricated in our lab, but literature values obtained by impedance spectroscopy for P3HT films in contact with deionized water are in the range of 3 to 6 μF cm^{-2},[2] which is around 2 times less in comparison to the value we obtained in deionized water.

1 G. Tarabella, C. Santato, S. Y. Yang, S. Iannotta, G. G. Malliaras and F. Cicoira , *Appl. Phys. Lett.*, 2010, **97**, 123304.
2 L. Kergoat, L. Herlogsson, D. Braga, B. Piro, M.-C. Pham, X. Crispin, M. Berggren and G. Horowitz, *Adv. Mater*, 2010, **22**, 2565–2569.

Andrew Mount remarked: The capacitance of the Pt electrode should be directly proportional to area. Has this been tested in these measurements *e.g.* by inserting wire of different defined lengths into the solution?

Katharina Melzer answered: Actually we did not perform measurements with different sizes of the gate electrode so far but we assured during all measurements shown in here that the length of the wire immersed in the electrolyte is constant to avoid any size-dependent influence of the gate electrode.

Fred Wudl opened the discussion of the paper by Donal Bradley: You mentioned PEDOT:PSS requires high conductivity to replace ITO. What are the latest conductivities? We have a sample from China claimed to exhibit 10^4 S cm^{-1}.

Donal Bradley answered: ITO has a conductivity of order 4 000 S cm^{-1} so ideally PEDOT:PSS should have this or higher conductivity. We have prepared vapour phase polymerized PEDOT (VPP–PEDOT) with conductivity above 1 700 S cm^{-1} and other groups have reported up to 3 400 S cm^{-1}.[1] If 10 000 S cm^{-1} has been achieved that would be an important advance. It will also be interesting to know what work-function the sample you mention has and what its transmission properties look like. One problem with increasing the conductivity of PEDOT is that it typically results in an increasing absorption in the red to near-infrared spectral range.

1 M. V. Fabretto, D. R. Evans, M. Mueller, K. Zuber, P. Hojati-Talemi, R. D. Short, G. G. Wallace and P. J. Murphy, *Chem. Mater.*, 2012, **24**, 3998–4003.

Faraday Discussions

PAPER

Organic electronics and photonics: concluding remarks

Donal D. C. Bradley*

Received 11th October 2014, Accepted 20th October 2014
DOI: 10.1039/c4fd00202d

Summing up the scientific content of a Faraday discussion meeting in a short paper is an impossible task and therefore, I have tried simply to draw-out a few more-general themes relating to the presentations made and the exciting research field that encompasses them.

It is in fact rather difficult to find a singular descriptor for the research field considered at Faraday discussion meeting 174; the label *organic* electronics and photonics is something of a misnomer since the devices fabricated generally also contain both metals and metal oxides and can contain inorganic molecules as well as organic ones. But equally, *molecular* electronics is incomplete, ditto for *printed* electronics; the materials used are not all molecular or necessarily printable. Should we simply say electronics and photonics and forget about the adjective? That would, I fear, be too general as there are many aspects of electronics that we do not (and cannot) address like central processing units (CPUs), graphical user interfaces (GUIs) and other components that require processing and data capacities far beyond our reach. *Large area* electronics is another possibility since this tends to rule out wafer-based materials. My own preference, however, is for *plastic* electronics and at Imperial College we have named our collective activities across physics, chemistry and materials science, the Centre for Plastic Electronics. It is so called because that adjective reflects, at least to us, a very practical combination of easy processing and functional properties and it is this same duality that has motivated and continues to motivate much of our interest, certainly in respect of applications, in solution processed semiconductors. Plastic also has the connotation, in biomedical science, of reconfigurable and that chimes with our general outlook as well. Plastics are moreover, in popular culture, often considered 'cheap and cheerful' and jolly useful; attributes we would surely wish for organic electronics and photonics as well.

The molecular nature of plastics means that they typically have strong bonding along the chain (*intra*-molecular) but relatively weak bonding between neighbouring chains (*inter*-molecular): the inter-molecular bonds, characteristically van der Waals, are roughly a twentieth as strong. This helps processing since the

Department of Physics and Centre for Plastic Electronics, Imperial College London, Exhibition Road, London SW7 2AZ, UK. E-mail: D.Bradley@imperial.ac.uk; Tel: +44 (0)2075942638

chains can be relatively easily separated, *via* heat (melt-processing) or a solvent (solution-processing), and then reassembled into new shapes, for instance conforming to the elaborate features of a mould or to a curved surface to thereby produce kitchen goods, drinks bottles or packaging bearing branding and labelling information. The same van der Waals solid or 'soft solid' properties apply to the solution processing of polymer and other (small molecule) organic and organo-metallic semiconductors, conductors and insulators that are widely used in our field. They are, moreover, at the centre of most that is of particular interest in respect to the optical and electrical properties of this class of materials. What's more they underpin the processing and some of the properties of colloidal metal and inorganic semiconductor particles, nanotubes and 2-D materials including the transition metal dichalcogenides and related systems. In the end the plastic electronics tag can reasonably encompass all of these material types especially if used in a way that helps us along the path to solution-based, low-temperature, high-throughput and (ideally) low-capital-cost fabrication of displays, lighting, solar cells, photodetectors, electronic circuits and photonic components.

Another reason to mention plastic electronics, perhaps, is that, as noted by Professor Fred Wudl in his introductory lecture (DOI: 10.1039/c4fd00191e), it is now twenty-five years since the 1989 demonstration of conjugated polymer electroluminescence from precursor-route poly(*p*-phenylenevinylene) diode structures. The associated 1989 patent[1] led to the 1992 founding of Cambridge Display Technology (now part of the Sumitomo Chemical Company) and the 1990 *Nature* paper[2] that followed has become one of the most highly cited articles in the physical sciences. This work can in many ways be considered the start of plastic electronics, acting as a trigger point for an explosion of interest in solution-processed semiconductors within both industrial labs and academic/government institutions. Its publication came at a critical juncture when interest in conjugated polymers was otherwise starting to wane following (i) their failure to live up to the expectations derived from the label "synthetic metals" and (ii) their inability to deliver the large, ultrafast non-resonant nonlinearities sought for all-optical switching.[3] The new directions opened by this paper transformed the field and it has gone from strength to strength with organic semiconductors now being considered mainstream. Ten years later, the significance of polymer-based synthetic metals was recognised by the 2000 Nobel Prize for Chemistry, awarded to Professors Heeger, MacDiarmid and Shirakawa.[4]

Professor Wudl focused his introductory talk on low band gap conjugated polymers and doping effects, reprising the development of charge transfer salts, the sulphur nitride polymer SN_x, poly(isothianapthene) (PITN) and related materials and discussing the development of new strong electron acceptor moieties such as benzobisthiadiazole (BBT) and their donor–acceptor copolymers with diketopyrrolopyrrole (DPP).[5] Professor Wudl also commented that (small) molecules are increasingly being used in preference to polymers because it is easier to control (limit) their structural polymorphs in order to obtain more reproducible thin film properties – essential for any commercialisation. This is certainly a direction of travel, with both solution processed and vacuum sublimed deposition protocols, but there are examples of polymers that have very precise and reproducible structures. I have a particular interest in poly(9,9-

dioctylfluorene) (see Fig. 1) for which the β-phase chain segment conformation is especially well defined.[6]

The patterning of β-phase chain segments *via* a simple dip-pen nanolithography procedure then offers a novel approach to the fabrication of nanophotonic structures.[7] Generation of the β-phase yields a conformational metamaterial; the sub-wavelength physical structuring strongly alters the material's interaction with electromagnetic radiation. Such avenues of exploration might be thought of as being analogous to the "silicon-approach" to semiconductor device fabrication in which desired properties are generated by processing (oxidizing, doping, etching) an otherwise uninteresting (from a semiconductor device perspective) intrinsic Si wafer. Much more also needs to be learned, in general, about how to tune weak interactions to thereby select specific molecular architectures and packing motifs. Professor Perepichka described (DOI: 10.1039/c4fd00133h), for instance, the use of complementary H-bonding in the formation of p–n heterojunction structures comprising p-type dipyrrolopyridines and n-type napthalenediimides.[8]

As already noted above, the molecular nature of many of our materials defines their functional properties. Wavefunctions localised to molecular units lead to large oscillator strengths (absorption coefficients $\sim 10^5$ cm^{-1}), coupling between electronic and vibrational degrees of freedom (polaronic charge states and vibronic structure in absorption and emission spectra – see Fig. 1), a distinct manifold of triplet states and, in combination with a low dielectric constant, to strongly bound excitons. These in turn typically lead to radiative decay times $\sim$ 1 ns, to 1/e absorption depths $\sim$ 100 nm that allow thin film structures to efficiently absorb solar energy, to broad emission bandwidths that can support short pulse optical gain, to hopping transport and to selection-rule-limited electroluminescence quantum efficiency. Nanometric length scales are also important *e.g.*

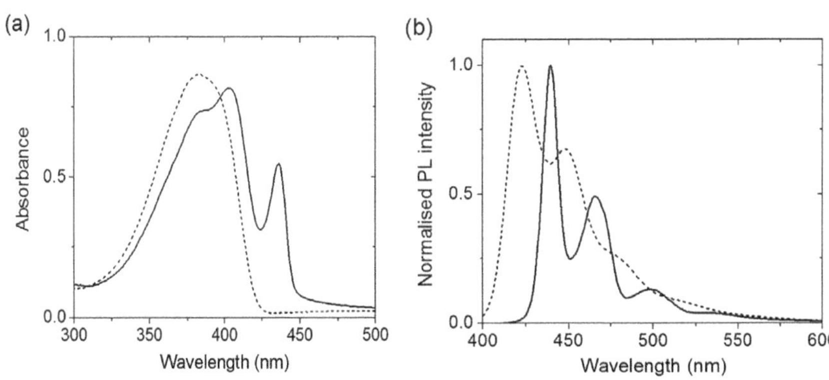

Fig. 1 (a) Absorbance and (b) photoluminescence (PL) spectra of a hot spin-coated (toluene, 100 °C), 100 nm thickness PFO glassy film before (dashed line) and after (solid line) flooding the film surface with decalin to generate a high fraction of β-phase chain segments. The β-phase PL vibronic structure is particularly well-resolved and characteristic of a rigid, chain-extended, planar structure. The absorbance spectrum for the decalin-immersed sample comprises contributions from both β-phase and glassy (α-phase) chain segments whilst the PL spectrum is dominated by β-phase segment contributions, populated *via* both direct excitation and following rapid energy transfer from the α-phase (unpublished data, courtesy of Alexandr Perevedentsev).

Kuhn segment lengths for conjugated polymers ~ 10–30 nm, exciton diffusion lengths ~ 5–50 nm, and Förster transfer radii ~ 1–10 nm but that is more a happenstance of the fact that we are dealing with molecular materials; nanometers are the natural length scale for molecules. Molecular orientation also matters (see Fig. 2);[9] it exposes the intrinsic anisotropy of many physical processes and can significantly affect the observed charge transport and light absorption and emission properties. State of order (glass/crystal/liquid crystal; mono-/polydomain) has also been used to good effect in establishing and understanding trade-offs between processing protocols and ultimate properties, especially for blends. An interesting example is the effect that solution pre-ordering has on the optical properties of poly(*p*-phenylenevinylene) thin films.[10]

The properties of individual molecules (in terms of geometry and electronic structure) are readily perturbed by the environment in which they sit; the 'usefulness' of a molecular material then often depends on the details of its processing and the form in which it is used. Six presentations (Solomon *et al.* (DOI: 10.1039/c4fd00083h), Bâldea *et al.* (DOI: 10.1039/c4fd00101j), Venkatramani *et al.* (DOI: 10.1039/c4fd00106k), Venkataraman *et al.* (DOI: 10.1039/c4fd00093e), Xu *et al.* (DOI: 10.1039/c4fd00080c) and Ottosson *et al.* (DOI: 10.1039/c4fd00084f)) at the Faraday discussion meeting addressed single- or few-molecule electronics, from both an experimental and a theoretical perspective, a topic that presents serious challenges for measurement and understanding but that can, for instance, offer important perspectives on the question: for how long will Moore's law continue to be obeyed? Exacting sample fabrication and measurement make this topic one for the specialist but it should also have a much broader significance in helping to understand the function of self-assembled monolayers and other ultrathin interlayer materials in modifying the injection properties of electrodes.[11,12] Many of the other presentations at the meeting considered more conventional electronics and photonics based on samples that are essentially macroscopic within the film-plane but nevertheless equally reliant on molecular properties and controlled by local environments.

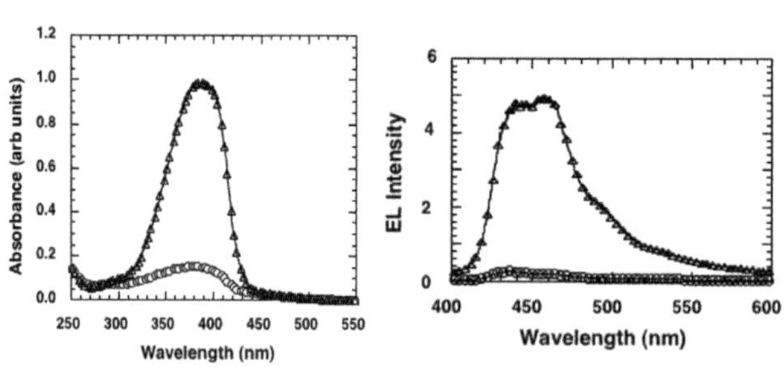

Fig. 2 Polarised absorption (left) and electroluminescence (right) emission spectra (triangles – parallel polarised, circles – perpendicular polarised) for poly(9,9-dioctylfluorene) homogeneously aligned in the nematic liquid crystal phase on a rubbed precursor-route poly(*p*-phenylenevinylene) (PPV) alignment layer. The absorption anisotropy is limited (≤7) by the liquid crystal order parameter and the $\pi-\pi^*$ transition dipole moment orientation relative to the chain axis projection. The higher (≤25) EL anisotropy is consistent with an emission zone localized close to the PPV alignment layer interface where the order parameter is enhanced relative to the bulk that is measured in absorption.

One good example of the influence of the environment is the class of molecules used as laser dyes that have close to unity quantum yield in dilute solution but whose fluorescence is drastically reduced in powder form by so-called concentration quenching. The use of low concentrations of laser dye 4-(dicyanomethylene)-2-methyl-6-(4-dimethylaminostyryl)-4*H*-pyran (DCM) dispersed in a tris(8-hydroxyquinolinato)aluminium (Alq_3) host as an optical gain medium in the research reported by Professor Lemmer (DOI: 10.1039/c4fd00077c) and Professor Leo, presented by Dr Brückner (DOI: 10.1039/c4fd00094c) represents the solid state equivalent; the Alq_3 host material keeps the DCM laser dye molecules far enough apart that they function as if in dilute solution. In Professor Lemmer's case the interest was in distributed feedback (DFB) lasers for use as light sources in lab-on-a-chip-based microfluidic analysis. Other approaches to the instrumentation of such structures include the use of OLED light sources and organic photodetectors.[13,14] The narrow laser linewidth in Professor Lemmer's devices, without the characteristic longer-wavelength vibronic tail found for most OLEDs, is beneficial to avoid overlap between excitation light and analyte fluorescence in biomedical assays.[15] In Professor Leo's structures the organic gain medium was interacting with a metal-containing vertical distributed-Bragg-reflector-encased photon environment that sustained Tamm plasmon polaritons. Related DBR-controlled photon environments include microcavities used to study strongly- and ultrastrongly-coupled exciton–photon mixed states (polaritons). A second example, not, however, discussed extensively at the meeting, is the pentacene doped *p*-terphenyl crystal that recently allowed fabrication of the first room temperature Maser[16] *via* optically pumped population inversion amongst the triplet sublevels of pentacene. In both of these cases, dilution prevents unwanted inter-molecular interactions and allows the properties of the individual molecule to shine through. Interestingly, films of many conjugated polymers do not greatly suffer from such concentration quenching effects (see Fig. 3),[17] although some do.[18]

H- and J-aggregates,[19] excimers and exciplexes also depend for their formation on intermolecular interactions, with interactions in the excited state key for the

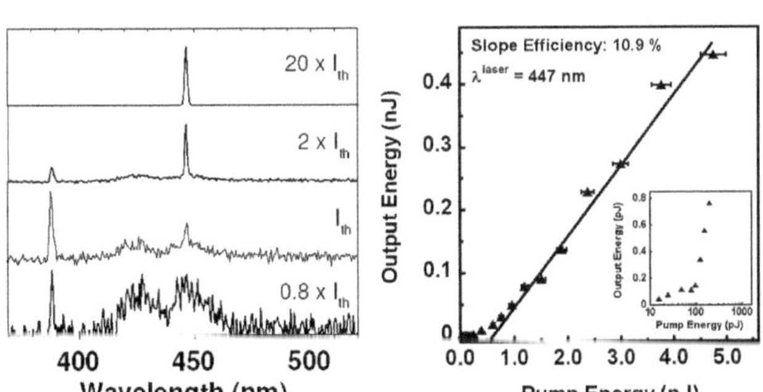

Fig. 3 Normalised laser spectra (left) and input–output characteristics (right) for a 447 nm one-dimensional DFB laser comprising a 90 nm thickness film of poly((9,9-dioctylfluorene)$_{0.8}$-*co*-(9,9-di(2-methyl)butylfluorene)$_{0.2}$) (Y80F8:20F5) spin coated on top of an etched silica grating (period $\Lambda = 290$ nm, 75% fill factor). The laser was pumped with 10 ns, 10 Hz laser pulses at $\lambda_{ext} = 390$ nm. The lasing threshold, $I_{th} \approx 100$ pJ (see inset to right panel) corresponds to 0.3 mJ cm^{-2} and combined with the measured slope efficiency of ~ 11% represents state-of-the-art performance for a conjugated polymer laser.

latter two. J-aggregated cyanine dyes have an enhanced absorption strength (and suppressed vibronic coupling) that was used for many years to good effect in sensitizing silver halide based photographic emulsions. More recently this large oscillator strength was shown to allow the fabrication of strongly coupled DBR-based microcavities that emit from hybrid exciton–polariton states.[20] The observed room temperature Rabi splittings are very substantial for this and a range of other molecular materials.

Intermolecular interactions are likewise key for charge transport through the bulk of a film which requires charges to transfer between neighbouring units, even when long chain polymers are used.[17] This can be strongly affected by the state-of-order and inter-molecular packing as discussed by Professor Nelson (DOI: 10.1039/c4fd00153b) and Dr McDouall (DOI: 10.1039/c4fd00082j). Crystallinity is not a pre-requisite with some of the highest mobilities in polymers arising for non-crystalline materials.[17] Charge trap states tailing into the optical gap can strongly affect the overall carrier mobility and its temperature dependence. In many cases an exponential gap state distribution provides a good model to explain a variety of experimental observations.[21]

Chemistry allows the tuning of many molecular properties including highest occupied and lowest unoccupied molecular orbitals (HOMOs and LUMOs) whose energy difference determines optical gap and thence absorption and emission spectra. Ionization potential and electron affinity, thermal properties *e.g.* glass and isotropic melt transition temperatures and the existence of any liquid crystal phases, solubility and the tendency to pack in, for example, herring-bone or cofacial stack arrangements are also controllable through the chemical structure design of both the conjugated core and peripheral solubilising groups. Starburst compounds such as the triphenylamine core/oligothiophene arm materials reported by Professor Ponomarenko (DOI:10.1039/c4fd00142g) provide an interesting example of this approach. Not all synthesis routes are, however, equivalent and for example the difference between the Wittig and Gilch syntheses of dialkoxy-*p*-phenylenevinylene polymers had an enormous impact on progress in OLED device performance with Gilch-route polymers[22] substantially outperforming on-paper-similar Wittig-synthesised materials.[23] Likewise, for fluorene-based polymers the Suzuki route proved especially useful[24] and current state of the art fluorene-based complex copolymers are Suzuki-coupled.

Differences in defect type and density, often at the detection limits of standard chemical analyses can have a major influence on device performance. Moreover, different device types and measurements may be more or less susceptible, for example the high carrier densities in transistors allow for significant trap filling that does not happen in time-of-flight photocurrent measurements performed at much lower densities.

Defect concentrations of a few percent, that might simply be what is present at the chain ends of a moderate molecular weight polymer, would be considered high in conventional semiconductor terms. Add to this the potential for catalyst residues, ionic impurities leached from the glassware, simultaneous synthesis of isomeric compounds, non-uniform coupling reactions, differing regio-regularities and oxidative products formed during storage and processing and it is small wonder that there can be a spread in reported performance for nominally the same materials. The influence of defects can also depend on environment, for instance the fluorenone oxidative degradation products in poly(9,9-

dialkylfluorene) polymers only modify the emission spectrum when they are able to form excimers that are then responsible for the characteristic green-band emission; if the polymer chains are isolated in a matrix then the only effect is a decrease in overall PL efficiency due to the quenching properties of the fluorenone moiety (see Fig. 4).[25] It should thus be evident that developing 'good' synthesis routes, effective procedures for impurity control and material-safe processing methods are necessary pre-requisites to undertaking true structure–property relationship studies.

Added complexities may occur when one looks to develop hybrid/composite materials such as the novel II–VI inorganic semiconductor/conjugated polymer systems discussed by Dr Kim (DOI: 10.1039/c4fd00141a) for use in solar energy generation. Here the favoured (for charge generation and collection) bulk heterojunction geometry is being formed *in situ* by chemical synthesis of the II–VI colloidal particles within the polymer matrix. The potential benefits of such hybrid systems can, however, be great as for example the newly developed use of methylammonium lead iodide and related perovskites for solar cells has demonstrated.[26] Hybrids can also encompass photonic structures, for instance those that dipole–dipole couple (In,Ga)N quantum well Mott–Wannier excitons with polyfluorene Frenkel excitons and that lead to efficient non-radiative energy transfer between the two.[27]

Professor Sasaki described a different composite system (DOI: 10.1039/c4fd00068d), in his case for photorefractive data storage, comprising mixtures of several ferroelectric LCs and 2,4,7-trinitrofluorenone (TNF) as an oxidative dopant. Photorefractive composites comprise a field in which one of this year's Nobel Laureates in Chemistry, W. E. Moerner, made seminal contributions at IBM.[28] Photorefraction is also found in conjugated polymers but it and other nonlinear optical effects are little studied at present.[29] W. E. Moerner shared the

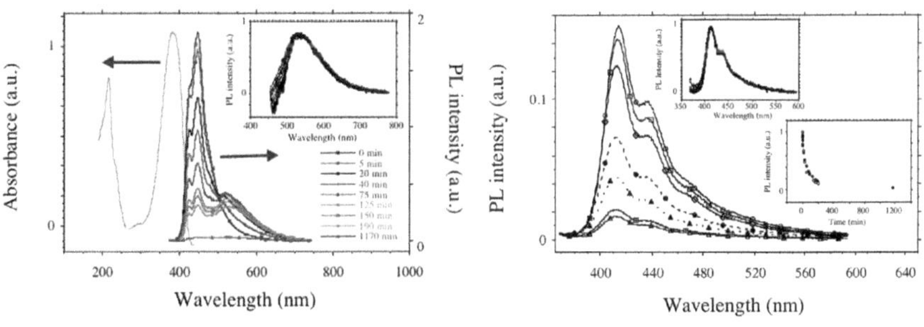

Fig. 4 Influence of local environment on the PL emission spectral changes induced by photo-oxidation of PFO thin films by exposure to UV radiation in air. Left panel shows the PFO absorbance (left ordinate) before irradiation and a sequence of PL spectra (right ordinate) recorded over time during irradiation. The inset shows the normalised green-band excimer emission spectra extracted from each of the irradiated sample spectra by subtraction of a scaled pre-irradiation PFO PL spectrum. Right panel shows the equivalent experiment performed for a PFO sample dispersed at low concentration in a polystyrene matrix film that prevents excimer formation. No green-band emission is observed; the PFO spectrum simply decreases in magnitude (see upper inset for peak normalised data). The lower inset shows the temporal decay of the integrated PL for this sample and superimposed data for the green-band-subtracted blue emission component of an undiluted film irradiated under the same conditions.

2014 Nobel Prize for his work on super-resolution fluorescence microscopy, a technique that might in the future be usefully applied to organic electronic and photonic materials systems to better understand their film microstructure.

Biomedical applications represent another interesting arena for the application of organic electronic and photonic devices as already mentioned above in connection with Professor Lemmer's presentation. This arena is sometimes referred to as bioelectronics and is seen as a promising outlet for the application of organic electronic and photonic devices.[30] In addition, within this meeting Dr Fraboni described work on organic single crystal X-ray detectors (DOI: 10.1039/c4fd00102h), Professor Torsi on ZnO transistors (DOI: 10.1039/c4fd00081a), Dr Melzer on electrolyte gated transistors (DOI: 10.1039/c4fd00095a), Dr Laurand on truxene-based DFB lasers (DOI: 10.1039/c4fd00091a) and Dr Dennany on solution phase electrochemiluminescence (DOI: 10.1039/c4fd00090k); all for use in imaging or sensing applications.

Dr Zysman-Colman also discussed electrochemiluminescence but in his case (DOI: 10.1039/c4fd00107a), the context was for use in light emitting structures for displays, more specifically focused on achieving efficient blue-light emission. This of course is the topic that has been recognized by the 2014 Nobel Prize for Physics, awarded to Professors Akasaki, Amano and Nakamura for their invention of the GaN-based blue LED that has produced a breakthrough in high efficiency solid-state lighting.

For those of us working in the organic LED (or OLED) field, it is perhaps fortunate that GaN LEDs were unknown at the time of the invention of OLEDs based on (i) vacuum sublimed small molecules, including Alq_3, by Tang and van Slyke at Kodak Research Labs in Rochester[31] and (ii) solution processed conjugated polymers (by Burroughes, Bradley and Friend[2] at the Cavendish Laboratory, Cambridge). Otherwise, this technology would surely have been harder to establish. It has taken twenty-five years but we have now seen OLEDs commercialized for use in mobile phone, tablet and television displays (a sector in which GaN LEDs are used to backlight liquid crystal displays) and they are starting to enter the lighting arena in direct competition to GaN-based LEDs. OLEDs can be fabricated as large area emitters whilst GaN LEDs are point source emitters and that difference in form factor will likely influence how they are used.

It is interesting finally to reflect on how many times statements that have acquired *accepted wisdom* status turn out to be incorrect. Good advice to someone entering a research field would be to beware such strongly held statements of belief. For example, in my own research career I have been told '*It is well known that…*

… polymer semiconductors do not have bound excitons'.

… organic semiconductors are intrinsically unstable and organic light emitting diodes (OLEDs) will, therefore, never be reliable'.

… organic semiconductors can't simultaneously transport electrons and holes'.

… organic semiconductors will never show strong exciton–photon coupling'.

… high charge carrier mobility and luminescence efficiency are mutually exclusive properties for organic semiconductors'.

Each one of these statements has proven to be wrong. There are plenty more statements that could also have been included in the list and one that we, and others, are currently working to challenge is that:

... electrically pumped organic lasers are impossible.

The latter device, the organic laser diode, remains the one major device type that has not been achieved. The closest[32] anyone has come to a compact structure has been to use a high power pulsed InGaN LED to optically pump an organic DFB laser (like those described by Professor Lemmer and Dr Laurand). Hopefully it will not be too long before this hybrid approach becomes unnecessary. Another promising result is the demonstration that electrically pumped resonant cavity OLED structures can emit very narrow linewidths.[33] Even if these are not lasers, just as in the case of inorganic resonant cavity (RC-) LEDs, such RC-OLEDs offer many desirable features that for specific applications are likely to negate the need for a laser.

In conclusion, organic electronics and photonics remains an exciting research field in which to work with plenty of fundamental issues to consider and many opportunities to discover new phenomena and to develop new materials, devices and applications. The Faraday 174 discussion meeting organizers Professors Peter Skabara, Simon Higgins, Iain McCulloch, George Malliaras, Andy Mount and Ifor Samuel should be congratulated for assembling an excellent programme and all of the attendees and the RSC support staff should be thanked for helping to create a very lively and stimulating three days in a warm and sunny Glasgow. I very much enjoyed this, my first, Faraday discussion meeting.

References

1 D. D. C. Bradley, J. H. Burroughes and R. H. Friend, Electroluminescent Devices, patent, WO 1990013148.

2 J. H. Burroughes, D. D. C. Bradley, A. R. Brown, R. N. Marks, K. D. Mackay, R. H. Friend, P. L. Burn and A. B. Holmes, *Nature*, 1990, **347**, 539.

3 B. I. Greene, J. Orenstein and S. Schmitt-Rink, *Science*, 1990, **247**, 679.

4 A. G. MacDiarmid, *Rev. Mod. Phys.*, 2001, **73**, 701.

5 J. D. Yuen, J. Fan, J. Seifter, B. Lim, R. Hufschmid, A. J. Heeger and F. Wudl, *J. Am. Chem. Soc.*, 2011, **133**, 20799.

6 G. Ryu, P. N. Stavrinou and D. D. C. Bradley, *Adv. Funct. Mater.*, 2009, **19**, 3237.

7 A. Perevedentsev, Y. Sonnefraud, S. Sharma, A. E. G. Cass, S. A. Maier, J. S. Kim, P. N. Stavrinou and D. D. C. Bradley, *Nat. Commun.*, 2014, submitted.

8 H. T. Black and D. F. Perepichka, *Angew. Chem.*, 2014, **53**, 2138.

9 K. S. Whitehead, M. Grell, D. D. C. Bradley, M. Jandke and P. Strohriegl, *Appl. Phys. Lett.*, 2000, **76**, 2946.

10 D. A. Halliday, P. L. Burn, D. D. C. Bradley, R. H. Friend, O. Gelsen, A. B. Holmes, A. Kraft, J. H. F. Martens and K. Pichler, *Adv. Mater.*, 1993, **5**, 40.

11 S. Khodabakhsh, D. Poplavskyy, S. Heutz, J. Nelson, D. D. C. Bradley, H. Murata and T. S. Jones, *Adv. Funct. Mater.*, 2004, **14**, 1205.

12 R. Xia, D.-S. Leem, T. Kirchartz, S. Spencer, C. Murphy, Z. He, H. Wu, S. Su, Y. Cao, J.-S. Kim, J. C. deMello, D. D. C. Bradley and J. Nelson, *Adv. Energy Mater.*, 2013, **3**, 718.

13 O. Hofmann, X. Wang, J. C. deMello, D. D. C. Bradley and A. J. deMello, *Lab Chip*, 2005, **5**, 863; O. Hofmann, P. Miller, P. Sullivan, T. S. Jones, J. C. de Mello, D. D. C. Bradley and A. J. de Mello, *Sens. Actuators, B*, 2005, **106**, 878.

14 J. Shinar and R. Shinar, *J. Phys. D: Appl. Phys.*, 2008, **41**, 133001.

15 G. Ryu, J.-S. Huang, O. Hofmann, C. Walshe, J. Y. Y. Sze, G. D. McClean, A. Mosley, J. C. de Mello, A. J. de Mello and D. D. C. Bradley, *Lab Chip*, 2011, **11**, 1664.

16 M. Oxborrow, J. D. Breeze and N. M. Alford, *Nature*, 2012, **488**, 353.

17 B. K. Yap, R. Xia, M. Campoy-Quiles, P. N. Stavrinou and D. D. C. Bradley, *Nat. Mater.*, 2008, **7**, 376.

18 R. Xia, P. N. Stavrinou, D. D. C. Bradley and Y. Kim, *J. Appl. Phys.*, 2012, **111**, 123107.

19 F. C. Spano and C. Silva, *Annu. Rev. Phys. Chem.*, 2014, **65**, 477.

20 D. G. Lidzey, D. D. C. Bradley, A. Armitage, S. Walker and M. S. Skolnick, *Science*, 2000, **288**, 1620.

21 T. Kirchartz, B. E. Pieters, J. Kirkpatrick, U. Rau and J. Nelson, *Phys. Rev. B: Condens. Matter Mater. Phys.*, 2011, **83**, 115209.

22 H. Becker, H. Spreitzer, K. Ibrom and W. Kreuder, *Macromolecules*, 1999, **32**, 4925.

23 T. Ohnishi, S. Doi, Y. Tsuchida and T. Noguchi, *IEEE Trans. Electron Devices*, 1997, **44**, 1253.

24 M. T. Bernius, M. Inbasekaran, J. O'Brien and W. S. Wu, *Adv. Mater.*, 2000, **12**, 1737.

25 T. A. M. Ferenczi, M. Sims and D. D. C. Bradley, *J. Phys.: Condens. Matter*, 2008, **20**, 045220.

26 M. Graetzel, *Nat. Mater.*, 2014, **13**, 838.

27 G. Heliotis, G. Itskos, R. Murray, M. D. Dawson, I. M. Watson and D. D. C. Bradley, *Adv. Mater.*, 2006, **18**, 334.

28 W. E. Moerner and S. M. Silence, *Chem. Rev.*, 1994, **94**, 127.

29 D. D. C. Bradley and O. M. Gelsen, *Phys. Rev. Lett.*, 1991, **67**, 2589; S. J. Martin, D. D. C. Bradley, P. A. Lane, H. Mellor and P. L. Burn, *Phys. Rev. B: Condens. Matter Mater. Phys.*, 1999, **59**, 15133.

30 M. Berggren and A. Richter-Dahlfors, *Adv. Mater.*, 2007, **19**, 3201.

31 C. W. Tang and S. A. van Slyke, *Appl. Phys. Lett.*, 1987, **51**, 913.

32 Y. Yang, G. A. Turnbull and I. D. W. Samuel, *Appl. Phys. Lett.*, 2008, **92**, 163306.

33 X. Liu, H. B. Li, C. Y. Song, Y. Q. Liao and M. M. Tian, *Opt. Lett.*, 2009, **34**, 503.

Poster titles

Spectroscopic measurement of critical lengthscales for free charge generation in organic photovoltaics, **J. Hodgkiss, K. Chen and A. J. Barker**, *Victoria University of Wellington, New Zealand*

Development of novel fused bis(thieno[3,2-b]thiophene) polymers for use in OFETs, **J. Shaw, H. Zhong, C. Nielsen, C. P. Yau, A. Casey, E. Buchaca-Domingo, S. E. Watkins, D. Sparrow, W. Mitchell and M. Heeney**, *Imperial College London, UK*

From linear to tripodal push-pull chromophores featuring N,N'-dibutylbarbituric acid and malondinitrile acceptors, **M. Klikar, D. Cvejn and F. Bureš**, *University of Pardubice, Czech Republic*

Donor-disubstituted 2,5-dihydropyrrolo[3,4-c]pyrrole-1,4-diones, **F. Bures and J. Podlesný**, *University of Pardubice, Czech Republic*

2,4-Bis(4-aryl-1,2,3-triazol-1-yl)-7H-pyrrolo[2,3-d]pyrimidines: optical properties and quantum chemical calculations, **J. Bucevičius, L. Skardžiūtė, K. Kazlauskas, S. Juršėnas and S. Tumkevičius**, *Vilnius University, Lithuania*

Polymer brush layer as an organic coating on ZnO nanocrystalline film towards hybrid photovoltaic systems, **A. Pomorska, K. Wolski and S. Zapotoczny**, *Jagiellonian University, Poland*

Engineered supramolecular semiconductors; a combined thin-film transistor and NEXAFS study, **C. Finlayson, R. Cross, S. Cooil and A. Evans**, *Aberystwyth University, UK*

Computational study of aggregation in TIPS-pentacene, **L. Weston and J. McDouall**, *University of Manchester, UK*

New N-annulated perylene and carbazole derivatives for OLEDs, **A. Bucinskas, A. Tomkeviciene, G. Bagdziunas, J. Vidas Grazulevicius, D. Volyniuk, P. Stakhira, V. Cherpak, M. Chapran and Z. Hotra**, *Kaunas University of Technology, Lithuania*

An organic down-converting material for white-light emission from hybrid LEDs, **N. Findlay, J. Bruckbauer, A. R. Inigo, B. Breig, S. Arumugam, D. J. Wallis, R. W. Martin and P. Skabara**, *University of Strathclyde, UK*

Nanoporous poly(3-hexylthiophene) thin films based on "click" prepared diblock copolymers, **G. Grancharov, V. Gancheva, P. Petrov, J. De Winter, P. Gerbaux, Ol. Coulembier and P. Dubois**, *Bulgarian Academy of Sciences, Bulgaria*

4,8-Benzobisthiazole as a weak donor for conjugated copolymers exhibiting high mobility in organic thin film transistors, **G. Conboy, N. Findlay, A. L. Kanibolotsky, A. R. Inigo and P. Skabara**, *University of Strathclyde, UK*

Synthesis and characterization of new donor-acceptor dyes for vacuum-deposited organic solar cells, **L. Fang, T. Matulaitis, O. Zeika, C. Koerner, R. Meerheim and K. Leo**, *TU Dresden, Germany*

Investigation of optoelectronic properties of selenophene and benzodithiophene based conjugated polymer, **H. Ünay, N. Akbasoglu Unlu, G. Hizalan, L. Toppare and A. Cirpan**, *Middle East Technical University, Turkey*

Energy transfer between F8BT polymer and photosynthetic LH2 complex, **K. Smolarek, D. Buczyńska, I. Samuel, R. Cogdell and S. Maćkowski**, *Nicolaus Copernicus University, Poland*

Borylative Fusion for producing low band gap donor acceptor materials, **D. Crossley, I. A. Cade, E. R. Clark, A. Escande, M. Humphries, M. Ingleson and M. Turner**, *University of Manchester, UK*

Synthesis of poly(arylene ethynylene)s with clickable side-chain groups for biosensor platforms, **Y. Braeken, L. Lutsen, D. Vanderzande and W. Maes**, *Hasselt university, Belgium*

Efficient dinuclear copper(I)-complexes for solution- and vacuum-deposited organic Light-emitting Diodes (OLEDs), **D. Volz, A. Jacob, C. Fléchon, J. Navarro, D. M. Zink, J. Friedrichs, D. Ambrosek, G. Liaptsis, S. Bräse and T. Baumann**, *cynora GmbH, Germany*

Conformational control via intramolecular non-covalent interactions in new benzobisthiazole-based materials, **E. Angioni, C. Risko, G. Conboy, N. Findlay, J. Brédas and P. Skabara**, *University of Strathclyde, UK*

Photovoltaic properties of silafluorene bearing polymer: effect of processing conditions on morphology, **G. Hizalan, O. Erlik, N. Akbasoglu Unlu, A. Cirpan and L. Toppare**, *Middle East Technical University, Turkey*

Synthesis of a benzotriazole based alternating copolymer for organic photovoltaic application, **C. Istanbulluoglu, S. Goker, G. Hizalan, S. O. Hacioglu, A. Cirpan and L. Toppare**, *Middle East Technical University, Turkey*

Fluorescence characterization of single gold nanoclusters, **D. Buczyńska, Ł. Bujak, F. Aldeek, H. Mattoussi and S. Maćkowski**, *Nicolaus Copernicus University, Poland*

Long pulse polymer laser dynamics & photophysics, **G. Whitworth, A. Kanibolotsky, P. Skabara, I. Samuel and G. Turnbull**, *University of St Andrews, UK*

Highly regioregular P3HT by Suzuki-Miyaura polycondensation of a novel MIDA boronate-derived monomer, **J. Ayuso-Carrillo, M. J. Ingleson and M. L. Turner**, *University of Manchester, UK*

Charge transport properties of the highly efficient organic photovoltaic material PTB7, **B. Ebenhoch, S. Thomson, K. Genevičius, G. Juška and I. Samuel**, *University of St Andrews, UK*

Novel donor-acceptor small molecules for use in bulk heterojunction organic photovoltaic devices, **J. Cameron, P. Skabara, N. Blouin and L. Nanson**, *University of Strathclyde*, *UK*

Excited state dynamics in self-assembled porphyrin nanotubes, **A. Stradomska-Szymczak and J. Knoester**, *University of Glasgow*, *UK*

Development of novel germanium-centred cruciform oligomers for organic photovoltaic devices, **A. Yeats, I. A. Wright, R. G. D. Taylor, A. L. Kanibolotsky, C. T. Howells, S. Gambino, I. Samuel and P. Skabara**, *University of Strathclyde*, *UK*

Novel extended carbazoles for organic electronics and optoelectronics, **P. Gawrys, A. Al-Muhammad, R. Whitby and B. Witulski**, *University of Southampton*, *UK*

Synthesis of novel polyaromatic molecules for applications in organic electronics, **A. Al-Muhammad, P. Gawrys, A. Mifleur, H. El-Siblani, J. Howe, B. Witulski and R. Whitby**, *University of Southampton*, *UK*

An organic optoelectronic visible light communication link operating at over 20 Mb/s, **S. Zhang, D. Tsonev, S. Ghosh, S. Videv, P. Manousiadis, M. T. Sajjad, D. C. V. Amarasinghe, J. D. D. McKendry, E. Xie, E. Gu, M. D. Dawson, H. Haas, G. Turnbull and I. Samuel**, *University of St Andrews*, *UK*

Conventional and inverted hybrid ultraviolet photodetector investigations based on solution processed PFE:ZnO blends, **C. Varlikli and G. Memisoglu**, *Ege University*, *Turkey*

QDLED applications of blue emitting conjugated polymer:CuInS2/ZnS QD nanoparticles, **N. Demir, I. Oner and C. Varlikli**, *Ege University*, *Turkey*

Single emissive layer white organic light-emitting diodes based on new green-emitting iridium complexes, **I. Oner, C. Sahin and C. Varlikli**, *Ege University*, *Turkey*

Synthesis and characterization of tetraphenylsilane based, solution processable host materials for organic light emitting diode (OLED) applications, **S. Oner, I. Oner and C. Varlikli**, *Ege University*, *Turkey*

Micro-size distributed feedback laser array, **B. Guilhabert, N. Laurand and M. D. Dawson**, *University of Strathclyde*, *UK*

Solution-processed high index polymer/inorganic photonic structures, **G. Richardson, P. N. Stavrinou and N. Stingelin**, *Imperial College London*, *UK*

Rhodanine flanked small molecules as non-fullerene acceptors for organic photovoltaics, **S. Holliday, R. Shahid Ashraf, C. Nielsen, M. Kirkus, C.-H. Tang, J. Durrant and I. McCulloch**, *Imperial College London*, *UK*

Passivated carbon nanotube field effect transistors as aptamer based biosensors, **H. Zheng, C. Wood and N. Plank**, *Victoria University of Wellington*, *New Zealand*

Charge separation in organic photovoltaic blends, **A. Matheson, A. Ruseckas and I. Samuel**, *University of St Andrews*, *UK*

Electrically detected magnetic resonance of polymer:fullerene solar cells, **S. Thomson, S. C. Hogg, D. J. Keeble and I. Samuel**, *University of St Andrews, UK*

Tetrathiafulavalene (TTF) incorporated macromolecules for field effect transistor and photovoltaic applications, **S. Arumugam, A. R. Inigo, N. J. Findlay, A. L. Kanibolotsky and P. Skabara**, *University of Strathclyde, UK*

Environmental gating of conductance and temperature dependence in oligothiophene single molecule junctions, **A. Vezzoli, I. Grace, E. Pires, S. J. Higgins, M. Elliott, C. J. Lambert, R. J. Nichols and J. E. MacDonald**, *University of Liverpool, UK*

Correlation between the structure of the dielectric monolayer and the performance of low-voltage transistors based on pentacene, **S. Gupta, S. Hannah and H. Gleskova**, *University of Strathclyde, UK*

Synthesis and characterization of soluble polyisobutylene based perylene diimides for organic solar cells, **A. Putta, N. Agnihotra, M. Al-Hashimi, E. Deniz and S. Al-Qaradawi**, *Qatar University, Qatar*

Covering the visible spectrum with organic light emitting materials, **C. Orofino-Peña, D. Cortizo-Lacalle, A. Kanibolotsky and P. Skabara**, *University of Strathclyde, UK*

Multidirectional waveguide lattices (MWGLs) for wide-angle light capture, **H. Lin, I. D. Hosein, M. R. Pont, D. B. Kumar and K. Saravanamuttu**, *McMaster University, Canada*

Statistical assessment of normal force testing to measure adhesion at organic-inorganic interfaces in organic optoelectronic devices, **H. Das Gupta, J. Yang, M. Kamperman and A. Turak**, *McMaster University, Canada*

Modelling nonlinear optical properties in nanophotonics, **A. Dekhira and O. Ouamerali**, *USTHB, Algeria*

The Skinner Prize for the best poster was awarded to Miss Sarah Holliday of Imperial College London, UK, for her poster on Rhodanine flanked small molecules as non-fullerene acceptors for organic photovoltaics.

List of participants

Dr Salima Alem, *National Research Council Canada, Canada*
Mr Ayham Al-Muhammad, *University of Southampton, United Kingdom*
Mr Enrico Angioni, *University of Strathclyde, United Kingdom*
Mr Josue Ayuso-Carrillo, *University of Manchester, United Kingdom*
Dr Ioan Baldea, *Heidelberg University, Germany*
Professor Donal Bradley, *Imperial College London, United Kingdom*
Miss Yasmine Braeken, *Hasselt university, Belgium*
Miss Roisin Brown, *University of Strathclyde, United Kingdom*
Dr Robert Brückner, *TU Dresden, Germany*
Mr Jonas Bucevičius, *Vilnius University, Lithuania*
Mr Audrius Bucinskas, *Kaunas University of Technology, Lithuania*
Ms Dorota Buczyńska, *Nicolaus Copernicus University, Poland*
Professor Filip Bures, *University of Pardubice, Czech Republic*
Mr Joseph Cameron, *University of Strathclyde, United Kingdom*
Mr Michael Clark, *Bentham Instruments Ltd, United Kingdom*
Mr Gary Conboy, *University of Strathclyde, United Kingdom*
Prof Graeme Cooke, *University of Glasgow, United Kingdom*
Mr Paul Cowper, *University of Bath, United Kingdom*
Mr Daniel Crossley, *University of Manchester, United Kingdom*
Dr Lynn Dennany, *University of Strathclyde, United Kingdom*
Mr Bernd Ebenhoch, *University of St Andrews, United Kingdom*
Miss Yasmin Elkhatib, *Qatar University, Qatar*
Mrs Lijia Fang, *TU Dresden, Germany*
Dr Neil Findlay, *University of Strathclyde, United Kingdom*
Dr Christopher Finlayson, *Aberystwyth University, United Kingdom*
Professor Beatrice Fraboni, *University of Bologna, Italy*
Professor Gitti Frey, *Technion, Israel*
Professor Ian Galbraith, *Heriot-Watt University, United Kingdom*
Mr Pawel Gawrys, *University of Southampton, United Kingdom*
Mrs Seza Goker, *Middle East Technical University, Turkey*
Mr Chris Goodall, *Royal Society of Chemistry, United Kingdom*
Dr Georgy Grancharov, *Bulgarian Academy of Sciences, Bulgaria*
Dr Benoit Guilhabert, *University of Strathclyde, United Kingdom*
Miss Abigail Hallowes, *Royal Society of Chemistry, United Kingdom*
Mr Stuart Hannah, *Univeristy of Strathclyde, United Kingdom*
Professor Simon Higgins, *University of Liverpool, United Kingdom*
Mrs Gonul Hizalan, *Middle East Technical University, Turkey*
Dr Justin Hodgkiss, *University of Wellington, New Zealand*
Miss Sarah Holliday, *Imperial College London, United Kingdom*
Professor Graham Hutchings, *Cardiff University, United Kingdom*
Dr Jesuraj Inigo, *University of Strathclyde, United Kingdom*
Miss Cagla Istanbulluoglu, *Middle East Technical University, Turkey*
Dr Andreas Jacob, *cynora GmbH, Germany*

Professor Rene Janssen, *Eindhoven University of Technology, Netherlands*
Miss Hasna Kandel, *Doha / Qatar University, Qatar*
Dr Alexander Kanibolotsky, *University of Strathclyde, United Kingdom*
Dr Ji-Seon Kim, *Imperial College London, United Kingdom*
Mr Milan Klikar, *University of Pardubice, Czech Republic*
Miss Rose Krawczuk, *University of Manchester, United Kingdom*
Dr Nicolas Laurand, *University of Strathclyde, United Kingdom*
Professor Uli Lemmer, *Karlsruhe Institute of Technology, Germany*
Professor Karl Leo, *TU Dresden, Germany*
Dr Felix Leyssner, *SPECS Surface Nano Analysis GmbH, Germany*
Mr Yuandi Li, *Royal Society of Chemistry, United Kingdom*
Miss Kirsty Lunnon, *Royal Society of Chemistry, United Kingdom*
Professor George Malliaras, *École Nationale Supérieure des Mines de Saint-Étienne, France*
Dr Luigi Martiradonna, *Nature Materials - Nature Publishing Group, United Kingdom*
Mr Andrew Matheson, *University of St Andrews, United Kingdom*
Mr Glenn McConnell, *University of Strathclyde, United Kingdom*
Mr Andrew McCrone, *University of Strathclyde, United Kingdom*
Professor Iain McCulloch, *Imperial College London, United Kingdom*
Dr Joseph McDouall, *University of Manchester, United Kingdom*
Mr Luke McLellan, *University of Strathclyde, United Kingdom*
Ms Katharina Melzer, *TU München, Germany*
Professor Andrew Mount, *University of Edinburgh, United Kingdom*
Miss Fiona Nalden, *Royal Society of Chemistry, United Kingdom*
Professor Jenny Nelson, *Imperial College London, United Kingdom*
Mr Ilker Oner, *Ege University, Turkey*
Mrs Saliha Oner, *Ege University, Turkey*
Dr Henrik Ottosson, *Uppsala University, Sweden*
Professor Ourida Ouamerali, *University STHB, Algeria*
Professor Dmitrii Perepichka, *McGill University, Canada*
Dr Natalie Plank, *Victoria University of Wellington, New Zealand*
Dr Agata Pomorska, *Jagiellonian University, Poland*
Professor Sergei Ponomarenko, *Enikolopov Institute of Synthetic Polymeric Materials of Russian Academy of Sciences, Russia*
Mr George Richardson, *Imperial College London, United Kingdom*
Professor Marc Salle, *University of Angers, CNRS, France*
Professor Ifor Samuel, *University of St Andrews, United Kingdom*
Professor Kalai Saravanamuttu, *McMaster University, Canada*
Professor Takeo Sasaki, *Tokyo University of Science, Japan*
Dr Volodymyr Senkovskyy, *Novaled GmbH, Germany*
Miss Jessica Shaw, *Imperial College London, United Kingdom*
Miss Debbie Shields, *Royal Society of Chemistry, United Kingdom*
Professor Peter Skabara, *University of Strathclyde, United Kingdom*
Miss Karolina Smolarek, *Nicolaus Copernicus University, Poland*
Dr Gemma Solomon, *University of Copenhagen, Denmark*
Dr Anna Stradomska-Szymczak, *University of Glasgow, United Kingdom*
Dr Rupert Taylor, *University of Strathclyde, United Kingdom*
Mr Stuart Thomson, *University of St Andrews, United Kingdom*
Professor Luisa Torsi, *Università degli Studi di Bari "Aldo Moro", Italy*
Ms Laura Tropf, *University of St Andrews, United Kingdom*
Dr Ayse Turak, *McMaster University, Canada*
Miss Hande Ünay, *Middle East Technical University, Turkey*

Dr Canan Varlikli, *Ege University*, *Turkey*
Professor Latha Venkataraman, *Columbia University*, *U.S.A.*
Dr Ravindra Venkatramani, *Tata Institute of Fundamental Reseach*, *India*
Mr Andrea Vezzoli, *University of Liverpool*, *United Kingdom*
Mr Daniel Volz, *cynora GmbH*, *Germany*
Professor John Wallis, *Nottingham Trent University*, *United Kingdom*
Dr Ian Watson, *University of Strathclyde*, *United Kingdom*
Ms Laura Weston, *University of Manchester*, *United Kingdom*
Professor Richard Whitby, *University of Southampton*, *United Kingdom*
Mr Guy Whitworth, *University of St Andrews*, *United Kingdom*
Mr Jack Wildman, *Heriot-Watt University*, *United Kingdom*
Dr Iain Alexander Wright, *Durham University*, *United Kingdom*
Professor Fred Wudl, *University of California*, *Santa Barbara*, *U.S.A.*
Professor Bingqian Xu, *University of Georgia*, *U.S.A.*
Mr Adam Yeats, *University of Strathclyde*, *United Kingdom*
Dr Shuyu Zhang, *University of St Andrews*, *United Kingdom*
Dr Eli Zysman-Colman, *University of St Andrews*, *United Kingdom*